CONDUCTION
OF
HEAT IN SOLIDS

CONDUCTION
OF
HEAT IN SOLIDS

BY

H. S. CARSLAW

EMERITUS PROFESSOR OF MATHEMATICS IN
THE UNIVERSITY OF SYDNEY

AND

J. C. JAEGER

PROFESSOR OF GEOPHYSICS IN THE
AUSTRALIAN NATIONAL UNIVERSITY

SECOND EDITION

OXFORD
AT THE CLARENDON PRESS

*Oxford University Press, Ely House, London W.*1

GLASGOW NEW YORK TORONTO MELBOURNE WELLINGTON
CAPE TOWN SALISBURY IBADAN NAIROBI LUSAKA ADDIS ABABA
BOMBAY CALCUTTA MADRAS KARACHI LAHORE DACCA
KUALA LUMPUR HONG KONG TOKYO

FIRST EDITION 1946
SECOND EDITION 1959

REPRINTED LITHOGRAPHICALLY IN GREAT BRITAIN
AT THE UNIVERSITY PRESS, OXFORD
BY VIVIAN RIDLER
PRINTER TO THE UNIVERSITY
FROM SHEETS OF THE SECOND EDITION
1960, 1962, 1965, 1967

PREFACE TO THE SECOND EDITION

THE death of Professor Carslaw in 1954 has left me with the task of preparing a new edition of this book. In doing this, I have attempted, while preserving so far as possible the form and spirit of Carslaw's mathematics, to provide as complete an account as possible of the exact solutions and soluble problems of the subject. To this end, a great many new results have been added and some parts of the discussion have been greatly expanded, for example, those on heat generation, surface heating, melting and freezing, geophysical applications, anisotropic media, moving media, and substances with variable thermal properties.

A number of new tables and text-figures giving numerical information on fundamental problems has been added. The number of references has now grown to over seven hundred; it is quite impossible to refer to all works on the subject, and the papers listed are largely confined to those which I have been able to consult, but I have attempted to give an adequate coverage of all branches of the subject.

Two short survey chapters have been added. The first of these gives an introduction to the integral transform notation and its relationship to the classical Fourier methods. The second gives an account of the numerical methods which have assumed great importance in the last decade and their relationship to the body of exact solutions given earlier in the text.

It is a pleasure to acknowledge the assistance of my wife and Mrs. A. Davidson in the preparation of the manuscript and the numerical calculations, and that of the staff of the Clarendon Press in the production of the book.

J. C. J.

PREFACE TO THE FIRST EDITION

CARSLAW'S *Introduction to the Theory of Fourier's Series and Integrals and the Mathematical Theory of the Conduction of Heat* was published at the end of 1906. In 1920 and 1921 the work was completely revised and rewritten in two volumes, the second of which, entitled *Introduction to the Mathematical Theory of the Conduction of Heat in Solids*, appeared in 1921. It became out of print in 1940.

In the last twenty-five years so many developments have been made, both in the theory and applications of the subject, that a new book embodying these advances seemed called for rather than a new and revised edition of the old one. This work, based on the earlier one and intended to supersede it, brings the discussion of the theory and applications up to date. In particular it contains a full treatment of the Laplace transformation method of dealing with problems in the Conduction of Heat. This takes the place of the method by contour integrals given in Chapters X and XI of the 1921 book. The Laplace transformation method, though similar in principle to that by contour integration, is much simpler, more direct and powerful.

In planning this book we have tried to make it as useful as possible to engineers and physicists without altering its character as a mathematical work. Explicit solutions of many problems of practical interest are included and much numerical information in the form of tables and text-figures is given. The discussion of the theory of systems used in experimental work has been greatly extended, and other subjects of practical importance are briefly noticed, such as the theory of automatic temperature control, which have hitherto not appeared in mathematical textbooks.

The earlier book, except in its final chapters, could be looked on as a treatise on the Fourier mathematics, developing the subject along the classical lines. The new book in Chapters I–X follows the same design. In these chapters it covers and often reproduces verbatim most of what is contained in Chapters I–IX of the old one, while giving fuller attention to the needs of the engineer and physicist.

In Chapters XII–XV the Laplace transformation method is introduced and applied in the main to more difficult problems. The reader, after the general discussion in Chapter XII, will see that its use would have simplified much of the preceding chapters, and he will probably, in the solution of problems as they arise, become accustomed to use it for himself.

A large number of interesting results has been given in small print, many without proof. These may be taken as examples for solution.

Over four hundred selected references to the mathematical and physical aspects of the various topics discussed here have been given as footnotes in the text. It is hoped that these will prove an adequate introduction to the literature of the subject. This has grown so much in recent years that it seemed impossible to give a complete bibliography.

Almost all the numerical material in the tables and text-figures has been calculated specially for this book. We are greatly indebted to Miss M. E. Clarke for her assistance in this computation and in many other ways.

H. S. C.
J. C. J.

CONTENTS

REFERENCES

THE following abbreviations for works frequently referred to are used throughout this volume.

H. S. Carslaw, *Fourier's Series and Integrals* (Macmillan, edn. 3, 1930): referred to as *F.S.*

H. S. Carslaw, *Introduction to the Mathematical Theory of the Conduction of Heat in Solids* (Macmillan, edn. 2, 1921): referred to as *C.H.*

G. N. Watson, *A Treatise on the Theory of Bessel Functions* (Cambridge, edn. 2, 1944): referred to as *W.B.F.*

Gray and Mathews, *Treatise on Bessel Functions* (Macmillan, edn. 2, 1922): referred to as *G. and M.*

J. Crank, *The Mathematics of Diffusion* (Oxford, 1956): referred to as *M.D.*

I

GENERAL THEORY

1.1. Introductory

WHEN different parts of a body are at different temperatures heat flows from the hotter parts to the cooler. There are three distinct methods by which this transference of heat takes place: (i) Conduction, in which the heat passes through the substance of the body itself, (ii) Convection, in which heat is transferred by relative motion of portions of the heated body, and (iii) Radiation, in which heat is transferred direct between distant portions of the body by electromagnetic radiation.

In liquids and gases convection and radiation are of paramount importance, but in solids convection is altogether absent and radiation usually negligible. In this book we shall consider conduction of heat only, and usually speak of the body as solid, though in certain circumstances the results will be valid for liquids or gases.

In this chapter the general theory of conduction of heat is developed; the subsequent chapters are devoted to special problems and methods.

1.2. Conductivity

The Mathematical Theory of the Conduction of Heat may be said to be founded upon a hypothesis suggested by the following experiment:

A plate of some solid is given, bounded by two parallel planes of such an extent that, so far as points well in the centre of the planes are concerned, these bounding surfaces may be supposed infinite. The two planes are kept at different temperatures, the difference not being so great as to cause any sensible change in the properties of the solid. For example, the upper surface may be kept at the temperature of melting ice by a supply of pounded ice packed upon it, and the lower at a fixed temperature by having a stream of warm water continually flowing over it. When these conditions have endured for a sufficient time the temperature of the different points of the solid settles down towards its steady value, and at points well removed from the ends the temperature will remain the same along planes parallel to the surfaces of the plate.

Consider the part of the solid bounded by an imaginary cylinder of cross-section S whose axis is normal to the surface of the plate. This cylinder is supposed so far in the centre of the plate that no flow of heat takes place across its generating lines. Let the temperature of the lower surface be $v_0{}^\circ$C and of the upper $v_1{}^\circ$C $(v_0 > v_1)$, and let the thickness of the plate be d centimetres. The results of experiments upon different solids suggest that, when the steady state of temperature has been reached, the quantity Q of heat which flows up through the plate in t seconds over the surface S is equal to

$$Q = \frac{K(v_0 - v_1)St}{d}, \tag{1}$$

where K is a constant, called the Thermal Conductivity of the substance, depending upon the material of which it is made. In other words, the flow of heat between these two surfaces is proportional to the difference of temperature of the surfaces.

This result must not be regarded as proved by these experiments. They suggest the law rather than verify it. The more exact verification is to be found in the agreement of experiment with calculations obtained from the mathematical theory based on the assumption of the truth of this law.

The reciprocal of the thermal conductivity of a substance is called its Thermal Resistivity.

Strictly speaking, the conductivity K is not constant for the same substance, but depends upon the temperature. However, when the range of temperature is limited, this change in K may be neglected, and in the ordinary mathematical theory it is assumed that the conductivity does not vary with the temperature. A nearer approximation to the actual state may be obtained by making K a linear function of the temperature v, e.g.
$$K = K_0(1 + \beta v),$$

where β is small, and, in fact, is negative for most substances.

From (1), the thermal conductivity is given by

$$K = \frac{Qd}{(v_0 - v_1)St}, \tag{2}$$

and from this its dimensions and the nature of the units in which it is expressed follow.

The system of units most frequently chosen in physical work uses the c.g.s. units of length, mass, and time, measures temperature in $^\circ$C, and takes as the unit quantity of heat the calorie, which is the quantity of heat required to raise the temperature of 1 gm of water† by 1° C. In this system, values of K are expressed in cal/(sec) (cm²) (°C/cm). This system will be used throughout this book when numerical values are given: values of the thermal properties of a few typical substances‡ are given in Appendix VI to give an idea of the orders of magnitude involved.

The other important system of units, which is the one commonly employed by engineers and used in works on heat transfer, takes the foot, pound, hour, and °F as units, and defines the unit quantity of heat as the British Thermal Unit (Btu.), which is the quantity of heat required to raise the temperature of 1 lb of water at its maximum density (39° F) by 1° F. The connexion between the two units is

$$1 \text{ Btu.} = 252 \cdot 0 \text{ cal.}$$

In this system, numerical values of the conductivity are given in Btu./(hr)(ft.²)(°F/ft.): to express these in terms of cal/(sec) (cm²)(°C/cm) multiply by 0·00413.

The dimensions of K in these systems in which the unit of heat is that which causes unit rise in temperature in unit mass of water may be seen from (2) to be

$$[K] = [M][L^{-1}][T^{-1}], \tag{3}$$

since those of $Q/(v_0 - v_1)$ are just those of mass.

If it is desired to measure quantity of heat by the work necessary to produce it, the unit would be the erg or joule. The number of joules in a calorie, J, is known as the mechanical equivalent of heat. For the 15° calorie defined above $J = 4 \cdot 184$.

In the fundamental experiment from which our definition of the conductivity is derived, the solid is supposed to be homogeneous and of such a material that, when a point within it is heated, the heat spreads out equally well in all directions. Such a solid is said to be isotropic, as opposed to crystalline and anisotropic solids, in which certain directions are more favourable for the conduction of heat than others. There are also heterogeneous solids, in which the conditions of conduction vary from point to point as well as in direction at each point.

---

† Experiments show that the quantity of heat required to raise the temperature of 1 gm of water by 1° C is not quite the same at different temperatures, and in an exact definition of the calorie the temperature of the water needs to be specified. Usually, this specified temperature is taken to be 15° C so that the '15° calorie' is the quantity of heat needed to raise 1 gm of water from 14·5° C to 15·5° C.

‡ For more extensive information see the *International Critical Tables* (McGraw-Hill, 1929), Vol. V, or, for rocks and minerals, Birch, Schairer, and Spicer, *Handbook of Physical Constants*, Geol. Soc. of America, Special papers, Number 36 (1942).

1.3. The flux of heat across any surface

The rate at which heat is transferred across any surface S at a point P, per unit area per unit time, is called the flux of heat† at that point across that surface, and we shall denote it by f.

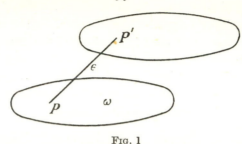

Fig. 1

First we show that the flux across a plane through a point P varies continuously with the position of the point P if the direction of the normal to the plane remains constant. Suppose an infinitesimal area ω enclosing P is taken in the plane, and a cylinder is formed on this area as base with generators equal and parallel to a line PP' whose length ϵ is an infinitesimal of lower order than the linear dimensions of ω (Fig. 1).

Let $f_1\omega$ and $f_2\omega$ be the rates of flow per unit time across the plane surfaces of the cylinder through P and P'. The flow across the curved surface is negligible compared with these. The rate of flow of heat into the cylinder is thus $\omega(f_1-f_2)$. Also if v is the average temperature in the cylinder, σ the distance between its plane faces, and ρ and c the average density and specific heat of its material, the rate at which the cylinder gains heat is

$$\rho c \omega \sigma \frac{\partial v}{\partial t}.$$

Equating these two expressions we have

$$f_1 - f_2 = \rho c \sigma \frac{\partial v}{\partial t},$$

and as $\sigma \to 0$ the expression on the right tends to zero and so $f_1 \to f_2$.

It is important to notice that this argument does not require the thermal properties of the medium to be continuous, only that they be finite. Thus it enables us to assert in § 1.9 that the flux is continuous at the surface of separation of two media.

Next we show that, if the values of f are given for three mutually

---

† Numerical values are usually given in cal/(cm²)(sec) or in Btu./(ft.²)(hr). The connexion between the two is 1 cal/(cm²)(sec) = 13,270 Btu./(ft.²)(hr).

perpendicular planes meeting at a point, its value for any other plane through the point may be written down.

Consider the elementary tetrahedron $PABC$, Fig. 2, whose three faces PBC, PCA, PAB are parallel to the coordinate planes, while the perpendicular from the point P to the face ABC has direction cosines (λ, μ, ν) and is of length p. Let the area of ABC be Δ; then the areas of PBC, PCA, and PAB are respectively $\lambda\Delta$, $\mu\Delta$, $\nu\Delta$.

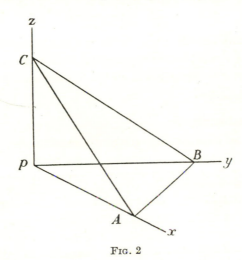

FIG. 2

If we denote the rates of flow of heat per unit time per unit area over the elementary areas PBC, PCA, PAB, and ABC by f_x, f_y, f_z, and f, the rate at which heat is gained by the tetrahedron is given by

$$(\lambda f_x + \mu f_y + \nu f_z - f)\Delta.$$

However, if ρ and c are the density and specific heat of the solid, and v the average temperature over the tetrahedron, this rate of gain of heat is equal to

$$\tfrac{1}{3}\Delta p\rho c \frac{\partial v}{\partial t}.$$

It follows that
$$\lambda f_x + \mu f_y + \nu f_z - f = \tfrac{1}{3}p\rho c \frac{\partial v}{\partial t}. \tag{1}$$

Now as p tends to zero, the right-hand side of (1) tends to zero, and f_x, f_y, f_z, and f become respectively the fluxes at the point P across planes parallel to the coordinate planes and across a plane through P which has λ, μ, ν for the direction cosines of its normal. Thus we have

$$f = \lambda f_x + \mu f_y + \nu f_z. \tag{2}$$

If the three fluxes f_x, f_y, f_z at a point P across planes parallel to the coordinate planes are known, the flux across any other plane through P can be determined from (2).

At every point P of the solid a vector $\mathbf{f}$ is defined whose components are f_x, f_y, f_z. Its magnitude is

$$f_m = \sqrt{(f_x^2 + f_y^2 + f_z^2)}, \tag{3}$$

and it lies along the line whose direction cosines are

$$f_x/f_m, \quad f_y/f_m, \quad f_z/f_m. \tag{4}$$

This vector may be called the *flux vector* at the point P. The flux of heat at P across a plane whose normal lies in the direction (4) is just f_m, while the flux at P across a plane whose normal makes an angle θ with this direction is $f_m \cos \theta$.

1.4. Isothermal surfaces

Consider a solid with a distribution of temperature at time t given by

$$v = f(x, y, z, t).$$

We may suppose a surface described in this solid, such that at every point upon it the temperature at this instant is the same, say V. Such a surface is called the isothermal surface for the temperature V, and it may be looked upon as separating the parts of the body which are hotter than V from the parts which are cooler than V. We may imagine the isothermals drawn for this instant for different degrees and fractions of a degree. These surfaces may be formed in any way, but no two isothermals can cut each other, since no part of the body can have two temperatures at the same time. The solid is thus pictured as divided up into thin shells by its isothermals.

1.5. Conduction of heat in an isotropic solid

In future, unless expressly stated, we shall consider only isotropic media, that is media whose structure and properties in the neighbourhood of any point are the same relative to all directions through the point. Because of this symmetry, the flux vector at a point must be along the normal to the isothermal surface through the point, and in the direction of falling temperature.

The relation between the rate of change of temperature along the normal to an isothermal and the flux vector in that direction may be deduced from the fundamental experiment described in § 1.2. In that case the isothermals are planes parallel to the faces of the slab. Suppose the isothermals for temperatures v and $v + \delta v$ are distant δx apart.

Then by 1.2 (1) the rate of flow of heat, per unit time per unit area, in the direction of x increasing is

$$-K\frac{\delta v}{\delta x}.$$

Thus in the limit as $\delta x \to 0$ we have

$$f_x = -K\frac{\partial v}{\partial x}. \tag{1}$$

We extend this to any isothermal surface and take as our *fundamental hypothesis for the Mathematical Theory of Conduction of Heat that the rate at which heat crosses from the inside to the outside of an isothermal surface per unit area per unit time at a point is equal to*

$$-K\frac{\partial v}{\partial n},$$

where K is the thermal conductivity of the substance, and $\partial/\partial n$ denotes differentiation along the outward-drawn normal to the surface.

We now proceed to find the flux at a point P across any surface, not necessarily isothermal. Let the tangent plane at P to the isothermal through P be taken as the XY-plane, so that the fluxes across planes through P parallel to the coordinate planes are

$$f_x = f_y = 0, \qquad f_z = -K\frac{\partial v}{\partial z}.$$

Then if the normal at P to the given surface has direction cosines (λ, μ, ν) relative to these axes, the flux across it is, by 1.3 (2),

$$-K\nu\frac{\partial v}{\partial z} = -K\frac{\partial v}{\partial h},$$

where $\partial/\partial h$ denotes differentiation in the direction (λ, μ, ν), since

$$\frac{\partial v}{\partial h} = \lambda\frac{\partial v}{\partial x} + \mu\frac{\partial v}{\partial y} + \nu\frac{\partial v}{\partial z}, \quad \text{and} \quad \frac{\partial v}{\partial x} = \frac{\partial v}{\partial y} = 0.$$

Thus *the flux of heat at a point across any surface is*

$$-K\frac{\partial v}{\partial h}, \tag{2}$$

where $\partial/\partial h$ denotes differentiation in the direction of the outward normal.

In particular, the fluxes across three planes parallel to the axes of coordinates are

$$f_{\dot{x}} = -K\frac{\partial v}{\partial x}, \qquad f_y = -K\frac{\partial v}{\partial y}, \qquad f_z = -K\frac{\partial v}{\partial z}. \tag{3}$$

Using the vector $\mathbf{f}$ introduced in § 1.3, the results of this section may be expressed by the formula

$$\mathbf{f} = -K \operatorname{grad} v. \tag{4}$$

1.6. The differential equation of conduction of heat in an isotropic solid

We first consider the case of a solid through which heat is flowing, but within which no heat is generated. The temperature v at the point $P(x, y, z)$ will be a continuous function of x, y, z, and t, and, as shown in § 1.3, the same is true of the flux.

Consider an element of volume of the solid at the point P, namely, the rectangular parallelepiped with this point as centre, its edges being parallel to the coordinate axes and of lengths $2dx$, $2dy$, and $2dz$. Let $ABCD$ and $A'B'C'D'$ be the faces in the planes $x-dx$ and $x+dx$, respectively, then the rate at which heat flows into the parallelepiped over the face $ABCD$ will ultimately be given by

$$4\left(f_x - \frac{\partial f_x}{\partial x}\, dx\right) dydz,$$

where f_x is the flux at P over a parallel plane. Similarly, the rate at which heat flows out over the face $A'B'C'D'$ is given by

$$4\left(f_x + \frac{\partial f_x}{\partial x}\, dx\right) dydz.$$

Thus the rate of gain of heat from flow across these two faces is equal to

$$-8\, dxdydz\, \frac{\partial f_x}{\partial x}.$$

There are similar expressions for the rates of gain from flow across the other pairs of faces, and, adding these, the total rate of gain of heat of the parallelepiped from flow across its faces is found to be

$$-8\left(\frac{\partial f_x}{\partial x} + \frac{\partial f_y}{\partial y} + \frac{\partial f_z}{\partial z}\right) dxdydz = -8\, dxdydz\, \operatorname{div} \mathbf{f}, \tag{1}$$

where $\mathbf{f}$ is the vector defined in 1.3 (4).

This rate of gain of heat is also given by

$$8\rho c\, \frac{\partial v}{\partial t}\, dxdydz, \tag{2}$$

where ρ is the density and c the specific heat† (at temperature v) of the solid. Equating (1) and (2) gives‡

$$\rho c \frac{\partial v}{\partial t} + \left(\frac{\partial f_x}{\partial x} + \frac{\partial f_y}{\partial y} + \frac{\partial f_z}{\partial z} \right) = 0. \tag{3}$$

This equation holds at any point of the solid, provided no heat is supplied at the point; the solid need not be homogeneous or isotropic. It corresponds to the equation of continuity in hydrodynamics.

For a homogeneous isotropic solid whose thermal conductivity is independent of the temperature, f_x, f_y, and f_z are given by 1.5 (3), and (3) becomes

$$\frac{\partial^2 v}{\partial x^2} + \frac{\partial^2 v}{\partial y^2} + \frac{\partial^2 v}{\partial z^2} - \frac{1}{\kappa} \frac{\partial v}{\partial t} = 0, \tag{4}$$

where

$$\kappa = \frac{K}{\rho c}. \tag{5}$$

The constant κ was called by Kelvin the Diffusivity§ of the substance, and by Clerk Maxwell its Thermometric Conductivity.‖

(4) is the equation commonly known as the equation of conduction of heat. In the case of steady temperature in which v does not vary with the time, it becomes Laplace's equation

$$\nabla^2 v \equiv \frac{\partial^2 v}{\partial x^2} + \frac{\partial^2 v}{\partial y^2} + \frac{\partial^2 v}{\partial z^2} = 0. \tag{6}$$

† The specific heat c of a substance at temperature v is defined as $\delta Q/\delta v$ where δQ is the quantity of heat necessary to raise the temperature of unit mass of the substance through the small temperature range from v to $v + \delta v$. It depends on both the temperature and the assumed mode of heating, which is taken here to be at constant strain. In the units employed in this book it is expressed in (cal)/(gm)(°C), and the specific heat of water at 15°C is 1 (cal)/(gm)(°C). It should be noticed that there is a considerable variety of usage in this matter. Some writers regard the above definition as that of heat capacity, or heat capacity per unit mass, of the substance, and define the specific heat of a substance as the ratio of its heat capacity per unit mass to that of water.

For solids, the effect of the method of heating on the specific heat is usually negligible and c may be replaced by c_p, the specific heat at constant pressure. The question is discussed in III at the end of this section.

‡ In transparent, as well as in fibrous or other materials with large pore spaces, transport of heat by radiation may be of importance, resulting in the appearance of an additional term in (3). Cf. van der Held, *Appl. Sci. Res.* A, **3** (1953) 237–49; A, **4** (1954) 77–99.

§ Some values for the diffusivities of various substances are given in Appendix VI. To find the dimensions of diffusivity we note that, writing $[Q]$ and $[v]$ for those of quantity of heat and temperature, respectively, $[K] = [Q][L^{-1}][T^{-1}][v^{-1}]$, $[c] = [Q][M^{-1}][v^{-1}]$, $[\rho] = [M][L^{-3}]$, so that $[\kappa] = [L^2][T^{-1}]$. It follows that if the units of length and time are the foot and the hour, as in many engineering tables, the value of κ for these units will have to be multiplied by $(30\cdot48)^2/3600 = 0\cdot258$ to reduce it to the c.g.s. system.

‖ Because it measures the change of temperature which would be produced in unit volume of the substance by the quantity of heat which flows in unit time through unit area of a layer of the substance of unit thickness with unit difference of temperature between its faces.

If heat is produced in the solid, so that at the point P (x, y, z) heat is supplied at the rate $A(x, y, z, t)$ per unit time per unit volume, a term $8A\,dxdydz$ has to be added to (1), and, for the case in which K is a constant, (4) is replaced by

$$\nabla^2 v - \frac{1}{\kappa}\frac{\partial v}{\partial t} = -\frac{A(x, y, z, t)}{K}. \tag{7}$$

For the case of steady flow, in which $\partial v/\partial t = 0$, equation (7) reduces to Poisson's equation.

In almost all the problems for which exact solution is possible, and in those discussed in this book unless otherwise stated, the thermal properties K, ρ, c are constants, independent of both position and temperature. If this is not the case, (3) still holds (with $A(x, y, z, t)$ added in the right-hand side if there is heat generation) but (7) is replaced by

$$\rho c\frac{\partial v}{\partial t} = \frac{\partial}{\partial x}\left(K\frac{\partial v}{\partial x}\right) + \frac{\partial}{\partial y}\left(K\frac{\partial v}{\partial y}\right) + \frac{\partial}{\partial z}\left(K\frac{\partial v}{\partial z}\right) + A. \tag{8}$$

If K and A are functions of position only, the solution of (8) offers no great difficulty in principle and a number of solutions are available for discontinuous thermal properties (composite solids) and for simple laws of variation of K with position. If the thermal properties depend on the temperature, the situation is more complicated since the equation becomes non-linear: few such cases have been studied in connexion with conduction of heat since the variation of the thermal properties with temperature is relatively slow and the information available about it is scanty and inaccurate. Nevertheless, they are becoming increasingly important when large ranges of temperature are involved, as in the solidification of castings; also the same equations arise in the theory of diffusion, where, because of the more rapid variation of diffusion coefficients with concentration, they are of much greater importance.† In most cases numerical methods have to be used, but a few general results, and cases in which exact solution is possible, will be noted below.

I. *The case of thermal properties varying with the temperature but independent of position*‡

In this case (8) becomes

$$\rho c\frac{\partial v}{\partial t} = K\nabla^2 v + A + \frac{\partial K}{\partial v}\left\{\left(\frac{\partial v}{\partial x}\right)^2 + \left(\frac{\partial v}{\partial y}\right)^2 + \left(\frac{\partial v}{\partial z}\right)^2\right\}, \tag{9}$$

which shows the non-linearity clearly.

† Cf. *M.D.*, Chaps. IX–XI.
‡ Some other methods for the one-dimensional case are given in § 2.16.

(8) may be reduced to a simpler form† by introducing the new variable

$$\Theta = \frac{1}{K_0} \int_0^v K \, dv, \tag{10}$$

where K_0 is the value of K when $v = 0$. This, and the lower limit of integration, are merely introduced to give Θ the dimensions of temperature and a definite value.

It follows from (10) that‡

$$\frac{\partial \Theta}{\partial t} = \frac{K}{K_0} \frac{\partial v}{\partial t}, \qquad \frac{\partial \Theta}{\partial x} = \frac{K}{K_0} \frac{\partial v}{\partial x}, \qquad \frac{\partial \Theta}{\partial y} = \frac{K}{K_0} \frac{\partial v}{\partial y}, \qquad \frac{\partial \Theta}{\partial z} = \frac{K}{K_0} \frac{\partial v}{\partial z},$$

and (8) becomes

$$\nabla^2 \Theta - \frac{1}{\kappa} \frac{\partial \Theta}{\partial t} = -\frac{A}{K_0}, \tag{11}$$

where, in (11), A and $\kappa = K/\rho c$ are expressed as functions of the new variable Θ. Thus, in terms of this new variable, the form (7) of the equation of conduction of heat is preserved, but with a diffusivity κ which depends on Θ. It is a fact that in many cases the variation of κ with temperature is much less important than that of K, so that, to a reasonable approximation, it may be taken to be constant; for example, for metals near absolute zero, both K and c are approximately proportional to the absolute temperature. In such cases, if A is independent of v, equation (11) becomes of type (7) and solutions for the case of constant conductivity may be taken over immediately by replacing v by Θ, provided that the boundary conditions prescribe only v or $K \, \partial v/\partial n$: if they are of the form

$$(\partial v/\partial n) + hv = 0,$$

where h is a constant, this remark does not hold.

The case of steady flow is of particular importance since (11) becomes Poisson's equation if A is constant, or Laplace's equation if $A = 0$. Thus solutions of problems of steady heat flow with conductivity any function of the temperature, and boundary conditions consisting of prescribed temperature or flux, may be derived immediately from the corresponding solutions for constant conductivity.

Another useful form may be obtained by introducing W, the heat content per unit mass of the material (measured from some arbitrary zero of temperature). In terms of this quantity, (8) becomes

$$\rho \frac{\partial W}{\partial t} = \frac{\partial}{\partial x}\left(K \frac{\partial v}{\partial x}\right) + \frac{\partial}{\partial y}\left(K \frac{\partial v}{\partial y}\right) + \frac{\partial}{\partial z}\left(K \frac{\partial v}{\partial z}\right) + A, \tag{12}$$

or, in terms of Θ defined by (10),

$$\frac{\rho}{K_0} \frac{\partial W}{\partial t} = \nabla^2 \Theta + \frac{A}{K_0}, \tag{13}$$

where W is related to Θ in a known manner. The introduction of W has advantages in problems involving latent heat.

II. *Heat production in the solid*

Cases in which heat is produced in the solid are becoming increasingly important in technical applications. Heat may be produced by (i) the passage of an electric

† van Dusen, *Bur. Stand. J. Res.* **4** (1930) 753–6; Eyres, Hartree, Ingham, Jackson, Sarjant, and Wagstaff, *Phil. Trans. Roy. Soc.* A, **240** (1946) 1–58. For steady flow, the method dates back to Kirchhoff's *Vorlesungen über die Theorie der Wärme* (1894).

‡ Θ is thus essentially a potential whose gradient is proportional to the flux, cf. Vernotte, *Comptes Rendus*, **218** (1944) 39–41.

current, (ii) dielectric or induction heating,[†] (iii) radioactive decay,[‡] (iv) absorption from radiation,[§] (v) mechanical generation in viscous or plastic flow, (vi) chemical reaction,[||] including in this such diverse matters as the hydration of cement[††] and the ripening of apples.[‡‡]

In all cases but the last, the rate of heat production is independent of v to a first approximation and, to a better approximation, often takes the form

$$A = a+bv, \tag{14}$$

where a and b are constants which may have either sign.[§§] It may be noted that (7), with A of this form, may be solved exactly by many of the methods used below, cf. §§ 1.14, 15.7.

When heat is produced by a zero-order chemical reaction, the rate of heat production may usually be represented by the Arrhenius relation

$$A_0 e^{-k/T}, \tag{15}$$

where A_0 and k are constants and T is the absolute temperature.[||||] For higher-order reactions, a more complicated expression of the same general type has to be used. In some cases it has been found experimentally that a law of type

$$A = A_0 \exp(bv)$$

holds.[†††] In all these cases no analytical solution exists, and they have to be treated

† Curtis, *High Frequency Induction Heating* (McGraw-Hill, 1944); Brown, *Proc. Inst. Radio Engrs.* **31** (1943) 537–48; Maddock, *J. Sci. Instr.* **23** (1946) 165–73; Brown, Hoyler, and Bierwirth, *Radio Frequency Heating* (van Nostrand, 1947).

‡ Published work on this subject is confined at present to studies of the effect of radioactivity on temperatures within the Earth, cf. Jeffreys, *The Earth* (Camb., edn. 3, 1952) p. 276; Lowan, *Phys. Rev.* (2) **44** (1933) 769–75; and further references in §§ 2.14, 9.14.

§ For example, infra-red rays are strongly absorbed in a layer of the retina of the eye and may cause a rise in temperature large enough to damage it.

|| For heat generation by chemical reactions of various orders see *M.D.*, Chap. VIII, also § 15.7 below.

†† There is a large literature on this subject, mostly in the *Proc. Amer. Concr. Inst.*, e.g. **31** (1934) 113; **34** (1937) 89, 105, 117, 477, 497. See also *Temperature, its Measurement and Control in Science and Industry* (Amer. Inst. Phys., 1941) pp. 905–16; Davey and Fox, *Building Research Technical Paper, No. 15* (London, H.M.S.O., 1933); Rawhouser, *J. Amer. Concr. Inst.* **16** (1945) 305. The rate of heat production is roughly of the form ke^{-at}, the heat generated in three days being of the order of 50–100 cal per gramme of cement; this is sufficient to have important technical consequences, particularly in the design of large dams.

‡‡ Awbery, *Phil. Mag.* (7) **4** (1927) 629–38.

§§ In the case of heat production by electric current, the temperature coefficient of electrical resistance is positive for most substances so that b in (14) is positive, but for some materials, notably carbon and fused salts, it is negative. Negative values of b also appear in problems on simultaneous diffusion and chemical reaction. (14), with a negative value of b, also gives a crude approximation to the conditions in a body from which heat is removed by fluid circulating in a network of pipes, cf. Jaeger, *Brit. J. Appl. Phys.* **3** (1952) 221–2.

|||| Crank and Nicholson, *Proc. Camb. Phil. Soc.* **43** (1947) 50–67, discuss the law (15) for linear flow, and Nichols and Presson, *J. Appl. Phys.* **25** (1954) 1469–72, discuss this and the case of a first-order reaction for radial flow in a cylinder. See also Goheen, *J. Math. Phys.* **28** (1949) 107–16.

††† This occurs in some chemical problems, cf. Rice, *J. Chem. Phys.* **8** (1940) 727, and also in the theory of the thermal breakdown of a dielectric, cf. Goodlet, Edwards, and Perry, *J. Instn. Elect. Engrs.* **69** (1931) 695; Hartree, *Mem. Manchr. Lit. Phil. Soc.* **80** (1935) 85. See also § 15.7 IV.

by numerical methods. The law (14) may be used as a very crude first approximation, but it appears that the exact solutions may have very different properties from those of these linear approximations.

III. *The effects of thermal expansion*

Equation (3) has been derived on the assumption that no work is done by deformation of the solid so that c is, in fact, the specific heat at constant strain. If the stresses in the body do work, (3) must be modified to take this into account. If unrestricted expansion at constant pressure is possible, (3) still holds if c is interpreted as c_p, the specific heat at constant pressure. If this expansion is restricted, additional terms appear: for example, if the stress-system consists of a hydrostatic pressure p, a term

$$\frac{3\alpha T}{J}\frac{\partial p}{\partial t} \tag{16}$$

has to be added to the right-hand side of (3), where T is the absolute temperature, α is the coefficient of linear thermal expansion, and c in (3) is to be interpreted as c_p, the specific heat at constant pressure. The case of a general stress-system has been treated by Jeffreys.[†]

1.7. The differential equation of conduction of heat in a moving medium

We consider first a solid medium moving with a velocity whose components are (u_x, u_y, u_z). When calculating the rate at which heat crosses any plane, a convective term of components $(\rho c v u_x, \rho c v u_y, \rho c v u_z)$ must be added to the part due to conduction. Thus the components of the heat-flux vector are now

$$f_x = -K\frac{\partial v}{\partial x}+\rho c v u_x, \qquad f_y = -K\frac{\partial v}{\partial y}+\rho c v u_y, \qquad f_z = -K\frac{\partial v}{\partial z}+\rho c v u_z, \tag{1}$$

and using these values in 1.6 (3) gives

$$\frac{\partial v}{\partial t}+u_x\frac{\partial v}{\partial x}+u_y\frac{\partial v}{\partial y}+u_z\frac{\partial v}{\partial z}-\kappa\nabla^2 v = 0, \tag{2}$$

for the case in which K is constant and there is no heat generation.[‡]

This may be written
$$\frac{Dv}{Dt}-\kappa\nabla^2 v = 0, \tag{3}$$

where D/Dt denotes the 'differentiation following the motion' of hydrodynamics.[§] If heat is produced in the solid at the rate A per unit time per unit volume, a term $A/\rho c$ is to be added to the right-hand side of (3).

[†] Jeffreys, *Proc. Camb. Phil. Soc.* **26** (1930) 101–6. An application to the determination of temperature in a cooling contracting sphere has been made by Lapwood, *Mon. Not. R. Astr. Soc. Geophys. Suppl.* **6** (1952) 402–7. The effect of gain of potential energy by contraction was first studied by Duhamel, *J. Éc. polyt. Paris*, **15**, Cah. 25 (1837) 1–57.

[‡] A discussion of this type for a moving fluid was given by Wilson, *Proc. Camb. Phil. Soc.* **12** (1904) 406–23, who solves a number of interesting problems on steady temperature distributions. See also Boussinesq, *Théorie analytique de la chaleur* (Gauthier-Villars, 1903) T. II; Jeffreys, *Phil. Mag.* (6) **35** (1918) 270–80; Goldstein, *Modern Developments in Fluid Dynamics* (Oxford, 1938) Vol. II, p. 603. A number of exact solutions is known for heat conduction in a viscous fluid in laminar motion, in particular for flow through a tube with the Poiseuille distribution of velocity: for these, see works on heat transfer, e.g. Gröber, Erk, and Grigull, *Wärmeübertragung* (Springer, edn. 3, 1955) p. 179; Jakob, *Heat Transfer* (Wiley, 1949) p. 451.

[§] Cf. Lamb, *Hydrodynamics* (Cambridge, edn. 4, 1916) § 4.

It may be remarked that (2) follows, as it should, by transformation from a coordinate system moving with velocity u_x, u_y, u_z, with reference to which the ordinary equation of conduction of heat 1.6 (4) holds.

As a matter of interest, and because of its similarity to the analysis in the case of conduction of heat in a deformable solid, the derivation of the equations of conduction of heat in a moving compressible fluid will be indicated briefly.† Clearly, the hydrodynamic and thermodynamic quantities involved must now be specified quite precisely, and for the latter we use the density ρ, absolute temperature T, and internal energy per unit mass U. We shall use the suffix notation, x_i, $i = 1, 2, 3$, being the coordinates, and a repeated suffix implying summation over the values 1 to 3, for example, $l_i \xi_i$ in (4) is $l_1 \xi_1 + l_2 \xi_2 + l_3 \xi_3$. We need Green's theorem‡ which states that if ξ_i, $i = 1, 2, 3$, and their first derivatives are continuous functions of x_i inside a closed surface S, then

$$\iint l_i \xi_i \, dS = \iiint \frac{\partial \xi_i}{\partial x_i} \, d\tau, \tag{4}$$

where l_i, $i = 1, 2, 3$, are the direction cosines of the outward normal to S, and the surface integral is taken over the surface of S, and the volume integral over its interior.

Now let S be a small closed surface in the fluid which is always bounded by the same particles of fluid. We write dS for an element of this surface and $d\tau$ for an element of volume contained in it. The fact that the mass $\rho \, d\tau$ in any element of volume $d\tau$ of S is conserved implies that

$$\frac{D(\rho \, d\tau)}{Dt} = 0, \tag{5}$$

where D/Dt implies differentiation following the motion of the fluid. (5) leads, as in Lamb, loc. cit., § 7, to the equation of continuity.

If X_i is the body force per unit mass, E_{ij} is the stress-tensor, and u_i are the components of velocity, the equation of motion of the element is

$$\frac{D}{Dt} \iiint \rho u_i \, d\tau = \iiint \rho X_i \, d\tau + \iint E_{ij} l_j \, dS, \qquad i = 1, 2, 3, \tag{6}$$

where the surface and volume integrals are taken over S and its interior.

Using (4) and (5), (6) may be written

$$\rho \frac{Du_i}{Dt} = \rho X_i + \frac{\partial}{\partial x_j} E_{ij}. \tag{7}$$

Also, applying the first law of thermodynamics to the element gives

$$\frac{D}{Dt} \iiint \tfrac{1}{2} \rho u_i u_i \, d\tau + \frac{D}{Dt} \iiint \rho U \, d\tau$$

$$= \iiint \rho X_i u_i \, d\tau + \iint E_{ij} u_i l_j \, dS - \iint f_i l_i \, dS, \tag{8}$$

† This discussion is based on that of Synge, *Quart. Appl. Math.* **13** (1955) 271–8, who gives a critical discussion of the thermodynamics involved. See also Eckart, *Phys. Rev.* (2) **58** (1940) 267–9.

‡ Cf. Lamb, loc. cit., § 42; Goursat, *Cours d'Analyse* (Gauthier-Villars, edn. 3, 1917) Vol. 1, § 143.

where in (8) f_i are the components of the heat-flux vector at S, and so

$$f_i = -K\frac{\partial T}{\partial x_i}, \qquad i = 1, 2, 3. \tag{9}$$

Using (4), (5), and (7) in (8), gives

$$\rho\frac{DU}{Dt} = E_{ij}e_{ij} - \frac{\partial f_i}{\partial x_i}, \tag{10}$$

where e_{ij} are the components of the rate of strain tensor, viz.

$$e_{ij} = \tfrac{1}{2}\left(\frac{\partial u_i}{\partial x_j} + \frac{\partial u_j}{\partial x_i}\right). \tag{11}$$

The equations of viscosity imply connexions between E_{ij} and e_{ij}, namely

$$E_{ij} = -p\delta_{ij} + 2\mu(e_{ij} - \tfrac{1}{3}\delta_{ij}e_{kk}), \tag{12}$$

where p is the hydrostatic pressure, μ is the coefficient of viscosity, and δ_{ij} is the Kronecker delta which is zero unless $i = j$, in which case it is unity. Using (12) and (9) in (10), gives finally

$$\rho\frac{DU}{Dt} = -pe_{ii} + 2\mu(e_{ij}e_{ij} - \tfrac{1}{3}e_{ii}^2) + \frac{\partial}{\partial x_i}\left(K\frac{\partial T}{\partial x_i}\right). \tag{13}$$

This is of the same form as (3). The first two terms on the right-hand side correspond to heat generation by processes of compression and friction in the fluid; the other term on the right-hand side is simply the usual div **f**; and on the left-hand side $\rho DU/Dt$ replaces the rather vaguely defined $\rho c Dv/Dt$ of (3). Further transformations of (13) based on thermodynamics can of course be made and, in particular, (13) may be expressed in terms of the entropy.

1.8. The transformation of coordinates

These equations may be easily transformed into other systems of orthogonal coordinates,† the most useful being the Spherical Polar System, in which the position of the point is determined by its distance r from the origin, its latitude θ, and its azimuth ϕ, and the Cylindrical System, in which its position is determined by the polar coordinates r, θ of its projection on the plane of x, y, and the coordinate z.

These are special cases of the general system of orthogonal coordinates, in which the position of a point is given by the intersection of the three orthogonal surfaces,

$$\xi = \text{const}, \qquad \eta = \text{const}, \qquad \zeta = \text{const}.$$

We proceed to show how this transformation may most easily be effected.

Consider the element of volume bounded by the surfaces $\xi \pm d\xi$, $\eta \pm d\eta$, $\zeta \pm d\zeta$, and let $A'B'C'D'$ and $ABCD$ be the faces $\xi \pm d\xi$.

† Elliptic coordinates will not be considered in this book. The elliptic cylinder is discussed by McLachlan, *Phil. Mag.* (7) **36** (1945) 600; ibid. (7) **37** (1946) 216. For the region bounded internally by an elliptic cylinder, see Tranter, *Quart. J. Mech. Appl. Math.* **4** (1951) 461–5. For ellipsoids see Mathieu, *Cours de Physique Mathématique* (Paris, 1873), Chaps. 8, 9; Niven, *Phil. Trans. Roy. Soc.* **171** (1879) 117.

Let
$$ds^2 = \lambda^2\,d\xi^2 + \mu^2\,d\eta^2 + \nu^2\,d\zeta^2$$

be the equation giving the length of the elementary arc joining the points (ξ, η, ζ) and $(\xi+d\xi, \eta+d\eta, \zeta+d\zeta)$.

Then the area of the section of the ξ surface through $P(\xi, \eta, \zeta)$ cut off by the surfaces $\eta\pm d\eta$, $\zeta\pm d\zeta$ is given by

$$4\mu\nu\,d\eta d\zeta,$$

and the rate at which heat flows across this section per unit time is

$$4\mu\nu\,d\eta d\zeta f_\xi,$$

f_ξ being the rate of flow of heat at P across the surface ξ.

Therefore the rate at which heat flows into the element across the face $ABCD$ is ultimately

$$4\left\{\mu\nu f_\xi - \frac{\partial}{\partial\xi}(\mu\nu f_\xi)\,d\xi\right\} d\eta d\zeta,$$

and the rate at which heat flows out across the face $A'B'C'D'$ is

$$4\left\{\mu\nu f_\xi + \frac{\partial}{\partial\xi}(\mu\nu f_\xi)\,d\xi\right\} d\eta d\zeta.$$

Hence the total rate of gain of heat from these two faces is

$$-8\frac{\partial}{\partial\xi}(\mu\nu f_\xi)\,d\xi d\eta d\zeta.$$

The other faces give respectively

$$-8\frac{\partial}{\partial\eta}(\nu\lambda f_\eta)\,d\xi d\eta d\zeta, \qquad -8\frac{\partial}{\partial\zeta}(\lambda\mu f_\zeta)\,d\xi d\eta d\zeta.$$

Inserting the values of f_ξ, f_η, f_ζ, namely,

$$f_\xi = -\frac{K}{\lambda}\frac{\partial v}{\partial\xi}, \qquad f_\eta = -\frac{K}{\mu}\frac{\partial v}{\partial\eta}, \qquad f_\zeta = -\frac{K}{\nu}\frac{\partial v}{\partial\zeta},$$

and equating the expression we thus obtain to

$$8\lambda\mu\nu\,d\xi d\eta d\zeta\, c\rho\frac{\partial v}{\partial t},$$

we have

$$\lambda\mu\nu c\rho\frac{\partial v}{\partial t} = \frac{\partial}{\partial\xi}\left(\frac{\mu\nu}{\lambda}K\frac{\partial v}{\partial\xi}\right) + \frac{\partial}{\partial\eta}\left(\frac{\nu\lambda}{\mu}K\frac{\partial v}{\partial\eta}\right) + \frac{\partial}{\partial\zeta}\left(\frac{\lambda\mu}{\nu}K\frac{\partial v}{\partial\zeta}\right),$$

which reduces to

$$\lambda\mu\nu\frac{\partial v}{\partial t} = \kappa\left[\frac{\partial}{\partial\xi}\left(\frac{\mu\nu}{\lambda}\frac{\partial v}{\partial\xi}\right) + \frac{\partial}{\partial\eta}\left(\frac{\nu\lambda}{\mu}\frac{\partial v}{\partial\eta}\right) + \frac{\partial}{\partial\zeta}\left(\frac{\lambda\mu}{\nu}\frac{\partial v}{\partial\zeta}\right)\right], \qquad (1)$$

when K is constant, and as usual we have written $\kappa = K/c\rho$.

Spherical polar coordinates

In this system
$$x = r \sin \theta \cos \phi,$$
$$y = r \sin \theta \sin \phi,$$
$$z = r \cos \theta,$$

and
$$ds^2 = dr^2 + r^2 \, d\theta^2 + r^2 \sin^2\theta \, d\phi^2.$$

Therefore the equation for v becomes

$$\frac{\partial v}{\partial t} = \frac{\kappa}{r^2}\left[\frac{\partial}{\partial r}\left(r^2\frac{\partial v}{\partial r}\right) + \frac{1}{\sin \theta}\frac{\partial}{\partial \theta}\left(\sin \theta \frac{\partial v}{\partial \theta}\right) + \frac{1}{\sin^2\theta}\frac{\partial^2 v}{\partial \phi^2}\right], \tag{2}$$

which may be written

$$\frac{\partial v}{\partial t} = \kappa\left[\frac{\partial^2 v}{\partial r^2} + \frac{2}{r}\frac{\partial v}{\partial r} + \frac{1}{r^2}\frac{\partial}{\partial \mu}\left\{(1-\mu^2)\frac{\partial v}{\partial \mu}\right\} + \frac{1}{r^2(1-\mu^2)}\frac{\partial^2 v}{\partial \phi^2}\right], \tag{3}$$

where $\mu = \cos \theta$.

Cylindrical coordinates

In this system $\quad x = r \cos \theta, \quad\quad y = r \sin \theta,$

and $\quad\quad\quad\quad\quad ds^2 = dr^2 + r^2 \, d\theta^2 + dz^2.$

Therefore the equation for v becomes

$$\frac{\partial v}{\partial t} = \frac{\kappa}{r}\left[\frac{\partial}{\partial r}\left(r\frac{\partial v}{\partial r}\right) + \frac{\partial}{\partial \theta}\left(\frac{1}{r}\frac{\partial v}{\partial \theta}\right) + \frac{\partial}{\partial z}\left(r\frac{\partial v}{\partial z}\right)\right], \tag{4}$$

which may be written

$$\frac{\partial v}{\partial t} = \kappa\left[\frac{\partial^2 v}{\partial r^2} + \frac{1}{r}\frac{\partial v}{\partial r} + \frac{1}{r^2}\frac{\partial^2 v}{\partial \theta^2} + \frac{\partial^2 v}{\partial z^2}\right]. \tag{5}$$

1.9. Initial and boundary conditions

Before we can proceed to the mathematical discussion of the problems of Conduction, it is necessary to determine the formulae which will express the Initial and Boundary Conditions which the temperature satisfies. These are partly the direct expression of the results of experiment and partly the mathematical statement of hypotheses founded upon these results.

We assume that in the interior of the solid v is a continuous function of x, y, z, and t; and that this holds also for the first differential coefficient with regard to t and for the first and second differential coefficients with regard to x, y, and z. At the boundary of the solid, and at the instant at which flow of heat is supposed to start, these assumptions are not made.

I. *Initial conditions*

The temperature throughout the body is supposed given arbitrarily at the instant which we take as the origin of the time coordinate t. If this arbitrary function is continuous, we require to find a solution of our problem which shall, as t tends to zero, tend to the given value. In other words, if the initial temperature is given by

$$v = f(x, y, z),$$

our solution of the equation

$$\frac{\partial v}{\partial t} = \kappa \nabla^2 v,$$

must be such that $\qquad \lim_{t \to 0} (v) = f(x, y, z)$

at all points of the solid.

If the initial distribution is discontinuous at points or surfaces, these discontinuities must disappear after ever so short a time, and in this case our solution must converge to the value given by the initial temperature at all points where this distribution is continuous.

II. *Boundary or surface conditions*

The surface conditions usually arising in the mathematical theory of Conduction of Heat are the following:

A. *Prescribed surface temperature.* This temperature may be constant, or a function of time, or position, or both. This is the easiest boundary condition to work with and that which has been most studied, but it must be remarked that in practice it is often difficult to prescribe surface temperature, and actual conditions may be better represented by a boundary condition of type D below.

B. *No flux across the surface, i.e.*

$$\frac{\partial v}{\partial n} = 0, \quad \text{at all points of the surface,}$$

where $\partial/\partial n$ denotes differentiation in the direction of the outward normal to the surface.

C. *Prescribed flux across the surface.*

D. *Linear heat transfer at the surface. The 'radiation' boundary condition.* If the flux across the surface is proportional to the temperature difference between the surface and the surrounding medium, so that it is given by

$$H(v - v_0), \tag{1}$$

where v_0 is the temperature of the medium and H is a constant, the boundary condition is

$$K\frac{\partial v}{\partial n}+H(v-v_0) = 0, \quad \text{or} \quad \frac{\partial v}{\partial n}+h(v-v_0) = 0, \tag{2}$$

where
$$h = H/K. \tag{3}$$

As† $h \to 0$ this tends to the boundary condition B, and as $h \to \infty$ it tends to the condition A.

The quantity H has been called the 'Outer' or 'Surface' Conductivity, but is now usually referred to as the *surface conductance* or the *coefficient of surface heat transfer*. It is often simpler to specify the *surface thermal resistance* per unit area, $R = 1/H$.

If, in addition, there is prescribed flux F into the surface, (2) is replaced by

$$\frac{\partial v}{\partial n}+h\left(v-v_0-\frac{F}{H}\right) = 0, \tag{4}$$

which is of the same form as (2) with v_0 replaced by $v_0+(F/H)$.

This boundary condition is referred to in the classical works as the 'radiation' boundary condition because of the application referred to in (12) below. This description is a little misleading since, in fact, when heat is transferred by radiation, the flux depends on the fourth powers of the absolute temperatures, but for shortness it will be used here so that the boundary condition (2) will usually be described as 'radiation into a medium at v_0' instead of the clumsier but more accurate 'linear heat transfer into a medium at v_0'. It is also sometimes referred to as 'Newton's Law', since Newton's Law of Cooling stated that for a body cooling 'in a draught' (that is, by forced convection, see (i) below) the rate of loss of heat is given by (1).

The flux across a boundary surface is given by (1) in several different physical connexions which are described briefly below. In each case a few numerical values are given which will enable the order of magnitude of H in a practical problem to be estimated roughly; accurate values are given in works on heat transmission.‡

(i) *Forced convection.* If fluid (gas or liquid) at temperature v_0 is forced rapidly past the surface of the solid, it is found experimentally that the rate of loss of heat from the surface is given by (1) with a value of the coefficient H which depends on the velocity and nature of the fluid and the shape of the surface. Most of the

† It is always understood that $h > 0$; $h < 0$ would correspond to *supply* of heat over the surface at a rate proportional to its temperature. The mathematical solutions for the case $h > 0$ frequently are not valid for the case $h = 0$.

‡ e.g. Fishenden and Saunders, *Heat Transfer* (Oxford, 1950); Jakob, *Heat Transfer* (Wiley, 1949); McAdams, *Heat Transmission* (McGraw-Hill, edn. 2, 1942); Gröber, Erk, and Grigull, *Wärmeübertragung* (Springer, 1955).

experiments have been made on fluid flowing inside circular pipes or flowing outside and perpendicular to circular cylinders, and the results have been expressed in terms of approximate power laws of which the following are typical:

For turbulent flow of air with velocity u cm/sec inside a circular pipe of diameter d cm

$$H \doteqdot 5 \cdot 5 \times 10^{-6} u^{0 \cdot 8} d^{-0 \cdot 2} \, \text{cal}/(\text{cm})^2(\text{sec})(^\circ \text{C}). \tag{5}$$

For water flowing in the same way the result is of the same general form but 500 to 1,000 times larger.

For turbulent flow of air with velocity u perpendicular to a circular cylinder of diameter d

$$H = 8 \times 10^{-5} (u/d)^{1/2} \, \text{cal}/(\text{cm})^2(\text{sec})(^\circ \text{C}). \tag{6}$$

For water the results are about 100 times larger.

It appears that in all cases the rate of loss of heat increases considerably with decrease of the diameter of the cylinder. For a plane surface H is of the same order as for a fairly large cylinder.

A law of type (1) holds approximately for heat transfer by combined forced convection and evaporation.

(ii) *A thin surface skin of poor conductor.* It frequently happens that on the surface of a body there is a thin skin of a poor conductor such as scale, grease, or oxide. Also, if the body is being cooled by gas or liquid flowing over the surface, there is usually a thin layer of undisturbed fluid in contact with the surface and, since this layer is not in motion, it has a relatively low thermal conductivity.

If the conductivity of the skin is K' and its thickness is d, and if as a first approximation we neglect the heat capacity of the skin,† the rate of flow of heat through it, per unit area per unit time, is

$$\frac{K'}{d}(v - v_0), \tag{7}$$

where v and v_0 are the temperatures inside and outside the skin. This is equivalent to the boundary condition

$$K \frac{\partial v}{\partial n} + \frac{K'}{d}(v - v_0) = 0, \tag{8}$$

for the medium inside the skin.

Medium Conductivity K'	Air 0·000053	Vaseline 0·00044	Water 0·00144	Mercury* 0·020
$d = 0 \cdot 1$ cm	0·00053	0·0044	0·0144	0·20
0·01	0·0053	0·044	0·144	2·0
0·001	0·053	0·44	1·44	20

* A layer of mercury or an amalgam is often used to secure good thermal contact.

Values‡ of K'/d for various substances are shown in the table, units being c.g.s. and $^\circ$ C. A typical value for a layer of boiler scale is 0·1.

If the outside of the skin, instead of being kept at v_0, loses heat at a rate H

---

† This is equivalent to neglecting the term in $\partial v / \partial t$ in the equation of conduction, and we then have a case of steady flow in which 1.2 (1) holds. Fox, *Phil. Mag.* (7) **18** (1934) 209–27, shows that a second approximation, taking into account the heat capacity of the skin, leads to a boundary condition of type (14) below; see also § 12.8.

‡ The reciprocal of this, d/K', is called the thermal resistance of the skin. For examples of the use of this concept see § 3.2.

times its temperature difference from a surrounding medium which is at v_1, the boundary condition becomes

$$K\frac{\partial v}{\partial n}+\frac{1}{(1/H)+(d/K')}\,(v-v_1) = 0.\tag{9}$$

E. *Non-linear heat transfer.* In most practical cases the flux of heat from the surface is not a linear function of the temperature difference between the surface and its surroundings, though it may be approximated to by a law of type (1) for small ranges of temperature. Typical examples are:

(i) *Black-body radiation.* A body at absolute temperature T surrounded by a black body at temperature T_0 will lose heat at the rate

$$\sigma E(T^4-T_0^4),\tag{10}$$

where σ is the Stefan–Bolzmann constant and E is the emissivity of the surface, that is the ratio of the heat emitted by it to that emitted by a black body at the same temperature.

For polished metals E runs from 0·02 to 0·05, for oxidized metals it is of the order of 0·6 or 0·7, for common substances such as paint, glass, paper, or wood 0·7 to 0·9, while for soot, lampblack, etc., it may rise as high as 0·98.

Introducing the numerical value of σ, (10) becomes

$$1\!\cdot\!37\times 10^{-12}E(T^4-T_0^4)\ \text{cal/(cm)}^2(\text{sec}).\tag{11}$$

If $T-T_0$ is not large this is approximately

$$5\!\cdot\!48\times 10^{-12}ET_0^3(T-T_0),\tag{12}$$

and if $T_0 = 300$ it becomes $1\!\cdot\!48\times 10^{-4}E(T-T_0)$.

Thus in this case the result (1) with constant H only holds approximately, and if $T-T_0$ is large will be seriously wrong. Few exact solutions of problems on conduction of heat in the variable state with the accurate boundary condition (10) have been found.†

In ordinary circumstances a body loses heat both by radiation and convection, and the coefficient H in (1) will be a compound one which takes both effects into account.

(ii) *Natural convection.* When a hot body is surrounded by fluid, the hotter fluid in the neighbourhood of the body tends to rise and in this way convection currents are set up. This process is called natural convection. In this case it is found experimentally that the rate of loss of heat from the body is proportional, not to the temperature difference $v-v_0$ between the body and the surrounding fluid, but very nearly to the 5/4 power of it.

For example, for surfaces a few centimetres or more in width the rate of loss of heat in air is approximately

$$5\times 10^{-5}(v-v_0)^{5/4}\ \text{cal/(cm)}^2(\text{sec}).\tag{13}$$

For very fine wires in air the value may be twenty times as large. In water the values are roughly 100 times as large.

As in the case of radiation exact solutions cannot be given for the accurate law (13) and an approximate value of H must be used in (1). It should be remarked that for bodies exposed in air at normal temperatures (12) and (13) give values of the same order of magnitude so both must be taken into account.

† The semi-infinite solid with a boundary condition involving (11) or (13) is discussed by Jaeger, *Proc. Camb. Phil. Soc.* **46** (1950) 634–41. A treatment of the general case using integral equations is given by Mann and Wolf, *Quart. Appl. Math.* **9** (1951) 163–84.

F. *Contact with a well-stirred fluid or perfect conductor.* In calorimetry, and elsewhere, it frequently happens that the surface of a solid is in contact with fluid which is so well stirred that its temperature may be taken to be constant throughout. Suppose the solid has conductivity K, surface area S, and surface temperature v, which must be constant over its surface. Let mass M of well-stirred fluid of specific heat c' be in contact with it and suppose V is the temperature of the fluid. For generality we suppose that heat is supplied to the mass M from an external source at the rate Q per unit time and that it loses heat by radiation into a medium at v_0 at the rate $H_1(V-v_0)$. If δV is the increase in temperature of the mass M in the small time δt, we have

$$Q\,\delta t - H_1(V-v_0)\,\delta t - K\,\delta t \iint \frac{\partial v}{\partial n}\,dS = Mc'\,\delta V,$$

i.e.
$$K \iint \frac{\partial v}{\partial n}\,dS + Mc'\frac{dV}{dt} + H_1(V-v_0) - Q = 0. \tag{14}$$

If we assume that the surface temperature of the solid equals that of the fluid for $t > 0$ (they need not, of course, be equal at $t = 0$), we have, in addition to (14),

$$v = V, \qquad t > 0. \tag{15}$$

If, instead, there is heat transfer between the solid and fluid by a law of type D, we have in place of (15)

$$K\frac{\partial v}{\partial n} + H(v-V) = 0. \tag{16}$$

Other physical conditions lead to boundary conditions of type (14); for example, if the surface of the solid is in contact with mass M of well-stirred fluid of which mass m per unit time is withdrawn and replaced by the same mass of fluid at v_0, we find

$$Mc'\frac{dV}{dt} + K \iint \frac{\partial v}{\partial n}\,dS + mc'(V-v_0) = 0. \tag{17}$$

The same boundary conditions (14), (15), (16) arise if the mass M is a perfectly conducting solid. When a metallic conductor is in contact with a non-metal, the conductivity of the metal is so much larger that a good approximation is often obtained by treating it as a perfect conductor. The resulting problem is much easier to solve than that for the composite region.

The essential difference between such boundary conditions and those of A to D above is the appearance of the term dV/dt or, in some cases,

$\partial v/\partial t$. The classical methods cannot always be used without modification, but the Laplace transformation procedure of Chapter XII applies equally well to both types. Some examples† are given in that chapter.

G. *The surface of separation of two media of different conductivities* K_1 *and* K_2. Let v_1 and v_2 denote the temperatures in the two media. It has been shown in § 1.3 that the flux is continuous over the surface of separation, that is

$$K_1 \frac{\partial v_1}{\partial n} = K_2 \frac{\partial v_2}{\partial n}, \tag{18}$$

where $\partial/\partial n$ denotes differentiation along the normal to the surface of separation.

If we can assume that at the surface of separation the temperatures in the two media are the same we have, in addition to (18),

$$v_1 = v_2. \tag{19}$$

This assumption will only be valid for very intimate contact, such as a soldered joint; in all other cases, even for optically flat surfaces pressed lightly together,‡ heat transfer between the two media takes place largely by the mechanisms D (ii) and E (i) described above, that is, the rate of transfer between the two surfaces is proportional to their temperature difference, so that

$$-K_1 \frac{\partial v_1}{\partial n} = H(v_1 - v_2). \qquad h = \frac{H}{k} \tag{20}$$

For this case (18) and (20) are the boundary conditions. $H = Kh$

H. *Contact with a thin skin of much better conductor.* This case arises, for example, if a thin metal sheet or wire is in contact with a relatively poor conductor such as soil, foodstuff, or insulating material. It also arises in problems of surface

† Problems leading to boundary conditions of these types are discussed in the following: Peddie, *Proc. Edin. Math. Soc.* **19** (1901) 34–35; March and Weaver, *Phys. Rev.* (2) **31** (1928) 1072–82; Schumann, *Phys. Rev.* **37** (1931) 1508–15; Peek, *Ann. Math. Princeton,* (2) **30** (1929) 265; Langer, *Tôhoku Math. J.* **35** (1932) 260–75; Lowan, *Phil. Mag.* (7) **17** (1934) 849–54; Jaeger, *J. Proc. Roy. Soc. N.S.W.* **74** (1940) 342–52, and **75** (1941) 130–9; *Aust. J. Phys.* **9** (1956) 167–79; Blackwell, *J. Appl. Phys.* **25** (1954) 137–44; Gaskell, *Amer. J. Math.* **64** (1942) 447–55. The last four authors use the Laplace transformation method, the others extensions of the classical methods. A case in which the quantity of fluid increases linearly with the time is discussed by Chao and Weiner, *Quart. Appl. Math.* **14** (1956) 214–17. These problems are also of great importance in connexion with diffusion, cf. *M.D.,* § 4.35, also Barrer, *Diffusion in and through Solids* (Cambridge, 1941), Chap. i.

‡ Jacobs and Starr, *Rev. Sci. Instr.* **10** (1941) 140, have measured H for optically flat surfaces pressed together *in vacuo*. If the surfaces are just touching H is very small, heat transfer being due to radiation only; as the pressure is increased H increases, for example for silver surfaces with a pressure of 2 kg/cm², $H = 0.07$.

or grain boundary diffusion.† The assumption made is that the skin is so thin that the temperature at any point of it is constant across its thickness d. If V is the temperature at any point of the skin and K_1 and κ_1 are the conductivity and diffusivity of its material, the equation of conduction of heat in the skin, derived by considering the heat balance of an element of area of the skin, is

$$\frac{\partial^2 V}{\partial \xi^2} + \frac{\partial^2 V}{\partial \eta^2} - \frac{1}{\kappa_1}\frac{\partial V}{\partial t} - \frac{K}{dK_1}\frac{\partial v}{\partial n} = 0, \tag{21}$$

where $\partial/\partial n$ denotes differentiation in the direction of the outward normal to the solid, and $\partial/\partial \xi$ and $\partial/\partial \eta$ correspond to two perpendicular directions.

If there is perfect thermal contact between the skin and the solid, the boundary conditions are (21) and $v = V$. If there is linear heat transfer between the skin and the interior or exterior, a slight further complication occurs.

For the case of a wire of radius a along the z-axis in solid $r > a$, (21) becomes

$$\frac{\partial^2 V}{\partial z^2} - \frac{1}{\kappa_1}\frac{\partial V}{\partial t} + \frac{2K}{aK_1}\frac{\partial v}{\partial r} = 0. \tag{22}$$

If the skin is so thin that its heat capacity is negligible, (21) reduces to

$$\frac{\partial^2 V}{\partial \xi^2} + \frac{\partial^2 V}{\partial \eta^2} - \frac{K}{dK_1}\frac{\partial v}{\partial n} = 0. \tag{23}$$

1.10. Dimensionless parameters

The solutions of problems in conduction of heat can always be expressed in terms of a number of dimensionless quantities. For example, consider the equation of linear flow of heat

$$\frac{\partial^2 v}{\partial x^2} - \frac{1}{\kappa}\frac{\partial v}{\partial t} = 0 \tag{1}$$

in the region $-l < x < l$ with boundary conditions of type 1.9 (2), namely

$$\frac{\partial v}{\partial x} + hv = 0, \quad x = l; \qquad \frac{\partial v}{\partial x} - hv = 0, \quad x = -l. \tag{2}$$

The position of the point x can be specified by the position ratio

$$\xi = x/l, \tag{3}$$

and the time can be specified by the quantity

$$\tau = \kappa t/l^2 \tag{4}$$

† It has been introduced by Jaeger, *Quart. J. Mech. Appl. Math.* **8** (1955) 101–6, in connexion with thermal problems and by Whipple, *Phil. Mag.* (7) **45** (1954) 1225–36, in connexion with grain boundary diffusion. The latter gives an alternative derivation of (21). Jaeger discusses the case of a wire of good conductor, and Whipple that of a plane sheet. See also § 14.11.

which was shown in § 1.6 to be dimensionless. Finally, in place of h in the boundary condition (2), the dimensionless quantity

$$L = lh \tag{5}$$

may be used.†

The solutions of problems on (1) with the boundary conditions (2) can always be expressed in terms of the three variables ξ, τ, L in place of the original quantities l, x, κ, t, h. It is always desirable to make this change before making numerical computations from the solutions. It will appear later that in very many cases alternative forms of solution, suitable for small, moderate, and large values of τ, can be found.

The range of the values of L and τ which occur in practice is enormous. With a value of $\kappa = 0{\cdot}01$ appropriate to a poor conductor, τ becomes large after a minute or so for a thin sheet 1 mm or less in thickness: on the other hand, for a body of the size of the Earth, τ has always been small throughout geological time.

It is possible to express the differential equation and boundary conditions in terms of dimensionless variables. For example, making the change of variables (3), (4), (5), in (1) and (2), these become

$$\frac{\partial^2 v}{\partial \xi^2} - \frac{\partial v}{\partial \tau} = 0, \qquad -1 < \xi < 1, \tag{6}$$

with $\quad \dfrac{\partial v}{\partial \xi} + Lv = 0, \quad \xi = 1; \qquad \dfrac{\partial v}{\partial \xi} - Lv = 0, \quad \xi = -1. \tag{7}$

This procedure is attractive from the pure-mathematical point of view and makes the mathematics a little more concise. It will not be adopted here since the physical significance of a formula is clearer if it is expressed in the original physical variables.

1.11. Experimental methods for the determination of thermal conductivity

A large number of methods‡ has at one time or another been used for measuring thermal conductivity: some of these must now be regarded as obsolete, but their theory remains of interest as they are based on solutions of the equations of conduction of heat for simple systems which often occur in practice.

Firstly it should be remarked that the thermal properties of any material occur in various combinations which may be regarded as characteristic of, and measured by, different experimental situations. These are: (a) the conductivity K which is measured by steady state experiments; (b) the heat capacity per unit volume ρc

† τ is occasionally referred to as the Fourier modulus or number, and L as the Nusselt number. Other important numbers of this sort appear in the study of heat transfer by convection.

‡ General surveys of the methods are given by Ingersoll, *J. Opt. Soc. Amer. and Rev. Sci. Instr.* **9** (1924) 495; Griffiths, *Proc. Phys. Soc.* **41** (1928) 151; and in the article on Conduction of Heat in the *Dictionary of Applied Physics*.

which is measured by calorimetry; (c) the quantity $(K\rho c)^{\frac{1}{2}}$ which is measured by some simple steady periodic experiments; (d) the diffusivity κ which is measured by the simplest variable state experiments. In fact, most variable state experiments, in principle, allow both K and κ to be determined.

We give here a brief classification of the commoner methods with references to the sections in which they are discussed. They divide essentially into Steady State, Periodic Heating, and Variable State methods, and subdivide again into methods suitable for poor conductors and for metals.

(i) *Steady state methods: Poor conductors.* The usual method consists of carrying out accurately the fundamental experiment of § 1.2 on a slab of the material.†
Alternatively the material may be used in the form of a hollow cylinder, § 7.2, or a hollow sphere, § 9.2. The flow of heat in a thick rod of the material has occasionally been used, but the theory is complicated, cf. §§ 6.1, 6.2, 8.3.

(ii) *Steady state thermal methods: Metals.* The metal is usually in the form of a rod whose ends are kept at different temperatures. The semi-infinite rod is considered in § 4.3, the rod of finite length in § 4.5.

(iii) *Steady state electrical methods: Metals.* The metal in the form of a wire is heated by passing electric current through it, the ends being kept at prescribed temperatures, §§ 4.11, 8.3 IX. Radial flow in a wire heated electrically has also been used, § 7.2 V.

(iv) *Steady state flow methods: Liquids.* The temperature in liquid flowing between two reservoirs at different temperatures is measured, § 4.9.

(v) *Periodic heating methods.* In these the conditions at the ends of a rod or slab are varied with period T; when steady conditions have been established the temperatures at certain points are studied. The semi-infinite rod is considered in § 4.4 and the rod of finite length in § 4.8. A similar method is used to find the diffusivity of soil from the temperature fluctuations caused by solar heating, § 2.12.

These methods have recently become of great importance for measurements at low temperatures: they have also the advantage that the theory of relatively complicated systems may be discussed by methods developed for the study of electrical wave-guides, cf. § 2.6.

(vi) *Variable state methods.* In the past, these have been used rather less than steady state methods. They suffer from the disadvantage that it is difficult to know how nearly the actual boundary conditions in an experiment agree with those postulated in the theory; the effect of a discrepancy of this sort, such as a contact resistance at a boundary, is more difficult to allow for, and may be more important than in steady state experiments; for discussion of this matter see § 2.10. On the other hand, variable state experiments have advantages of their own: for example, some methods allow for very rapid measurements and involve only small changes of temperature, also, some methods may be used *in situ* without removing a sample to the laboratory, this is very desirable in connexion with materials such as soils and rocks. In most of the older methods only the latter part of a curve of temperature against time, in which the solution consists of a single exponential term, is used; the case of a body of simple geometrical form cooling by linear heat transfer from its surface is discussed in §§ 4.7, 6.5, 8.5, 9.5. The use of the variable temperature in a wire heated by electric current is discussed in § 4.14. In some cases the complete curve of temperature variation at a point has been used, cf. §§ 2.10, 3.3.

† Griffiths and Kaye, *Proc. Roy. Soc.* A, **104** (1923) 71.

In recent years much attention has been paid to the possibility of determining both K and κ from a single experiment. In one method, a quantity which ultimately increases linearly with the time (for example, the quantity of heat which flows in time t through a slab with its faces at constant temperature) is measured and from the parameters of this linear asymptote both K and κ can be determined, cf. §§ 12.6 II, 15.6. In another method, a heating probe is used whose temperature ultimately varies linearly with $\ln t$, cf. § 13.7.

1.12. The mathematical interpretation of the initial and boundary conditions

In the mathematical treatment the boundary and initial conditions are not regarded as conditions which the temperature v must satisfy on the surface itself or at the instant $t = 0$. They are taken as limiting conditions. The boundary conditions are to be understood in the sense that, for fixed $t > 0$, the given combination of the temperature and its derivatives is to tend to the prescribed value as we approach a point of the surface. The initial conditions are to be understood in the sense that, for a fixed point within the region, the temperature is to tend to the prescribed value as $t \to 0$.

For example, for the problem which may be stated briefly as '*Conduction of heat in the region* $0 < x < l$. $x = 0$ *is maintained at zero temperature for* $t > 0$, *and at* $x = l$ *there is radiation into a medium at zero. The initial temperature of the solid is unity*', the equations to be satisfied are

$$\kappa \frac{\partial^2 v}{\partial x^2} = \frac{\partial v}{\partial t}, \quad 0 < x < l, \quad t > 0,$$

$$v \to 0 \quad \text{as } x \to 0, \text{ for fixed } t > 0,$$

$$\frac{\partial v}{\partial x} + hv \to 0 \quad \text{as } x \to l-0, \text{ for fixed } t > 0,$$

$$v \to 1 \quad \text{as } t \to 0, \text{ for fixed } x \text{ in } 0 < x < l.$$

1.13. Related differential equations

In essence, the whole of this book is concerned with the solution of the equations

$$\operatorname{div} \mathbf{f} + \rho c \frac{\partial v}{\partial t} = A, \tag{1}$$

$$\mathbf{f} = -K \operatorname{grad} v, \tag{2}$$

or

$$\nabla^2 v - \frac{1}{\kappa} \frac{\partial v}{\partial t} = -\frac{A}{K}, \tag{3}$$

in regions of various shapes with boundary conditions which are usually expressed in terms of v and $\mathbf{f}$. Ordinarily, K, ρ, and c are constants, and A may be a function of position or time.

A number of generalizations of (3), notably

$$\nabla^2 v - \frac{1}{\kappa} \frac{\partial v}{\partial t} = Av + B \tag{4}$$

and

$$\nabla^2 v - \frac{U}{\kappa} \frac{\partial v}{\partial x} - \frac{1}{\kappa} \frac{\partial v}{\partial t} = Av + B, \tag{5}$$

where U, A, and B are constants, have been noted in §§ 1.6 and 1.7 and may be treated by the same methods. As remarked earlier, cases in which the thermal parameters, K, c, etc., are functions of v are becoming of increasing importance but have not yet been studied to any great extent.

The same sets of equations and boundary conditions arise in many other physical contexts and many of the solutions given here can be taken over with a simple change of notation. On the other hand, there are many small changes in the types of problem which are of practical importance. Some of the main applications are noted briefly below.

I. *Diffusion*

This subject is now covered very thoroughly in the treatises[†] by Crank, Barrer, and Jost. If C is the concentration of the diffusing substance and $\mathbf{F}$ is its rate of transfer, (1) and (2) are replaced by

$$\operatorname{div}\mathbf{F} = -\frac{\partial C}{\partial t} \quad \text{and} \quad \mathbf{F} = -D\operatorname{grad}C. \tag{6}$$

If D is constant, the equations of diffusion are thus the same as those of conduction of heat with $K = \kappa = D$, so that a slight simplification results. At the surface of separation of two media, 1.9 (18), (19) are replaced by

$$D_1(\partial C_1/\partial n) = D_2(\partial C_2/\partial n) \quad \text{and} \quad C_1 = kC_2, \tag{7}$$

where k is a constant. The results given later for composite media thus apply also to this case after a change of notation.

Equations of types (4) and (5) arise much more frequently. They occur, for example, in problems on simultaneous diffusion and chemical reaction,[‡] in connexion with diffusion in tissue containing cells,[§] in the theory of consolidation of soils,[||] and in genetics. Problems of simultaneous diffusion of heat and moisture lead to a pair of simultaneous differential equations.[††]

II. *Diffusion in a field of force. Sedimentation*

In such problems an equation of type

$$\nabla^2 C = \frac{1}{D}\frac{\partial C}{\partial t} + A\operatorname{div}(\mathbf{U}C)$$

appears,[‡‡] where $\mathbf{U}$ is a constant vector. This is, essentially, the equation of conduction of heat (5) in a moving medium.

III. *The slowing down of neutrons*

On certain assumptions, this problem is reduced to the solution of problems in conduction of heat,[§§] mostly involving Green's functions.

[†] Crank, *The Mathematics of Diffusion* (Oxford, 1956), referred to as *M.D.*; Barrer, *Diffusion in and through Solids* (Cambridge, 1941); Jost, *Diffusion* (Acad. Press, 1952). Many problems leading to the diffusion equation are discussed by Babbitt, *Canad. J. Res.* A, **28** (1950) 449–74. Many physiological applications are given by Hill, *Proc. Roy. Soc.* B, **104** (1929) 39–96, and by Thews, *Acta Biotheoretica*, A, **10** (1953) 105–38.

[‡] *M.D.*, Chap. VIII; Danckwerts, *Trans. Faraday Soc.* **47** (1951) 1014–23.

[§] Keynes, *Proc. Roy. Soc.* B, **142** (1954) 359–82.

[||] Gibson and Henkel, *Geotechnique*, **4** (1954) 6–15.

[††] *M.D.*, Chap. XIII; Henry, *Proc. Roy. Soc.* A, **171** (1939) 215.

[‡‡] Riemann–Weber, *Die Differential und Integralgleichungen der Mechanik und Physik*, ed. Frank (Vieweg, 1927) Vol. 2, Chap. 7; Fürth, *Geofis. Pura é Appl.* **31** (1955) 80–89; Davies, *Proc. Roy. Soc.* A, **200** (1949) 100–13; Hulburt, *Phys. Rev.* (2) **31** (1928) 1018 gives equations for diffusion of ions in the ionosphere.

[§§] Marshak, *Rev. Mod. Phys.* **19** (1947) 185–238; Sneddon, *Fourier Transforms* (McGraw-Hill, 1951) Chap. 6.

IV. *Viscous motion*

The equation of conduction of heat appears in two related but slightly different contexts. Firstly, many problems of one-dimensional laminar motion ·lead directly to (3) in one variable.† Secondly, the differential equations satisfied by the vorticity are of 'Diffusion' type.‡

V. *Electrical problems*

The differential equation for the potential in a non-inductive transmission line§ takes the form (4) in one space-variable if the line is leaky, or (3) if there is no leakage. See, also, § 2.6.

VI. *Flow of fluids through porous media*

The differential equation governing the flow of a compressible liquid‖ through a porous medium is precisely (3). It, and also (4) and (5), arise in many other problems of this type.††

1.14. Simplification of the general problem of conduction

In this section several classical methods of reducing general problems in conduction to simpler ones will be noted. It should be remarked that if the Laplace transformation method of Chapters XII to XV is used, none of these devices is necessary, all problems being treated in the same manner.

I. *Surface conditions independent of the time*

Suppose we have to satisfy

$$\nabla^2 v - \frac{1}{\kappa}\frac{\partial v}{\partial t} = A(x, y, z) \tag{1}$$

throughout the solid, with $v = f(x, y, z)$, initially, and $v = \phi(x, y, z)$ at the surface. Put

$$v = u + w, \tag{2}$$

where u is a function of x, y, z only which satisfies

$$\nabla^2 u = A(x, y, z) \tag{3}$$

throughout the solid and

$$u = \phi(x, y, z) \tag{4}$$

† Lamb, *Hydrodynamics* (Camb., edn. 4, 1916) § 345.
‡ Lamb, ibid., § 328; Goldstein, *Proc. Lond. Math. Soc.* (2) **34** (1932) 51; McEwen, *J. Marine Res.* **7** (1948) 188–216.
§ Carslaw and Jaeger, *Operational Methods in Applied Mathematics* (Oxford, edn. 2, 1947) Chap. 9.
‖ Muskat, *The Flow of Homogeneous Fluids through Porous Media* (McGraw-Hill, 1937) Chap. 10, gives many solutions of problems of this sort, mostly involving line sources and cylindrical boundaries.
†† Rosenhead and Miller, *Proc. Roy. Soc.* A, **163** (1937) 298–317; Hantush and Jacob, *Trans. Amer. Geophys. Union*, **36** (1955) 95–112.

at its surface, while w is a function of x, y, z, t, which satisfies

$$\nabla^2 w - \frac{1}{\kappa}\frac{\partial w}{\partial t} = 0, \tag{5}$$

throughout the solid,

$$w = f(x, y, z) - u, \text{ initially,} \tag{6}$$

and $\qquad\qquad w = 0, \text{ at the surface.} \tag{7}$

Clearly v given by (2) satisfies all the conditions of the problem, so that the solution is reduced to that of two problems, one of steady temperature, and one of variable temperature with prescribed initial temperature and zero surface temperature. The case of radiation at the surface into a medium at constant temperature may be treated in the same way.

II. *Surface conditions prescribed functions of the time. No heat production*

The solution for this case can be deduced from that for constant surface conditions by the use of *Duhamel's theorem*,† namely:

If $v = F(x, y, z, \lambda, t)$ represents the temperature at (x, y, z) at the time t in a solid in which the initial temperature is zero, while its surface temperature is $\phi(x, y, z, \lambda)$, then the solution of the problem in which the initial temperature is zero, and the surface temperature is $\phi(x, y, z, t)$, is given by

$$v = \int_0^t \frac{\partial}{\partial t} F(x, y, z, \lambda, t - \lambda)\, d\lambda.$$

When the surface temperature is zero from $t = -\infty$ to $t = 0$, and $\phi(x, y, z, \lambda)$ from $t = 0$ to $t = t$, we may say that the initial temperature is zero and the surface temperature is $\phi(x, y, z, \lambda)$, so that the temperature at the time t is given by

$$v = F(x, y, z, \lambda, t), \quad \text{when } t > 0.$$

Therefore, when the surface temperature is zero from $t = -\infty$ to $t = \lambda$ and $\phi(x, y, z, \lambda)$ from $t = \lambda$ to $t = t$, we have

$$v = F(x, y, z, \lambda, t - \lambda), \quad \text{when } t > \lambda.$$

Also, when the surface temperature is zero from $t = -\infty$ to $t = \lambda + d\lambda$ and $\phi(x, y, z, \lambda)$ from $t = \lambda + d\lambda$ to $t = t$, we have

$$v = F(x, y, z, \lambda, t - \lambda - d\lambda), \quad \text{when } t > \lambda + d\lambda.$$

Hence, when the surface temperature is zero from $t = -\infty$ to $t = \lambda$,

† *Mémoire sur la méthode générale relative au mouvement de la chaleur dans les corps solides plongés dans les milieux dont la température varie avec le temps*, cf. J. Éc. polyt. Paris, **14** (1833) Cah. 22, p. 20.

$\phi(x, y, z, \lambda)$ from $t = \lambda$ to $t = \lambda + d\lambda$, and zero from $t = \lambda + d\lambda$ to $t = t$, we have

$$v = F(x, y, z, \lambda, t - \lambda) - F(x, y, z, \lambda, t - \lambda - d\lambda),$$

or ultimately

$$v = \frac{\partial}{\partial t} F(x, y, z, \lambda, t - \lambda)\, d\lambda \quad (t > \lambda).$$

In this way, by breaking up the interval $t = 0$ to $t = t$ into these small intervals, and then summing the results thus obtained, we find the solution of the problem for the surface temperature $\phi(x, y, z, t)$ in the form

$$v = \int_0^t \frac{\partial}{\partial t} F(x, y, z, \lambda, t - \lambda)\, d\lambda. \tag{8}$$

The corresponding theorem for the case of radiation is as follows:

If $v = F(x, y, z, \lambda, t)$ represents the temperature at (x, y, z) at the time t in a solid in which the initial temperature is zero, while radiation takes place at its surface into a medium at $\phi(x, y, z, \lambda)$, then the solution of the problem in which the initial temperature is zero, and the temperature of the medium is $\phi(x, y, z, t)$, is given by

$$v = \int_0^t \frac{\partial}{\partial t} F(x, y, z, \lambda, t - \lambda)\, d\lambda. \tag{9}$$

When the surface temperature, or the temperature of the medium into which radiation takes place, does not vary from point to point but changes only with time, these results may be stated in a slightly simpler form as follows:

If $v = F(x, y, z, t)$ represents the temperature at (x, y, z) at the time t in a solid in which the initial temperature is zero, while its surface is kept at temperature unity [or, in the case of radiation, while radiation takes place into a medium at temperature unity], then the solution of the problem when the surface is kept at temperature $\phi(t)$ [or, in the case of radiation, while radiation takes place into a medium at temperature $\phi(t)$], is given by

$$v = \int_0^t \phi(\lambda) \frac{\partial}{\partial t} F(x, y, z, t - \lambda)\, d\lambda. \tag{10}$$

This follows at once since $F(x, y, z, \lambda, t)$ now takes the simpler form

$$F(x, y, z, t) \phi(\lambda).$$

Returning, now, to the general problem with varying surface temperature, this requires the solution of (1) with $v = f(x, y, z)$, initially, and $v = \phi(x, y, z, t)$ at the surface.

This is satisfied by $v = u+w$, where

$$\nabla^2 u - \frac{1}{\kappa}\frac{\partial u}{\partial t} = A(x,y,z), \tag{11}$$

$$u = f(x,y,z), \text{ initially,} \tag{12}$$

$$u = 0, \text{ at the surface,} \tag{13}$$

and
$$\nabla^2 w - \frac{1}{\kappa}\frac{\partial w}{\partial t} = 0, \tag{14}$$

$$w = 0, \text{ initially,} \tag{15}$$

$$w = \phi(x,y,z,t), \text{ at the surface.} \tag{16}$$

The equations for w we have just discussed, while those for u have been considered in I above. Problems involving other boundary conditions may be reduced in the same way.

In fact equation (8) holds in much more general cases than those given above: for example, consider the differential equation

$$\frac{\partial}{\partial x}\Big(K_1\frac{\partial v}{\partial x}\Big)+\frac{\partial}{\partial y}\Big(K_2\frac{\partial v}{\partial y}\Big)+\frac{\partial}{\partial z}\Big(K_3\frac{\partial v}{\partial z}\Big)+K_4 v+A(x,y,z,t) = \rho c\frac{\partial v}{\partial t}, \tag{17}$$

where K_1, K_2, K_3, K_4, and ρc may be functions of x, y, z. This by §§ 1.6, 1.17 is the equation of conduction of heat† in a non-homogeneous anisotropic solid in which heat is generated at the rate $A(x,y,z,t)$ per unit volume. Suppose the boundary condition is

$$k_1\frac{\partial v}{\partial x} + k_2\frac{\partial v}{\partial y}+k_3\frac{\partial v}{\partial z}+k_4 v = g(x,y,z,t), \tag{18}$$

where $k_1,..., k_4$ are functions of x, y, z only. Finally, suppose the initial condition is

$$v \to \phi(x,y,z) \quad \text{as } t \to 0. \tag{19}$$

Now let $F(x,y,z,\lambda,t)$ be the solution of the same problem except that $A(x,y,z,t)$ and $g(x,y,z,t)$ are replaced by $A(x,y,z,\lambda)$ and $g(x,y,z,\lambda)$, the values of these functions at time λ. Then‡ the solution of (17), (18), and (19) is

$$v(x,y,z,t) = \frac{\partial}{\partial t}\int_0^t F(x,y,z,\lambda,t-\lambda)\,d\lambda = \phi(x,y,z)+ \int_0^t \frac{\partial}{\partial t} F(x,y,z,\lambda,t-\lambda)\,d\lambda, \tag{20}$$

which is of the same form as (8) if $\phi(x,y,z) = 0$. It may be added that the result still holds if there are discontinuities in the thermal properties of the medium.

III. *The case of the equation*

$$\nabla^2 v - bv - \frac{1}{\kappa}\frac{\partial v}{\partial t} = A(x,y,z), \tag{21}$$

---

† The term $K_4 v$ is added for greater generality. Such a term appears when each element of volume of the solid loses heat at a rate proportional to its temperature, as in the problems of Chapter IV and in other physical applications.

‡ A proof using the Laplace Transformation method is given by Bartels and Churchill, *Bull. Amer. Math. Soc.* **48** (1942) 276. It may be extended to cover more general boundary conditions such as those of § 1.9 F.

where b is a constant which may have either sign. The substitution

$$v = ue^{-\kappa bt} \tag{22}$$

reduces this to

$$\nabla^2 u - \frac{1}{\kappa}\frac{\partial u}{\partial t} = e^{\kappa bt}A(x,y,z), \tag{23}$$

which may be treated by the preceding methods.

Another important case for which even simpler treatment† is possible is that of

$$\nabla^2 v - bv - \frac{1}{\kappa}\frac{\partial v}{\partial t} = 0, \tag{24}$$

where b is a constant (which may have either sign), with zero initial temperature and boundary conditions of either constant temperature or 'radiation' at the surface. In this case, if u is the solution for the case $b = 0$ and the same boundary conditions, it may be verified by differentiation that

$$v = \kappa b \int_0^t e^{-\kappa bt'}u(t')\,dt' + ue^{-\kappa bt} \tag{25}$$

satisfies (24) and the boundary conditions. Thus solutions for this case may be obtained by simple integration from those given later.

1.15. Problems whose solutions can be expressed as a product of solutions of simpler problems

Consider the equation of conduction of heat

$$\frac{\partial^2 v}{\partial x_1^2} + \frac{\partial^2 v}{\partial x_2^2} + \frac{\partial^2 v}{\partial x_3^2} = \frac{1}{\kappa}\frac{\partial v}{\partial t}, \quad t > 0, \tag{1}$$

in the rectangular parallelepiped

$$a_1 < x_1 < b_1, \qquad a_2 < x_2 < b_2, \qquad a_3 < x_3 < b_3. \tag{2}$$

For certain important types of initial and boundary conditions the solution of this is the product of the solutions of three one-variable problems, and thus can be written down immediately if these are known.

Suppose $v_r(x_r, t)$, $r = 1, 2, 3$, is the solution of

$$\frac{\partial^2 v_r}{\partial x_r^2} = \frac{1}{\kappa}\frac{\partial v_r}{\partial t}, \quad a_r < x_r < b_r, \quad t > 0, \tag{3}$$

with boundary conditions

$$\alpha_r \frac{\partial v_r}{\partial x_r} - \beta_r v_r = 0, \quad x_r = a_r, \quad t > 0, \tag{4}$$

$$\alpha_r' \frac{\partial v_r}{\partial x_r} + \beta_r' v_r = 0, \quad x_r = b_r, \quad t > 0, \tag{5}$$

(where the α_r and β_r, etc., are constants, either of which may be zero, so that the cases of zero surface temperature and of no flow of heat at

† Danckwerts, *Trans. Faraday Soc.* **47** (1951) 1014–23 gives explicit formulae for the slab, sphere, cylinder, and semi-infinite solid.

the surface are included) and with initial conditions

$$v_r(x_r, t) = V_r(x_r), \quad t = 0, \quad a_r < x_r < b_r. \tag{6}$$

Then the solution of (1) in the region (2), with

$$v = V_1(x_1)V_2(x_2)V_3(x_3), \quad \text{when } t = 0, \tag{7}$$

and with boundary conditions

$$\alpha_r \frac{\partial v}{\partial x_r} - \beta_r v = 0, \quad x_r = a_r, \quad t > 0, \quad r = 1, 2, 3, \tag{8}$$

$$\alpha_r' \frac{\partial v}{\partial x_r} + \beta_r' v = 0, \quad x_r = b_r, \quad t > 0, \quad r = 1, 2, 3, \tag{9}$$

is
$$v = v_1(x_1, t)v_2(x_2, t)v_3(x_3, t). \tag{10}$$

For substituting (10) in (1) gives

$$v_2 v_3 \frac{\partial^2 v_1}{\partial x_1^2} + v_3 v_1 \frac{\partial^2 v_2}{\partial x_2^2} + v_1 v_2 \frac{\partial^2 v_3}{\partial x_3^2} - \frac{1}{\kappa}\left(v_2 v_3 \frac{\partial v_1}{\partial t} + v_3 v_1 \frac{\partial v_2}{\partial t} + v_1 v_2 \frac{\partial v_3}{\partial t}\right) = 0,$$

using (3). And clearly the initial and boundary conditions (7), (8), and (9) are satisfied.

A similar result holds for combined radial and axial flow in a solid or hollow cylinder. Here the differential equation 1.8 (5) becomes, since we are assuming that all quantities are independent of θ,

$$\frac{1}{r}\frac{\partial}{\partial r}\left(r\frac{\partial v}{\partial r}\right) + \frac{\partial^2 v}{\partial z^2} = \frac{1}{\kappa}\frac{\partial v}{\partial t}. \tag{11}$$

Suppose it has to be solved in the region

$$a < r < b, \quad z_1 < z < z_2. \tag{12}$$

Let $v_1(r, t)$ be the solution of

$$\frac{1}{r}\frac{\partial}{\partial r}\left(r\frac{\partial v_1}{\partial r}\right) = \frac{1}{\kappa}\frac{\partial v_1}{\partial t}, \quad t > 0, \quad a < r < b,$$

with

$$\alpha_1 \frac{\partial v_1}{\partial r} - \beta_1 v_1 = 0, \quad r = a, \quad t > 0,$$

$$\alpha_1' \frac{\partial v_1}{\partial r} + \beta_1' v_1 = 0, \quad r = b, \quad t > 0,$$

and $v_1 = V_1(r)$ when $t = 0$.

Also let $v_2(z, t)$ be the solution of

$$\frac{\partial^2 v_2}{\partial z^2} = \frac{1}{\kappa}\frac{\partial v_2}{\partial t}, \quad t > 0, \quad z_1 < z < z_2,$$

with

$$\alpha_2 \frac{\partial v_2}{\partial z} - \beta_2 v_2 = 0, \quad z = z_1, \quad t > 0,$$

$$\alpha_2' \frac{\partial v_2}{\partial z} + \beta_2' v_2 = 0, \quad z = z_2, \quad t > 0,$$

and
$$v_2 = V_2(z) \quad \text{when } t = 0.$$

Then $v = v_1(r, t) v_2(z, t)$ is the solution of (11) in the region (12) with boundary conditions

$$\alpha_1 \frac{\partial v}{\partial r} - \beta_1 v = 0, \quad r = a, \quad z_1 < z < z_2, \quad t > 0,$$

$$\alpha_1' \frac{\partial v}{\partial r} + \beta_1' v = 0, \quad r = b, \quad z_1 < z < z_2, \quad t > 0,$$

$$\alpha_2 \frac{\partial v}{\partial z} - \beta_2 v = 0, \quad z = z_1, \quad a < r < b, \quad t > 0,$$

$$\alpha_2' \frac{\partial v}{\partial z} + \beta_2' v = 0, \quad z = z_2, \quad a < r < b, \quad t > 0,$$

and with initial condition

$$v = V_1(r) V_2(z), \quad \text{when } t = 0.$$

The same procedure may be applied to other regions such as the infinite rectangular corner, $x > 0$, $y > 0$, the semi-infinite cylinder, etc. Examples are given in §§ 5.6, 6.4, 8.4. If the solid is anisotropic and the axes are chosen so that the differential equation takes the form 1.18 (4) the method may still be applied if the bounding surfaces are perpendicular to the axes.

These results are of practical importance[†] because they allow numerical values for the temperatures in the simple solids mentioned above to be written down very simply if the solid has constant initial temperature and at the surface there is radiation into a medium at constant temperature.

1.16. The uniqueness of the solution of the problem

We consider the problem of conduction of heat in a finite closed region with prescribed initial and surface temperatures.

† Newman, *Ind. Eng. Chem.* **28** (1936) 545; Olson and Schultz. ibid. **34** (1942) 874.

If possible, let there be two independent solutions, v_1, v_2 of the equations

$$\frac{\partial v}{\partial t} = \kappa \nabla^2 v \quad \text{in the solid}$$

$$v = f(x, y, z) \quad \text{for } t = 0 \text{ in the solid} \left.\begin{array}{c} \\ \\ \\ \end{array}\right\} . \qquad (1)$$

$$v = \phi(x, y, z, t) \quad \text{at the surface}$$

Let $V = v_1 - v_2$. Then V satisfies

$$\frac{\partial V}{\partial t} = \kappa \nabla^2 V \quad \text{in the solid}$$

$$V = 0 \quad \text{for } t = 0 \text{ in the solid} \left.\begin{array}{c} \\ \\ \\ \end{array}\right\} . \qquad (2)$$

$$V = 0 \quad \text{at the surface}$$

If the equations (1) have a unique solution, we must have $v_1 \equiv v_2$, i.e. $V \equiv 0$.

Consider the volume integral

$$J = \tfrac{1}{2} \iiint V^2 \, dx \, dy \, dz \geqslant 0, \qquad (3)$$

the integration being taken through the solid.

Then $\quad \dfrac{\partial J}{\partial t} = \iiint V \dfrac{\partial V}{\partial t} \, dx \, dy \, dz = \kappa \iiint V \nabla^2 V \, dx \, dy \, dz. \qquad (4)$

But putting

$$\xi = V \frac{\partial V}{\partial x}, \qquad \eta = V \frac{\partial V}{\partial y}, \qquad \zeta = V \frac{\partial V}{\partial z}$$

in Green's theorem, 1.7 (4), and assuming that V satisfies sufficient conditions for that theorem to hold (e.g. continuity of V and its first and second derivatives), we have

$$\iint V \frac{\partial v}{\partial n} \, dS$$

$$= \iiint V \nabla^2 V \, dx \, dy \, dz + \iiint \left\{ \left(\frac{\partial V}{\partial x} \right)^2 + \left(\frac{\partial V}{\partial y} \right)^2 + \left(\frac{\partial V}{\partial z} \right)^2 \right\} dx \, dy \, dz,$$

the integrals being taken over the surface and through the volume of the solid. Therefore

$$\frac{\partial J}{\partial t} = \kappa \iint V \frac{\partial V}{\partial n} \, dS - \kappa \iiint \left\{ \left(\frac{\partial V}{\partial x} \right)^2 + \left(\frac{\partial V}{\partial y} \right)^2 + \left(\frac{\partial V}{\partial z} \right)^2 \right\} dx \, dy \, dz. \quad (5)$$

Since V is zero over the surface, the first integral vanishes provided $(\partial V/\partial n)$ is bounded on the surface, and we obtain

$$\frac{\partial J}{\partial t} = -\kappa \iiint \left\{ \left(\frac{\partial V}{\partial x} \right)^2 + \left(\frac{\partial V}{\partial y} \right)^2 + \left(\frac{\partial V}{\partial z} \right)^2 \right\} dx \, dy \, dz.$$

Therefore
$$\frac{\partial J}{\partial t} \leqslant 0. \tag{6}$$

If we could assert that
$$J = 0 \quad \text{when } t = 0, \tag{7}$$

it would follow from (6) that $J \leqslant 0$ when $t > 0$. From this and (3) we must have $J = 0$ and, since V is continuous, it follows that $V = 0$. This is the classical proof of uniqueness which has been reproduced in many places.

It was remarked by Doetsch† that further assumptions as to V are needed to ensure the validity of (7). In fact it does not follow that, if $V(x, y, z, t) \to 0$ as $t \to 0$ for any fixed x, y, z in the volume, then $J \to 0$ as $t \to 0$. The simplest illustration of this arises in the one-dimensional semi-infinite region $x > 0$.

Consider the function
$$V(x, t) = xt^{-\frac{3}{2}}e^{-x^2/4\kappa t}. \tag{8}$$

This satisfies the equation of Conduction of Heat in one dimension,
$$\kappa \frac{\partial^2 V}{\partial x^2} = \frac{\partial V}{\partial t}.$$

Also $\qquad V(x, t) = 0 \quad \text{for } x = 0, t > 0,$

and $\qquad V(x, t) \to 0 \quad \text{as } t \to 0 \text{ for any fixed } x > 0.$

Thus the function (8) satisfies the equation of Conduction of Heat and vanishes for $t = 0$, and on the boundary of the region, but it does not vanish identically. Physically it is the temperature due to a doublet and will be discussed further in § 10.8. It does not tend to zero uniformly in x as $t \to 0$, and is unbounded in the neighbourhood of $x = 0$, $t = 0$, e.g. for $x = t^{\frac{1}{2}}$.

The integral corresponding to (3) for this function is
$$J = \frac{1}{2t^3} \int\limits_0^\infty x^2 e^{-x^2/2\kappa t} \, dx = \left(\frac{\pi\kappa^3}{8t^3}\right)^{\frac{1}{2}},$$

and so $J \to \infty$ as $t \to 0$. Thus (7) is not satisfied, and the above uniqueness proof fails in this case as it should.

If (7) is to be valid, further assumptions about the function V must be made which will exclude such functions. Clearly (7) holds if we assume either

(i) $V \to 0$ as $t \to 0$ uniformly for x, y, z in the volume,

or (ii) $\partial V/\partial x$, $\partial V/\partial y$, $\partial V/\partial z$ are all less than a constant M which is independent of x, y, z in the volume, and of t in $0 < t \leqslant t_0$.

† *Math. Z.* **22** (1925) 293; ibid. **25** (1926) 608. *Enseign. math.* **35** (1936) 43.

But few attempts have yet been made to establish broad conditions under which the solution is unique, and the tendency has been to prove the uniqueness of the solution of each special problem.

A similar discussion† can be applied to other boundary conditions, to anisotropic media,‡ and to the case of steady temperature.

To prove that the equations have a solution is an even more difficult matter than that of proving uniqueness. The physical interpretation of the equations requires that there be a solution: the mathematical demonstration of such existence theorems belongs to pure analysis.

1.17. Conduction of heat in an anisotropic solid

Anisotropic media§ are of considerable importance in practice. The commonest examples are crystals, naturally occurring non-crystalline substances such as sedimentary rocks or wood, and the laminated materials such as transformer cores used in engineering practice.

For such substances the results of §§ 1.3, 1.4 hold unchanged, but it will appear that it is not usually true that the direction of the flux vector at a point is normal to the isothermal through the point. The simplest fundamental assumption, generalizing that of 1.5 (3) for the isotropic solid, is that each component of the flux vector at a point is a linear function of the components of the temperature gradient at the point, that is

$$
\left.
\begin{aligned}
-f_x &= K_{11}\frac{\partial v}{\partial x} + K_{12}\frac{\partial v}{\partial y} + K_{13}\frac{\partial v}{\partial z} \\
-f_v &= K_{21}\frac{\partial v}{\partial x} + K_{22}\frac{\partial v}{\partial y} + K_{23}\frac{\partial v}{\partial z} \\
-f_z &= K_{31}\frac{\partial v}{\partial x} + K_{32}\frac{\partial v}{\partial y} + K_{33}\frac{\partial v}{\partial z}
\end{aligned}
\right\}.
\tag{1}
$$

† More careful discussions are given by Goursat, *Cours d'Analyse* (Gauthier–Villars, 1927), Vol. III, Chap. 29; Titchmarsh, *Fourier Integrals* (Oxford, 1937) pp. 281–3; Tychonoff, *Rec. Math.* (*Mat. Sbornik*), **42** (1935) 199–216; Churchill, *Amer. J. Math.* **61** (1939) 651; Mersman, *Bull. Amer. Math. Soc.* **47** (1941) 956; Widder, *Trans. Amer. Math. Soc.* **55** (1944) 85–95, **75** (1953) 510–25; Cooper, *J. Lond. Math. Soc.* **25** (1950) 173–80; Birkhoff and Kotik, *Proc. Amer. Math. Soc.* **5** (1954) 162–8; Kampé de Fériet, *C.R. Acad. Sci. Paris*, **236** (1953) 1527–9; Fulks, *Pacific J. Math.* **2** (1952) 141–5, **3** (1953) 387–91, 567–83; Hartman and Wintner, *Amer. J. Math.* **72** (1950) 367–95; Rayner, *Quart. J. Mech. Appl. Math.* **6** (1953) 385–90.

‡ In this case we use Green's theorem 1.7 (4) with $\xi = Vf_x$, etc., and use 1.6 (3), 1.17 (1), and the fact that the expression 2Φ defined in 1.20 (3) is always positive.

§ The mathematical theory of conduction of heat in crystals was first developed by Duhamel, *J. Éc. polyt. Paris*, **13**, Cah. 21 (1832) 356, **19**, Cah. 32 (1848) 155, and Lamé, *Leçons sur la théorie de la chaleur* (Paris, 1861) on the hypothesis of a mechanism of 'molecular radiation'. The modern treatment in essentially the form given here is due to Stokes, *Camb. and Dublin Math. J.* **6** (1851) 215–38. A very full analytical treatment is given in Boussinesq, *Théorie analytique de la chaleur* (Paris, 1901). From the point of view of crystal physics, the subject is developed at length in Voigt, *Lehrbuch der Krystallphysik* (Leipzig, 1910), and, briefly but in a more modern form, in Wooster, *Text-book on Crystal Physics* (Cambridge, 1949). Because of the difficulty of making accurate measurements on conduction of heat, and, in particular, on crystals, little accurate experimental information is available even now, and few special problems have been solved.

The quantities K_{rs} are called the conductivity coefficients, they are the components of a second-order tensor. The equations (1) may be solved for $\partial v/\partial x$, etc., to give

$$
\left.
\begin{aligned}
-\frac{\partial v}{\partial x} &= R_{11}f_x + R_{12}f_y + R_{13}f_z \\
-\frac{\partial v}{\partial y} &= R_{21}f_x + R_{22}f_y + R_{23}f_z \\
-\frac{\partial v}{\partial z} &= R_{31}f_x + R_{32}f_y + R_{33}f_z
\end{aligned}
\right\}, \tag{2}
$$

where the R_{rs}, which can be written down as determinants involving the K_{rs}, are called resistivity coefficients. For example,

$$
R_{11} = (K_{22}K_{33} - K_{23}K_{32})/\Delta, \qquad R_{12} = (K_{13}K_{32} - K_{12}K_{33})/\Delta, \tag{3}
$$

$$
R_{21} = (K_{31}K_{23} - K_{21}K_{33})/\Delta, \qquad R_{22} = (K_{11}K_{33} - K_{31}K_{13})/\Delta, \tag{4}
$$

where

$$
\Delta = \begin{vmatrix} K_{11} & K_{12} & K_{13} \\ K_{21} & K_{22} & K_{23} \\ K_{31} & K_{32} & K_{33} \end{vmatrix}, \tag{5}
$$

and, similarly, the K_{rs} can be expressed in terms of the R_{rs}. It should be noted that some problems can be formulated most naturally in terms of the K_{rs} and some in terms of the R_{rs}. If the equations (2) are regarded as fundamental they may be solved for f_x, f_y, f_z in terms of $\partial v/\partial x$, etc., and it may be noted that the determinant which then appears, namely

$$
\Delta' = \begin{vmatrix} R_{11} & R_{12} & R_{13} \\ R_{21} & R_{22} & R_{23} \\ R_{31} & R_{32} & R_{33} \end{vmatrix}, \tag{6}
$$

is $(1/\Delta^3)$ times the adjoint of Δ so that, by a general theorem,[†]

$$
\Delta' = 1/\Delta, \tag{7}
$$

and the minors of Δ' are $1/\Delta$ times the algebraic complements of the corresponding minors of Δ, for example,

$$
R_{11}R_{22} - R_{12}R_{21} = K_{33}/\Delta. \tag{8}
$$

It should first be noticed that it is implied in (1) that the signs of f_x, f_y, and f_z are changed if those of all the components of the temperature gradient are changed. That is, essentially, that the conductivity of the material in opposite directions is the same. For the case of a crystal with a centre of symmetry, this follows from symmetry: this is the case for 21 of the 32 crystal classes. The remaining 11 classes do not have centres of symmetry, and for these the form (1) must be regarded as being justified by experiments[‡] which have shown that the conductivities in opposite directions are approximately the same.

The form (1) is that which frequently relates two vectors in anisotropic media. Because of crystalline symmetry, it frequently simplifies if the axes are chosen in appropriate crystallographic directions. The results for various crystal systems are given below; details may be found in Wooster's *Crystal Physics*, chap. I, § 4.

[†] Bôcher, *Higher Algebra* (Macmillan, 1924).

[‡] Experiments have been made on polar crystals such as tourmaline. They were at first believed to have indicated a substantial difference in conductivity but it was subsequently shown by Thomson and Lodge, *Phil. Mag.* (5) **8** (1879) 18, and Stenger, *Wied. Ann.* **22** (1884) 522, that the difference, if any, is small.

Triclinic system. There is no simplification.

Monoclinic system. All classes of this system have either a diad axis (such that a rotation of 180° about it brings the crystal into a position congruent with its original one) or a plane of reflection symmetry. If the z-axis is either the diad axis or is normal to the plane of reflection symmetry, the scheme of conductivities is

$$\begin{array}{ccc} K_{11} & K_{12} & 0 \\ K_{21} & K_{22} & 0 \\ 0 & 0 & K_{33}. \end{array} \qquad (9)$$

It will be remarked below that it is probably true also that $K_{21} = K_{12}$, but this does not follow from considerations of symmetry.

Orthorhombic system. All classes of this system have either two perpendicular diad axes, or a diad axis with a plane of symmetry through it. If one of the axes is taken along a diad axis, and another along a second diad axis or in the plane of symmetry, the scheme of conductivities is

$$\begin{array}{ccc} K_{11} & 0 & 0 \\ 0 & K_{22} & 0 \\ 0 & 0 & K_{33}. \end{array} \qquad (10)$$

Cubic system. In this case a cyclical interchange of the axes of the orthorhombic system is possible, so that the scheme of conductivities becomes

$$\begin{array}{ccc} K_{11} & 0 & 0 \\ 0 & K_{11} & 0 \\ 0 & 0 & K_{11}. \end{array} \qquad (11)$$

Tetragonal, Trigonal, and Hexagonal systems. If the z-axis is a tetrad, triad, or hexad axis (corresponding to rotations of 90°, 120°, or 60° respectively) the scheme of conductivities reduces to

$$\begin{array}{ccc} K_{11} & K_{12} & 0 \\ -K_{12} & K_{11} & 0 \\ 0 & 0 & K_{33}. \end{array} \qquad (12)$$

In some classes of these systems there is, in addition, a diad axis perpendicular to the z-axis or a plane of symmetry through it, and, if the x-axis is taken along this axis or in this plane, $K_{12} = 0$ in (12). There are some classes of these systems for which it does not follow from considerations of symmetry alone that $K_{12} = 0$, though, as remarked below, this relation probably holds.

The results above are all that can be deduced from considerations of macroscopic symmetry. But, in fact, it is probably true that the K_{rs} in (1) are symmetrical, that is, that $K_{rs} = K_{sr}$ for all r and s. This implies that $K_{12} = 0$ in (12) and $K_{21} = K_{12}$ in (9).

In many branches of crystal physics in which a fundamental law of type (1) appears, it follows from classical thermodynamics that the coefficients are symmetrical, that is, $K_{rs} = K_{sr}$. In the present case no such general proof is possible, so that it has been necessary to rely on experiment to show that the coefficients are symmetrical. For this reason, the mathematical theory has usually been developed without using the assumption of symmetry so that effects of asymmetry can be calculated and compared with experiment, cf. § 1.19. Recently, proofs of the symmetry law, based on Onsager's principle of microscopic reversibility have been given.†

Formulae for change of axes are frequently needed. Suppose that we wish to transfer to a new system of rectangular axes x', y', z', whose direction cosines relative to the old x, y, z-system are (c_{11}, c_{21}, c_{31}), (c_{12}, c_{22}, c_{32}), (c_{13}, c_{23}, c_{33}), respectively, then the conductivity coefficients K'_{ik} relative to the x', y', z'-system are given by

$$K'_{ik} = \sum_{r=1}^{3} \sum_{s=1}^{3} c_{ri} c_{sk} K_{rs}, \tag{13}$$

while the K_{rs} are expressed in terms of the K'_{ik} by

$$K_{rs} = \sum_{i=1}^{3} \sum_{k=1}^{3} c_{ri} c_{sk} K'_{ik}. \tag{14}$$

These are the transformation laws for a second order tensor; for a proof and application in the present context, see Wooster's *Crystal Physics*. The same transformation laws apply to the R_{rs}.

The most important case of a non-crystalline anisotropic material is that of the *orthotropic solid* which has different conductivities K_1, K_2, K_3 in three mutually perpendicular directions. Taking these as the axes of x, y, and z, we have

$$f_x = -K_1 \frac{\partial v}{\partial x}, \qquad f_y = -K_2 \frac{\partial v}{\partial y}, \qquad f_z = -K_3 \frac{\partial v}{\partial z}. \tag{15}$$

Again, for a substance such as wood which has different conductivities K_1, K_2, K_3, in the directions r, θ, z, of a particular system of cylindrical coordinates‡ (i.e. in the direction of the rays, the rings, and the axis of the tree) the fluxes in these directions are

$$f_r = -K_1 \frac{\partial v}{\partial r}, \qquad f_\theta = -K_2 \frac{\partial v}{r \, \partial \theta}, \qquad f_z = -K_3 \frac{\partial v}{\partial z}. \tag{16}$$

1.18. The differential equation of conduction of heat in an aniso-tropic solid

We now proceed to develop the theory on the general assumption 1.17 (1). Equation 1.6 (3) remains valid in the present case, and substituting 1.17 (1) in it gives the equation of conduction of heat

$$\rho c \frac{\partial v}{\partial t} = K_{11} \frac{\partial^2 v}{\partial x^2} + K_{22} \frac{\partial^2 v}{\partial y^2} + K_{33} \frac{\partial^2 v}{\partial z^2} + (K_{23} + K_{32}) \frac{\partial^2 v}{\partial y \partial z} +$$
$$+ (K_{31} + K_{13}) \frac{\partial^2 v}{\partial z \partial x} + (K_{12} + K_{21}) \frac{\partial^2 v}{\partial x \partial y}, \quad (1)$$

provided that the medium is homogeneous and that no heat is generated in it. The extension to other cases is made as in § 1.6.

† Crandall, *Physica*, **21** (1955) 251–2; Casimir, *Rev. Mod. Phys.* **17** (1945) 343–50; Onsager, *Phys. Rev.* (2) **37** (1931) 405, ibid. (2) **38** (1931) 2265.

It should be noticed that the equation of conduction of heat, 1.18 (1), which is essentially the expression of the first law of thermodynamics, only involves the K_{rs} in the combination $(K_{rs} + K_{sr})$, so that their symmetry or lack of it does not affect deductions from this equation. It is only in calculations of the direction of the flux vector that the effects of anti-symmetry appear.

‡ Griffiths and Kaye, *Proc. Roy. Soc.* A, **104** (1923) 71.

Now consider the quadric†

$$K_{11}x^2 + K_{22}y^2 + K_{33}z^2 + (K_{23}+K_{32})yz + (K_{31}+K_{13})zx + (K_{12}+K_{21})xy = \text{const.} \quad (2)$$

It is known that a transformation to a new system of rectangular coordinates, ξ, η, ζ, can be found which reduces the left-hand side of (2) to a sum of squares

$$K_1\xi^2 + K_2\eta^2 + K_3\zeta^2. \quad (3)$$

In terms of the same variables, (1) becomes

$$\rho c\frac{\partial v}{\partial t} = K_1\frac{\partial^2 v}{\partial \xi^2} + K_2\frac{\partial^2 v}{\partial \eta^2} + K_3\frac{\partial^2 v}{\partial \zeta^2}. \quad (4)$$

These new axes are called the *principal axes of conductivity* and the coefficients K_1, K_2, K_3 are the *principal conductivities*. It follows from § 1.17, (9) to (12), that when the crystal has axes of symmetry these will be principal axes of conductivity. If we make the additional transformation

$$\xi_1 = \xi(K/K_1)^{\frac{1}{2}}, \qquad \eta_1 = \eta(K/K_2)^{\frac{1}{2}}, \qquad \zeta_1 = \zeta(K/K_3)^{\frac{1}{2}}, \quad (5)$$

where K may be chosen arbitrarily, (4) becomes

$$\frac{\partial v}{\partial t} = \frac{K}{\rho c}\left(\frac{\partial^2 v}{\partial \xi_1^2} + \frac{\partial^2 v}{\partial \eta_1^2} + \frac{\partial^2 v}{\partial \zeta_1^2}\right), \quad (6)$$

which has the same form as the equation 1.6 (4) for the isotropic solid. Thus this transformation reduces problems on the anisotropic solid to the solution of corresponding problems on the isotropic solid when the solid is infinite, or when it is bounded by planes perpendicular to the principal axes of conductivity, or, in the case $K_2 = K_3$ by planes perpendicular to the axis of ξ and by circular cylinders with this as axis. In most other cases the bounding surfaces are distorted: for example, a circular cylinder with its axis along one of the principal axes becomes an elliptic cylinder.

For the homogeneous orthotropic solid of 1.17 (15) the equation of conduction of heat is

$$K_1\frac{\partial^2 v}{\partial x^2} + K_2\frac{\partial^2 v}{\partial y^2} + K_3\frac{\partial^2 v}{\partial z^2} - \rho c\frac{\partial v}{\partial t} = 0, \quad (7)$$

while for the solid of 1.17 (16) with cylindrical symmetry it is

$$\frac{K_1}{r}\frac{\partial}{\partial r}\left(r\frac{\partial v}{\partial r}\right) + \frac{K_2}{r^2}\frac{\partial^2 v}{\partial \theta^2} + K_3\frac{\partial^2 v}{\partial z^2} - \rho c\frac{\partial v}{\partial t} = 0. \quad (8)$$

Various important special cases in which the differential equation contains only one or two space variables may be noted:

I. *The temperature a function of x only*

By symmetry, this is the case of flow into a semi-infinite solid or slab with faces perpendicular to the x-axis and surface conditions independent of y and z. In this case $\partial v/\partial y = \partial v/\partial z = 0$ and 1.17 (1) gives

$$-f_x = K_{11}\frac{\partial v}{\partial x}, \qquad -f_y = K_{21}\frac{\partial v}{\partial x}, \qquad -f_z = K_{31}\frac{\partial v}{\partial x}. \quad (9)$$

Also the differential equation (1) becomes

$$K_{11}\frac{\partial^2 v}{\partial x^2} - \rho c\frac{\partial v}{\partial t} = 0. \quad (10)$$

† It will be shown in § 1.20 that $K_{11} \geqslant 0$, $K_{22} \geqslant 0$, $K_{33} \geqslant 0$.

Thus the whole of the theory of Chapters II and III holds for the anisotropic solid with

$$\kappa = K_{11}/\rho c. \tag{11}$$

If the x-axis has direction cosines l, m, n relative to the principal axes of conductivity, it follows from 1.17 (13) that

$$K_{11} = l^2 K_1 + m^2 K_2 + n^2 K_3. \tag{12}$$

When v has been found, the fluxes f_x, f_y, f_z follow from (9) and it appears that the direction of the flux vector is not normal to the isothermals. These results follow from the general theory leading to 1.20 (15).

II. *Flow of heat in the x-direction only*

This is the case of a thin rod in the direction of the x-axis. We now have $f_y = f_z = 0$ and, from 1.17 (2),

$$-\frac{\partial v}{\partial x} = R_{11} f_x, \qquad -\frac{\partial v}{\partial y} = R_{21} f_x, \qquad -\frac{\partial v}{\partial z} = R_{31} f_x. \tag{13}$$

The differential equation (1) now becomes by 1.6 (3)

$$\frac{1}{R_{11}} \frac{\partial^2 v}{\partial x^2} - \rho c \frac{\partial v}{\partial t} = 0, \tag{14}$$

and so it is the equation of linear flow of heat with

$$\kappa = 1/R_{11}\rho c. \tag{15}$$

It should be noted that R_{11} is not equal to $1/K_{11}$ but is given by 1.17 (3). If the x-axis has direction cosines (l, m, n) relative to the principal axes, we get in fact from 1.17 (13) for R_{rs},

$$R_{11} = \frac{l^2}{K_1} + \frac{m^2}{K_2} + \frac{n^2}{K_3}, \tag{16}$$

which is derived in a general way in 1.20 (25).

The theory of Chapter IV on the propagation of heat in rods may thus be taken over for crystalline rods with the value (15) of κ.

III. *The temperature a function of x and y only*

This is the case of flow into an infinite cylinder parallel to the z-axis with surface conditions independent of z. Since $\partial v/\partial z = 0$, (1) becomes

$$K_{11} \frac{\partial^2 v}{\partial x^2} + (K_{12} + K_{21}) \frac{\partial^2 v}{\partial x \partial y} + K_{22} \frac{\partial^2 v}{\partial y^2} - \rho c \frac{\partial v}{\partial t} = 0 \tag{17}$$

and the fluxes are given by 1.17 (1) with $\partial v/\partial z = 0$. It will be noticed that $f_z \neq 0$. This case is much less important than that of two-dimensional flow in a thin plate ($f_z = 0$) which is discussed in detail in § 1.19.

1.19. Conduction in a thin crystal plate

Conduction in two dimensions is of interest both as illustrating the main features of the general case and because several important methods of determining thermal conductivity involve the use of thin crystal plates. In this section the general theory of flow in such a plate, without any assumptions of symmetry, will be developed.

Taking x- and y-axes in the plane of the plate, we assume that there is no flow of heat in the z-direction so that $f_z = 0$ and 1.17 (2) gives

$$-\frac{\partial v}{\partial x} = R_{11} f_x + R_{12} f_y, \qquad -\frac{\partial v}{\partial y} = R_{21} f_x + R_{22} f_y. \tag{1}$$

Solving for f_x and f_y, and using 1.17 (3), (4), (8), gives

$$-f_x = K'_{11}\frac{\partial v}{\partial x}+K'_{12}\frac{\partial v}{\partial y}, \qquad -f_y = K'_{21}\frac{\partial v}{\partial x}+K'_{22}\frac{\partial v}{\partial y}, \tag{2}$$

where

$$K'_{11} = (K_{11}K_{33}-K_{31}K_{13})/K_{33}, \qquad K'_{12} = (K_{12}K_{33}-K_{13}K_{32})/K_{33}, \tag{3}$$

$$K'_{21} = (K_{21}K_{33}-K_{31}K_{23})/K_{33}, \qquad K'_{22} = (K_{22}K_{33}-K_{23}K_{32})/K_{33}. \tag{4}$$

The four quantities K'_{11}, etc., may be called the conductivity coefficients for a thin plate in the xy-plane: they reduce to K_{11}, etc., only if one of K_{13} and K_{31} and also one of K_{23} and K_{32} vanish.

Substituting (2) in 1.6 (3) gives as the equation of conduction of heat in the steady state

$$K'_{11}\frac{\partial^2 v}{\partial x^2}+(K'_{12}+K'_{21})\frac{\partial^2 v}{\partial x \partial y}+K'_{22}\frac{\partial^2 v}{\partial y^2} = 0. \tag{5}$$

The alternative form

$$R_{22}\frac{\partial^2 v}{\partial x^2}-(R_{12}+R_{21})\frac{\partial^2 v}{\partial x \partial y}+R_{11}\frac{\partial^2 v}{\partial y^2} = 0 \tag{6}$$

may be noted.

Referred to principal axes ξ, η as in § 1.18, (5) becomes

$$K'_1\frac{\partial^2 v}{\partial \xi^2}+K'_2\frac{\partial^2 v}{\partial \eta^2} = 0, \tag{7}$$

where K'_1 and K'_2 may be called the principal conductivities in the plane (the prime is used to distinguish them from the principal conductivities in three dimensions of § 1.18 with which they may coincide in special cases). Referred to these principal axes, the components of the flux, f_ξ, f_η, must take the form

$$-f_\xi = K'_1\frac{\partial v}{\partial \xi}+A\frac{\partial v}{\partial \eta}, \tag{8}$$

$$-f_\eta = -A\frac{\partial v}{\partial \xi}+K'_2\frac{\partial v}{\partial \eta}, \tag{9}$$

since the equation of conduction of heat (7) in these coordinates contains no term in $\partial^2 v/\partial\xi\partial\eta$, and, by comparison with (5), this requires that the sum of the co-efficients of $(\partial v/\partial\eta)$ in (8) and $(\partial v/\partial\xi)$ in (9) must vanish. K'_1, K'_2, and A can, in principle, be determined from the coefficients in (1) and (2).

As a specific example of considerable practical importance, we determine the isothermals and lines of flow of heat for the case of steady supply of heat to the plate at the origin. Writing, as in 1.18 (5),

$$\xi_1 = \xi(K/K'_1)^{\frac{1}{2}}, \qquad \eta_1 = \eta(K/K'_2)^{\frac{1}{2}}, \tag{10}$$

(7) becomes
$$\frac{\partial^2 v}{\partial \xi_1^2}+\frac{\partial^2 v}{\partial \eta_1^2} = 0, \tag{11}$$

of which a solution with radial symmetry is

$$v = -m\ln(\xi_1^2+\eta_1^2)/K = -m\ln\left(\frac{\xi^2}{K'_1}+\frac{\eta^2}{K'_2}\right), \tag{12}$$

where m is a constant. The flux is, by (8) and (9),

$$f_\xi = \frac{2m(K'_2\xi+A\eta)}{K'_2(\xi^2/K'_1+\eta^2/K'_2)}, \qquad f_\eta = -\frac{2m(A\xi-K'_1\eta)}{K'_1(\xi^2/K'_1+\eta^2/K'_2)}. \tag{13}$$

The total quantity of heat Q crossing a circle of radius a about the origin (per unit thickness of the plate) is

$$Q = \int_0^{2\pi} (f_\xi \cos\theta + f_\eta \sin\theta) a \, d\theta,$$

and, using (13) with $\xi = a\cos\theta$, $\eta = a\sin\theta$, this gives

$$Q = 4\pi m (K_1' K_2')^{\frac{1}{2}}, \tag{14}$$

which is independent of both a and A. With the value (14) of m, (12) gives the steady temperature due to supply of heat at the origin at the rate Q.

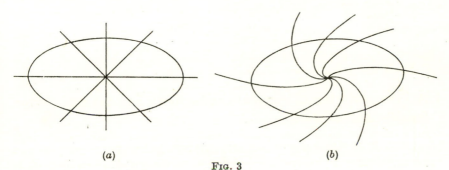

(a)   (b)

Fig. 3

The isothermals are the family of ellipses

$$\frac{\xi^2}{K_1'} + \frac{\eta^2}{K_2'} = \text{const.} \tag{15}$$

The direction of the flux vector is given by

$$\frac{f_\eta}{f_\xi} = \frac{K_2'(K_1'\eta - A\xi)}{K_1'(K_2'\xi + A\eta)}. \tag{16}$$

If $A = 0$, this direction is radially from the origin (and not normal to the equipotentials) so that the equipotentials and lines of flow of heat (curves whose direction at each point is in the direction of the flux vector) are as shown in Fig. 3 (a). If $A \neq 0$, the differential equation of the lines of flow of heat is

$$\frac{d\eta}{d\xi} = \frac{K_2'(K_1'\eta - A\xi)}{K_1'(K_2'\xi + A\eta)}. \tag{17}$$

The solution of this is

$$(K_1' K_2')^{\frac{1}{2}} \tan^{-1}(\eta K_1'^{\frac{1}{2}}/\xi K_2'^{\frac{1}{2}}) + \tfrac{1}{2} A \ln(K_1'\eta^2 + K_2'\xi^2) = \text{const.} \tag{18}$$

The curves (18) are the family of spirals shown in Fig. 3 (b): if $K_1' = K_2'$, or in the ξ_1, η_1 plane, these spirals are equiangular. Thus if the so-called 'rotatory' term A is not zero, the directions of flow of heat from a point source in an infinite plate are as shown in Fig. 3 (b). It follows that if a slit is cut in a radial direction in the plate, heat will not be able to flow in these spirals so that there should be a difference in temperature between the two sides of the slit. The fact that no such temperature difference has been observed in the few experiments of this type which have been carried out indicates that A is small. Other methods are described by Voigt (loc. cit.); it is concluded† that A is less than one thousandth of K_1 or K_2.

† Voigt, *Gött. Nach.* (1896) 223.

The elliptical form (15) of the isothermals has been used by de Senarmont[†] to determine the ratio $K_1':K_2'$ of the principal conductivities, the isothermals being made visible by melting a thin film of wax on the surface of the crystal plate.

For the case in which there is linear heat transfer between the surface of the plate and an external medium at zero temperature at the rate Hv per unit time per unit area, the equation of steady flow is found as in § 5.4 to be

$$K_1'\frac{\partial^2 v}{\partial \xi^2}+K_2'\frac{\partial^2 v}{\partial \eta^2}-\frac{2H}{D}v = 0, \tag{19}$$

referred to principal axes in the plane, where D is the thickness of the plate. The solution of (19) corresponding to steady flow from a point is now

$$v = mK_0\left\{\left(\frac{\xi^2}{K_1'}+\frac{\eta^2}{K_2'}\right)^{\frac{1}{2}}\left(\frac{2H}{D}\right)^{\frac{1}{2}}\right\}, \tag{20}$$

where K_0 is the Bessel function defined in Appendix III. The isothermals are still the ellipses (15), and the discussion above still holds.

1.20. The variation of thermal conductivity and the flux vector in anisotropic solids

In an anisotropic medium the direction of the flux vector f_m at a point in general does not lie along the normal to the isothermal through the point. Let $\partial v/\partial n$ and $\partial v/\partial m$ be the rates of change of temperature along the normal to the isothermal through the point P, and along the direction of the flux vector at P, respectively.

Considering first the direction of the flux vector at P, its direction cosines are

$$f_x/f_m, \qquad f_y/f_m, \qquad f_z/f_m, \tag{1}$$

and thus

$$\frac{\partial v}{\partial m} = \frac{f_x}{f_m}\frac{\partial v}{\partial x}+\frac{f_y}{f_m}\frac{\partial v}{\partial y}+\frac{f_z}{f_m}\frac{\partial v}{\partial z} = -\frac{2\Phi}{f_m}, \tag{2}$$

where 1.17 (1) has been used, and

$$2\Phi = K_{11}\left(\frac{\partial v}{\partial x}\right)^2+K_{22}\left(\frac{\partial v}{\partial y}\right)^2+K_{33}\left(\frac{\partial v}{\partial z}\right)^2+(K_{23}+K_{32})\frac{\partial v}{\partial y}\frac{\partial v}{\partial z}+$$
$$+(K_{31}+K_{13})\frac{\partial v}{\partial z}\frac{\partial v}{\partial x}+(K_{12}+K_{21})\frac{\partial v}{\partial x}\frac{\partial v}{\partial y}. \tag{3}$$

The homogeneous quadratic form (3) has the physical significance $-f_m(\partial v/\partial m)$ and thus its value is independent of the choice of axes. Further, since $(\partial v/\partial m)$ must be negative if f_m is positive, the form (3) is positive definite: this requires

$$K_{11} \geqslant 0, \qquad K_{22} \geqslant 0, \qquad K_{33} \geqslant 0. \tag{4}$$

It follows from (2) that

$$f_m = -K_m\frac{\partial v}{\partial m}, \quad \text{where } K_m = \frac{2\Phi}{(\partial v/\partial m)^2}, \tag{5}$$

and K_m may be called the conductivity in the direction of the flux vector at P.

Considering next the normal to the isothermal through P, its direction cosines λ, μ, ν are given by

$$\lambda = \frac{1}{\Delta_1 v}\frac{\partial v}{\partial x}, \qquad \mu = \frac{1}{\Delta_1 v}\frac{\partial v}{\partial y}, \qquad \nu = \frac{1}{\Delta_1 v}\frac{\partial v}{\partial z}, \tag{6}$$

[†] *Comptes rendus*, **21** (1847) 459, 707, 829; von Lang, *Pogg. Ann.* **135** (1868) 29; Roentgen, *Pogg. Ann.* **151** (1874) 603.

where
$$\Delta_1 v = \left\{ \left(\frac{\partial v}{\partial x}\right)^2 + \left(\frac{\partial v}{\partial y}\right)^2 + \left(\frac{\partial v}{\partial z}\right)^2 \right\}^{\frac{1}{2}}. \tag{7}$$

It follows that
$$\frac{\partial v}{\partial n} = \lambda \frac{\partial v}{\partial x} + \mu \frac{\partial v}{\partial y} + \nu \frac{\partial v}{\partial z} = \Delta_1 v. \tag{8}$$

Also from 1.3 (2) the flux f_n in the direction of the normal to the isothermal at P is
$$f_n = \lambda f_x + \mu f_y + \nu f_z = -\frac{2\Phi}{\Delta_1 v}, \tag{9}$$
which may be written
$$f_n = -K_n \frac{\partial v}{\partial n}, \quad \text{where } K_n = \frac{2\Phi}{(\Delta_1 v)^2}, \tag{10}$$

and K_n may be called the conductivity normal to the isothermals at P. The two quantities K_m and K_n defined by (5) and (10) are independent of the choice of axes.

We now consider the way in which K_n varies with the direction cosines λ, μ, ν of the normal to the isothermal surface. Substituting the values (6) of these in (3) and (10) gives

$$K_n = K_{11}\lambda^2 + K_{22}\mu^2 + K_{33}\nu^2 + (K_{23} + K_{32})\mu\nu + (K_{31} + K_{13})\nu\lambda + (K_{12} + K_{21})\lambda\mu. \tag{11}$$

Now if we measure off a length $kK_n^{-\frac{1}{2}}$, where k is a constant, in the direction (λ, μ, ν), the locus of its end point Q,
$$x = k\lambda K_n^{-\frac{1}{2}}, \qquad y = k\mu K_n^{-\frac{1}{2}}, \qquad z = k\nu K_n^{-\frac{1}{2}}, \tag{12}$$
is the ellipsoid

$$K_{11}x^2 + K_{22}y^2 + K_{33}z^2 + (K_{23} + K_{32})yz + (K_{31} + K_{13})zx + (K_{12} + K_{21})xy = k^2, \tag{13}$$

which arose in 1.18 (2) and is now seen to have the property that the square of the radius vector in any direction is inversely proportional to the conductivity normal to isothermals at points where their normals are in this direction.

As in § 1.18 a system of rectangular axes ξ, η, ζ can be found relative to which (13) takes the form
$$K_1 \xi^2 + K_2 \eta^2 + K_3 \zeta^2 = k^2, \tag{14}$$

where K_1, K_2, K_3 are the principal conductivities. Then the conductivity normal to isothermals whose normals have direction cosines (l, m, n) relative to the principal axes of conductivity is k^2/r^2, where r is the radius vector of (14) in the direction (l, m, n). Alternatively, it is given by the formula

$$K_n = l^2 K_1 + m^2 K_2 + n^2 K_3. \tag{15}$$

It is this conductivity K_n which would be measured, for example, by the fundamental experiment of § 1.2 applied to a plane slice of a crystal cut so that its normal has direction cosines (l, m, n) relative to the principal axes of conductivity of the crystal. In the important special case $K_1 = K_2$ and $n = \cos\theta$, that is, any direction making an angle θ with the axis of symmetry of the ellipsoid, (15) becomes

$$K_n = K_1 + (K_3 - K_1)\cos^2\theta. \tag{16}$$

Next, we study the fluxes of heat over planes perpendicular to the principal axes of conductivity. If the transformation from the x, y, z of (13) to the ξ, η, ζ of (14) were written out explicitly, the values of the fluxes f_ξ, f_η, f_ζ across the new coordinate planes could be obtained from 1.17 (1) and 1.3 (2). Without doing this, however, it is clear that the new fluxes must be of the form

$$\left.\begin{aligned}
-f_\xi &= K_1\frac{\partial v}{\partial \xi} + A\frac{\partial v}{\partial \eta} + B\frac{\partial v}{\partial \zeta}\\
-f_\eta &= -A\frac{\partial v}{\partial \xi} + K_2\frac{\partial v}{\partial \eta} + C\frac{\partial v}{\partial \zeta}\\
-f_\zeta &= -B\frac{\partial v}{\partial \xi} - C\frac{\partial v}{\partial \eta} + K_3\frac{\partial v}{\partial \zeta}
\end{aligned}\right\}, \tag{17}$$

since, if we repeat the argument which led from 1.17 (1) to (13), we must now arrive at (14) which consists of a sum of squares. Thus, when referred to principal axes of conductivity, it appears that there are at most six independent coefficients in the linear connexions between the fluxes and the rates of change of temperature. Further, as in the two-dimensional case of § 1.19, there are good reasons for supposing the 'rotatory' terms A, B, C to vanish so that (17) becomes

$$-f_\xi = K_1\frac{\partial v}{\partial \xi}, \qquad -f_\eta = K_2\frac{\partial v}{\partial \eta}, \qquad -f_\zeta = K_3\frac{\partial v}{\partial \zeta}. \tag{18}$$

It is easy to show from the formulae for change of axes that $A = B = C = 0$ implies symmetry in the conductivity coefficients, that is, in the formulae 1.17 (1) for any rectangular axes we must have

$$K_{12} = K_{21}, \qquad K_{23} = K_{32}, \qquad K_{31} = K_{13}. \tag{19}$$

Finally, we consider the direction of the flux vector, and the way in which the conductivity K_m in this direction varies, on the assumption that (19) holds so that the flux is given by (18).

The point (ξ, η, ζ) in (14) was chosen on a radius vector normal to the isothermal at P, that is

$$\xi : \eta : \zeta = \frac{\partial v}{\partial \xi} : \frac{\partial v}{\partial \eta} : \frac{\partial v}{\partial \zeta}.$$

Using this result and (18) gives

$$\frac{\partial v/\partial \xi}{\xi} = \frac{\partial v/\partial \eta}{\eta} = \frac{\partial v/\partial \zeta}{\zeta} = -\frac{f_\xi}{K_1\xi} = -\frac{f_\eta}{K_2\eta} = -\frac{f_\zeta}{K_3\zeta}. \tag{20}$$

The square of each ratio in (20) is equal to

$$-\frac{f_\xi(\partial v/\partial \xi) + f_\eta(\partial v/\partial \eta) + f_\zeta(\partial v/\partial \zeta)}{K_1\xi^2 + K_2\eta^2 + K_3\zeta^2} = \frac{f_\xi^2 + f_\eta^2 + f_\zeta^2}{K_1^2\xi^2 + K_2^2\eta^2 + K_3^2\zeta^2}. \tag{21}$$

Using (2) and (14) this becomes

$$f_m = -\frac{OR^2}{k^2}\frac{\partial v}{\partial m}, \tag{22}$$

where OR is the radius vector from the origin to the point $X = K_1\xi$, $Y = K_2\eta$, $Z = K_3\zeta$. By (20), OR is the direction of the flux vector. The conductivity K_m in this direction is thus

$$K_m = OR^2/k^2. \tag{23}$$

Now, by (14), X, Y, Z lies on the ellipsoid

$$\frac{X^2}{K_1} + \frac{Y^2}{K_2} + \frac{Z^2}{K_3} = k^2. \tag{24}$$

Thus, if the flux vector has direction cosines (l, m, n) relative to the principal axes of conductivity, the conductivity K_m in this direction is given by

$$\frac{1}{K_m} = \frac{l^2}{K_1} + \frac{m^2}{K_2} + \frac{n^2}{K_3}. \tag{25}$$

This is the conductivity which would be measured by experiments on a long thin rod cut in the direction (l, m, n). If $K_1 = K_2$ and $n = \cos\theta$, (25) becomes†

$$\frac{1}{K_m} = \frac{1}{K_1} + \left(\frac{1}{K_3} - \frac{1}{K_1}\right)\cos^2\theta. \tag{26}$$

The ellipsoid (24) was called by Lamé the Principal Ellipsoid; it is sometimes referred to as the Thermal Ellipsoid. It is of the same general shape as the variation of conductivity.

Numerous geometrical properties of the conductivities and the directions of the flux vector and the normal to the isothermal at a point follow from consideration of the ellipsoids (14) and (24). For example, it follows from the last three equations of (20) that if the normal to the isothermal at a point is in the direction (ξ, η, ζ), the direction of the flux vector at the point is perpendicular to the tangent plane to (14) at the point (ξ, η, ζ). Again, if the direction of the flux vector (X, Y, Z) at a point is known, the isothermal through the point has its normal perpendicular to the tangent plane to (24) at the point (X, Y, Z).

† This relation has been verified experimentally by Bridgman, *Proc. Amer. Acad. Arts Sci.* **61** (1925) 101. See also Hume-Rothery, *The Metallic State* (Oxford, 1931) Chap. IV.

LINEAR FLOW OF HEAT: THE INFINITE AND SEMI-INFINITE SOLID

2.1. Introductory. Simple solutions of the equation of linear flow of heat

IN this chapter we shall examine various problems in which the isothermal surfaces are planes parallel to $x = 0$ and the flow of heat is linear, the lines of flow being parallel to the axis of x. The results obtained in this way also serve for the flow of heat along straight rods of small cross-section, when there is no loss of heat from the surface; problems in which this is not the case are considered in Chapter IV.

After obtaining the solution for the *infinite solid*, we proceed to examine, in detail, the many important problems of linear flow of heat in the *semi-infinite solid*, or the solid which is bounded by the plane $x = 0$ and extends to infinity in the direction of x positive. In all cases the thermal properties will be assumed to be independent of position and temperature: the extension to variable thermal properties is discussed in § 2.16.

The equation of linear flow of heat is

$$\frac{\partial^2 v}{\partial x^2} - \frac{1}{\kappa} \frac{\partial v}{\partial t} = 0. \tag{1}$$

We proceed to indicate a number of solutions of this equation which have simple mathematical forms.† These will all appear in many contexts later, together with their physical significance.

I. *The source solution*

Consider the expression

$$u = t^{-\frac{1}{2}} e^{-x^2/4\kappa t}. \tag{2}$$

Since
$$\frac{\partial u}{\partial t} = -\frac{1}{2t^{\frac{3}{2}}} e^{-x^2/4\kappa t} + \frac{x^2}{4\kappa t^{\frac{5}{2}}} e^{-x^2/4\kappa t},$$

and
$$\frac{\partial^2 u}{\partial x^2} = -\frac{1}{2\kappa t^{\frac{3}{2}}} e^{-x^2/4\kappa t} + \frac{x^2}{4\kappa^2 t^{\frac{5}{2}}} e^{-x^2/4\kappa t},$$

---

† The relationship between various types of solution of (1) is discussed by **Gray**, *Proc. Roy. Soc. Edin.* **45** (1924–5) 230–44.

it follows that (2) is a particular solution of (1). The solution (2) has the properties

$$u \to 0, \quad \text{as } t \to 0, \quad \text{for fixed } x \neq 0,$$

$$u \to \infty, \quad \text{as } t \to 0, \quad \text{if } x = 0,$$

$$\int_{-\infty}^{\infty} u \, dx = 2(\pi\kappa)^{\frac{1}{2}}, \quad \text{for all } t > 0,$$

it may thus be regarded physically as the solution corresponding to the release of the quantity of heat $2\rho c(\pi\kappa)^{\frac{1}{2}}$ per unit area over the plane $x = 0$ at time $t = 0$.

Clearly, further solutions of (1) are obtained by differentiating, or in some cases by integrating, (2) with respect to either x or t.

II. *The error function solution*

It follows as in I that (1) is satisfied by

$$\int_{0}^{x} t^{-\frac{1}{2}} e^{-x^2/4\kappa t} \, dx = 2\kappa^{\frac{1}{2}} \int_{0}^{x/2(\kappa t)^{\frac{1}{2}}} e^{-\xi^2} \, d\xi.$$

The notation

$$\operatorname{erf} x = \frac{2}{\sqrt{\pi}} \int_{0}^{x} e^{-\xi^2} \, d\xi \tag{3}$$

will always be used here, so that we have just shown that

$$A \operatorname{erf} \frac{x}{2(\kappa t)^{\frac{1}{2}}}, \tag{4}$$

where A is an arbitrary constant, is a solution of (1).

The 'error function' defined by (3) has the properties

$$\operatorname{erf} 0 = 0, \quad \operatorname{erf} \infty = 1, \quad \operatorname{erf}(-x) = -\operatorname{erf} x. \tag{5}$$

Further results and a short table of numerical values are given in Appendix II. The notation

$$\operatorname{erfc} x = 1 - \operatorname{erf} x, \tag{6}$$

$$\operatorname{ierfc} x = \operatorname{i}^1 \operatorname{erfc} x = \int_{x}^{\infty} \operatorname{erfc} \xi \, d\xi, \tag{7}$$

$$\operatorname{i}^n \operatorname{erfc} x = \int_{x}^{\infty} \operatorname{i}^{n-1} \operatorname{erfc} \xi \, d\xi, \quad n = 2, 3, 4,..., \tag{8}$$

will also be used frequently.

III. *Solutions of the form $t^m f[x/(4\kappa t)^{\frac{1}{2}}]$*

It may be verified that an expression of this type satisfies (1) if $f(z)$ is a solution of the differential equation

$$\frac{d^2f}{dz^2} + 2z\frac{df}{dz} - 4mf = 0. \tag{9}$$

This is, in fact, the equation (16) of Appendix II so that it follows that, if n is integral,

$$t^{\frac{1}{2}n}\mathrm{i}^n\mathrm{erfc}[x/(4\kappa t)^{\frac{1}{2}}] \tag{10}$$

is a solution† of (1).

IV. *Solution in exponentials*

It follows immediately by differentiating that

$$P\exp(\kappa A^2t \pm Ax), \tag{11}$$

where P and A are constants, real or complex, satisfies (1).

V. *The steady state solution*

For the case in which v is independent of time, the solution of (1) is

$$Ax + B, \tag{12}$$

where A and B are constants.

It has been shown‡ that (4), (11), and (12) are (apart from trivial modifications such as replacing x by $x+a$) the *only* solutions of (1) of the form $f[\phi(x)\psi(t)]$.

VI. *Double power series solution*

It is easy to verify by substitution that

$$v = a_0 + a_1x + a_2(x^2 + 2\kappa t) + a_3(x^3 + 6\kappa xt) + a_4(x^4 + 12\kappa x^2t + 12\kappa^2t^2) +$$
$$+ a_5(x^5 + 20\kappa x^3t + 60\kappa^2xt^2) + a_6(x^6 + 30\kappa x^4t + 180\kappa^2x^2t^2 + 120\kappa^3t^3) + ...,$$
$$\tag{13}$$

where $a_0, a_1,...$ are constants, satisfies (1).

VII. *A solution involving two arbitrary functions of the time*

$$v = \phi + \frac{x^2}{\kappa 2!}\dot{\phi} + \frac{x^4}{\kappa^24!}\ddot{\phi} + ... + x\psi + \frac{x^3}{\kappa 3!}\dot{\psi} + \frac{x^5}{\kappa^25!}\ddot{\psi} + ..., \tag{14}$$

where ϕ and ψ are any functions of the time and dots denote differentiation with respect to t, satisfies (1). This solution has the property that $v = \phi(t)$ and $\partial v/\partial x = \psi(t)$ when $x = 0$.

---

† Ribaud, *C.R. Acad. Sci. Paris*, **226** (1948) 140–2, 204–6, 449–51. Nordon, ibid. **228** (1949), 167–8, remarks that for all values of n the solutions of (9) are Weber functions. These and other questions of this section are discussed by Appell, *J. Math. Pure Appl.* (4) **8** (1892) 187–216 and Goursat, *Cours d'Analyse* (Gauthier–Villars. edn. 5, 1942) Vol. 3.

‡ Paterson, *Proc. Glasgow Math. Ass.* **1** (1952–3) 48–52.

2.2. The infinite solid: Laplace's solution

It is required to find the solution of the equation of linear flow of heat 2.1 (1) in the infinite region $-\infty < x < \infty$ with the initial condition

$$v = f(x), \quad \text{when } t = 0.$$

The usual formal discussion of this problem is as follows: by 2.1 (2),

$$\frac{1}{2(\pi\kappa t)^{\frac{1}{2}}} e^{-(x-x')^2/4\kappa t}$$

is a particular integral of 2.1 (1).

Further, the equation being linear, the sum of any number of particular integrals is also an integral, and thus

$$v = \frac{1}{2\sqrt{(\pi\kappa t)}} \int_{-\infty}^{\infty} f(x') e^{-(x-x')^2/4\kappa t} \, dx'$$

satisfies the equation, assuming that this integral is convergent.

Putting $\qquad x' = x + 2\sqrt{(\kappa t)}\xi,$

we find that $\qquad v = \frac{1}{\sqrt{\pi}} \int_{-\infty}^{\infty} f\{x + 2\sqrt{(\kappa t)}\xi\} e^{-\xi^2} \, d\xi.$

In the limit when $t \to 0$, $f\{x + 2\sqrt{(\kappa t)}\xi\} = f(x)$, if this function is continuous; and it is assumed that the limiting value of this integral is given by

$$\frac{1}{\sqrt{\pi}} \int_{-\infty}^{\infty} f(x) e^{-\xi^2} \, d\xi,$$

which is equal to $f(x)$.

Therefore the temperature in the infinite solid at time t, due to the initial temperature $v = f(x)$, is given by

$$v = \frac{1}{2\sqrt{(\pi\kappa t)}} \int_{-\infty}^{\infty} f(x') e^{-(x-x')^2/4\kappa t} \, dx'. \tag{1}$$

Since† $\qquad \int_{0}^{\infty} e^{-a^2 x^2} \cos 2bx \, dx = \frac{\sqrt{\pi}}{2a} e^{-b^2/a^2},$

and therefore

$$\int_{0}^{\infty} e^{-\kappa\alpha^2 t} \cos \alpha(x'-x) \, d\alpha = \frac{\sqrt{\pi}}{2\sqrt{(\kappa t)}} e^{-(x'-x)^2/4\kappa t},$$

† Cf. *F.S.*, p. 213, Ex 13.

we may transform the expression for v into

$$\frac{1}{\pi} \int_{-\infty}^{\infty} dx' \int_{0}^{\infty} f(x')\cos \alpha(x'-x)e^{-\kappa\alpha^2 t}\, d\alpha, \tag{2}$$

a form which would be suggested by Fourier's integral 2.3 (2) for $f(x)$.

The above discussion has been formal only, and it is necessary to consider the conditions on $f(x)$ under which it is valid. It is shown in *C.H.*, § 18 that if $f(x)$ is bounded and integrable in any given interval, and $\int_{-\infty}^{\infty} |f(x')|\, dx'$ converges, v given by (1) has the limit $f(x)$ as $t \to 0$ if $f(x)$ is continuous at x, or $\frac{1}{2}[f(x+0)+f(x-0)]$ if $f(x)$ has an ordinary discontinuity at x. Also, if $f(x)$ is continuous in a closed interval $[\alpha,\beta]$, v tends uniformly to $f(x)$ in this interval as $t \to 0$.

These results are, in fact, true under much less stringent conditions on $f(x)$, for example, if $f(x)$ is any polynomial, or if $f(x)$ is of exponential type, i.e. if

$$|f(x)| < Ke^{c|x|},$$

where K and c are constants.†

Finally, some results of practical importance which follow from, or extend, (1) may be noted.

(i) If the region $-a < x < a$ is initially at constant temperature V and the region $|x| > a$ is initially at zero,‡

$$v = \tfrac{1}{2}V\left\{\operatorname{erf}\frac{a-x}{2\sqrt{(\kappa t)}} + \operatorname{erf}\frac{a+x}{2\sqrt{(\kappa t)}}\right\}, \quad -\infty < x < \infty. \tag{3}$$

Some numerical values of v/V given by (3) for various values of the parameter $\kappa t/a^2$ are given in Fig. 4 (a). Corresponding results for a cylinder of radius a and a sphere of radius a, initially at constant temperatures, are given in Figs. 4 (b) and 4 (c), respectively.

(ii) If the region $-a < x < a$ is initially at zero and $|x| > a$ is initially at V,

$$v = \tfrac{1}{2}V\left\{\operatorname{erfc}\frac{a-x}{2\sqrt{(\kappa t)}} + \operatorname{erfc}\frac{a+x}{2\sqrt{(\kappa t)}}\right\}, \quad -\infty < x < \infty. \tag{4}$$

(iii) If initially the region $|x| > a$ is at zero, and $v = V(a-x)/a$ if $a > x > 0$, and $v = V(a+x)/a$ if $-a < x < 0$,

$$v = \frac{V}{2a}\left\{(a-x)\operatorname{erf}\frac{a-x}{2(\kappa t)^{\frac{1}{2}}} + (a+x)\operatorname{erf}\frac{a+x}{2(\kappa t)^{\frac{1}{2}}} - 2x\operatorname{erf}\frac{x}{2(\kappa t)^{\frac{1}{2}}} + \right.$$
$$\left. + 2(\kappa t/\pi)^{\frac{1}{2}}[e^{-(x+a)^2/4\kappa t} + e^{-(x-a)^2/4\kappa t} - 2e^{-x^2/4\kappa t}]\right\}. \tag{5}$$

† For a more general discussion see Goursat, *Cours d'Analyse* (Gauthier–Villars, edn. 5, 1942) Vol. 3, Chap. 29; Titchmarsh, *Theory of Fourier Integrals* (Oxford, 1937) § 10.6.
‡ Lovering, *Bull. Geol. Soc. Amer.* **46** (1935) 69–93, uses (3) as an approximation to the temperature when a flat sheet of molten rock (dike or sill) is intruded into other rock at zero temperature. He gives graphs of (3) in a form suitable for practical application. He also uses (9) and (10) for the case of cylindrical and compact intrusions, and 2.4 (14) for a lava flow on the surface of other rock. See also Lovering, *Econ. Geol.* fiftieth anniv. vol. (1955) 249. For the effect of the heat of solidification of the rocks see § 11.2.

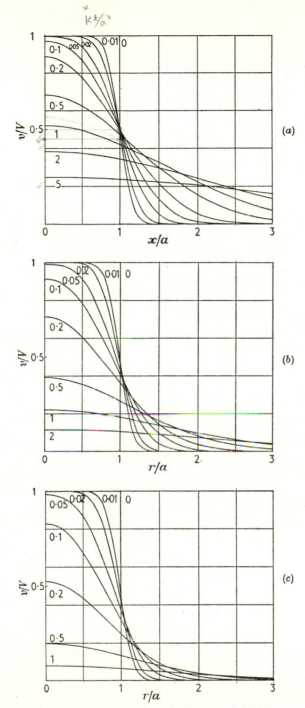

Fig. 4. Temperatures in an infinite region of which part is initially at constant temperature V and the remainder at zero: (a) the region $|x| < a$ initially at V; (b) the interior of the cylinder $r = a$ initially at V; (c) the interior of the sphere $r = a$ initially at V. In all cases the numbers on the curves are the values of $\kappa t/a^2$.

(iv) If the plane $x = 0$ is impervious to heat, the solution takes the form

$$v = \frac{1}{2\sqrt{(\pi\kappa t)}} \int_0^\infty f(x')\{e^{-(x-x')^2/4\kappa t} + e^{-(x+x')^2/4\kappa t}\}\, dx' \tag{6}$$

$$= \frac{2}{\pi} \int_0^\infty dx' \int_0^\infty f(x')\cos\alpha x' \cos\alpha x\, e^{-\kappa\alpha^2 t}\, d\alpha.$$

(v) The results corresponding to (1) in two and three dimensions are

$$v = \frac{1}{4\pi\kappa t} \int_{-\infty}^\infty \int_{-\infty}^\infty f(x',y')e^{-[(x-x')^2+(y-y')^2]/4\kappa t}\, dx'dy', \tag{7}$$

$$v = \frac{1}{8(\pi\kappa t)^{\frac{3}{2}}} \int_{-\infty}^\infty \int_{-\infty}^\infty \int_{-\infty}^\infty f(x',y',z')e^{-[(x-x')^2+(y-y')^2+(z-z')^2]/4\kappa t}\, dx'dy'dz'. \tag{8}$$

(vi) If the infinite cylinder $|x| < a$, $|y| < b$ is initially at constant temperature V, and the infinite region outside it is initially at zero,

$$v = \tfrac{1}{4}V\left\{\operatorname{erf}\frac{a-x}{2(\kappa t)^{\frac{1}{2}}} + \operatorname{erf}\frac{a+x}{2(\kappa t)^{\frac{1}{2}}}\right\}\left\{\operatorname{erf}\frac{b-y}{2(\kappa t)^{\frac{1}{2}}} + \operatorname{erf}\frac{b+y}{2(\kappa t)^{\frac{1}{2}}}\right\}. \tag{9}$$

(vii) If the parallelepiped $|x| < a$, $|y| < b$, $|z| < c$ is initially at constant temperature V, and the infinite region outside it is initially at zero,

$$v = \tfrac{1}{8}V\left\{\operatorname{erf}\frac{a-x}{2(\kappa t)^{\frac{1}{2}}} + \operatorname{erf}\frac{a+x}{2(\kappa t)^{\frac{1}{2}}}\right\}\left\{\operatorname{erf}\frac{b-y}{2(\kappa t)^{\frac{1}{2}}} + \operatorname{erf}\frac{b+y}{2(\kappa t)^{\frac{1}{2}}}\right\}\left\{\operatorname{erf}\frac{c-z}{2(\kappa t)^{\frac{1}{2}}} + \operatorname{erf}\frac{c+z}{2(\kappa t)^{\frac{1}{2}}}\right\}. \tag{10}$$

2.3. The use of Fourier integrals and Fourier transforms

It was remarked in § 2.2 that Laplace's solution could be put in the form 2.2 (2) which was related to Fourier's integral for $f(x)$. Alternatively the solution can be deduced from Fourier's Integral Theorem. Probably the most satisfactory way of doing this is to use the Fourier Transform. We give here a brief résumé of the method for occasional reference, and indicate how it leads to Laplace's solution.

Fourier's integral theorem [*F.S.*, § 119] states that, if $f(x)$ is defined for all x and satisfies Dirichlet's conditions† in any finite interval, and if

$$\int_{-\infty}^\infty |f(x)|\, dx \quad \text{exists,‡} \tag{1}$$

then

$$f(x) = \frac{1}{\pi} \int_0^\infty d\xi \int_{-\infty}^\infty f(x')\cos\xi(x-x')\, dx' \tag{2}$$

$$= \frac{1}{2\pi} \int_{-\infty}^\infty d\xi \int_{-\infty}^\infty f(x')\cos\xi(x-x')\, dx', \tag{3}$$

† For a statement of these cf. *F.S.*, § 93. They are satisfied by a function with only a finite number of maxima, minima, and ordinary discontinuities.

‡ This condition, which is an extremely restrictive one, greatly reduces the value of Fourier's integral theorem in this form for practical applications. Results for much wider classes of function can be obtained by the use of Generalized Fourier Integrals (Titchmarsh, *Theory of Fourier Integrals*, Oxford, 1937) or the related Laplace Transformation method; for most applications in conduction of heat the latter is more convenient.

at every point of continuity of $f(x)$, with a corresponding result, namely,
$$\tfrac{1}{2}[f(x+0)+f(x-0)],$$
at points of ordinary discontinuity. Since

$$\frac{1}{2\pi} \int_{-\infty}^{\infty} d\xi \int_{-\infty}^{\infty} f(x')\sin\xi(x-x')\,dx' = 0, \tag{4}$$

it follows from (3) and (4) that

$$f(x) = \frac{1}{2\pi} \int_{-\infty}^{\infty} e^{-i\xi x}\,d\xi \int_{-\infty}^{\infty} f(x')e^{i\xi x'}\,dx'. \tag{5}$$

This is usually known as the complex form of Fourier's integral theorem. It may be stated in the form that, if

$$F(\xi) = \frac{1}{\sqrt{(2\pi)}} \int_{-\infty}^{\infty} e^{i\xi x'}f(x')\,dx', \tag{6}$$

then

$$f(x) = \frac{1}{\sqrt{(2\pi)}} \int_{-\infty}^{\infty} e^{-i\xi x}F(\xi)\,d\xi. \tag{7}$$

Here $F(\xi)$ and $f(x)$ are called the Fourier transforms of each other: if either is known, the other follows from the appropriate formula (6) or (7).

Two other cases are of importance in applications. If $f(x)$ is an odd function of x, (2) reduces to

$$f(x) = \frac{2}{\pi} \int_{0}^{\infty} \sin\xi x\,d\xi \int_{0}^{\infty} f(x')\sin x'\xi\,dx'. \tag{8}$$

This may be stated in the form that either of

$$F_s(\xi) = \sqrt{\left(\frac{2}{\pi}\right)} \int_{0}^{\infty} f(x')\sin x'\xi\,dx', \tag{9}$$

$$f(x) = \sqrt{\left(\frac{2}{\pi}\right)} \int_{0}^{\infty} F_s(\xi)\sin\xi x\,d\xi \tag{10}$$

implies the other. The functions $F_s(\xi)$ and $f(x)$ are called Fourier Sine Transforms of each other.

Again, if $f(x)$ is an even function of x, (2) reduces to

$$f(x) = \frac{2}{\pi} \int_{0}^{\infty} \cos\xi x\,d\xi \int_{0}^{\infty} f(x')\cos x'\xi\,dx', \tag{11}$$

and this may be stated in the form that either of

$$F_c(\xi) = \sqrt{\left(\frac{2}{\pi}\right)} \int_{0}^{\infty} f(x')\cos x'\xi\,dx', \tag{12}$$

$$f(x) = \sqrt{\left(\frac{2}{\pi}\right)} \int_{0}^{\infty} F_c(\xi)\cos\xi x\,d\xi \tag{13}$$

implies the other. $F_c(\xi)$ and $f(x)$ are Fourier Cosine Transforms of each other.

We now show how (6) and (7) can be applied formally to the problem of § 2.2, that is to the solution of

$$\frac{\partial^2 v}{\partial x^2} = \frac{1}{\kappa}\frac{\partial v}{\partial t}, \quad -\infty < x < \infty, \quad t > 0, \tag{14}$$

with
$$v = f(x), \quad -\infty < x < \infty, \quad t = 0. \tag{15}$$

We notice that
$$e^{-i\xi x - \kappa \xi^2 t} \tag{16}$$

satisfies (14) for any ξ. Thus we assume for the general solution of (14) and (15)

$$v(x, t) = \frac{1}{\sqrt{(2\pi)}} \int_{-\infty}^{\infty} e^{-i\xi x - \kappa \xi^2 t}\phi(\xi)\, d\xi. \tag{17}$$

For $t = 0$ we must have

$$f(x) = \frac{1}{\sqrt{(2\pi)}} \int_{-\infty}^{\infty} e^{-i\xi x}\phi(\xi)\, d\xi.$$

Therefore by (6) it follows that

$$\phi(\xi) = \frac{1}{\sqrt{(2\pi)}} \int_{-\infty}^{\infty} e^{i\xi x'}f(x')\, dx'.$$

Thus, using this value in (17),

$$v(x, t) = \frac{1}{2\pi} \int_{-\infty}^{\infty} e^{-\kappa \xi^2 t - i\xi x}\, d\xi \int_{-\infty}^{\infty} e^{i\xi x'}f(x')\, dx'$$

$$= \frac{1}{2\pi} \int_{-\infty}^{\infty} f(x')\, dx' \int_{-\infty}^{\infty} e^{-\kappa \xi^2 t - i\xi(x-x')}\, d\xi$$

$$= \frac{1}{2\sqrt{(\pi\kappa t)}} \int_{-\infty}^{\infty} f(x')e^{-(x-x')^2/4\kappa t}\, dx'.$$

Clearly, to make this analysis rigorous further discussion is needed; Titchmarsh, loc. cit., § 10.6, discusses the problem for the case in which $f(x)$ is of exponential type.

2.4. The semi-infinite solid. Initial temperature $f(x)$. Surface temperature zero

Let the solid be bounded by the plane $x = 0$ and extend to infinity in the direction of x positive, the initial temperature being given by $v = f(x)$, and the plane $x = 0$ being kept at zero temperature. The solution of this problem may be deduced from that of the infinite solid.

We suppose the solid continued on the negative side of the plane $x = 0$, and the initial temperature at $-x'$ ($x' > 0$) to be $-f(x')$, the initial temperature at x' being $f(x')$. With this distribution the plane $x = 0$ will remain at zero.

Then we have from 2.2 (1)

$$v = \frac{1}{2\sqrt{(\pi\kappa t)}}\left\{\int_0^\infty f(x')e^{-(x-x')^2/4\kappa t}\,dx' + \int_{-\infty}^0 \{-f(-x')\}e^{-(x-x')^2/4\kappa t}\,dx'\right\}$$

$$= \frac{1}{2\sqrt{(\pi\kappa t)}}\int_0^\infty f(x')\{e^{-(x-x')^2/4\kappa t} - e^{-(x+x')^2/4\kappa t}\}\,dx'. \tag{1}$$

It is clear that this value of v satisfies all the conditions of the problem of the semi-infinite solid whose bounding plane is kept at zero temperature.

The expression (1) for the temperature may be transformed as in § 2.2 into

$$v = \frac{1}{\pi}\int_0^\infty dx' \int_0^\infty f(x')[\cos\alpha(x'-x) - \cos\alpha(x'+x)]e^{-\kappa\alpha^2 t}\,d\alpha$$

$$= \frac{2}{\pi}\int_0^\infty dx' \int_0^\infty f(x')\sin\alpha x' \sin\alpha x\, e^{-\kappa\alpha^2 t}\,d\alpha, \tag{2}$$

a form suggested by Fourier's sine integral, 2.3 (8). The result may also be obtained by using Fourier sine transforms, 2.3 (9) and (10), along the lines developed in § 2.3 for the infinite solid.

When the initial temperature is a constant, V, the expression (1) may be simplified by substituting $x' = x + 2\xi\sqrt{(\kappa t)}$ in the first part, and $x' = -x + 2\xi\sqrt{(\kappa t)}$ in the second. We thus obtain

$$v = \frac{V}{\sqrt{\pi}}\int_{-x/2\sqrt{(\kappa t)}}^{x/2\sqrt{(\kappa t)}} e^{-\xi^2}d\xi = \frac{2V}{\sqrt{\pi}}\int_0^{x/2\sqrt{(\kappa t)}} e^{-\xi^2}\,d\xi.$$

This is the integral 2.1 (3), so that the solution of the problem of the semi-infinite solid, whose surface is kept at zero temperature, the initial temperature being V, is given by

$$v = V\,\mathrm{erf}\left\{\frac{x}{2\sqrt{(\kappa t)}}\right\}. \tag{3}$$

This result could have been written down immediately from 2.1 (3), since it follows from the properties of the error function that it satisfies the differential equation and initial and boundary conditions.

It is important to notice that in this case of constant initial temperature V and zero surface temperature, the result (3) depends on the single dimensionless parameter

$$\frac{x}{2\sqrt{(\kappa t)}}. \tag{4}$$

This makes it easy to compare temperatures at different times and places in solids of different diffusivity. Similar results hold for the rate of cooling and the temperature gradient at any point, both of which are often needed.

The rate of cooling at any point is

$$-\frac{\partial v}{\partial t} = \frac{Vx}{2\sqrt{(\pi\kappa t^3)}}e^{-x^2/4\kappa t}. \tag{5}$$

The temperature gradient at any point is

$$\frac{\partial v}{\partial x} = \frac{V}{\sqrt{(\pi\kappa t)}}e^{-x^2/4\kappa t}. \tag{6}$$

Thus in terms of the parameter $\dfrac{x}{2\sqrt{(\kappa t)}}$ we have

$$\frac{v}{V} = \mathrm{erf}\frac{x}{2\sqrt{(\kappa t)}}, \tag{7}$$

$$-\frac{t\sqrt{\pi}}{V}\frac{\partial v}{\partial t} = \frac{x\sqrt{\pi}}{2V}\frac{\partial v}{\partial x} = \frac{x}{2\sqrt{(\kappa t)}}e^{-x^2/4\kappa t}. \tag{8}$$

Numerical values of these quantities are given in Appendix II, they are also plotted in curves I and II of Fig. 5. Curve II has a maximum value of

$$\frac{1}{\sqrt{2}}e^{-\frac{1}{2}} = 0{\cdot}4288\ldots \quad \text{when} \quad \frac{x}{2\sqrt{(\kappa t)}} = \frac{1}{\sqrt{2}}. =0{\cdot}707 \tag{9}$$

It follows from (7) that in any substance the time required for any point to reach a given temperature is proportional to the square of its distance from the surface. Also that the time required for a given point to attain a given temperature varies inversely as the diffusivity.

For example, it follows from Fig. 5 or Table I in Appendix II that $v = \frac{1}{2}V$ when

$$\frac{x}{2\sqrt{(\kappa t)}} = 0{\cdot}477.$$

Thus in silver, $\kappa = 1{\cdot}72$, the time taken to reach this temperature at a depth of 1 cm is $0{\cdot}64$ sec; in bismuth, $\kappa = 0{\cdot}07$, it is $15{\cdot}7$ sec; and in soil, $\kappa = 0{\cdot}0047$, it is 234 sec. At a depth of 10 cm the corresponding times would be 100 times the above.

Finally we give some results of practical importance which follow easily from (1).

(i) If the boundary $x = 0$ is kept at constant temperature V and the initial temperature is zero,

$$v = V\left\{1 - \mathrm{erf}\frac{x}{2\sqrt{(\kappa t)}}\right\} = V\,\mathrm{erfc}\frac{x}{2\sqrt{(\kappa t)}}. \tag{10}$$

This follows by adding to $v = V$, $x > 0$, $t > 0$, which is a solution of the differential equation of conduction of heat, the solution (3) for initial temperature $-V$ and zero surface temperature.

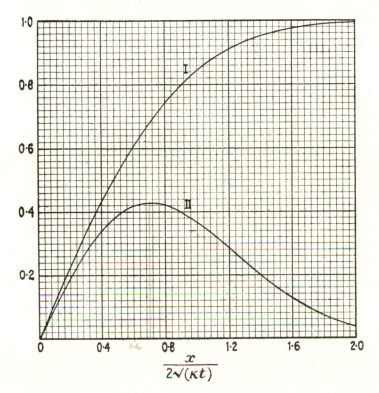

FIG. 5. The functions $\mathrm{erf}[x/2\sqrt{(\kappa t)}]$, curve I, and $[x/2\sqrt{(\kappa t)}]\exp[-x^2/4\kappa t]$, curve II.

The flux of heat at the surface is

$$-K\left[\frac{\partial v}{\partial x}\right]_{x=0} = \frac{KV}{\sqrt{(\pi\kappa t)}},\tag{11}$$

and this tends to infinity as $t \to 0$.

(ii) If the region $x > 0$ is initially at constant temperature V, and the region $x < 0$ is initially at zero temperature,

$$v = \tfrac{1}{2}V\left\{1 + \mathrm{erf}\frac{x}{2\sqrt{(\kappa t)}}\right\}, \quad -\infty < x < \infty.\tag{12}$$

(iii) If the region $x > 0$ is initially at temperature $V + kx$ and the plane $x = 0$ is kept at zero temperature,

$$v = V\,\mathrm{erf}\frac{x}{2\sqrt{(\kappa t)}} + kx.\tag{13}$$

(iv) If the region $0 < x < d$ is initially at constant temperature V, and the region $x > d$ is at zero, the surface $x = 0$ being maintained at zero for $t > 0$,

$$v = \tfrac{1}{2}V\Big\{2\,\mathrm{erf}\frac{x}{2(\kappa t)^{\frac{1}{2}}} - \mathrm{erf}\frac{x-d}{2(\kappa t)^{\frac{1}{2}}} - \mathrm{erf}\frac{x+d}{2(\kappa t)^{\frac{1}{2}}}\Big\}. \tag{14}$$

(v) If the region $a < x < b$ is initially at constant temperature V, and the regions $0 < x < a$ and $x > b$ are at zero, the surface $x = 0$ being maintained at zero for $t > 0$,

$$v = \tfrac{1}{2}V\Big\{\mathrm{erf}\frac{x-a}{2(\kappa t)^{\frac{1}{2}}} + \mathrm{erf}\frac{x+a}{2(\kappa t)^{\frac{1}{2}}} - \mathrm{erf}\frac{x-b}{2(\kappa t)^{\frac{1}{2}}} - \mathrm{erf}\frac{x+b}{2(\kappa t)^{\frac{1}{2}}}\Big\}. \tag{15}$$

2.5. Semi-infinite solid. Initial temperature zero. Surface at temperature $\phi(t)$

We have seen in § 1.14 that, when the surface temperature varies with the time, the solution may be deduced, by Duhamel's theorem, from the case in which this temperature is constant.

Now, in the semi-infinite solid, where v has to satisfy

$$\frac{\partial v}{\partial t} = \kappa\,\frac{\partial^2 v}{\partial x^2},$$

$$v = 0 \quad \text{when } t = 0,$$

and
$$v = 1 \quad \text{at } x = 0,$$

the solution, given in 2.4 (10), is

$$v = 1 - \frac{2}{\sqrt{\pi}}\int_0^{x/2\sqrt{(\kappa t)}} e^{-\xi^2}\,d\xi = \frac{2}{\sqrt{\pi}}\int_{x/2\sqrt{(\kappa t)}}^{\infty} e^{-\xi^2}\,d\xi.$$

Therefore, if
$$\frac{\partial v}{\partial t} = \kappa\,\frac{\partial^2 v}{\partial x^2},$$

$$v = 0 \quad \text{for } t = 0$$

and
$$v = \phi(t) \quad \text{at } x = 0,$$

it follows from 1.14 (10) that the solution is given by

$$v = \int_0^t \phi(\lambda)\frac{\partial}{\partial t}F(x, t-\lambda)\,d\lambda,$$

where
$$F(x, t-\lambda) = \frac{2}{\sqrt{\pi}}\int_{x/2\sqrt{\{\kappa(t-\lambda)\}}}^{\infty} e^{-\xi^2}\,d\xi.$$

In this case

$$\frac{\partial}{\partial t}F(x, t-\lambda) = -\frac{2}{\sqrt{\pi}}e^{-x^2/4\kappa(t-\lambda)}\frac{\partial}{\partial t}\bigg[\frac{x}{2\sqrt{\{\kappa(t-\lambda)\}}}\bigg]$$

$$= \frac{x}{2\sqrt{\{\pi\kappa(t-\lambda)^3\}}}e^{-x^2/4\kappa(t-\lambda)}.$$

Therefore the solution of our problem is

$$v = \frac{x}{2\sqrt{(\pi\kappa)}} \int_0^t \phi(\lambda) \frac{e^{-x^2/4\kappa(t-\lambda)}}{(t-\lambda)^{\frac{3}{2}}} \, d\lambda.$$

Putting

$$\frac{x}{2\sqrt{\{\kappa(t-\lambda)\}}} = \mu,$$

we have

$$t-\lambda = \frac{x^2}{4\kappa\mu^2},$$

and

$$v = \frac{2}{\sqrt{\pi}} \int_{x/2\sqrt{(\kappa t)}}^\infty \phi\!\left(t - \frac{x^2}{4\kappa\mu^2}\right) e^{-\mu^2} \, d\mu. \tag{1}$$

In this form it is clear that our solution satisfies the differential equation and the initial and boundary conditions.

Special cases of some practical interest are the following:†

(i) $\phi(t) = V_0$, constant, $0 < t < T$
             $= V_1$, constant, $t > T$.

$$v = V_0 \operatorname{erfc} \frac{x}{2\sqrt{(\kappa t)}}, \quad 0 < t < T. \tag{2}$$

$$v = V_0 \operatorname{erfc} \frac{x}{2\sqrt{(\kappa t)}} + (V_1 - V_0)\operatorname{erfc} \frac{x}{2\sqrt{[\kappa(t-T)]}}, \quad t > T. \tag{3}$$

(ii) $\phi(t) = kt$, where k is a constant.‡

$$v = kt\left\{\left(1 + \frac{x^2}{2\kappa t}\right)\operatorname{erfc} \frac{x}{2\sqrt{(\kappa t)}} - \frac{x}{\sqrt{(\pi\kappa t)}} e^{-x^2/4\kappa t}\right\} \tag{4}$$

$$= 4kt \, i^2\operatorname{erfc} \frac{x}{2\sqrt{(\kappa t)}}. \tag{5}$$

For the notation used in (5), (7), and (8) see Appendix II.

(iii) $\phi(t) = kt^{\frac{1}{2}}$, where k is a constant.

$$v = kt^{\frac{1}{2}}\left\{e^{-x^2/4\kappa t} - \frac{x\sqrt{\pi}}{2\sqrt{(\kappa t)}} \operatorname{erfc} \frac{x}{2\sqrt{(\kappa t)}}\right\} \tag{6}$$

$$= k(\pi t)^{\frac{1}{2}} \operatorname{ierfc} \frac{x}{2\sqrt{(\kappa t)}}. \tag{7}$$

It will be seen in § 2.9 that this is the surface temperature corresponding to constant flux at the surface.

(iv) $\phi(t) = kt^{\frac{1}{2}n}$, where n is any positive integer, even or odd.

$$v = k\Gamma(\tfrac{1}{2}n+1)(4t)^{\frac{1}{2}n} \, i^n\operatorname{erfc} \frac{x}{2\sqrt{(\kappa t)}}. \tag{8}$$

† These results are most easily obtained by the Laplace transformation method of Chapter XII; see § 12.4.

‡ Linearly increasing surface temperature is of importance in some practical applications, cf. Williamson and Adams, *Phys. Rev.* **14** (1919) 99; Taylor, *Phil. Trans. Roy. Soc.* A, **215** (1915) 1.

By the use of (8) the temperature at any depth for the case in which the surface temperature is a polynomial in t, or in $t^{\frac{1}{2}}$, can be written down in terms of tabulated functions. It may be remarked that a polynomial in $t^{\frac{1}{2}}$ may be useful for representing an observed surface temperature empirically, since the term in $t^{\frac{1}{2}}$ is that corresponding to constant flux at the surface (cf. § 2.9).

(v) $\phi(t) = e^{\lambda t}$, where λ is a constant, positive or negative,

$$v = \tfrac{1}{2}e^{\lambda t}\left\{e^{-x\sqrt{(\lambda/\kappa)}}\operatorname{erfc}\left[\frac{x}{2\sqrt{(\kappa t)}} - \sqrt{(\lambda t)}\right] + e^{x\sqrt{(\lambda/\kappa)}}\operatorname{erfc}\left[\frac{x}{2\sqrt{(\kappa t)}} + \sqrt{(\lambda t)}\right]\right\}. \qquad (9)$$

The error functions of complex argument which are needed for the important case in which λ is negative have only recently been tabulated, cf. Appendix II. The solution for positive values of λ is needed in Chapter IV.

We may solve the problem of the semi-infinite solid with initial temperature $f(x)$ and surface temperature $\phi(t)$ by putting

$$v = u + w,$$

where u satisfies

$$\frac{\partial u}{\partial t} = \kappa\frac{\partial^2 u}{\partial x^2},$$

with

$$u = 0, \quad \text{initially,}$$

and

$$u = \phi(t), \quad \text{at } x = 0,$$

and w satisfies

$$\frac{\partial w}{\partial t} = \kappa\frac{\partial^2 w}{\partial x^2},$$

with

$$w = f(x), \quad \text{initially,}$$

and

$$w = 0, \quad \text{at } x = 0.$$

That is, u is the solution of the problem considered above, and w is the solution of the problem of § 2.4.

2.6. Semi-infinite solid. Surface temperature a harmonic function of the time

Problems on the conduction of heat in solids with periodic surface temperature are of great practical importance. They arise: (i) in the study of the fluctuations in temperature of the Earth's crust due to periodic heating by the Sun, cf. § 2.12; (ii) in various experimental arrangements for the determination of diffusivity, §§ 2.12, 4.4, 4.8; (iii) in the calculation of the periodic temperatures (and thus of periodic thermal stresses) in the cylinder walls of steam† and internal combustion engines; (iv) in the theory of automatic temperature control systems.

If the surface temperature in the semi-infinite solid $x > 0$ is given by $v = A\cos(\omega t - \epsilon)$, and the initial temperature is zero, the solution is by 2.5 (1)

$$v = \frac{2A}{\sqrt{\pi}}\int\limits_{x/2\sqrt{(\kappa t)}}^{\infty}\cos\left\{\omega\left(t - \frac{x^2}{4\kappa\mu^2}\right) - \epsilon\right\}e^{-\mu^2}\,d\mu. \qquad (1)$$

† Kirsch, *Die Bewegung der Wärme in den Cylinderwandungen der Dampfmaschine* (Leipzig, 1886). Dahl, *Trans. Amer. Soc. Mech. Engrs.* **46** (1924) 161.

Since by a known definite integral,

$$\frac{2}{\pi^{\frac{1}{2}}} \int_0^\infty \cos\left\{\omega\left(t-\frac{x^2}{4\kappa\mu^2}\right)-\epsilon\right\}e^{-\mu^2}\,d\mu = e^{-x(\omega/2\kappa)^{\frac{1}{2}}}\cos\left\{\omega t-x\left(\frac{\omega}{2\kappa}\right)^{\frac{1}{2}}-\epsilon\right\},$$

it follows that (1) can be written in the form

$$v = Ae^{-x(\omega/2\kappa)^{\frac{1}{2}}}\cos\left\{\omega t-x\left(\frac{\omega}{2\kappa}\right)^{\frac{1}{2}}-\epsilon\right\} -$$

$$-\frac{2A}{\pi^{\frac{1}{2}}} \int_0^{x/2(\kappa t)^{\frac{1}{2}}} \cos\left\{\omega\left(t-\frac{x^2}{4\kappa\mu^2}\right)-\epsilon\right\}e^{-\mu}\,d\mu. \quad (2)$$

The second term in (2) is a transient disturbance† caused by starting the oscillations of surface temperature at time $t = 0$; it dies away as t increases, leaving the first term which is a steady oscillation of period $2\pi/\omega$. Before discussing it further, we give an alternative derivation‡ starting from the differential equation

$$\frac{\partial^2 v}{\partial x^2}-\frac{1}{\kappa}\frac{\partial v}{\partial t} = 0. \quad (3)$$

We seek a solution of (3) of type

$$v = ue^{i(\omega t-\epsilon)}, \quad (4)$$

where u is a function of x only. This solution v will have period $2\pi/\omega$. Substituting (4) in (3) it follows that u must satisfy

$$\frac{d^2u}{dx^2} = \frac{i\omega}{\kappa}u. \quad (5)$$

The solution of (5) which is finite as $x \to \infty$ is

$$u = Ae^{-x\sqrt{(i\omega/\kappa)}} = Ae^{-x(1+i)\sqrt{(\omega/2\kappa)}}.$$

Thus, the solution of (4) of period $2\pi/\omega$ is

$$v = Ae^{-kx}\frac{\cos}{\sin}\{\omega t-\epsilon-kx\}, \quad (6)$$

where

$$k = (\omega/2\kappa)^{\frac{1}{2}}, \quad (7)$$

and the solution which has the value $A\cos(\omega t-\epsilon)$ at $x = 0$ is

$$v = Ae^{-kx}\cos(\omega t-kx-\epsilon). \quad (8)$$

† An alternative form is given in 12.7 (8). If the initial temperature of the solid is $f(x)$ instead of zero, a term comprising the solution of the problem of the semi-infinite solid with this initial temperature and zero surface temperature must be added to (1), see § 2.4, and the time must be so large that this term, also, is negligible.

‡ Stokes, *Scientific Papers*, iii. 1; Lamb, *Hydrodynamics*, edn. 4, § 345.

Equation (8) represents a temperature wave of wave number $\mathbf{k}$ and wavelength λ given by

$$\lambda = \frac{2\pi}{\mathbf{k}} = \left(\frac{4\pi\kappa}{n}\right)^{\frac{1}{2}}, \tag{9}$$

where n is the frequency $\omega/2\pi$.

For typical rock material with $\kappa = 0\cdot01$, the wavelength is about $2\cdot7$ cm for a frequency of 1 cycle/min, 1 m for 1 cycle/day, and 20 m for 1 cycle/year. For a metallic conductor with $\kappa = 1$ the wavelength is $3\cdot5$ cm at 1 cycle/sec, and 27 cm at 1 cycle/min, while for metals at temperatures near absolute zero, where κ is of the order 10^4, the wavelength is 11 cm at 1,000 cycle/sec. These figures are of importance in connexion with measurements by periodic methods: ordinarily, observations will be made over a distance of the order of a wavelength, and this determines the frequency to be used. The fact that for metals at very low temperatures frequencies in the audio range can be used, has recently led to a considerable development of these methods.[†]

The important properties of the steady periodic temperature are the following:

(i) The amplitude of the temperature oscillation diminishes like

$$e^{-\mathbf{k}x} = e^{-x\sqrt{(\omega/2\kappa)}} = e^{-2\pi x/\lambda}, \tag{10}$$

and thus falls off more rapidly for large ω. If the surface temperature is given by a Fourier series as in (17), the higher harmonics disappear most rapidly as we move into the solid. At a distance of one wavelength the amplitude is reduced by a factor of $\exp[-2\pi] = 0\cdot0019$, so the waves are very strongly attenuated. This implies that the present solution for the semi-infinite solid can, in fact, be used for a conductor whose thickness is one or two wavelengths.

(ii) There is a progressive lag

$$\mathbf{k}x = x(\omega/2\kappa)^{\frac{1}{2}} \tag{11}$$

in the phase of the temperature wave. This lag increases with ω.

(iii) The temperature fluctuations, e.g. the positions of the maxima and minima of temperature, are propagated into the solid with velocity

$$(2\kappa\omega)^{\frac{1}{2}}. \tag{12}$$

It follows from (10), (11), and (12) that a measurement of the amplitude or phase at depth x, or of the velocity of propagation, is sufficient to determine the diffusivity κ.

† Howling, Mendoza, and Zimmerman, *Proc. Roy. Soc.* A, **229** (1955) 86–109.

The flux of heat F at the surface is

$$F = -K\left[\frac{\partial v}{\partial x}\right]_{x=0} = 2^{\frac{1}{2}}\mathbf{k}KA\cos(\omega t - \epsilon + \tfrac{1}{4}\pi). \qquad (13)$$

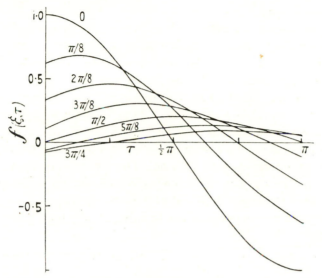

Fig. 6. Oscillations of temperature at various depths due to harmonic surface temperature

Thus the steady temperature in the semi-infinite solid which is heated at its face $x = 0$ by periodic flux $F_0\cos(\omega t - \epsilon)$ is

$$v = \frac{F_0}{K\mathbf{k}\sqrt{2}}\cos(\omega t - \mathbf{k}x - \epsilon - \tfrac{1}{4}\pi). \qquad (14)$$

The amplitude of this involves the thermal constants in the combination $K\rho c$, and so provides a means of measuring this quantity.

The expression (8) involves the dimensionless quantities

$$\xi = x\sqrt{(\omega/2\kappa)} \quad \text{and} \quad \tau = \omega t. \qquad (15)$$

Then, if we choose the origin of time so that $\epsilon = 0$, (8) involves the function
$$f(\xi, \tau) = e^{-\xi}\cos(\tau - \xi). \qquad (16)$$

In Fig. 6, $f(\xi, \tau)$ is plotted against τ for the values $0, \pi/8, \ldots, 6\pi/8$ of ξ; as ξ increases these curves show the progressive lag in phase and diminution in amplitude which has been described above.

In Fig. 7, $f(\xi, \tau)$ is plotted against ξ for the values $0, \pi/8, \ldots, \pi$ of τ. These curves show the way in which the temperature varies with depth at different times. Families of curves of the types shown in Figs. 6

and 7 often appear as the result of experiment; for example, in many of the works referred to in § 2.12.

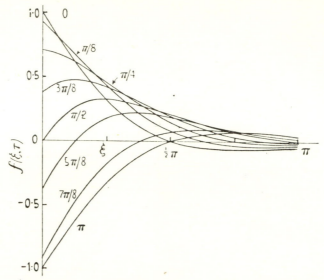

FIG. 7. Variation of temperature with depth when the surface temperature is harmonic

When the surface temperature is a periodic function of the time of period $2\pi/\omega$, we can obtain the solution by using the Fourier series for $\phi(t)$:

$$\phi(t) = A_0 + A_1 \cos(\omega t - \epsilon_1) + A_2 \cos(2\omega t - \epsilon_2) + \dots. \qquad (17)$$

With this value of $\phi(t)$ we have from (8)

$$v = A_0 + \sum_{n=1}^{\infty} A_n e^{-x\sqrt{(n\omega/2\kappa)}} \cos\left\{ n\omega t - \epsilon_n - x\left(\frac{n\omega}{2\kappa}\right)^{\frac{1}{2}} \right\}, \qquad (18)$$

provided, as before, that $x/2\sqrt{(\kappa t)}$ is small.

For example if the surface temperature is

$$\left. \begin{aligned} \phi(t) &= V, & 2rT < t < (2r+1)T, & \quad r = 0, 1, 2,\dots \\ \phi(t) &= -V, & (2r+1)T < t < (2r+2)T \end{aligned} \right\} \qquad (19)$$

so that

$$\phi(t) = \frac{4V}{\pi} \sum_{n=0}^{\infty} \frac{1}{(2n+1)} \sin\frac{(2n+1)\pi t}{T},$$

the steady state periodic temperature is

$$v = \frac{4V}{\pi} \sum_{n=0}^{\infty} \frac{1}{(2n+1)} e^{-x\sqrt{[(2n+1)\pi/2\kappa T]}} \sin\left[\frac{(2n+1)\pi t}{T} - x\left\{ \frac{(2n+1)\pi}{2\kappa T} \right\}^{\frac{1}{2}} \right]. \qquad (20)$$

In Fig. 8, v/V is plotted against t for values 0, $0{\cdot}5$, $1{\cdot}0$, $2{\cdot}0$ of $x\sqrt{(\pi/2\kappa T)}$. The way in which the higher harmonics disappear and the square wave gradually becomes sinusoidal as we move into the solid shows up clearly.

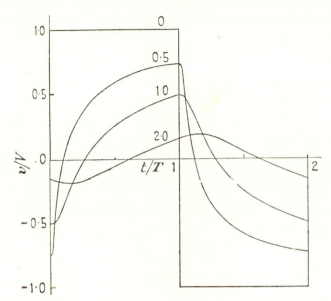

FIG. 8. Oscillations of temperature at various depths caused by a 'square wave' surface temperature

Finally, the analogy with the theory of electric transmission lines must be noted. For generality, we consider a rod of area Ω with no loss of heat from its surface.† Then the temperature v and rate of flow of heat I in the rod satisfy

$$\Omega\rho c\,\frac{\partial v}{\partial t} = -\frac{\partial I}{\partial x}, \qquad I = -K\Omega\,\frac{\partial v}{\partial x}. \tag{21}$$

These are precisely the equations satisfied by the potential v and current I in a transmission line with series resistance $1/K\Omega$ and shunt capacitance $\Omega\rho c$, per unit length, and with zero inductance and leakance (the 'submarine cable'). The theory of the steady periodic behaviour of such a line is well known and may be taken over immediately. This implies that the behaviour of a complicated thermal network may be written down immediately by the methods of circuit theory. In this notation,‡ the series impedance per unit length of the line is $Z = 1/\Omega K$ and its shunt admittance is $Y = i\omega\Omega\rho c$. Its characteristic impedance Z_0 is given by

$$Z_0 = [Z/Y]^{\frac{1}{2}} = \frac{1}{\Omega(iK\omega\rho c)^{\frac{1}{2}}} = \frac{1}{\Omega\mathbf{k}K(1+i)}, \tag{22}$$

† If there is loss of heat at a rate proportional to the temperature, the analogy is with the leaky submarine cable.

‡ For the theory see any work on circuit theory, e.g. Slater, *Microwave Transmission* (McGraw-Hill, 1942) § 2.

and the propagation constant γ is

$$\gamma = (YZ)^{\frac{1}{2}} = \mathbf{k}(1+i). \tag{23}$$

I and v at any point of a semi-infinite line are given by

$$I = v/Z_0, \qquad v = e^{-\gamma x}. \tag{24}$$

These are the results already obtained in (8) and (13). This method has been very fully developed by Marcus† who uses methods developed in the theory of wave-guides to study the effect on temperature waves of a change in cross-section of a conductor.

2.7. Semi-infinite solid. Radiation at the surface into a medium at zero temperature. Initial temperature constant‡

When the initial temperature is constant and equal to V, the equations for v are as follows:

$$\frac{\partial v}{\partial t} = \kappa \frac{\partial^2 v}{\partial x^2},$$

$$v = V \quad \text{when } t = 0,$$

$$-\frac{\partial v}{\partial x} + hv = 0, \quad \text{when } x = 0.$$

Let

$$\phi = v - \frac{1}{h}\frac{\partial v}{\partial x}.$$

Then we have

$$\frac{\partial \phi}{\partial t} = \kappa \frac{\partial^2 \phi}{\partial x^2},$$

$$\phi = V \quad \text{when } t = 0,$$

$$\phi = 0 \quad \text{when } x = 0.$$

Therefore, from 2.4 (3),

$$\phi(x,t) = \frac{2V}{\sqrt{\pi}} \int_0^{x/2\sqrt{(\kappa t)}} e^{-u^2}\, du,$$

and it will be noticed that, when $x \to \infty$, $\phi(x,t)$ has the limit V.

To determine v we have the equation

$$\frac{\partial v}{\partial x} - hv = -h\phi(x,t).$$

Thus

$$v = Ce^{hx} - he^{hx} \int_\infty^x \phi(\xi,t)e^{-h\xi}\, d\xi,$$

on integrating this equation in the usual way.

Therefore

$$v = Ce^{hx} + h\int_0^\infty \phi(x+\eta,t)e^{-h\eta}\, d\eta,$$

on putting $\xi = x + \eta$.

† Marcus, Carnegie Institute of Technology Report (1953).
‡ The case of initial temperature $f(x)$ is treated in § 14.2.

But as $x \to \infty$, $\phi(x, t)$ has the limit V, and, as v must be finite, it follows that C must be zero.

Hence the solution of our problem is given by

$$v = h \int_0^\infty \phi(x+\eta, t)e^{-h\eta}\, d\eta = \frac{2Vh}{\sqrt{\pi}} \int_0^\infty e^{-h\eta} \left[\int_0^{(x+\eta)/2\sqrt{(\kappa t)}} e^{-u^2}\, du \right] d\eta.$$

Therefore

$$v = \frac{2V}{\sqrt{\pi}} \left[-e^{-h\eta} \int_0^{(x+\eta)/2\sqrt{(\kappa t)}} e^{-u^2}\, du \right]_0^\infty + \frac{2V}{\sqrt{\pi}} \int_0^\infty e^{-h\eta} \frac{d}{d\eta} \left(\int_0^{(x+\eta)/2\sqrt{(\kappa t)}} e^{-u^2}\, du \right) d\eta$$

$$= \frac{2V}{\sqrt{\pi}} \int_0^{x/2\sqrt{(\kappa t)}} e^{-u^2}\, du + \frac{V}{\sqrt{(\pi\kappa t)}} \int_0^\infty e^{-h\eta - (x+\eta)/4\kappa t}\, d\eta$$

$$= \frac{2V}{\sqrt{\pi}} \int_0^{x/2\sqrt{(\kappa t)}} e^{-u^2}\, du + \frac{V}{\sqrt{(\pi\kappa t)}} e^{hx+h^2\kappa t} \int_0^\infty e^{-(x+\eta+2h\kappa t)^2/4\kappa t}\, d\eta.$$

In the second integral put

$$x + \eta + 2h\kappa t = 2\sqrt{(\kappa t)}u.$$

Then

$$v = \frac{2V}{\sqrt{\pi}} \int_0^{x/2\sqrt{(\kappa t)}} e^{-u^2}\, du + \frac{2V}{\sqrt{\pi}} e^{hx+h^2\kappa t} \int_{(x+2h\kappa t)/2\sqrt{(\kappa t)}}^\infty e^{-u^2}\, du$$

$$= \frac{2V}{\sqrt{\pi}} \int_0^{x/2\sqrt{(\kappa t)}} e^{-u^2}\, du + \frac{2V}{\sqrt{\pi}} e^{hx+h^2\kappa t} \left(\int_0^\infty e^{-u^2}\, du - \int_0^{(x+2h\kappa t)/2\sqrt{(\kappa t)}} e^{-u}\, du \right).$$

Therefore

$$\frac{v}{V} = \operatorname{erf}\frac{x}{2\sqrt{(\kappa t)}} + e^{hx+h^2\kappa t} \operatorname{erfc}\left\{ \frac{x}{2\sqrt{(\kappa t)}} + h\sqrt{(\kappa t)} \right\}, \tag{1}$$

where $\operatorname{erf} x$ and $\operatorname{erfc} x$ are the integrals defined in Appendix II.

The surface temperature of the solid, v_s, obtained by putting $x = 0$ in (1), is given by

$$\frac{v_s}{V} = e^{h^2\kappa t} \operatorname{erfc} h\sqrt{(\kappa t)}. \tag{2}$$

A short table of this function is given in Appendix II. For large values of the time it follows from (5) of Appendix II that v_s/V is given approximately by

$$\frac{1}{h\sqrt{(\pi\kappa t)}} \left\{ 1 - \frac{1}{2h^2\kappa t} + \frac{3}{4h^4\kappa^2 t^2} - \cdots \right\}. \tag{3}$$

Thus, when the cooling has been going on for a very long time the surface temperature may be taken to be

$$\frac{V}{h\sqrt{(\pi\kappa t)}},$$

(4)

with an error less than $V/2h^3(\pi\kappa^3t^3)^{\frac{1}{2}}$.

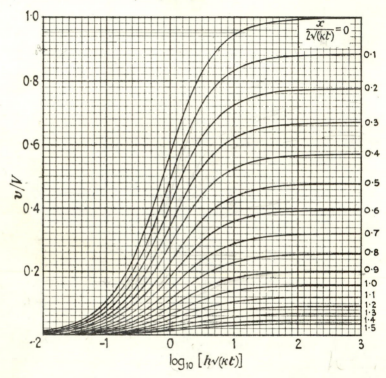

FIG. 9. Temperature distribution in the semi-infinite solid with radiation at its surface

For the problem of the *semi-infinite solid, initially at temperature zero, which is heated at $x = 0$ by radiation from a medium at temperature V,* the solution is

$$\frac{v}{V} = \operatorname{erfc}\frac{x}{2\sqrt{(\kappa t)}} - e^{hx+h^2\kappa t}\operatorname{erfc}\left\{\frac{x}{2\sqrt{(\kappa t)}} + h\sqrt{(\kappa t)}\right\}.$$

(5)

The solution (5) can be expressed in terms of any two of the dimensionless parameters

$$\frac{x}{2\sqrt{(\kappa t)}}, \quad h\sqrt{(\kappa t)}, \quad \text{or} \quad hx.$$

(6)

Each choice of two has its own advantages: here in Fig. 9 we plot v/V against $\log_{10} h(\kappa t)^{\frac{1}{2}}$ for the values $0, 0\cdot1,..., 1\cdot5$ of $x/2\sqrt{(\kappa t)}$.

It was remarked in § 1.9 that the radiation boundary condition can arise in several ways, notably in cases of heat transfer by forced convection or radiation, and of heat transfer through a thin surface skin. In the former cases values of h can be obtained from formulae such as those of § 1.9 and the temperature at any position and time in the semi-infinite solid can then be found from (5).

It is of importance to consider the effect of a thin surface film of poor conductor and to compare results deduced from (5) for this case with those obtained on the assumption of § 2.4 that the surface at $x = 0$ is maintained at V for all $t > 0$. The boundary condition of constant surface temperature has been used much more than any other in the theory of conduction of heat, but it is not usually realized in practice, since there is generally some sort of imperfect contact at $x = 0$ so that this surface does not attain the temperature V instantaneously. It is desirable to know what errors may arise from this source, so we represent the imperfect contact by surface films of likely substances and thicknesses.

We consider surface films of air, vaseline, water, and mercury, each 0·001 cm thick; values of the surface conductance H for these have been given in the table in § 1.9 D. As solids we consider a good conductor, silver, $K = 1·001$, $\kappa = 1·716$; a poor metallic conductor, bismuth, $K = 0·020$, $\kappa = 0·0699$; and a non-metal, glass, $K = 0·0028$, $\kappa = 0·0058$. We take $x = 1$ cm in all cases and $t = 0·64$, 15·7, and 189 sec, respectively, so that for all three solids $x/2\sqrt{(\kappa t)} = 0·477$ and the value of v/V is 0·5 for the case of constant surface temperature § 2.4. Values of v/V calculated from (5) for the various surface films are given in the table below.

	Constant temp. $H = \infty$	Air $H = 0·053$	Light pressure† $H = 0·07$	Vaseline $H = 0·44$	Water $H = 1·44$	Mercury $H = 20$
Silver	0·5	0·022	0·029	0·145	0·295	0·479
Bismuth	0·5	0·368	0·394	0·481	0·494	0·500
Glass	0·5	0·478	0·483	0·497	0·499	0·500

† Cf. § 1.9 G, footnote.

It appears that with a good conductor a surface film of poor conductor can have a very marked effect; for a poorly conducting solid the effect of surface films is much less important.

2.8. Semi-infinite solid. Radiation at the surface into a medium at temperature $f(t)$. Initial temperature zero

In this problem the temperature v has to satisfy

$$\frac{\partial v}{\partial t} = \kappa \frac{\partial^2 v}{\partial x^2},$$

$$-\frac{\partial v}{\partial x} + hv = hf(t) \quad \text{at } x = 0,$$

$$v = 0 \quad \text{when } t = 0.$$

Proceeding as in the last article, put

$$\phi = v - \frac{1}{h}\frac{\partial v}{\partial x}.$$

Then we have the following equations to determine ϕ:

$$\frac{\partial \phi}{\partial t} = \kappa \frac{\partial^2 \phi}{\partial x^2},$$

$$\phi = f(t) \quad \text{at } x = 0,$$

$$\phi = 0 \quad \text{when } t = 0.$$

These equations have already been discussed in § 2.5, and we have seen that

$$\phi = \frac{2}{\sqrt{\pi}} \int_{x/2\sqrt{(\kappa t)}}^{\infty} f\left(t - \frac{x^2}{4\kappa\mu^2}\right)e^{-\mu^2}\, d\mu.$$

Hence, as in § 2.7,

$$v = \frac{2h}{\sqrt{\pi}} \int_{0}^{\infty} e^{-h\eta}\, d\eta \int_{(x+\eta)/2\sqrt{(\kappa t)}}^{\infty} f\left(t - \frac{(x+\eta)^2}{4\kappa\mu^2}\right)e^{-\mu^2}\, d\mu. \tag{1}$$

Special cases of interest are

(i) $\qquad\qquad f(t) = A, \quad \text{constant, } 0 < t < T$

$$= B, \quad \text{constant, } t > T.$$

Here

$$v = A\, \mathrm{erfc}\,\frac{x}{2\sqrt{(\kappa t)}} - Ae^{hx+h^2\kappa t}\mathrm{erfc}\left\{\frac{x}{2\sqrt{(\kappa t)}}+h\sqrt{(\kappa t)}\right\}, \quad 0 < t < T, \tag{2}$$

$$v = A\, \mathrm{erfc}\,\frac{x}{2\sqrt{(\kappa t)}} + (B-A)\,\mathrm{erfc}\,\frac{x}{2\sqrt{[\kappa(t-T)]}} - Ae^{hx+h^2\kappa t}\,\mathrm{erfc}\left\{\frac{x}{2\sqrt{(\kappa t)}}+h\sqrt{(\kappa t)}\right\} -$$

$$- (B-A)e^{hx+h^2\kappa(t-T)}\mathrm{erfc}\left(\frac{x}{2\sqrt{[\kappa(t-T)]}}+h\sqrt{[\kappa(t-T)]}\right), \quad t > T. \tag{3}$$

(ii) $\qquad\qquad\qquad\qquad f(t) = \sin(\omega t + \epsilon).$

$$v = \frac{h}{\sqrt{[(h+\omega')^2+\omega'^2]}}\, e^{-\omega' x}\sin(\omega t + \epsilon - \omega' x - \delta) +$$

$$+ \frac{2\kappa h}{\pi}\int_{0}^{\infty}\frac{(\omega\cos\epsilon - \kappa u^2\sin\epsilon)(u\cos ux + h\sin ux)}{(\kappa^2 u^4 + \omega^2)(h^2 + u^2)}\, e^{-\kappa u^2 t}\, u\, du, \tag{4}$$

where $\omega' = \sqrt{(\omega/2\kappa)}$ and $\delta = \tan^{-1}[\omega'/(h+\omega')]$.

The first term of (4) is the steady periodic solution and is seen to have the same properties as 2.6 (6).

2.9. The semi-infinite solid. The flux of heat at $x = 0$ a prescribed function of the time. Zero initial temperature

(i) *Constant flux, F_0 per unit time per unit area.*

$$\text{The flux} \qquad f = -K\frac{\partial v}{\partial x}, \qquad (1)$$

satisfies the same differential equation as v, namely

$$\kappa \frac{\partial^2 f}{\partial x^2} = \frac{\partial f}{\partial t}, \quad x > 0,\, t > 0. \qquad (2)$$

The solution of (2) with

$$f = F_0, \quad \text{constant}, \, x = 0, \, t > 0, \qquad (3)$$

is, by 2.4 (10),

$$f = F_0 \operatorname{erfc} \frac{x}{2\sqrt{(\kappa t)}}. \qquad (4)$$

Thus, from (1), using Appendix II, (9) and (11),

$$v = \frac{F_0}{K} \int_x^\infty \operatorname{erfc} \frac{x}{2\sqrt{(\kappa t)}}\, dx \qquad (5)$$

$$= \frac{2F_0\sqrt{(\kappa t)}}{K} \operatorname{ierfc} \frac{x}{2\sqrt{(\kappa t)}} \qquad (6)$$

$$= \frac{2F_0}{K}\left\{ \left(\frac{\kappa t}{\pi}\right)^{\frac{1}{2}} e^{-x^2/4\kappa t} - \frac{x}{2}\operatorname{erfc} \frac{x}{2\sqrt{(\kappa t)}} \right\}. \qquad (7)$$

A table of values of the function $\operatorname{ierfc} x$ is given in Appendix II. The temperature at $x = 0$ is given by

$$\frac{2F_0}{K}\left(\frac{\kappa t}{\pi}\right)^{\frac{1}{2}}. \qquad (8)$$

The boundary condition of constant flux is of considerable practical importance. It appears if heat is generated by a flat heating element carrying electric current, if heat is generated by friction, and as an approximation in the early stages of heating a furnace or a room. It has also important applications to problems on diffusion. The cooling of the Earth's surface after sunset on a clear windless night† is very nearly that due to removal of heat at a constant rate per unit area per unit time, thus (8) gives the way in which the surface temperature falls after sunset.

The results above apply also to the case of the region $-\infty < x < \infty$ with heat supply $2F_0$ in the plane $x = 0$. The corresponding results for

† Cf. Brunt, *Quart. J. R. Met. Soc.* **58** (1932) 389.

the case in which the regions $x > 0$ and $x < 0$ are of different materials are given in § 2.15.

(ii) *The region $x > 0$ initially at zero temperature. Flux $f(t)$ per unit time per unit area at $x = 0$.*

$$v = \frac{\kappa^{\frac{1}{2}}}{K\pi^{\frac{1}{2}}} \int_0^t f(t-\tau) e^{-x^2/4\kappa t} \frac{d\tau}{\tau^{\frac{1}{2}}}. \tag{9}$$

This follows from (5) and Duhamel's theorem, § 1.14.

(iii) *The region $x > 0$ initially at zero temperature. Heat supplied at $x = 0$ at the constant rate F_0 per unit time per unit area for time T. At time T the supply of heat ceases and the end $x = 0$ is thermally insulated.*

The temperature at time t is given by (6) or (7) if $0 < t < T$, and if $t > T$ by

$$\frac{2F_0\kappa^{\frac{1}{2}}}{K}\left\{ t^{\frac{1}{2}}\operatorname{ierfc}\frac{x}{2\kappa^{\frac{1}{2}}t^{\frac{1}{2}}} - (t-T)^{\frac{1}{2}}\operatorname{ierfc}\frac{x}{2\kappa^{\frac{1}{2}}(t-T)^{\frac{1}{2}}} \right\}. \tag{10}$$

This gives the temperature if a flat heating element to which heat is supplied at the rate $2F_0$ per unit time per unit area is immersed in infinite medium and switched on for time T and then switched off.

(iv) *The region $x > 0$ is initially at zero temperature. Heat is supplied at $x = 0$ at the rate† $K(\kappa\pi t)^{-\frac{1}{2}}$ per unit time per unit area for time T. At time T the supply of heat ceases and the surface is insulated.*

The temperature at the end $x = 0$ at time t is

$$\left. \begin{array}{l} 1, \quad 0 < t \leqslant T \\ \frac{2}{\pi}\sin^{-1}(T/t)^{\frac{1}{2}}, \quad t > T \end{array} \right\}. \tag{11}$$

This gives the surface temperature of the semi-infinite solid for the case in which the end $x = 0$ is kept at unit temperature for time T and then insulated.

The temperature v at x for $t > T$ may be obtained from (9) as the definite integral

$$v = \frac{1}{\pi} \int_0^T e^{-x^2/4\kappa(t-\tau)} \frac{d\tau}{\tau^{\frac{1}{2}}(t-\tau)^{\frac{1}{2}}} = \frac{2}{\pi} \int_0^{[T/(t-T)]^{\frac{1}{2}}} e^{-x^2(1+u^2)/4\kappa t} \frac{du}{1+u^2}. \tag{12}$$

(v) *The region $x > 0$ is initially at zero temperature. The flux of heat at $x = 0$ for $t > 0$ is $\sin(\omega t + \epsilon)$.*

$$v = \frac{\kappa^{\frac{1}{2}}}{K\omega^{\frac{1}{2}}} e^{-x\sqrt{(\omega/2\kappa)}}\sin[\omega t + \epsilon - \tfrac{1}{4}\pi - x\sqrt{(\omega/2\kappa)}] -$$

$$- \frac{2\kappa}{K\pi} \int_0^\infty \frac{(\kappa u^2 \sin\epsilon - \omega\cos\epsilon)\cos ux}{\omega^2 + \kappa^2 u^4} e^{-\kappa u^2 t} \, du. \tag{13}$$

As before, the first term of (13) is the steady periodic solution, and the second a transient.

---

† This is the rate at which heat must be supplied to maintain the surface at unit temperature, cf. 2.4 (11). The integral (12) has been tabulated by Smith, *Aust. J. Phys.* **6** (1953) 127–30. The question of the minimum supply of heat necessary to give a specified rise in temperature at a specified point in heating by methods (iii) and (iv) is discussed by Lauwerier, *Appl. Sci. Res. A*, **4** (1954) 142–52.

(vi) *If the flux at $x = 0$ is*

$$\left.\begin{array}{ll} -(1/\pi), & 2r\pi/\omega < t < (2r+1)\pi/\omega, \\ -\sin\omega t - (1/\pi), & (2r+1)\pi/\omega < t < (2r+2)\pi/\omega, \end{array}\right\} \quad (14)$$

the periodic part of the temperature at x at time t is

$$-\frac{\kappa^{\frac{1}{2}}}{2K\omega^{\frac{1}{2}}} e^{-x\sqrt{(\omega/2\kappa)}} \sin[\omega t - \tfrac{1}{4}\pi - x\sqrt{(\omega/2\kappa)}] -$$

$$-\frac{(2\kappa)^{\frac{1}{2}}}{\pi K\omega^{\frac{1}{2}}} \sum_{n=1}^{\infty} \frac{e^{-x\sqrt{(n\omega/\kappa)}} \sin[2n\omega t + \tfrac{1}{4}\pi - x\sqrt{(n\omega/\kappa)}]}{n^{\frac{1}{2}}(4n^2-1)}. \quad (15)$$

This is an approximation† to the effect of solar heating on the Earth's surface at the equinox; the first line of (14) corresponding to nighttime and the second to daytime conditions.

(vii) *The region $x > 0$ initially at zero temperature. Flux $F_0 t^{\frac{1}{2}n}$ into the solid for $t > 0$ where n may be -1, 0 or a positive integer.*

$$v = \frac{F_0 \kappa^{\frac{1}{2}} \Gamma(\tfrac{1}{2}n+1)}{K} (4t)^{\frac{1}{2}(n+1)} i^{n+1} \text{erfc} \frac{x}{2(\kappa t)^{\frac{1}{2}}}, \quad (16)$$

where the function $i^n \text{erfc}\, x$ is defined in Appendix II. The surface temperature is

$$\frac{F_0 \kappa^{\frac{1}{2}} \Gamma(\tfrac{1}{2}n+1)}{K\Gamma(\tfrac{1}{2}n+\tfrac{3}{2})} t^{\frac{1}{2}(n+1)}. \quad (17)$$

2.10. Application to the determination of thermal conductivity

In the mathematical problem, where the semi-infinite solid is initially at zero temperature and the end $x = 0$ is kept at temperature unity, the temperature at time t is, by 2.4 (10),

$$v = 1 - \text{erf} \frac{x}{2\sqrt{(\kappa t)}}. \quad (1)$$

Thus from the observed temperature at any point x_1 at time t_1, the value of $x_1/2\sqrt{(\kappa t_1)}$ can be found from a Table of the Error Function (cf. Appendix II, Table I) and hence κ would be known. The difficulty in using this method depends on the fact that if the end of the bar, $x = 0$, is heated, say by a current of fluid at constant temperature, it has been shown by experiment that it is not true that the end of the bar immediately attains the temperature of the fluid; and thus the mathematical statement of the conditions of the experiment can only be accepted as an approximation.‡ It appears from the discussion at the end of § 2.7 that it is likely to be a fair approximation for poor conductors, but that for good conductors considerable errors may be introduced and care must be taken with the experimental arrangements, for example by greatly increasing the velocity with which the fluid flows over the surface of the solid.§

---

† Brunt, loc. cit. The extension to any latitude and season is made by Jaeger and Johnson, *Geofis. Pura é Appl.* **24** (1953) 104. The case in which the surface loses heat by black-body radiation (corresponding very nearly to the surface temperature of the moon) is discussed by Jaeger, *Proc. Camb. Phil. Soc.* **49** (1953) 355–9; *Aust. J. Phys.* **6** (1953) 10–21.

‡ Cf. Schulze, *Ann. Physik* (N.F.), **66** (1898) 207; Schaufelberger, ibid. (4) **7** (1902) 589.

§ Cf. Frazier, *Phys. Rev.* (2) **39** (1932) 515.

Various attempts to avoid this difficulty have been made. Kirchhoff and Hanse-mann† made the assumption that the temperature at $x = 0$ would be given by $C + \phi(t)$, where C was a constant and $\phi(t)$ was a function of the time which was negligible except for small values of t. The value of C was to be determined by temperature observations near the heated end and was not assumed to be equal to the temperature of the fluid by which the heat was supplied. A modification of this method assumes the surface temperature to be $C(1 - e^{-\alpha t})$, where α is large.

In another method‡ of treating the same problem the temperatures at two points are studied and the conditions at $x = 0$ are only used to suggest a suitable mathematical form for the solution.

Many of the mathematical solutions of the previous sections can be used as bases of experimental methods for measuring diffusivity. Thus if the solid be heated by a flat heating coil of negligible heat capacity, or by radiation from a very high temperature, the boundary condition of constant flux at the surface is approximately realized, and 2.9 (6) can be arranged to give κ from two observations of the temperature. Again, 2.7 (5) can be arranged to give h and κ from two observed temperatures.

2.11. The semi-infinite solid with heat generated within it

For linear flow of heat, the differential equation 1.6 (7) becomes

$$\frac{\partial^2 v}{\partial x^2} - \frac{1}{\kappa}\frac{\partial v}{\partial t} = -\frac{A}{K}, \tag{1}$$

where A, the rate of heat production per unit time per unit volume is, in general, a function of x, t, and v. Here we shall consider only the case in which A is independent of v; in this case§ three methods are available for the solution of (1): (i) integration of the source solutions of Chapters X, XIV; (ii) the Laplace transformation method of Chapter XII; (iii) reduction of (1) to a homogeneous form by a change of variable. The first two methods are the most powerful: here we shall merely illustrate the third briefly and give a few results which are of interest in connexion with radioactive heat generation in the Earth's crust (cf. §§ 2.13, 2.14). Some further results are given in §§ 12.4 and 15.7. It may be remarked that heat production at a rate which is a linear function of the temperature may be studied by the methods of §§ 1.14 and 15.7.

Since the solution for the case in which the rate of heat production is a function of the time can be obtained by Duhamel's theorem, § 1.14,

---

† *Ann. Physik* (N.F.), **9** (1880) 1, and **13** (1881) 406.

‡ Grüneisen, *Ann. Physik*, (4) **3** (1900) 43; Giebe, *Verh. dtsch. phys. Ges.* (1903) p. 60; Hobson and Diesselhorst, 'Wärmeleitung', *Enc. der Math. Wiss.*, Bd. V, Tl. I, (1905) pp. 224–7.

§ A large number of results for the semi-infinite solid, slab, sphere, and cylinder, with constant heat production and various surface conditions, is given by Fox, *Phil. Trans. Roy. Soc.* A, **232** (1934) 431. See also Paterson, *Phil. Mag.* (7) **32** (1941) 384; *Proc. Glasgow Math. Ass.* **1** (1953) 164–9.

from that for heat production independent of the time, it is only necessary to consider the latter, though explicit solutions for simple types
of variation with the time are easily obtained.†

(i) *The region $x > 0$ with initial temperature $a+bx$. Heat is produced
at a constant rate A_0 per unit time per unit volume for $t > 0$. The surface
$x = 0$ is maintained at zero temperature.*

Here we have to solve

$$\frac{\partial^2 v}{\partial x^2} - \frac{1}{\kappa}\frac{\partial v}{\partial t} + \frac{A_0}{K} = 0, \quad x > 0, \, t > 0,$$

with

$$v = a+bx, \quad \text{when } t = 0,$$

and

$$v = 0, \quad x = 0, \, t > 0.$$

Putting

$$v = u - \frac{A_0}{2K}x^2,$$

these equations become

$$\frac{\partial^2 u}{\partial x^2} - \frac{1}{\kappa}\frac{\partial u}{\partial t} = 0, \quad x > 0, \, t > 0$$

with

$$u = a+bx+\frac{A_0}{2K}x^2, \quad \text{when } t = 0,$$

and

$$u = 0, \quad x = 0, \, t > 0.$$

It follows from 2.4 (1) that

$$u = \left(a+\frac{\kappa t A_0}{K}+\frac{A_0 x^2}{2K}\right)\mathrm{erf}\frac{x}{2\sqrt{(\kappa t)}}+\frac{A_0 x}{K}\left(\frac{\kappa t}{\pi}\right)^{\frac{1}{2}} e^{-x^2/4\kappa t}+bx,$$

and thus

$$v = \left(a+\frac{\kappa t A_0}{K}+\frac{A_0 x^2}{2K}\right)\mathrm{erf}\frac{x}{2\sqrt{(\kappa t)}}+\frac{A_0 x}{K}\left(\frac{\kappa t}{\pi}\right)^{\frac{1}{2}} e^{-x^2/4\kappa t}+bx-\frac{A_0 x^2}{2K}. \quad (2)$$

(ii) *The problem of* (i) *except that the rate of heat production is $A_0 e^{-\alpha x}$.*

$$v = bx+\left(a-\frac{A_0}{K\alpha^2}\right)\mathrm{erf}\frac{x}{2\sqrt{(\kappa t)}}+\frac{A_0}{K\alpha^2}\left\{1-e^{-\alpha x}+\tfrac{1}{2}e^{\kappa\alpha^2 t-\alpha x}\,\mathrm{erfc}\left[\alpha\sqrt{(\kappa t)}-\frac{x}{2\sqrt{(\kappa t)}}\right]-\right.$$

$$\left.-\tfrac{1}{2}e^{\kappa\alpha^2 t+\alpha x}\,\mathrm{erfc}\left[\alpha\sqrt{(\kappa t)}+\frac{x}{2\sqrt{(\kappa t)}}\right]\right\}. \quad (3)$$

(iii) *The problem of* (i) *except that heat is produced only in the layer‡ $0 < x < l$.*

$$v = \frac{\kappa A_0 t}{K}\left\{1-4\,\mathrm{i}^2\mathrm{erfc}\frac{x}{2\sqrt{(\kappa t)}}+2\,\mathrm{i}^2\mathrm{erfc}\frac{l+x}{2\sqrt{(\kappa t)}}-2\,\mathrm{i}^2\mathrm{erfc}\frac{l-x}{2\sqrt{(\kappa t)}}\right\}+bx+a\,\mathrm{erf}\frac{x}{2\sqrt{(\kappa t)}},$$

$$0 < x < l, \quad (4)$$

† Some results for heat production at the rate $t^{\frac{1}{2}n}$, where $n = -1, 0, 1, 2,...$, are given
in § 12.4. See also Jaeger, *Quart. Appl. Math.* **4** (1946) 100–3.
‡ For the case in which the thermal constants in $0 < x < l$ and $x > l$ are different
see § 12.8.

$$v = \frac{\kappa A_0 t}{K}\left\{2\,\mathrm{i}^2\mathrm{erfc}\,\frac{x-l}{2\sqrt{(\kappa t)}} + 2\,\mathrm{i}^2\mathrm{erfc}\,\frac{x+l}{2\sqrt{(\kappa t)}} - 4\,\mathrm{i}^2\mathrm{erfc}\,\frac{x}{2\sqrt{(\kappa t)}}\right\} + bx + a\,\mathrm{erf}\,\frac{x}{2\sqrt{(\kappa t)}},$$
$$x > l. \quad (5)$$

The temperature gradient at the surface $x = 0$ is

$$b + \frac{a}{\sqrt{(\pi\kappa t)}} + \frac{2A_0\sqrt{(\kappa t)}}{K}\left\{\frac{1}{\sqrt{\pi}} - \mathrm{ierfc}\,\frac{l}{2\sqrt{(\kappa t)}}\right\}. \quad (6)$$

In the application to temperature in the earth's crust heat is generated only in a layer at the surface less than 50 km in thickness, so that in (5) l is small compared with $2\sqrt{(\kappa t)}$. Thus we may expand (5) in powers of l by Taylor's theorem and obtain†

$$v = bx + a\,\mathrm{erf}\,\frac{x}{2\sqrt{(\kappa t)}} + \frac{A_0 l^2}{2K}\,\mathrm{erfc}\,\frac{x}{2\sqrt{(\kappa t)}}, \quad x > l, \quad (7)$$

for depths below the radioactive layer.

Also the temperature gradient at the surface is given approximately by

$$b + \frac{a}{\sqrt{(\pi\kappa t)}} + \frac{A_0}{K}\left(l - \frac{l^2}{2\sqrt{(\pi\kappa t)}} + \dots\right). \quad (8)$$

(iv) *The region $x > 0$ with zero initial temperature. Heat is produced at a constant rate A_0 per unit time per unit volume for $t > 0$ in the region $0 < x < l$. No flow of heat at $x = 0$.*

$$v = \frac{\kappa A_0 t}{K}\left\{1 - 2\,\mathrm{i}^2\mathrm{erfc}\,\frac{l-x}{2\sqrt{(\kappa t)}} - 2\,\mathrm{i}^2\mathrm{erfc}\,\frac{l+x}{2\sqrt{(\kappa t)}}\right\}, \quad 0 < x < l, \quad (9)$$

$$v = \frac{2\kappa A_0 t}{K}\left\{\mathrm{i}^2\mathrm{erfc}\,\frac{x-l}{2\sqrt{(\kappa t)}} - \mathrm{i}^2\mathrm{erfc}\,\frac{x+l}{2\sqrt{(\kappa t)}}\right\}, \quad x > l. \quad (10)$$

This is also the solution for the case of heat production in a strip of thickness $2l$ in infinite solid.

(v) *The region $x > 0$ with zero initial and surface temperature. Heat is produced at a constant rate A_0 per unit time per unit volume for $t > 0$ in the region $a < x < b$.* The temperature gradient‡ at the surface is

$$\frac{2A_0(\kappa t)^{\frac{1}{2}}}{K}\left\{\mathrm{ierfc}\,\frac{a}{2(\kappa t)^{\frac{1}{2}}} - \mathrm{ierfc}\,\frac{b}{2(\kappa t)^{\frac{1}{2}}}\right\}, \quad (11)$$

(vi) *The region $x > 0$ with zero initial temperature and heat production at the rate $A_0 e^{-\lambda t}$. The surface $x = 0$ maintained at zero for $t > 0$*

$$v = \frac{A_0}{\rho c\lambda}\,\mathrm{erf}\,\frac{x}{2(\kappa t)^{\frac{1}{2}}} - \frac{A_0}{\rho c\lambda}e^{-\lambda t}\left\{1 - \mathbf{R}e^{-ix\sqrt{(\lambda/\kappa)}}\mathrm{erfc}\left[\frac{x}{2(\kappa t)^{\frac{1}{2}}} - i(\lambda t)^{\frac{1}{2}}\right]\right\}, \quad (12)$$

where $\mathbf{R}$ implies that the real part is taken. For the error function of complex argument, see Appendix II.

(vii) *The region $x > 0$ with initial temperature $a + bx$. Heat production at the rate $A_0 e^{-\alpha x}$ for $t > 0$. No flow§ of heat at $x = 0$.*

$$v = a + bx + \left(2b + \frac{2A_0}{K\alpha}\right)(\kappa t)^{\frac{1}{2}}\,\mathrm{ierfc}\,\frac{x}{2(\kappa t)^{\frac{1}{2}}} - \frac{A_0}{\alpha^2 K}e^{-\alpha x} +$$
$$+ \frac{A_0}{2K\alpha^2}e^{\alpha^2\kappa t - \alpha x}\,\mathrm{erfc}\left[\alpha(\kappa t)^{\frac{1}{2}} - \frac{x}{2(\kappa t)^{\frac{1}{2}}}\right] + \frac{A_0}{2K\alpha^2}e^{\alpha^2\kappa t + \alpha x}\mathrm{erfc}\left[\alpha(\kappa t)^{\frac{1}{2}} + \frac{x}{2(\kappa t)^{\frac{1}{2}}}\right] \quad (13)$$

† Jeffreys, *Beitr. Geophys.* **18** (1927) 1. For an extension to the case of heat production and thermal constants varying with the depth see Bullard, *Mon. Not. R. Astr. Soc. Geophys. Supp.* **4** (1939) 534; Jeffreys, *The Earth* (Cambridge, ed. 3, 1952) Chap. X.
‡ van Orstrand, *J. Wash. Acad. Sci.* **22** (1932) 529, considers the case of a thin region.
§ Cook, *Brit. J. Appl. Phys.* **3** (1952) 1–6. The problem appears in connexion with the heating of the body by microwaves. See also Jaeger, ibid. **3** (1952) 221–2.

2.12. Terrestrial temperature : surface oscillations

Observations of the temperature at points near the surface of the Earth have been carried out at a large number of meteorological stations in different parts of the world for many years. These results have established that the variations of the surface temperature from the heat by day to the cold by night do not affect the temperatures of points at a depth of more than 3–4 feet, while the yearly changes from the cold of winter to the heat of summer may be observed up to a depth of 60–70 feet. Below that depth the temperature remains practically constant from day to day and is not subject to alterations due to the changes at the surface. In other words, the heat waves due to the changes of the temperature at the surface die away before they penetrate to a depth of more than 60–70 feet, and the heat which is thus transferred to the Earth oscillates in the upper crust, and while it proceeds inwards at certain seasons of the year, at others it ascends and radiates into space at the surface.

The periodic oscillations in the temperature near the surface have been used by writers from Fourier and Poisson onwards for the determination of the thermal conductivity of rocks near the surface. Taking the surface to be the plane $x = 0$ with periodic temperature

$$v = V_0 + \sum_{n=1}^{\infty} V_n \cos(n\omega t - \epsilon_n), \tag{1}$$

the temperature at depth x is, by 2.6 (18),

$$v = V_0 + \sum_{n=1}^{\infty} V_n e^{-\mathbf{k}x\sqrt{n}} \cos(n\omega t - \epsilon_n - \mathbf{k}x n^{\frac{1}{2}}), \tag{2}$$

where $$\mathbf{k} = (\omega/2\kappa)^{\frac{1}{2}}. \tag{3}$$

As remarked in § 2.6, the theory shows that each partial wave is propagated with unaltered period inwards, and that the amplitudes of the waves of shorter period diminish more rapidly than those of greater period so that the periodic variation takes a simpler form as we descend, the principal wave with the longest period persisting to greatest depth. The depth at which the amplitude of the annual variation is reduced by a factor of $0 \cdot 1$ is approximately $\sqrt{365} = 19$ times that for the diurnal variation, corresponding to the statement above that these variations are noticeable to depths of 60 to 70, and 3 to 4 feet, respectively.

The classical work on the use of these observations is Kelvin's paper† on 'The Reduction of Observations of Underground Temperature'. Kelvin used Forbes's Edinburgh observations for a period of eighteen

---

† *Trans. Roy. Soc. Edin.* **22** (1861) 405.

years. From these a mean temperature curve for a year was found and harmonically analysed: thus, the temperatures v_1 and v_2 at depths x_1 and x_2 were found in the forms

$$v_1 = V_0' + \sum_{n=1}^{\infty} V_n' \cos(n\omega t - \epsilon_n'), \qquad v_2 = V_0'' + \sum_{n=1}^{\infty} V_n'' \cos(n\omega t - \epsilon_n''). \quad (4)$$

Comparing coefficients between (2) and (4) gives

$$V_0' = V_0'' = V_0, \qquad V_n' = V_n e^{-kx_1\sqrt{n}}, \qquad V_n'' = V_n e^{-kx_2\sqrt{n}},$$

$$\epsilon_n' = kx_1 n^{\frac{1}{2}} + \epsilon_n, \qquad \epsilon_n'' = kx_2 n^{\frac{1}{2}} + \epsilon_n.$$

Therefore
$$\frac{\ln V_n' - \ln V_n''}{x_2 - x_1} = \frac{\epsilon_n'' - \epsilon_n'}{x_2 - x_1} = kn^{\frac{1}{2}} = \left(\frac{n\omega}{2\kappa}\right)^{\frac{1}{2}}. \quad (5)$$

It appears from (5) that either the amplitude or the phase of any harmonic may be used to estimate κ. Kelvin found almost complete agreement between values of κ deduced from the amplitude and phase of the first harmonic and, as might be expected, less satisfactory values from the higher harmonics.

These methods are of interest as giving an average value for the diffusivity of soil, but it is recognized that the conduction of heat in soils is a very complicated process, being affected by the presence of water.[†] The addition of water to dry soil causes a great increase in the thermal conductivity, while the diffusivity usually rises to a maximum value of 2 to 3 times the dry value at a moisture content of 5 to 10 per cent. Further, when a soil is periodically heated, there will be periodic fluctuations in its moisture content as well as in its temperature, so that a theory based on the assumption of constant diffusivity can only give approximate results.

The theory given above only gives the temperature fluctuations at various depths in terms of those at the surface, it does not give absolute values. To obtain these, it is necessary to know the rate at which radiation is received from the sun, the rate at which it is lost from the Earth's surface, and the way in which it is absorbed in the atmosphere: the latter is particularly difficult to estimate on account of the predominant part played by water vapour in the process. However, for days with cloudless skies, Brunt[‡] has obtained curves agreeing well with observation by assuming that the Earth loses heat by radiation at a constant rate during

---

† A good account is given by Keen, *The Physical Properties of the Soil* (Rothamsted Monographs on Agricultural Science, 1931) Chap. IX. A large number of results for various soils is given by Patten, *Bull. U.S. Div. Soils*, No. 59 (1909). For the effect of water movement see de Vries, *Trans. Int. Congr. Soil Sci.* (1950) Vol. II.

‡ *Quart. J. R. Met. Soc.* **58** (1932) 389. His formula is given in 2.9 (15). See also Lettau, *Trans. Amer. Geophys. Union*, **32** (1951) 189–200.

both day and night, while in the daytime heat is received from the sun at a rate proportional to the cosine of the Sun's zenith distance. The mean surface temperature of the Earth is determined entirely by solar radiation, the flow of heat from the interior discussed in the next section being quite negligible from this point of view.

2.13. The geothermal gradient and the heat flux from the Earth

From the early days of mining it has been known that the temperature in the Earth increases with depth, the vertical distance between points whose temperatures differ by 1° C (sometimes called the *geothermal step*) being of the order of 80 feet. It appears that in deep mines relatively high temperatures will be attained, which add to the difficulties of deep mining.

Many measurements of temperatures in deep drill-holes have been made, with the result that the rate of increase of temperature with depth, which is called the *geothermal gradient*, has been found to vary between about 10° C per km and 50° C per km on land; a few measurements in the ocean floor have been made giving a value of about 40° C per km. The values referred to above, and all that will be said below, refer to regions far removed from volcanic activity; in thermal regions and near active volcanoes, observed temperatures are much higher. It will appear from the calculations below, that the differences mentioned above are due mainly to differences in the thermal conductivity of the rocks concerned, and that when these are taken into account observations† at all points of the Earth (including the ocean floor) are consistent with a flux which only varies from about $0 \cdot 6$ to $2 \cdot 0 \times 10^{-6}$ cal/cm² sec between different regions, the mean value being about $1 \cdot 2 \times 10^{-6}$. No systematic variation of the heat flux with position has yet been discovered.

To determine the theoretical form of the variation of temperature with depth, suppose that the conductivity K and the rate of heat production A are functions

---

† There is now a very large literature on this subject which is summarized in an article by Bullard in *The Earth as a Planet*, ed. Kuiper (University of Chicago Press, 1953), and by Jacobs in the *Encyclopaedia of Physics*, Vol. 47 (Springer, 1956). Many early results which give only temperatures and not rock conductivities are given in *Internal Constitution of the Earth*, ed. Gutenberg (Dover, 1951), *Temperature, its Measurement and Control in Science and Industry* (Amer. Inst. Phys., 1941) p. 1014. The modern practice of systematically measuring both temperature and conductivity at close intervals in bore-holes was initiated by Benfield, *Proc. Roy. Soc.* A, **173** (1939) 428–50, and Bullard, ibid. A, **173** (1939) 474–502. The use of temperature measurements in tunnels for this purpose is discussed by Birch, *Bull. Geol. Soc. Amer.* **61** (1950) 567–630. Effects which occur in Permafrost regions are described by Terzaghi, *J. Boston Soc. Ci il Engrs.* **39** (1952) 1–50, and Misener, *Trans. Amer. Geophys. Union* **36** (1955) 1055–60.

only of the depth x below the surface. The equation of steady flow of heat is

$$\frac{d}{dx}\left(K\frac{dv}{dx}\right) = -A. \tag{1}$$

This takes a simpler form if the new variable

$$\xi = \int_0^x \frac{dx}{K} \tag{2}$$

is introduced. Since $R = 1/K$ is the thermal resistivity of the solid, ξ is the total thermal resistance of the solid between the surface and depth x. By (2),

$$\frac{dv}{d\xi} = K\frac{dv}{dx}, \tag{3}$$

so that $dv/d\xi$ is a measure of the heat flux. Using (3) in (1) gives

$$\frac{d^2v}{d\xi^2} = -AK. \tag{4}$$

(i) *The case of no heat generation, $A = 0$.* Integrating (4) gives

$$v = V_0 + F\xi,$$

where V_0 and F are constants which may be interpreted as the surface temperature and heat flux. Thus, in the absence of heat supply, the graph of v against ξ is a straight line and, when observed temperatures are reduced in this way using measured conductivities, many apparent anomalies disappear. The value of V_0 found in this way would be expected to agree with the mean annual air temperature but is usually higher, this effect being attributed to evaporation.

(ii) *Heat supply at the rate Q per unit time per unit area in the plane ξ_1.*

$$\left.\begin{aligned} v &= V_0 + F\xi, \quad 0 < \xi < \xi_1 \\ v &= V_0 + Q\xi_1 + (F-Q)\xi, \quad \xi > \xi_1 \end{aligned}\right\}. \tag{5}$$

It follows that a sudden change in slope of the v, ξ curve is due to heat supply or removal at this depth, for example by flowing water.

(iii) *Heat supply at a constant rate in a region.* This may arise in three ways: (a) transport of heat by ground water, (b) radioactivity, (c) chemical reactions near ore bodies. As an example, consider the three-layer case in which $K = K_1, A = 0$ in $0 < x < x_1$; $K = K_2, A = A_2, x_1 < x < x_2$; $K = K_3, A = 0$, in $x > x_2$, where K_1, K_2, K_3, A_2 are constants. Then

$$\left.\begin{aligned} v &= V_0 + Fx/K_1, \quad 0 < x < x_1 \\ v &= V_0 + \frac{Fx_1}{K_1} + \frac{F(x-x_1)}{K_2} - \frac{A_2(x-x_1)^2}{2K_2}, \quad x_1 < x < x_2 \\ v &= V_0 + \frac{Fx_1}{K_1} + \frac{F(x_2-x_1)}{K_2} - \frac{A_2(x_2-x_1)^2}{2K_2} + \frac{[F-A_2(x_2-x_1)](x-x_2)}{K_3}, \quad x > x_3 \end{aligned}\right\}. \tag{6}$$

It appears from (4) and (6) that the curve of v against ξ is concave downwards in a region in which heat is being produced. Formulae such as (6), using the known value of the heat flux F at the surface and assumed distributions of radioactivity, have frequently been used to estimate temperatures in the crust.

These simple considerations are greatly modified by geological conditions, for example, the effects of uplift and denudation (cf. § 15.2), the effect of deviation from a horizontal surface (cf. § 16.3), the effect of lateral variations in conductivity, and the effect of variation in surface temperature. In particular, the effect of recent glacial epochs is quite marked and has been carefully discussed.† The simplest case, from which others can be built up, is that of a glacial epoch lasting for time T during which time the surface was maintained at zero, its temperature before and after this time being V_0. Taking the end of the glacial epoch as time $t = 0$, we assume that at time $t = -T$ the temperature was $v = V_0 + Gx$ corresponding to a constant geothermal gradient G. Then by 2.4 (10), the temperature at time $t > 0$ is

$$v = V_0 + Gx + V_0 \operatorname{erf} \frac{x}{2[\kappa(t+T)]^{\frac{1}{2}}} - V_0 \operatorname{erf} \frac{x}{2(\kappa t)^{\frac{1}{2}}}. \tag{7}$$

2.14. The age of the Earth. Kelvin's treatment

It was remarked by Fourier himself‡ that the measured value of the geothermal gradient might be used to obtain a rough estimate of the time which has elapsed since the Earth began to cool from an initially molten state. In the problem, as simplified by him for mathematical treatment, the curvature of the Earth is neglected and the diffusivity κ assumed to be constant. The surface is taken as the plane $x = 0$, and radiation takes place into a medium at temperature zero. The temperature when cooling began, taken as the time $t = 0$, is constant and equal to v_0. He obtained the result given in § 2.7 that for large values of t the temperature gradient near the surface is approximately $v_0(\pi\kappa t)^{-\frac{1}{2}}$.

Kelvin§ took the simpler problem of the semi-infinite solid bounded by the plane $x = 0$, the boundary being kept at zero temperature and the initial temperature being constant and equal to v_0. By 2.4 (3) the temperature v at the depth x at time t is given by

$$v = v_0 \operatorname{erf} \frac{x}{2(\kappa t)^{\frac{1}{2}}}, \tag{1}$$

and

$$\frac{\partial v}{\partial x} = \frac{v_0}{(\pi\kappa t)^{\frac{1}{2}}} e^{-x^2/4\kappa t}.$$

It follows that the value of $\partial v/\partial x$ at $x = 0$, that is, the geothermal gradient G, is given by

$$G = v_0(\pi\kappa t)^{-\frac{1}{2}}, \tag{2}$$

as in Fourier's problem. Taking $v_0 = 7,000°$ F as a suitable temperature for melting rock, $G = 1/2700$, and the reasonable average value $\kappa = 0\cdot0118$, Kelvin found from (2) a value of about 94×10^6 years for the time needed for the geothermal gradient to fall to its present value,

---

† Birch, *Amer. J. Sci.* **246** (1948) 729–60.

‡ 'Extrait d'un Mémoire sur le refroidissement du globe terrestre', *Bull. Sci. par la Société philomathique de Paris* (1820). Also *Œuvres de Fourier*, T. II (Paris, 1888), cf. p. 284.

§ 'The secular cooling of the Earth', *Trans. Roy. Soc. Edin.* **23** (1864) 157. It should be said that in his discussion Kelvin carefully excepted the possibility of heat generation by chemical action.

that is, for the 'age' of the Earth. The assumed initial temperature of $7,000°$ F he recognized[†] to be a high value, and later experiments upon the behaviour of rocks at high temperatures led him to believe that $1,200°$ C would be a fairer estimate. This change would reduce his estimate of the age to less than 10^7 years and he seems to have been somewhat of the opinion of King[‡] that we have no warrant in this argument for extending the Earth's age beyond 24,000,000 years.

The limits of the age of the Earth given by Kelvin in 1864 attracted much attention, since the geologists then, as now, demanded a much longer period of time for the cooling from the molten state, their arguments being based on the visible processes and effects of stratification. Following Kelvin's pronouncement much discussion took place between the physicists on the one hand and the geologists on the other,[§] and the controversy was only closed by the discovery of radioactivity early in the twentieth century. It should be said, however, that Kelvin's problem is essentially that of the extraction of heat from a thin surface layer—since $\mathrm{erf}\,2 = 0\cdot995$, it follows with the numerical values used above that after 10^8 years the temperature at a depth of 250 km will only have changed by half a per cent., and thus the vast amount of heat in the interior of the Earth is quite untouched. Perry[||] and Heaviside[††] pointed out that if physical conditions in the Earth were such as to make more of this heat available, much greater ages could be obtained.

It is now known that heat is generated by disintegration of radioactive substances[‡‡] in the rocks of the Earth's crust: the rate of heat production is difficult to estimate since there are wide variations between the amounts of radioactive matter in different specimens from the same type of rock as well as differences between the different rock types. Recent figures are $6\cdot3\times10^{-6}$, $1\cdot7\times10^{-6}$, $0\cdot04\times10^{-6}$ cal per gm per annum

[†] Cf. *Nature*, **59** (1895) 438; also *Phil. Mag.* (5) **47** (1899) 66.

[‡] *Amer. J. Sci.* **145** (1893) 1.

[§] Cf. Woodward, 'The mathematical theories of the Earth', *Amer. Ass. Adv. Sci.* (Toronto, 1889); 'The century's progress in applied mathematics', *Bull. Amer. Math. Soc.* **6** (1900) 147.

[||] *Nature*, **51** (1895) 224, 341, 582.

[††] *Electromagnetic Theory* (1899), Vol. II, Chap. V entitled 'Mathematics and the age of the Earth'. See also § 12.8 below.

[‡‡] The classical early references are Rutherford, Chadwick, and Ellis, *Radiation from Radioactive Substances* (Cambridge, 1930); Holmes, *The Age of the Earth* (Harper's, 1913); *Internal Constitution of the Earth,* ed. Gutenberg (McGraw-Hill, 1939); *The Age of the Earth,* published in the same series; Jeffreys, *Mon. Not. R. Astr. Soc. Geophys. Supp.* **5** (1942) 37; Bullard, ibid. **5** (1942) 41. Recent surveys are given by Bullard in *The Earth as a Planet,* ed. Kuiper (Univ. of Chicago Press, 1954) and by Jacobs in Vol. 47 (Geophysics I) of the *Encyclopaedia of Physics* (Springer, 1956). See also Birch, *Nuclear Geology,* ed. Faul (Wiley, 1954) Chap. 5.

for granitic, basic, and ultrabasic rocks respectively. Values for sediments are very variable, but 2×10^{-6} is not uncommon. The distribution of radioactive material with depth is unknown, but amounts of the order of those observed at the surface must be confined to a relatively thin layer below the Earth's surface of the order of a few tens of kilometres in thickness, otherwise more heat would be generated than can be accounted for by the observed loss from the surface. The problem can thus be represented by models such as those of § 2.11 (ii), (iii) and § 2.13 (iii), and these have frequently been used to discuss temperatures in the Earth's crust and possible ages for the Earth. An additional complication arises from the fact that the half-life times of some of the radioactive elements (in particular, potassium, $1 \cdot 3 \times 10^9$ years) are considerably shorter than the ages of the order of 4×10^9 years now considered likely for the Earth, so that exponential decay in the rate of heat production must be allowed for.

As early as 1893 it was recognized† that Kelvin's assumption of constant initial temperature should be replaced by one which allowed for the increase of melting-point with pressure. A linear increase $v_0 + bx$ was commonly assumed, where v_0 was about $1,400°$ C and b about $3°$ C/km. The problem is still that of 2.11 (iii), and in this way Jeffreys, loc. cit., deduced a value of $1 \cdot 6 \times 10^9$ years for the age of the Earth.

In recent years, interest has changed from attempts to calculate the age of the Earth by thermal methods to attempts to calculate temperatures within the Earth starting from an assumed age. These require consideration of the Earth as a sphere and are discussed in § 9.14.

2.15. The infinite composite solid

Suppose the region $x > 0$ is of one substance, K_1, ρ_1, κ_1, and $x < 0$ of another, K_2, ρ_2, κ_2, the boundary conditions at the plane of separation $x = 0$ being 1.9 (18), (19), namely,

$$v_1 = v_2, \qquad x = 0, t > 0, \tag{1}$$

$$K_1 \frac{\partial v_1}{\partial x} = K_2 \frac{\partial v_2}{\partial x}, \qquad x = 0, t > 0, \tag{2}$$

where we write v_1 for the temperature in the region $x > 0$, and v_2 for that in the region $x < 0$.

Many problems on such regions may be solved by the use of the solutions of § 2.4 for the semi-infinite solid.

† King, *Amer. J. Sci.* **145** (1893) 1; Barus, *Phil. Mag.* (5) **35** (1893) 173.

(i) *The initial temperature V, constant, in $x > 0$ and zero in $x < 0$.*
We seek solutions of type

$$v_1 = A_1 + B_1 \operatorname{erf} \frac{x}{2\sqrt{(\kappa_1 t)}}, \quad x > 0, \tag{3}$$

$$v_2 = A_2 + B_2 \operatorname{erf} \frac{|x|}{2\sqrt{(\kappa_2 t)}}, \quad x < 0. \tag{4}$$

It is known from § 2.4 that these satisfy the differential equations of conduction of heat in their respective regions. We choose the constants A_1, B_1, A_2, B_2 to satisfy the initial and boundary conditions. The former give

$$A_1 + B_1 = V, \qquad A_2 + B_2 = 0;$$

while the latter, (1) and (2), give

$$A_1 = A_2, \quad \text{and} \quad B_1 K_1 \kappa_1^{-\frac{1}{2}} = -B_2 K_2 \kappa_2^{-\frac{1}{2}}.$$

Solving these equations and substituting in (3) and (4) gives finally†

$$v_1 = \frac{K_1 \kappa_1^{-\frac{1}{2}} V}{K_1 \kappa_1^{-\frac{1}{2}} + K_2 \kappa_2^{-\frac{1}{2}}} \left\{ 1 + \frac{K_2 \kappa_2^{-\frac{1}{2}}}{K_1 \kappa_1^{-\frac{1}{2}}} \operatorname{erf} \frac{x}{2\sqrt{(\kappa_1 t)}} \right\}, \tag{5}$$

$$v_2 = \frac{K_1 \kappa_1^{-\frac{1}{2}} V}{K_1 \kappa_1^{-\frac{1}{2}} + K_2 \kappa_2^{-\frac{1}{2}}} \operatorname{erfc} \frac{|x|}{2\sqrt{(\kappa_2 t)}}. \tag{6}$$

(ii) *The initial temperature zero. Heat is supplied for $t > 0$ at the constant rate F_0 per unit time per unit area in the plane $x = 0$.*
Here, following 2.9 (6), we assume

$$v_1 = \frac{2F_1(\kappa_1 t)^{\frac{1}{2}}}{K_1} \operatorname{ierfc} \frac{x}{2\sqrt{(\kappa_1 t)}}, \quad x > 0, \tag{7}$$

$$v_2 = \frac{2F_2(\kappa_2 t)^{\frac{1}{2}}}{K_2} \operatorname{ierfc} \frac{|x|}{2\sqrt{(\kappa_2 t)}}, \quad x < 0, \tag{8}$$

where the unknown constants F_1 and F_2 are to be found from the boundary conditions at $x = 0$, which give

$$\frac{F_1 \kappa_1^{\frac{1}{2}}}{K_1} = \frac{F_2 \kappa_2^{\frac{1}{2}}}{K_2}, \quad \text{and} \quad F_1 + F_2 = F_0.$$

Therefore

$$v_1 = \frac{2F_0 \sqrt{(\kappa_1 \kappa_2 t)}}{K_1 \kappa_2^{\frac{1}{2}} + K_2 \kappa_1^{\frac{1}{2}}} \operatorname{ierfc} \frac{x}{2\sqrt{(\kappa_1 t)}}, \tag{9}$$

$$v_2 = \frac{2F_0 \sqrt{(\kappa_1 \kappa_2 t)}}{K_1 \kappa_2^{\frac{1}{2}} + K_2 \kappa_1^{\frac{1}{2}}} \operatorname{ierfc} \frac{|x|}{2\sqrt{(\kappa_2 t)}}. \tag{10}$$

The case in which there is a contact resistance between the surfaces is discussed by Schaaf.‡ He distinguishes two cases: (a) heat supply in one or both surfaces, corresponding to 'dry' friction, and (b) heat supply between the surfaces, corresponding to 'wet' friction or to a thin flat heating element.

† For other methods of solution and discussion see Tranter, *Phil. Mag.* (7) **28** (1939) 579; Carslaw, *Phil. Mag.* (7) **30** (1940) 414; Churchill, *Phil. Mag.* (7) **31** (1941) 81.
‡ Schaaf, *Quart. Appl. Math.* **5** (1947) 107–11.

(iii) *The problem of* (i) *except that at the plane* $x = 0$ *there is a contact resistance, so that in place of* (1) *we have*

$$K_1 \frac{\partial v_1}{\partial x} + H(v_2 - v_1) = 0, \tag{11}$$

while (2) *holds as before* (cf. § 1.9 G).

Proceeding as in (i), except that since we have a 'radiation' type of boundary condition at $x = 0$ we seek solutions of type appropriate to this, we have finally

$$v_1 = \frac{K_1 \kappa_1^{-\frac{1}{2}} V}{K_1 \kappa_1^{-\frac{1}{2}} + K_2 \kappa_2^{-\frac{1}{2}}} \left\{ 1 + \frac{K_2 \kappa_2^{-\frac{1}{2}}}{K_1 \kappa_1^{-\frac{1}{2}}} \left[\operatorname{erf} \frac{x}{2\sqrt{(\kappa_1 t)}} + e^{h_1 x + h_1^2 \kappa_1 t} \operatorname{erfc} \left(\frac{x}{2\sqrt{(\kappa_1 t)}} + h_1 \sqrt{(\kappa_1 t)} \right) \right] \right\}, \tag{12}$$

$$v_2 = \frac{K_1 \kappa_1^{-\frac{1}{2}} V}{K_1 \kappa_1^{-\frac{1}{2}} + K_2 \kappa_2^{-\frac{1}{2}}} \left\{ \operatorname{erfc} \frac{|x|}{2\sqrt{(\kappa_2 t)}} - e^{h_2 x + h_2^2 \kappa_2 t} \operatorname{erfc} \left(\frac{|x|}{2\sqrt{(\kappa_2 t)}} + h_2 \sqrt{(\kappa_2 t)} \right) \right\}, \tag{13}$$

where

$$h_1 = \frac{H(K_1 \kappa_1^{-\frac{1}{2}} + K_2 \kappa_2^{-\frac{1}{2}})}{K_1 K_2 \kappa_2^{-\frac{1}{2}}}, \qquad h_2 = \frac{H(K_1 \kappa_1^{-\frac{1}{2}} + K_2 \kappa_2^{-\frac{1}{2}})}{K_1 K_2 \kappa_1^{-\frac{1}{2}}}. \tag{14}$$

These and more complicated problems are most easily solved by the Laplace transformation method, Chapter XII.

2.16. The case of thermal properties varying with the temperature

It was shown in 1.6 (11) that if both K and c depend on the temperature, the equation of linear flow of heat

$$\frac{\partial}{\partial x} \left(K \frac{\partial v}{\partial x} \right) - \rho c \frac{\partial v}{\partial t} = 0, \tag{1}$$

or

$$\frac{\partial^2 v}{\partial x^2} + \frac{d(\ln K)}{dv} \left(\frac{\partial v}{\partial x} \right)^2 = \frac{1}{\kappa} \frac{\partial v}{\partial t}, \tag{2}$$

is reduced by the change of variable

$$\Theta = \int_0^v \frac{K}{K_0} \, dv \tag{3}$$

to the form

$$\frac{\partial^2 \Theta}{\partial x^2} - \frac{1}{\kappa} \frac{\partial \Theta}{\partial t} = 0, \tag{4}$$

where $\kappa = K/\rho c$ is a function of Θ. Either of the equations (1), (2), or (4) may be used as a basis for discussion.

Historically, a number of early papers discuss the case of variable thermal properties; subsequently, partly because of their difficulty and partly because of lack of information about the variation of thermal properties with temperature, little work was done on problems of this type, but recently, because of increasing knowledge of thermal properties and of the importance of similar problems in diffusion, they have attracted a great deal of attention. Usually, numerical methods have to be used, but there are some theoretical approaches of considerable interest. Only those which are of importance in connexion with conduction of heat will be discussed in detail here: a full account of others is given in *M.D.*, Chap. IX.

I. *Boltzmann's transformation*†

The solutions of a number of important problems with constant diffusivity are functions of

$$\xi = xt^{-\frac{1}{2}} \tag{5}$$

† *Ann. Physik* (N.F.), **53** (1894) 959.

only. This suggests that the possibility of finding solutions of (1) of this type should be examined. If it is assumed that v is a function of ξ only, (1) reduces to the ordinary differential equation

$$\frac{d}{d\xi}\left(K\frac{dv}{d\xi}\right) + \tfrac{1}{2}\rho c\xi \frac{dv}{d\xi} = 0, \tag{6}$$

or (4) becomes
$$\kappa\frac{d^2\Theta}{d\xi^2} + \tfrac{1}{2}\xi\frac{d\Theta}{d\xi} = 0. \tag{7}$$

The usefulness of (5) is limited by the fact that the initial and boundary conditions must also be expressible in terms of ξ only. For example, since $\xi \to \infty$ as $t \to 0$ for $x > 0$, and $\xi = 0$ when $x = 0$ for $t > 0$, it is clearly applicable to the region $x > 0$ with constant initial temperature and constant temperature at $x = 0$ for $t > 0$.

It should be remarked that the term 'transformation' as applied to (5) is a misnomer since it suggests that if the variables in (1) are changed to, say, ξ and t, the partial differential equation reduces to the ordinary differential equation (6), but it, in fact, merely transforms into a partial differential equation in ξ and t. The status of the method is that if the initial and boundary conditions to which (1) is subject can be expressed in terms of ξ only, the solution of (6), subject to these conditions, gives a solution of (1) and its boundary conditions which is a function of $xt^{-\frac{1}{2}}$ only and may be presumed to be the unique solution.

If κ is constant, (7) leads immediately to the usual solution

$$\Theta = A\operatorname{erf}\frac{\xi}{2\kappa^{\frac{1}{2}}} = A\operatorname{erf}\frac{x}{2(\kappa t)^{\frac{1}{2}}}. \tag{8}$$

Various integrals of (6) and (7) are available, for example

$$\int \xi\, dv = -2\kappa\frac{dv}{d\xi} \tag{9}$$

for the case of constant c, which has been made the basis of a numerical method by Philip.[†] Approximate solutions of (6) for cases in which K varies slowly are readily obtained.

Kirchhoff and Hansemann[‡] treat the case $c = c_0 + c_1 v$, $K = K_0 + K_1 v$ by seeking a solution $v = v(\xi)$, and using the values for the case $c_1 = K_1 = 0$ in small terms containing c_1 and K_1.

II. *Peek's transformation*[§]

Suppose that ψ is a function of x and t which satisfies

$$\frac{\partial\psi}{\partial t} = \frac{(\partial\psi/\partial x)^2}{f_1(\psi)} = \frac{\partial^2\psi/\partial x^2}{f_2(\psi)}, \tag{10}$$

where $f_1(\psi)$ and $f_2(\psi)$ are functions of ψ only. Then, if v is a function of ψ only, (2) becomes

$$\left[\frac{1}{\kappa} - f_2(\psi)\right]\frac{dv}{d\psi} - f_1(\psi)\left(\frac{dv}{d\psi}\right)^2\frac{d(\ln K)}{dv} = f_1(\psi)\frac{d^2v}{d\psi^2}. \tag{11}$$

It is also necessary that the boundary conditions should transform: for example, if the boundary conditions in v are $v = v_1$, constant, when $F_1(x, t) = 0$, and

† Philip, *Trans. Faraday Soc.* **51** (1955) 885–92. For an application to (1) with an additional term $A(v)\,\partial v/\partial x$ see *Aust. J. Phys.* **10** (1957) 29–42.
‡ *Ann. Phys. und Chem.* (N.F.), 9 (1880) 1–47.
§ Peek, *Phys. Rev.* (2) **35** (1930) 554–61, develops this method in three dimensions.

$v = v_2$, when $F_2(x,t) = 0$, the boundary conditions in ψ must be of the form $\psi = V_1$ when $F_1(x,t) = 0$, and $\psi = V_2$ when $F_2(x,t) = 0$.

For example, for the region $x > 0$ with $v = v_1$, when $x > 0, t = 0$, and $v = v_2$, when $x = 0, t > 0$, $\psi = \exp(-xt^{-\frac{1}{2}})$ satisfies (10), and (11) becomes

$$\left[\frac{1}{2\kappa}\ln\psi + 1\right]\frac{dv}{d\psi} + \psi\left(\frac{dv}{d\psi}\right)^2\frac{d(\ln K)}{dv} + \psi\frac{d^2v}{d\psi^2} = 0,$$

to be solved with $v = v_1$ when $\psi = 0$, and $v = v_2$ when $\psi = 1$.

III. *Exact solutions for particular cases*

The case in which the thermal properties are step functions of the temperature[†] may be treated exactly by the methods of § 11.2.

Conductivities of the form $K_0/(1-\lambda v)$, $K_0/(1-\lambda v)^2$ and $K_0/(1+2av+bv^2)$ have been treated[‡] by Fujita.

IV. *Other methods*

Various approximate methods for conductivities of types $a + bv$ and $\exp(av)$ are described in *M.D.*, Chap. IX. It may be noted that the most important law in the theory of diffusion, viz.

$$D = D_0\exp[-E/RT],$$

where T is the absolute temperature and D_0, E, R are constants, has not as yet been fully discussed.

Hopkins[§] has suggested a method of successive approximations in which the case of constant thermal properties is regarded as the zero approximation and the Green's function for this problem is used to obtain the first approximation.

V. *The steady state*

In this case (1) reduces to the ordinary differential equation

$$K\frac{dv}{dx} = \text{const},$$

which may be integrated immediately in many important special cases. For the case in which K is a function of v only, results follow immediately from those for constant K in many important special cases in one or more dimensions, çf. § 1.6 I. Barrer[||] has given graphs of the temperature distributions in linear and radial flow for K of the forms $K_0[1+f(v)]$ and $K_0[1+f(x)]$.

---

† See also *M.D.*, Chap. VII; Crank, *Trans. Faraday Soc.* **47** (1951) 450–61.
‡ Fujita, *Text. Res. J.* **22** (1952) 757, 823; *M.D.*, Chap. IX.
§ Hopkins, *Proc. Phys. Soc.* **50** (1938) 703.
|| Barrer, *Proc. Phys. Soc.* **58** (1946) 321–31.

III

LINEAR FLOW OF HEAT IN THE SOLID BOUNDED BY TWO PARALLEL PLANES

3.1. Introductory

In this chapter we shall examine various cases of linear flow of heat in a solid bounded by a pair of parallel planes, usually $x = 0$ and $x = l$. This region we shall usually refer to briefly as the 'slab $0 < x < l$'. The results apply also to a rod of length l with the same end conditions and with no loss of heat from its surface.

3.2. Steady temperature

In the case of steady flow in a slab of conductivity K and thickness l whose surfaces are kept at temperatures v_1 and v_2 the differential equation becomes

$$\frac{d^2v}{dx^2} = 0.$$

Thus

$$\frac{dv}{dx} = \text{constant} = \frac{v_2 - v_1}{l}.$$

Also the flux at any point is

$$f = -K\frac{dv}{dx} = -\frac{K(v_2 - v_1)}{l} = \frac{v_1 - v_2}{R}, \tag{1}$$

where

$$R = l/K. \tag{2}$$

The relation (1) is precisely analogous to Ohm's law for the steady flow of electric current: the flux f corresponds to the electric current, and the fall in temperature $v_1 - v_2$ to the fall in potential. Thus R may be called the thermal resistance of the slab.

Next suppose we have a composite wall composed of n slabs of thicknesses $l_1, ..., l_n$ and conductivities $K_1, ..., K_n$. If the slabs are in perfect thermal contact at their surfaces of separation, the fall of temperature over the whole wall will be the sum of the falls over the component slabs, and, since the flux is the same at every point, this sum is

$$\frac{fl_1}{K_1} + \frac{fl_2}{K_2} + ... + \frac{fl_n}{K_n} = (R_1 + R_2 + ... + R_n)f. \tag{3}$$

This is equivalent to the statement that the thermal resistance of a composite wall is the sum of the thermal resistances of the separate layers, assuming perfect thermal contact between them.

Finally, consider a composite wall as before, but with contact resistances between the layers such that the flux of heat between the surfaces of consecutive layers is H times the temperature difference between these surfaces (cf. 1.9 (20)). Here $1/H$ may be regarded as the thermal resistance of the contact, and the total thermal resistance of the composite wall will be the sum of the thermal resistances of the separate layers plus the sum of the thermal resistances of the contacts between them.

If the conductivity K is a function of the temperature, the differential equation is

$$\frac{d}{dx}\left(K\frac{dv}{dx}\right) = 0.$$

Thus the relation $\qquad -K\dfrac{dv}{dx} = f, \quad$ constant,

still holds. Integrating between the surface temperatures v_1 and v_2 of a slab of thickness l we have

$$-\int_{v_1}^{v_2} K\,dv = lf,$$

and thus
$$f = \frac{(v_1-v_2)K_{\mathrm{av}}}{l}, \tag{4}$$

where
$$K_{\mathrm{av}} = \frac{1}{v_2-v_1}\int_{v_1}^{v_2} K\,dv \tag{5}$$

is the average conductivity over the temperature range in the slab. Thus, if conductivity is a function of temperature, the previous results hold good with K_{av} in place of K.

3.3. The region $0 < x < l$. Ends kept at zero temperature. Initial temperature $f(x)$

The differential equation to be solved is

$$\frac{\partial v}{\partial t} = \kappa\frac{\partial^2 v}{\partial x^2}, \quad 0 < x < l, \tag{1}$$

with $\qquad v = 0, \quad$ when $x = 0$ and $x = l$, $\tag{2}$

and $\qquad v = f(x), \quad$ when $t = 0$. $\tag{3}$

If the initial distribution were

$$v = A_n \sin\frac{n\pi x}{l},$$

it is clear that $\qquad v = A_n \sin\dfrac{n\pi x}{l}\,e^{-\kappa n^2\pi^2 t/l^2}$

would satisfy all the conditions (1), (2), (3) of the problem.

Let us suppose that the initial temperature, $f(x)$, is a bounded function satisfying Dirichlet's conditions† ($F.S.$, § 93) in the interval $(0, l)$ so that it can be expanded in the sine series

$$\sum_{n=1}^{\infty} a_n \sin \frac{n\pi x}{l},$$

where

$$a_n = \frac{2}{l} \int_0^l f(x') \sin \frac{n\pi x'}{l} \, dx'. \tag{4}$$

Now consider the function v defined by the infinite series

$$v = \sum_{n=1}^{\infty} a_n \sin \frac{n\pi x}{l} e^{-\kappa n^2 \pi^2 t/l^2}. \tag{5}$$

This series, owing to the presence of the convergency factor $\exp[-(\kappa n^2 \pi^2 t/l^2)]$, is uniformly convergent‡ for any interval of x, when $t > 0$; and, regarded as a function of t, it is uniformly convergent when $t \geqslant t_0 > 0$, t_0 being any positive number.

Thus the function v, defined by the series (5), is a continuous function of x, and a continuous function of t, in these intervals.§

It is easy to show that the series obtained by term-by-term differentiation of (5) with respect to x and t are also uniformly convergent in these intervals of x and t respectively. Thus they are equal to the differential coefficients of the function v.

Hence

$$\frac{\partial v}{\partial t} = -\sum_1^{\infty} \frac{\kappa n^2 \pi^2}{l^2} a_n \sin \frac{n\pi x}{l} e^{-\kappa n^2 \pi^2 t/l^2}$$

and

$$\kappa \frac{\partial^2 v}{\partial x^2} = -\sum_1^{\infty} \frac{\kappa n^2 \pi^2}{l^2} a_n \sin \frac{n\pi x}{l} e^{-\kappa n^2 \pi^2 t/l^2},$$

when $t > 0$, and $0 < x < l$.

---

† This restriction is removed in $C.H.$ § 31 where it is shown that the results below hold if $f(x)$ is bounded and integrable in $0 \leqslant x \leqslant l$.

‡ Since $f(x)$ is bounded there is a positive number M such that $|f(x)| < M$ in $0 < x < l$. It follows that $|a_n| < 2M$ for all values of n. Therefore

$$\left| a_n \sin \frac{n\pi x}{l} e^{-\kappa n^2 \pi^2 t/l^2} \right| < 2M e^{-\kappa n^2 \pi^2 t_0/l^2}, \quad \text{where} \quad t \geqslant t_0 > 0.$$

Now the series

$$\sum_{n=1}^{\infty} e^{-\kappa n^2 \pi^2 t_0/l^2}$$

is convergent and its terms are independent of both x and t, and the results follow.

§ Regarded as a function of the two variables x, t, it is a continuous function of (x, t) in the regions $0 \leqslant x \leqslant l$, $t \geqslant t_0 > 0$. (Cf. $F.S.$, § 37.)

Thus the equation
$$\frac{\partial v}{\partial t} = \kappa \frac{\partial^2 v}{\partial x^2}$$

is satisfied at all points of the rod, when $t > 0$, by the function defined by (5).

We have now to see whether this function also satisfies the boundary and initial conditions.

Since the series is uniformly convergent with respect to x in the interval $0 \leqslant x \leqslant l$, when $t > 0$, it represents a continuous function of x in this interval.

Thus

$$\lim_{x \to 0} v = \text{the value of the sum of the series when } x = 0$$
$$= 0,$$

and
$$\lim_{x \to l} v = \text{the value of the sum of the series when } x = l$$
$$= 0.$$

Hence the *boundary conditions* are satisfied.

With regard to the *initial conditions*, we may use the extension of Abel's theorem contained in *F.S.*, § 73, I.

We have assumed that $f(x)$ is bounded and satisfies Dirichlet's conditions in $(0, l)$.

Therefore the sine series for $f(x)$,

$$a_1 \sin \frac{\pi x}{l} + a_2 \sin \frac{2\pi x}{l} + \dots,$$

converges, and its sum is $f(x)$ at every point between 0 and l where $f(x)$ is continuous, and $\frac{1}{2}\{f(x+0)+f(x-0)\}$ at all other points.† (Cf. *F.S.*, § 98.)

It follows from the extension of Abel's theorem referred to above that when v is defined by (5), we have

$$\lim_{t \to 0} v = \lim_{t \to 0} \sum_{1}^{\infty} a_n \sin \frac{n\pi x}{l} e^{-\kappa n^2 \pi^2 l/l^2}$$

$$= f(x) \quad \text{at a point of continuity}$$

$$= \tfrac{1}{2}\{f(x+0)+f(x-0)\} \quad \text{at all other points.}$$

Thus we have shown that if the initial temperature satisfies Dirichlet's

† If $f(x)$ is bounded and satisfies Dirichlet's conditions, it follows from *F.S.*, § 93, that it can only have ordinary discontinuities.

conditions, and is continuous from $x = 0$ to $x = l$, while $f(0) = f(l) = 0$, the function defined by (5)† satisfies all the conditions of the problem.

If the initial temperature has discontinuities, the function defined by (5) at these points tends to $\frac{1}{2}\{f(x+0)+f(x-0)\}$ as $t \to 0$. If t is taken small enough, v will bridge the gap from $f(x-0)$ to $f(x+0)$, and the temperature curve will pass close the point $\frac{1}{2}\{f(x+0)+f(x-0)\}$.

It must be remembered that the physical problem, as we have stated it for discontinuity, either at the ends of the rod or in the rod itself, is an ideal one. In nature there cannot be a discontinuity in the temperature in the rod initially. In the physical problem we must assume that a sudden change of temperature takes place at the instant from which our observations are measured, in the immediate neighbourhood of the point of discontinuity or the ends, if they are points of discontinuity. The gap in the temperature is thus smoothed over. The solution of the mathematical problem we have obtained satisfies these conditions, and it may be taken as representing the physical problem in this modified aspect.

The following special cases of (5) are of interest:‡

(i) *Constant initial temperature* $f(x) = V_0$, *constant.*

$$v = \frac{4V_0}{\pi} \sum_{n=0}^{\infty} \frac{1}{(2n+1)} e^{-\kappa(2n+1)^2\pi^2t/l^2} \sin\frac{(2n+1)\pi x}{l}. \tag{6}$$

(ii) *A linear initial distribution* $f(x) = kx.$

$$v = \frac{2lk}{\pi} \sum_{n=1}^{\infty} \frac{(-1)^{n-1}}{n} e^{-\kappa n^2\pi^2t/l^2} \sin\frac{n\pi x}{l}. \tag{7}$$

In general, it is a little more satisfactory to set out results for the symmetrical case of the slab $-l < x < l$ so that direct comparison with similar results for the sphere and cylinder is possible. Also, it is invariably found that series such as (6) and (7) converge slowly for small values of $\kappa t/l^2$, say $\kappa t/l^2 < 0.01$, but it will appear later (cf. § 12.5) that alternative series involving error functions or their integrals are rapidly convergent for such values. For convenience these alternative series will be given here, cf. (9), (11), their derivation will be discussed in § 12.5. All the results given below hold, of course, also for the slab $0 < x < l$, with no flow of heat at $x = 0$, and $x = l$ at zero temperature.

† This can be written as

$$v = \frac{2}{l} \int_0^l f(x') \sum_1^{\infty} \left(\sin\frac{n\pi x'}{l} \sin\frac{n\pi x}{l} e^{-\kappa n^2\pi^2t/l^2} \right) dx',$$

since the series under the integral is uniformly convergent. (*F.S.*, § 70.)

‡ Series such as (6) can also be expressed in terms of theta functions, cf. Whittaker and Watson, *Modern Analysis* (Cambridge, edn. 3, 1920) Chap. XXI.

(iii) *The slab* $-l < x < l$ *with constant initial temperature* V_0.

Changing the origin in (6) to the mid-point of the slab and replacing $\frac{1}{2}l$ by l gives

$$v = \frac{4V_0}{\pi} \sum_{n=0}^{\infty} \frac{(-1)^n}{(2n+1)} e^{-\kappa(2n+1)^2\pi^2t/4l^2} \cos \frac{(2n+1)\pi x}{2l} \tag{8}$$

$$= V_0 - V_0 \sum_{n=0}^{\infty} (-1)^n \left\{ \operatorname{erfc} \frac{(2n+1)l-x}{2(\kappa t)^{\frac{1}{2}}} + \operatorname{erfc} \frac{(2n+1)l+x}{2(\kappa t)^{\frac{1}{2}}} \right\}. \tag{9}$$

Some numerical results for this problem are given in Figs. 10 (a) and 11.
The average temperature v_{av} in the slab at time t is

$$v_{\mathrm{av}} = \frac{8V_0}{\pi^2} \sum_{n=0}^{\infty} \frac{1}{(2n+1)^2} e^{-\kappa(2n+1)^2\pi^2t/4l^2} \tag{10}$$

$$= V_0 - 2V_0 \left(\frac{\kappa t}{l^2}\right)^{\frac{1}{2}} \left\{ \pi^{-\frac{1}{2}} + 2 \sum_{n=1}^{\infty} (-1)^n \operatorname{ierfc} \frac{nl}{(\kappa t)^{\frac{1}{2}}} \right\}. \tag{11}$$

The quantity of heat per unit area of the slab at time t is just $2l\rho c v_{\mathrm{av}}$. Measurements of this quantity are frequently used to determine diffusivities or diffusion coefficients.[†]

The flux of heat f at the surface is

$$f = -K\left[\frac{\partial v}{\partial x}\right]_{x=l} = \frac{2KV_0}{l} \sum_{n=0}^{\infty} e^{-\kappa(2n+1)^2\pi^2t/4l^2} \tag{12}$$

$$= \frac{KV_0}{(\pi\kappa t)^{\frac{1}{2}}} \left\{ 1 + 2 \sum_{n=1}^{\infty} (-1)^n e^{-n^2l^2/\kappa t} \right\}. \tag{13}$$

Ingersoll and Koepp[‡] have used this solution for the determination of κ for earth materials; also Frazier[§] has used it for metal rods by observing the difference in temperature between the points $x = a$ and $x = b$ of the rod. He chooses a and b so that

$$\cos(3\pi a/2l) = \cos(3\pi b/2l).$$

In this case the second term of the series derived from (8) for the temperature difference vanishes, and the third term which has $\exp[-25\kappa\pi^2t/4l^2]$ as a factor disappears very rapidly.

(iv) *The region* $-l < x < l$ *with initial temperature* $V_0(l-|x|)/l$ *and zero surface temperature.*

$$v = \frac{8V_0}{\pi^2} \sum_{n=0}^{\infty} \frac{1}{(2n+1)^2} \cos \frac{(2n+1)\pi x}{2l} e^{-\kappa(2n+1)^2\pi^2t/4l^2} \tag{14}$$

$$= \frac{V_0(l-|x|)}{l} - \frac{2V_0(\kappa t)^{\frac{1}{2}}}{l} \sum_{n=0}^{\infty} (-1)^n \left\{ \operatorname{ierfc} \frac{2nl+|x|}{2(\kappa t)^{\frac{1}{2}}} - \operatorname{ierfc} \frac{(2n+2)l-|x|}{2(\kappa t)^{\frac{1}{2}}} \right\}. \tag{15}$$

[†] Anderson and Saddington, *J. Chem. Soc.* (1949) 381–6.
[‡] *Phys. Rev.* (2) **24** (1924) 92.
[§] *Phys. Rev.* (2) **39** (1932) 515.

(v) *The region* $-l < x < l$ *with initial temperature* $V_0(l^2-x^2)/l^2$ *and zero surface temperature.*

$$v = \frac{32V_0}{\pi^3} \sum_{n=0}^{\infty} \frac{(-1)^n}{(2n+1)^3} e^{-\kappa(2n+1)^2\pi^2t/4l^2} \cos\frac{(2n+1)\pi x}{2l} \tag{16}$$

$$= \frac{V_0(l^2-x^2)}{l^2} - \frac{2\kappa t V_0}{l^2} + \frac{8\kappa t V_0}{l^2} \sum_{n=0}^{\infty} (-1)^n \left\{ i^2\mathrm{erfc}\,\frac{(2n+1)l-x}{2(\kappa t)^{\frac{1}{2}}} + i^2\mathrm{erfc}\,\frac{(2n+1)l+x}{2(\kappa t)^{\frac{1}{2}}} \right\}. \tag{17}$$

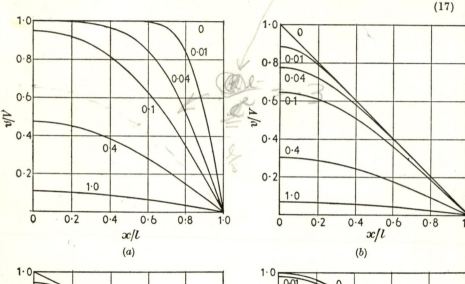

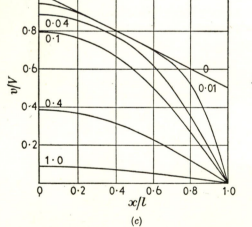

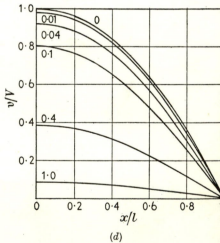

FIG. 10. Temperatures in the slab $0 < x < l$ with no flow at $x = 0$, zero temperature at $x = l$, and various initial distributions of temperature. The numbers on the curves are the values of $\kappa t/l^2$. (a) Constant initial temperature; (b) linear initial temperature, § 3.3 (iv); (c) 'linear+constant' initial temperature; (d) parabolic initial temperature, § 3.3 (v)

(vi) *The region* $-l < x < l$ *with initial temperature* $V_0 \cos(\pi x/2l)$ *and zero surface temperature.*

$$v = V_0 \cos\frac{\pi x}{2l}\, e^{-\kappa\pi^2 t/4l^2}. \tag{18}$$

These results are interesting since they give a qualitative idea of the way in which heat is extracted from a slab with a given initial distribution of temperature. It appears from (5) that the higher harmonics in the Fourier series for $f(x)$ disappear first, leaving the fundamental whose amplitude diminishes exponentially. This is, in effect, restated in (18). In Fig. 10 (a)–(d) the decay of temperature for four different initial distributions of temperature is shown, viz. constant, linear, 'linear+constant', and parabolic. It appears that heat is removed in such a way as to make the distribution approximate to a cosine: for a constant distribution heat is taken first from near the surface; for a linear distribution from near the centre; for a 'linear+constant' distribution from both centre and surface.

3.4. The region $0 < x < l$. Initial temperature $f(x)$. The ends at constant temperature or insulated

In the case in which the ends are kept at constant temperatures v_1 and v_2 we have the equations

$$\frac{\partial v}{\partial t} = \kappa\frac{\partial^2 v}{\partial x^2} \quad (0 < x < l),$$

$$v = v_1, \quad \text{when } x = 0,$$

$$v = v_2, \quad \text{when } x = l,$$

and $\qquad\qquad v = f(x), \quad \text{when } t = 0.$

As in § 1.14, we reduce this to a case of steady temperature, and a case where the ends are kept at zero temperature.

Put $\qquad\qquad\qquad v = u+w,$

where u and w satisfy the following equations:

$$\frac{d^2u}{dx^2} = 0 \quad (0 < x < l),$$

$$u = v_1, \quad \text{when } x = 0,$$

$$u = v_2, \quad \text{when } x = l,$$

and $\qquad\quad \dfrac{\partial w}{\partial t} = \kappa\dfrac{\partial^2 w}{\partial x^2} \quad (0 < x < l),$

$$w = 0, \quad \text{when } x = 0 \text{ and } x = l,$$

$$w = f(x)-u, \quad \text{when } t = 0.$$

We find at once that

$$u = v_1+(v_2-v_1)x/l,$$

and it follows from § 3.3 that

$$w = \sum_{1}^{\infty} a_n \sin\frac{n\pi x}{l} e^{-\kappa n^2 \pi^2 t/l^2},$$

where $\qquad a_n = \frac{2}{l} \int_{0}^{l} \left[f(x') - \left\{ v_1 + (v_2 - v_1)\frac{x'}{l} \right\} \right] \sin\frac{n\pi x'}{l} \, dx'.$

Thus

$$v = v_1 + (v_2 - v_1)\frac{x}{l} + \frac{2}{\pi}\sum_{1}^{\infty} \frac{v_2 \cos n\pi - v_1}{n} \sin\frac{n\pi x}{l} e^{-\kappa n^2 \pi^2 t/l^2} +$$

$$+ \frac{2}{l}\sum_{1}^{\infty} \sin\frac{n\pi x}{l} e^{-\kappa n^2 \pi^2 t/l^2} \int_{0}^{l} f(x')\sin\frac{n\pi x'}{l} \, dx'. \quad (1)$$

The simplest and most important case is that of *the region* $-l < x < l$ *with zero initial temperature and with the surfaces* $x = \pm l$ *kept at constant temperature V for* $t > 0$. The solution, which follows immediately from (1) or 3.3 (8), is

$$v = V - \frac{4V}{\pi}\sum_{n=0}^{\infty} \frac{(-1)^n}{2n+1} e^{-\kappa(2n+1)^2\pi^2 t/4l^2} \cos\frac{(2n+1)\pi x}{2l}. \quad (2)$$

Introducing the dimensionless parameters

$$T = \frac{\kappa t}{l^2}, \qquad \xi = \frac{x}{l}, \quad (3)$$

(2) may be written in the form

$$\frac{v}{V} = 1 - \frac{4}{\pi}\sum_{n=0}^{\infty} \frac{(-1)^n}{2n+1} e^{-(2n+1)^2\pi^2 T/4} \cos\frac{(2n+1)\pi\xi}{\cdot 2}, \quad (4)$$

and the solution for all values of κ, l, t, and x may be obtained from a family of curves in two dimensions. In Fig. 11, v/V is plotted against ξ for various values of T.

In Fig. 12 values of the centre temperature,[†] $x = 0$, and the average temperature in the slab are shown, in curves II and I respectively, together with corresponding curves for the cylinder and sphere.

For the case in which *the end* $x = 0$ *is insulated and the end* $x = l$ *is kept at V, the initial temperature being* $f(x)$, the solution, obtained as

[†] Olson and Schultz, *Ind. Eng. Chem.* **34** (1942) 874, give extensive numerical values for the centre temperatures of slabs and spheres.

in §§ 3.3, 3.4 except that now a cosine series for $f(x)$ has to be used, is

$$v = V + \frac{2}{l} \sum_{n=0}^{\infty} e^{-\kappa(2n+1)^2\pi^2 t/4l^2} \cos \frac{(2n+1)\pi x}{2l} \left\{ \frac{2l(-1)^{n+1}V}{(2n+1)\pi} + \right.$$

$$\left. + \int_0^l f(x') \cos \frac{(2n+1)\pi x'}{2l}\, dx' \right\}. \quad (5)$$

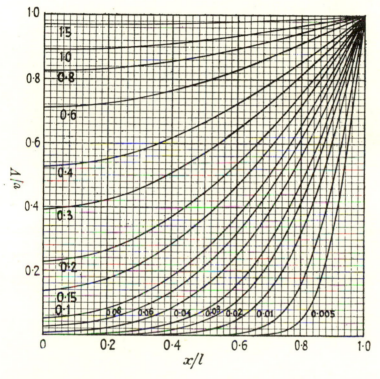

FIG. 11. Temperature distribution at various times in the slab $-l < x < l$ with zero initial temperature and surface temperature V. The numbers on the curves are the values of $\kappa t/l^2$

Some results for this case have already been given in § 3.3 (iii)–(vi).

If the initial temperature is $f(x)$ and both ends $x = 0$ and $x = l$ are thermally insulated the solution is

$$v = \frac{1}{l} \int_0^l f(x')\, dx' + \frac{2}{l} \sum_{n=1}^{\infty} e^{-\kappa n^2 \pi^2 t/l^2} \cos \frac{n\pi x}{l} \int_0^l f(x') \cos \frac{n\pi x'}{l}\, dx'. \quad (6)$$

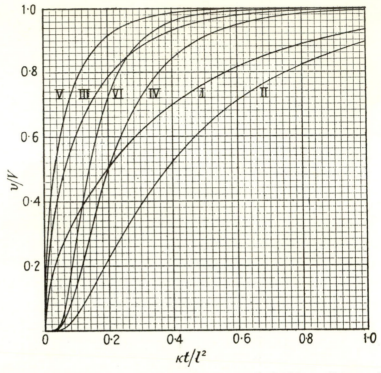

Fig. 12. Centre and average temperatures for a slab of thickness $2l$ (curves II and I); for an infinite circular cylinder of diameter $2l$ (IV, III); and for a sphere of diameter $2l$ (VI, V). Initial temperature zero and surface temperature V

3.5. The region $0 < x < l$. Ends at temperatures $\phi_1(t)$ and $\phi_2(t)$. Initial temperature $f(x)$

In this case we have the equations

$$\frac{\partial v}{\partial t} = \kappa \frac{\partial^2 v}{\partial x^2} \quad (0 < x < l),$$

$$v = \phi_1(t), \quad \text{when } x = 0,$$

$$v = \phi_2(t), \quad \text{when } x = l,$$

and $\qquad v = f(x), \quad \text{when } t = 0.$

Following the general method given in § 1.14, put

$$v = u + w,$$

where

$$\frac{\partial u}{\partial t} = \kappa \frac{\partial^2 u}{\partial x^2} \quad (0 < x < l),$$

$$u = 0, \quad \text{when } x = 0 \text{ and } x = l,$$

$$u = f(x), \quad \text{when } t = 0,$$

and

$$\frac{\partial w}{\partial t} = \kappa \frac{\partial^2 w}{\partial x^2} \quad (0 < x < l),$$

$$w = \phi_1(t), \quad \text{when } x = 0,$$

$$w = \phi_2(t), \quad \text{when } x = l,$$

$$w = 0, \quad \text{when } t = 0.$$

The value of u follows from § 3.3, and is given by

$$u = \frac{2}{l} \sum_1^\infty e^{-\kappa n^2 \pi^2 t / l^2} \sin\frac{n\pi x}{l} \int_0^l f(x')\sin\frac{n\pi x'}{l}\, dx'.$$

To obtain w we may use Duhamel's theorem† (§ 1.14), where the solution for the surface temperatures $\phi_1(t)$ and $\phi_2(t)$ is derived from that for the surface temperatures v_1 and v_2.

In this case the temperature at time t, when the temperature through the slab at $t = \lambda$ is zero, and the ends are kept at $\phi_1(\lambda)$ and $\phi_2(\lambda)$ from $t = \lambda$ to $t = t$, is given by

$$w = \phi_1(\lambda)\left[1 - \frac{x}{l} - \frac{2}{\pi}\sum_1^\infty \frac{1}{n} e^{-\kappa n^2 \pi^2 (t-\lambda)/l^2}\sin\frac{n\pi x}{l}\right] +$$

$$+ \phi_2(\lambda)\left[\frac{x}{l} + \frac{2}{\pi}\sum_1^\infty \frac{1}{n}\cos n\pi e^{-\kappa n^2\pi^2(t-\lambda)/l^2}\sin\frac{n\pi x}{l}\right].$$

Hence, when the surface temperatures are $\phi_1(t)$ and $\phi_2(t)$, we obtain

$$w = \int_0^t \left[\phi_1(\lambda)\frac{\partial}{\partial t}F_1(x, t-\lambda) + \phi_2(\lambda)\frac{\partial}{\partial t}F_2(x, t-\lambda)\right]d\lambda,$$

where

$$F_1(x, t-\lambda) = 1 - \frac{x}{l} - \frac{2}{\pi}\sum_1^\infty \frac{1}{n}e^{-\kappa n^2\pi^2(t-\lambda)/l^2}\sin\frac{n\pi x}{l},$$

$$F_2(x, t-\lambda) = \frac{x}{l} + \frac{2}{\pi}\sum_1^\infty \frac{1}{n}\cos n\pi e^{-\kappa n^2\pi^2(t-\lambda)/l^2}\sin\frac{n\pi x}{l}.$$

Thus

$$w = \frac{2\kappa\pi}{l^2}\sum_1^\infty n e^{-\kappa n^2\pi^2 t/l^2}\sin\frac{n\pi x}{l}\int_0^t e^{\kappa n^2\pi^2\lambda/l^2}[\phi_1(\lambda) - (-1)^n\phi_2(\lambda)]\, d\lambda. \quad (1)$$

† An alternative method due to Stokes is given by Mollison, *Messeng. Math.* **10** (1881) 170–4.

Therefore, finally,

$$v = \frac{2}{l} \sum_{1}^{\infty} e^{-\kappa n^2 \pi^2 t/l^2} \sin \frac{n\pi x}{l} \left[\int_0^l f(x') \sin \frac{n\pi x'}{l} \, dx' + \right.$$

$$\left. + \frac{n\kappa\pi}{l} \int_0^t e^{\kappa n^2 \pi^2 \lambda/l^2} \{\phi_1(\lambda) - (-1)^n \phi_2(\lambda)\} \, d\lambda \right]. \quad (2)$$

For *the region $0 < x < l$ with initial temperature $f(x)$, no flow of heat at $x = 0$, and $x = l$ kept at temperature $\phi_2(t)$,* the solution, obtained in the same way, is

$$\frac{2}{l} \sum_{n=0}^{\infty} e^{-\kappa(2n+1)^2\pi^2 t/4l^2} \cos \frac{(2n+1)\pi x}{2l} \left\{ \frac{(2n+1)\pi\kappa(-1)^n}{2l} \int_0^t e^{\kappa(2n+1)^2\pi^2\lambda/4l^2} \phi_2(\lambda) \, d\lambda + \right.$$

$$\left. + \int_0^l f(x') \cos \frac{(2n+1)\pi x'}{2l} \, dx' \right\}. \quad (3)$$

The following results are of some practical importance. They are given for the region $-l < x < l$, since, in symmetrical cases such as these, comparison with the corresponding results for the cylinder and sphere is easier.

(i) *The region $-l < x < l$ with zero initial temperature. The surfaces kept*[†] *at temperature kt for $t > 0$.*

$$v = kt + \frac{k(x^2 - l^2)}{2\kappa} + \frac{16kl^2}{\kappa\pi^3} \sum_{n=0}^{\infty} \frac{(-1)^n}{(2n+1)^3} e^{-\kappa(2n+1)^2\pi^2 t/4l^2} \cos \frac{(2n+1)\pi x}{2l}. \quad (4)$$

(ii) *The region $-l < x < l$ with zero initial temperature. The surfaces kept at temperature $V(1 - e^{-\beta t})$ for $t > 0$.*

$$v = V - Ve^{-\beta t} \frac{\cos x(\beta/\kappa)^{\frac{1}{2}}}{\cos l(\beta/\kappa)^{\frac{1}{2}}} -$$

$$- \frac{16\beta Vl^2}{\pi} \sum_{n=0}^{\infty} \frac{(-1)^n e^{-\kappa(2n+1)^2\pi^2 t/4l^2}}{(2n+1)[4\beta l^2 - \kappa\pi^2(2n+1)^2]} \cos \frac{(2n+1)\pi x}{2l}, \quad (5)$$

provided that β is not equal to any of the values of $\kappa(2n+1)^2\pi^2/4l^2$. The solution (5) is useful if the surface temperature is changed rapidly but not instantaneously.[‡]

[†] Cf. Williamson and Adams, *Phys. Rev.* (2) **14** (1919) 99; Gurney and Lurie, *Ind. Eng. Chem.* **15** (1923) 1170, who give some numerical results.
[‡] Austin, *J. Appl. Phys.* **3** (1932) 179.

(iii) *The region* $-l < x < l$ *with zero initial temperature. The surfaces* $x = \pm l$ *kept at temperature* $Ve^{\nu t}$ *for* $t > 0$.

$$v = Ve^{\nu t}\frac{\cosh x(\nu/\kappa)^{\frac{1}{2}}}{\cosh l(\nu/\kappa)^{\frac{1}{2}}} -$$

$$-\frac{4V}{\pi}\sum_{n=0}^{\infty}\frac{(-1)^n e^{-\kappa(2n+1)^2\pi^2 t/4l^2}}{(2n+1)[1+\{4\nu l^2/(2n+1)^2\pi^2\kappa\}]}\cos\frac{(2n+1)\pi x}{2l}. \quad (6)$$

3.6. The slab with periodic surface temperature

We consider first the problem of *the slab* $-l < x < l$ *with zero initial temperature and its surface maintained at temperature* $\sin(\omega t+\epsilon)$ *for* $t > 0$.

The solution, obtained from 3.5 (2) or § 12.6, is

$$v = A\sin(\omega t+\epsilon+\phi)+$$

$$+4\pi\kappa\sum_{n=0}^{\infty}\frac{(-1)^n(2n+1)[4l^2\omega\cos\epsilon-\kappa(2n+1)^2\pi^2\sin\epsilon]}{16l^4\omega^2+\kappa^2\pi^4(2n+1)^4}\times$$

$$\times e^{-\kappa(2n+1)^2\pi^2 t/4l^2}\cos\frac{(2n+1)\pi x}{2l}, \quad (1)$$

where

$$A = \left|\frac{\cosh kx(1+i)}{\cosh kl(1+i)}\right| = \left\{\frac{\cosh 2kx+\cos 2kx}{\cosh 2kl+\cos 2kl}\right\}^{\frac{1}{2}}, \quad (2)$$

$$\phi = \arg\left\{\frac{\cosh kx(1+i)}{\cosh kl(1+i)}\right\}, \quad (3)$$

$$k = \left(\frac{\omega}{2\kappa}\right)^{\frac{1}{2}}. \quad (4)$$

The first term of (1) is the periodic steady state solution and the second term the transient. The former might have been found from first principles by the argument used in 2.6 (4)–(6) for the semi-infinite solid.

The quantities A and ϕ which are the amplitude and phase of the steady temperature oscillation at the point x are functions of the two dimensionless quantities x/l and kl.

In Figs. 13 and 14 respectively, the way in which A and ϕ vary across the cross-section of the slab for the values 0·5, 1·0, 1·5, 2·0, 3·0, 5·0, 10·0 of kl are shown.

For the slab $0 < x < l$ *with zero initial temperature and with the planes* $x = 0$ *and* $x = l$ *kept at zero and* $\sin(\omega t+\epsilon)$ *respectively*

$$v = A\sin(\omega t+\epsilon+\phi)+2\pi\kappa\sum_{n=1}^{\infty}\frac{n(-1)^n(\kappa n^2\pi^2\sin\epsilon-\omega l^2\cos\epsilon)}{\kappa^2 n^4\pi^4+\omega^2 l^2}\sin\frac{n\pi x}{l}e^{-\kappa n^2\pi^2 t/l^2},$$

$$(5)$$

where

$$A = \left| \frac{\sinh \mathbf{k}x(1+i)}{\sinh \mathbf{k}l(1+i)} \right| = \left(\frac{\cosh 2\mathbf{k}x - \cos 2\mathbf{k}x}{\cosh 2\mathbf{k}l - \cos 2\mathbf{k}l} \right)^{\frac{1}{2}}, \tag{6}$$

$$\phi = \arg\left\{ \frac{\sinh \mathbf{k}x(1+i)}{\sinh \mathbf{k}l(1+i)} \right\},$$

and $\mathbf{k}$ is defined in (4).

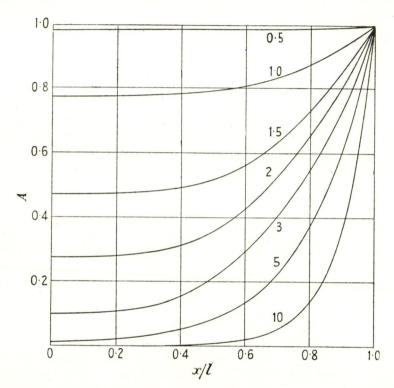

FIG. 13. Variation of amplitude of the steady oscillation of temperature in a slab caused by harmonic surface temperature

If the surface temperature can be represented by the Fourier series

$$\sum_{m=1}^{\infty} a_m \sin(m\omega t + \epsilon_m), \tag{7}$$

the steady periodic part of the solution is

$$\sum_{m=1}^{\infty} a_m A_m \sin(m\omega t + \epsilon_m + \phi_m), \tag{8}$$

where A_m and ϕ_m are defined by (2) and (3)—or, for the second problem considered, by the corresponding equations—with ω replaced by $m\omega$.

Since the amplitudes and phases, A_m and ϕ_m, in (8) are given by rather complicated expressions, the above method, though always available, is often clumsy to use. We shall now give an elegant alternative method,† in which the solution is expressed as a trigonometric series in x with coefficients which are functions of the time; this method is particularly useful for many simple types of surface temperature which often arise in practice, such as 'square' and 'saw-tooth' oscillations.

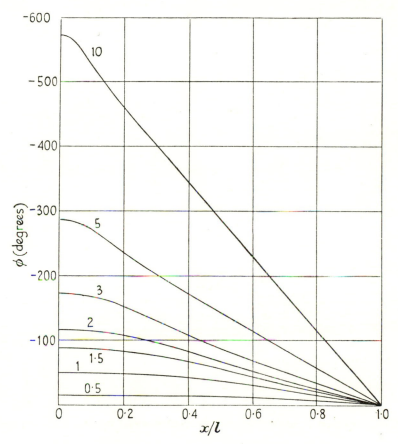

FIG. 14. Variation in phase of the steady oscillation of temperature in a slab caused by harmonic surface temperature

As a first example we consider *the steady periodic temperature in the slab* $0 < x < l$ *in which the surface* $x = 0$ *is maintained at zero, and* $x = l$ *at the temperature*

$$\phi_2(t) = V, \quad rT < t < rT+T_1, \quad r = 0, 1, 2,... \atop = 0, \quad rT+T_1 < t < (r+1)T \Big\}, \qquad (9)$$

† Weber, *Ann. Physik,* **146** (1872) 257. See also §§ 4.8, 15.5.

that is, the temperature V is 'on' for time T_1 and 'off' for time $T-T_1$, this cycle being repeated indefinitely.

We suppose this surface oscillation of temperature has been going on for so long that the steady periodic conditions are established and the influence of the initial temperature has disappeared, and we proceed to find the temperature at time t' after the beginning of an 'on' period.

In 3.5 (2) we take $f(x) = 0$, $\phi_1(\lambda) = 0$, and

$$t = rT+t', \tag{10}$$

where $0 < t' < T_1$, and r is large. Writing

$$\alpha_n = \frac{\kappa n^2 \pi^2}{l^2}, \tag{11}$$

it follows from 3.5 (2) that the solution is

$$v = \frac{2\pi\kappa}{l^2} \sum_{n=1}^{\infty} (-1)^{n+1} n e^{-\alpha_n t} \sin\frac{n\pi x}{l} \int_0^{} e^{\alpha_n \lambda} \phi_2(\lambda)\, d\lambda. \tag{12}$$

Introducing the values (9) and (10) we have

$$e^{-\alpha_n t} \int_0^t e^{\alpha_n \lambda} \phi_2(\lambda)\, d\lambda = Ve^{-\alpha_n t}\left\{ \int_0^{T_1} + \int_T^{T+T_1} + \ldots + \int_{(r-1)T}^{(r-1)T+T_1} + \int_{rT}^{rT+t'} e^{\alpha_n \lambda}\, d\lambda \right\}$$

$$= \frac{V}{\alpha_n} e^{-\alpha_n t}\left\{ (e^{\alpha_n T_1} - e^{\alpha_n T}) \sum_{s=0}^{r-1} e^{s\alpha_n T} - 1 + e^{\alpha_n T} \right\}$$

$$= \frac{V}{\alpha_n} e^{-\alpha_n t}\left\{ \frac{(e^{\alpha_n T_1} - e^{\alpha_n T})(1 - e^{r\alpha_n T})}{1 - e^{\alpha_n T}} - 1 + e^{\alpha_n T} \right\}. \tag{13}$$

For large values of r, (13) becomes

$$\frac{V}{\alpha_n}\left\{ 1 - \frac{e^{\alpha_n(T_1-t')} - e^{\alpha_n(T-t')}}{1 - e^{\alpha_n T}} \right\}. \tag{14}$$

Thus for large values of the time the solution is

$$v = \frac{2V}{\pi} \sum_{n=1}^{\infty} \frac{(-1)^{n+1}}{n} \sin\frac{n\pi x}{l}\left\{ 1 - \frac{e^{\alpha_n(T_1-t')} - e^{\alpha_n(T-t')}}{1 - e^{\alpha_n T}} \right\}. \tag{15}$$

Since

$$\frac{2}{\pi} \sum_{n=1}^{\infty} \frac{(-1)^{n+1}}{n} \sin\frac{n\pi x}{l} = \frac{x}{l}, \tag{16}$$

an alternative form of (15) is

$$v = \frac{Vx}{l} + \frac{2V}{\pi} \sum_{n=1}^{\infty} \frac{(-1)^n}{n} \sin\frac{n\pi x}{l} \frac{e^{\alpha_n(T_1-t')} - e^{\alpha_n(T-t')}}{1 - e^{\alpha_n T}}. \tag{17}$$

In the same way, putting $t = rT + T_1 + t''$, where r is large, we find

$$Ve^{-\alpha_n t} \int_0^t e^{\alpha_n \lambda} \phi_2(\lambda)\, d\lambda \to V \frac{e^{\alpha_n(T-T_1-t'')} - e^{\alpha_n(T-t'')}}{\alpha_n(1 - e^{\alpha_n T})}, \tag{18}$$

and the temperature at time t'' after the beginning of an 'off' interval is

$$v = \frac{2V}{\pi} \sum_{n=1}^{\infty} \frac{(-1)^{n+1}[e^{\alpha_n(T-T_1-t'')} - e^{\alpha_n(T-t'')}]}{n[1 - e^{\alpha_n T}]} \sin \frac{n\pi x}{l}. \tag{19}$$

These results are usually rather more suitable for numerical work than the Fourier series (8). Moreover, the method is quite general, and the results (14) and (18) are immediately available for use in any problem in which the solution for constant external conditions is given as a series of exponentials of $(-\alpha_n t)$, and the solution for external conditions represented by (9) is to be obtained by Duhamel's theorem. Thus, using the results §§ 3.12, 3.8 with the appropriate values of α_n, solutions for problems on a rod radiating into a medium at $\phi_2(t)$ or with supply of heat given by $\phi_2(t)$, etc., where $\phi_2(t)$ is given by (9), can be written down.

Again, using (14) and (18) in 3.5 (3), the solution of the problem of *steady periodic temperature in the rod* $0 < x < l$ *with no flow of heat at* $x = 0$, *and* $x = l$ *maintained at temperature* $\phi_2(t)$ *defined in* (9), *is*

$$\frac{4V}{\pi} \sum_{n=0}^{\infty} \frac{(-1)^n}{2n+1} \cos \frac{(2n+1)\pi x}{2l} \left\{ 1 - \frac{e^{\beta_n(T_1-t')} - e^{\beta_n(T-t')}}{1 - e^{\beta_n T}} \right\} \tag{20}$$

and

$$\frac{4V}{\pi} \sum_{n=0}^{\infty} \frac{(-1)^n}{(2n+1)} \frac{e^{\beta_n(T-T_1-t'')} - e^{\beta_n(T-t'')}}{1 - e^{\beta_n T}} \cos \frac{(2n+1)\pi x}{2l} \tag{21}$$

in the 'on' and 'off' intervals respectively, where now

$$\beta_n = \kappa(2n+1)^2\pi^2/4l^2. \tag{22}$$

3.7. Steady periodic temperature in composite slabs

Such problems are best treated by the matrix methods commonly used in electric circuit theory.† We first discuss periodic temperature in the slab in this notation. All quantities are supposed to be multiplied by a time-factor $\exp(i\omega t)$ which is omitted throughout, and only included at the end of the calculation if real or imaginary parts have to be taken. At each point we shall always be interested in two quantities, the temperature v and the flux f. Then, as in § 2.6, the general solution

---

† The essential theory is that of the four-terminal network, cf. *Electric Circuits* (M.I.T. Staff, Wiley, 1943) p. 452. For the present method, or similar ones, see van Gorcum, *Appl. Sci. Res.* A, **2** (1951) 272–80; Vodicka, *Appl. Sci. Res.* A, **5** (1955) 108–14; Caquot, *Comptes Rendus*, **222** (1946) 486–7; Shklover, *C.R.* (*Doklady*) *Acad. Sci. U.R.S.S.* **45** (1944) 106–10.

corresponding to steady periodic conditions (omitting the time-factor as explained above) is

$$v_x = P \sinh \mathbf{k}x(1+i) + Q \cosh \mathbf{k}x(1+i), \tag{1}$$

$$f_x = -K\mathbf{k}P(1+i)\cosh \mathbf{k}x(1+i) - K\mathbf{k}Q(1+i)\sinh \mathbf{k}x(1+i), \tag{2}$$

where
$$\mathbf{k} = (\omega/2\kappa)^{\frac{1}{2}}, \tag{3}$$

and P and Q are (complex) constants, and v_x and f_x are the temperature and flux at the point x.

Let v and f be the temperature and flux at the face $x = 0$ of the slab and let v' and f' be their values at the face $x = l$. Then, if any two of these four quantities are prescribed, P and Q can be determined from them, and thus the remaining two of v, v', f, f' can be expressed in terms of the original two. In particular,

$$\left.\begin{aligned} v' &= Av + Bf \\ f' &= Cv + Df \end{aligned}\right\}, \tag{4}$$

where
$$A = \cosh \mathbf{k}l(1+i), \qquad B = -\frac{\sinh \mathbf{k}l(1+i)}{K\mathbf{k}(1+i)}, \tag{5}$$

$$C = -K\mathbf{k}(1+i)\sinh \mathbf{k}l(1+i), \qquad D = \cosh \mathbf{k}l(1+i). \tag{6}$$

It follows from (5) and (6) that

$$AD - BC = 1. \tag{7}$$

The equations (4) may be solved to give

$$\left.\begin{aligned} v &= Dv' - Bf' \\ f &= -Cv' + Af' \end{aligned}\right\}. \tag{8}$$

The essential new point to be made is that (4) may be regarded as a matrix equation

$$\begin{pmatrix} v' \\ f' \end{pmatrix} = \begin{pmatrix} A & B \\ C & D \end{pmatrix}\begin{pmatrix} v \\ f \end{pmatrix} \tag{9}$$

connecting the two matrices (v', f') and (v, f), each of two rows and one column, by the ordinary law of matrix multiplication which states that if a_{rs} is the element in the rth row and sth column in a matrix (a_{rs}) of m rows and n columns, and b_{rs} is an element in a matrix (b_{rs}) of n rows and t columns, the product of the matrices (a_{rs}) and (b_{rs}) is a matrix (c_{rs}) of m rows and t columns such that the element c_{rs} in the rth row and sth column is†

$$c_{rs} = \sum_{j=1}^{n} a_{rj} b_{js}. \tag{10}$$

† It should be noted that matrix multiplication is not commutative.

For example,

$$\begin{pmatrix} A_1 & B_1 \\ C_1 & D_1 \end{pmatrix}\begin{pmatrix} A_2 & B_2 \\ C_2 & D_2 \end{pmatrix} = \begin{pmatrix} A_1 A_2 + B_1 C_2 & A_1 B_2 + B_1 D_2 \\ C_1 A_2 + D_1 C_2 & C_1 B_2 + D_1 D_2 \end{pmatrix}. \qquad (11)$$

Suppose, now, that we have a composite wall of n layers, the rth being of thickness l_r, conductivity K_r, and diffusivity κ_r, and with v_r, f_r and v'_r, f'_r at its left-hand and right-hand faces, respectively. Then, if there is perfect thermal contact between the faces of the slabs, repeated application of (9) gives

$$\begin{pmatrix} v'_n \\ f'_n \end{pmatrix} = \begin{pmatrix} A_n & B_n \\ C_n & D_n \end{pmatrix}\begin{pmatrix} A_{n-1} & B_{n-1} \\ C_{n-1} & D_{n-1} \end{pmatrix} \cdots \begin{pmatrix} A_1 & B_1 \\ C_1 & D_1 \end{pmatrix}\begin{pmatrix} v_1 \\ f_1 \end{pmatrix}, \qquad (12)$$

where the A_r, B_r, etc., are the quantities (5), (6) for the individual slabs. The multiplication of matrices in (12) can be carried out successively by (11). Explicit formulae for a slab of n layers may be written out in this way, but they are extremely complicated, and the essential value of the present method is that it allows numerical values for special cases to be studied very easily by inserting numerical values of A_r, B_r,... in (12) and multiplying these *numerical* matrices.

If there are contact resistances between the slabs or at the surfaces, they may also be expressed in matrix notation and included in the chain of products (12). For example, if there is contact resistance R_1 between the first and second slabs, we have

$$f'_1 = f_2 = (v'_1 - v_2)/R_1, \qquad (13)$$

or

$$\begin{pmatrix} v_2 \\ f_2 \end{pmatrix} = \begin{pmatrix} 1 & -R_1 \\ 0 & 1 \end{pmatrix}\begin{pmatrix} v'_1 \\ f'_1 \end{pmatrix}, \qquad (14)$$

so that, for example,

$$\begin{pmatrix} v'_2 \\ f'_2 \end{pmatrix} = \begin{pmatrix} A_2 & B_2 \\ C_2 & D_2 \end{pmatrix}\begin{pmatrix} 1 & -R_1 \\ 0 & 1 \end{pmatrix}\begin{pmatrix} A_1 & B_1 \\ C_1 & D_1 \end{pmatrix}\begin{pmatrix} v_1 \\ f_1 \end{pmatrix}. \qquad (15)$$

The final result of this calculation is a pair of linear relations between the temperatures and fluxes v_1, v'_n, f_1, f'_n at the two surfaces of the composite slab. The surface conditions will provide two more relations so that all four quantities may be determined. The temperature within any one of the slabs, if needed, can be found from (1).

As a simple example, consider the slab $0 < x < l$ with no flow of heat at $x = 0$ and with heat transfer at $x = l$ through a thermal resistance R into a medium at $V \cos \omega t$. Then, if v_1 and $f_1 = 0$ are the temperature and flux at $x = 0$, and V, f_2, those in the outside medium, we have by (14) and (9)

$$\begin{pmatrix} V \\ f_2 \end{pmatrix} = \begin{pmatrix} 1 & -R \\ 0 & 1 \end{pmatrix}\begin{pmatrix} A & B \\ C & D \end{pmatrix}\begin{pmatrix} v_1 \\ 0 \end{pmatrix} = \begin{pmatrix} Av_1 - RCv_1 \\ Cv_1 \end{pmatrix}.$$

It follows that

$$v_1 = \frac{V}{A-RC} = V\{\cosh \mathbf{k}l(1+i) + RK\mathbf{k}(1+i)\sinh \mathbf{k}l(1+i)\}^{-1}.$$

Inserting the time factor $\exp(i\omega t)$ and taking the real part, gives a result which agrees with 3.12 (7).

3.8. The slab with prescribed flux at its surface

Problems of this type are of increasing importance in technical applications. They divide into two classes. In the first of these, heat is supplied by a flat heater embedded in the solid; in this case there is no loss of heat at the surface, and the boundary condition is accurately satisfied if the thermal capacity of the heater is negligible—if it is not it may be regarded as a perfect conductor as in § 3.13. In the second class of problem, which appears in surface heating of metals by induction, heat can escape from the surface, and if linear heat transfer with surface conductance H into a medium at zero is assumed, the boundary condition at the surface is, by 1.9 (4),

$$-K\frac{\partial v}{\partial n} = -F + Hv, \qquad (1)$$

or

$$\frac{\partial v}{\partial n} + h(v - V) = 0, \qquad (2)$$

where F is the surface flux into the solid and $V = F/H$. Thus, for constant H, this problem is that of heating by radiation from a medium at V which is treated in § 3.11.

Some results for the case in which there is no loss of heat from the surface are given below; they are most easily obtained by the methods of Chapter XII.

(i) *The region $0 < x < l$. Zero initial temperature. Constant flux F_0 into the solid at $x = l$. No flow of heat over $x = 0$.*[†]

$$v = \frac{F_0 t}{\rho c l} + \frac{F_0 l}{K}\left\{\frac{3x^2 - l^2}{6l^2} - \frac{2}{\pi^2}\sum_{n=1}^{\infty}\frac{(-1)^n}{n^2}e^{-\kappa n^2\pi^2 t/l^2}\cos\frac{n\pi x}{l}\right\} \qquad (3)$$

$$= \frac{2F_0(\kappa t)^{\frac{1}{2}}}{K}\sum_{n=0}^{\infty}\left\{\text{ierfc}\frac{(2n+1)l-x}{2(\kappa t)^{\frac{1}{2}}} + \text{ierfc}\frac{(2n+1)l+x}{2(\kappa t)^{\frac{1}{2}}}\right\}. \qquad (4)$$

[†] Dufton, *Phil. Mag.* **34** (1943) 376; Smith, *J. Appl. Phys.* **12** (1941) 638; Macey, *Proc. Phys. Soc.* **54** (1942) 128; Brown, *Phil. Mag.* (7) **37** (1946) 318–22; Clarke and Kingston, *Aust. J. Appl. Sci.* **1** (1950) 172–88; Vernotte, *C.R. Acad. Sci. Paris,* **204** (1937) 563.

The temperature distribution (3) corresponds to a linear increase with time, $F_0 t/\rho c l$, together with a correcting term which depends on position and time. This latter term is plotted in Fig. 15.

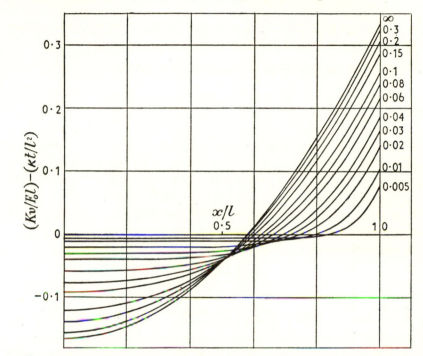

FIG. 15. Values of $(Kv/F_0 l)-(\kappa t/l^2)$ for a slab of thickness l with no flow of heat at $x = 0$ and constant flux F_0 at $x = l$. The numbers on the curves are values of $\kappa t/l^2$

(ii) *The region $0 < x < l$. Zero initial temperature. Constant flux F_0 into the region at $x = l$. $x = 0$ kept at zero temperature.*

$$v = \frac{F_0 x}{K} - \frac{8F_0 l}{K\pi^2} \sum_{n=0}^{\infty} \frac{(-1)^n}{(2n+1)^2} e^{-\kappa(2n+1)^2\pi^2 t/4l^2} \sin\frac{(2n+1)\pi x}{2l} \tag{5}$$

$$= \frac{2F_0(\kappa t)^{\frac{1}{2}}}{K} \sum_{n=0}^{\infty} (-1)^n \left\{ \mathrm{ierfc}\frac{(2n+1)l-x}{2(\kappa t)^{\frac{1}{2}}} - \mathrm{ierfc}\frac{(2n+1)l+x}{2(\kappa t)^{\frac{1}{2}}} \right\}. \tag{6}$$

(iii) *The flux a prescribed function of the time.* Results for this case may be written down by Duhamel's theorem. Two simple results for flux $F_0 t^{\frac{1}{2}m}$, where $m = -1, 0, 1,...$ may be noted: (4) and (6), respectively, are generalized to

$$v = \frac{2^{m+1}F_0 \kappa^{\frac{1}{2}} t^{\frac{1}{2}(m+1)}\Gamma(\tfrac{1}{2}m+1)}{K} \sum_{n=0}^{\infty} \left\{ i^{m+1}\mathrm{erfc}\frac{(2n+1)l-x}{2(\kappa t)^{\frac{1}{2}}} + i^{m+1}\mathrm{erfc}\frac{(2n+1)l+x}{2(\kappa t)^{\frac{1}{2}}} \right\}, \tag{7}$$

I

and†

$$v = \frac{2^{m+1}F_0\kappa^{\frac{1}{2}}t^{\frac{1}{2}(m+1)}\Gamma(\frac{1}{2}m+1)}{K} \sum_{n=0}^{\infty} (-1)^n \left\{ i^{m+1}\mathrm{erfc}\,\frac{(2n+1)l-x}{2(\kappa t)^{\frac{1}{2}}} - i^{m+1}\mathrm{erfc}\,\frac{(2n+1)l+x}{2(\kappa t)^{\frac{1}{2}}} \right\}.$$

(8)

3.9. The region $0 < x < l$. Radiation at the ends into a medium at zero temperature. Initial temperature $f(x)$

In this case the equations for the temperature are

$$\frac{\partial v}{\partial t} = \kappa\frac{\partial^2 v}{\partial x^2} \quad (0 < x < l),$$

(1)

$$-\frac{\partial v}{\partial x} + hv = 0 \quad \text{at } x = 0,$$

(2)

$$\frac{\partial v}{\partial x} + hv = 0 \quad \text{at } x = l,$$

(3)

and

$$v = f(x), \quad \text{when } t = 0.$$

(4)

The expression $e^{-\kappa\alpha^2 t}(A\cos\alpha x + B\sin\alpha x)$ satisfies (1).
It also satisfies (2) and (3), provided that

$$-\alpha B + hA = 0,$$

and

$$\alpha(B\cos\alpha l - A\sin\alpha l) + h(B\sin\alpha l + A\cos\alpha l) = 0.$$

From these we obtain $A/B = \alpha/h$, and

$$\tan\alpha l = \frac{2\alpha h}{\alpha^2 - h^2}.$$

(5)

Hence the expression

$$A\left(\cos\alpha x + \frac{h}{\alpha}\sin\alpha x\right)e^{-\kappa\alpha^2 t}$$

satisfies (1), (2), and (3), where A is an arbitrary constant and α is any root other than zero of the equation

$$\tan\alpha l = \frac{2h\alpha}{\alpha^2 - h^2}.$$

To form an idea of the distribution of the real roots‡ of (5), it is only necessary to note that they correspond to the abscissae of the common points of the curves

$$\eta = 2\cot\xi \quad \text{and} \quad \eta = \frac{\xi}{hl} - \frac{hl}{\xi},$$

where we have put $\alpha l = \xi$.

† For the case $m = 2$ see Newman, *Trans. Amer. Inst. Chem. Engrs.* **30** (1934) 598.
‡ For an alternative discussion of these roots, and numerical values, see § 3.10.

The second of these curves is a hyperbola, whose centre is at the origin, and whose asymptotes are

$$\xi = 0 \quad \text{and} \quad \eta = \frac{\xi}{hl}.$$

If this hyperbola and the cotangent curve are drawn as in Fig. 16, it is clear from the figure that the positive roots lie one in each of the intervals $(0, \pi)$, $(\pi, 2\pi)$,..., and the negative roots are equal in absolute value to the positive ones. Also there are no repeated roots.

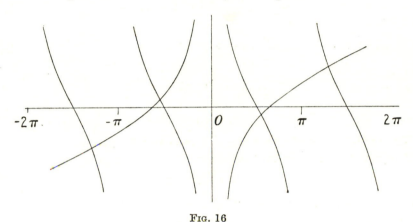

FIG. 16

Further, it is clear that (5) cannot have a pure imaginary root ib, since we would have

$$\tanh lb + \frac{2hb}{b^2 + h^2} = 0,$$

which is impossible as both terms are of the same sign.

Also we shall show at the end of this section that it cannot have a complex root of the form $a \pm ib$; therefore its roots are all real.

Let us assume that $f(x)$ can be developed in an infinite series

$$f(x) = A_1 X_1 + A_2 X_2 + ..., \tag{6}$$

where

$$X_n = \cos \alpha_n x + \frac{h}{\alpha_n} \sin \alpha_n x, \tag{7}$$

α_n being the nth positive root of (5).

Then the solution of our problem is

$$v = \sum_{1}^{\infty} A_n X_n e^{-\kappa \alpha_n^2 t}. \tag{8}$$

The possibility of the expansion (6) and the question of the validity of this solution will be referred to again, cf. § 14.1, but if we assume that such an expansion exists, and that we may integrate the series term by term, the value of the coefficients may be obtained in a similar way to that in which the coefficients in Fourier series, with similar assumptions, may be found.

This depends upon the fact that

$$\int_0^l X_m X_n \, dx = 0 \quad (m \neq n), \tag{9}$$

$$\int_0^l X_n^2 \, dx = \frac{(\alpha_n^2 + h^2)l + 2h}{2\alpha_n^2}, \tag{10}$$

which we shall now prove.

Since $\dfrac{d^2 X_m}{dx^2} + \alpha_m^2 X_m = 0$ and $\dfrac{d^2 X_n}{dx^2} + \alpha_n^2 X_n = 0,$

$$(\alpha_m^2 - \alpha_n^2) \int_0^l X_m X_n \, dx = \int_0^l \left(X_m \frac{d^2 X_n}{dx^2} - X_n \frac{d^2 X_m}{dx^2} \right) dx$$

$$= \left[X_m \frac{dX_n}{dx} - X_n \frac{dX_m}{dx} \right]_0^l.$$

But $\qquad -\dfrac{dX_r}{dx} + hX_r = 0, \quad$ when $x = 0,$

and $\qquad \dfrac{dX_r}{dx} + hX_r = 0, \quad$ when $x = l,$

whatever positive integer r may be.

Thus $\qquad (\alpha_m^2 - \alpha_n^2) \displaystyle\int_0^l X_m X_n \, dx = 0,$

and, when m is not equal to n,

$$\int_0^l X_m X_n \, dx = 0.$$

To obtain the value of $\displaystyle\int_0^l X_n^2 \, dx$, we note that

$$\alpha_n^2 \int_0^l X_n^2 \, dx = -\int_0^l X_n \frac{d^2 X_n}{dx^2} \, dx.$$

Thus $\qquad \alpha_n^2 \displaystyle\int_0^l X_n^2 \, dx = -\left[X_n \frac{dX_n}{dx} \right]_0^l + \int_0^l \left(\frac{dX_n}{dx} \right)^2 dx.$

But
$$\alpha_n X_n = \alpha_n \cos \alpha_n x + h \sin \alpha_n x$$

and
$$\frac{dX_n}{dx} = -\alpha_n \sin \alpha_n x + h \cos \alpha_n x.$$

Therefore
$$\alpha_n^2 X_n^2 + \left(\frac{dX_n}{dx}\right)^2 = \alpha_n^2 + h^2$$

and
$$\alpha_n^2 \int_0^l X_n^2\, dx + \int_0^l \left(\frac{dX_n}{dx}\right)^2 dx = (\alpha_n^2 + h^2)l.$$

But we have seen that

$$\alpha_n^2 \int_0^l X_n^2\, dx - \int_0^l \left(\frac{dX_n}{dx}\right)^2 dx = -\left[X_n \frac{dX_n}{dx}\right]_0^l.$$

Therefore

$$2\alpha_n^2 \int_0^l X_n^2\, dx = l(\alpha_n^2 + h^2) - \left[X_n \frac{dX_n}{dx}\right]_0^l.$$

But
$$-\frac{dX_n}{dx} + hX_n = 0, \quad \text{when } x = 0,$$

and
$$\frac{dX_n}{dx} + hX_n = 0, \quad \text{when } x = l.$$

Therefore
$$X_n \frac{dX_n}{dx} = -hX_n^2, \quad \text{when } x = l,$$

and
$$X_n \frac{dX_n}{dx} = hX_n^2, \quad \text{when } x = 0.$$

But
$$\alpha_n^2 X_n^2 + \left(\frac{dX_n}{dx}\right)^2 = \alpha_n^2 + h^2.$$

Therefore $\quad X_n^2 = 1, \quad$ both when $x = 0$ and $x = l.$

Thus
$$\left[X_n \frac{dX_n}{dx}\right]_0^l = -2h$$

and
$$\int_0^l X_n^2\, dx = \frac{(\alpha_n^2 + h^2)l + 2h}{2\alpha_n^2}.$$

Hence, if we assume the possibility of the expansion and that we may integrate term by term, we have

$$A_n \int_0^l X_n^2\, dx = \int_0^l f(x) X_n\, dx$$

and
$$A_n = \frac{2\alpha_n^2}{(\alpha_n^2+h^2)l+2h} \int_0^l f(x)X_n \, dx.\tag{11}$$

Thus†
$$v = 2 \sum_{n=1}^\infty e^{-\kappa\alpha_n^2 l} \frac{\alpha_n \cos \alpha_n x + h \sin \alpha_n x}{(\alpha_n^2+h^2)l+2h} \int_0^l f(x)(\alpha_n \cos \alpha_n x + h \sin \alpha_n x) \, dx.$$
$$\tag{12}$$

If radiation takes place at $x = 0$ and $x = l$ into media at temperatures v_1 and v_2, the problem can be reduced to the above as usual by putting
$$v = u+w,$$
where u is a function of x only which satisfies the equations
$$\frac{d^2u}{dx^2} = 0 \quad (0 < x < l)$$

$$-\frac{du}{dx}+h(u-v_1) = 0, \quad \text{when } x = 0,$$

and
$$\frac{du}{dx}+h(u-v_2) = 0, \quad \text{when } x = l,$$

so that
$$u = \frac{(v_2-v_1)hx+v_1(1+lh)+v_2}{lh+2};$$

and w is a function of x and t which satisfies the equations
$$\frac{\partial w}{\partial t} = \kappa\frac{\partial^2 w}{\partial x^2} \quad (0 < x < l),$$

$$-\frac{\partial w}{\partial x}+hw = 0, \quad \text{when } x = 0,$$

$$\frac{\partial w}{\partial x}+hw = 0, \quad \text{when } x = l,$$

and
$$w = f(x)-u, \quad \text{when } t = 0.$$

The problems where one end of the rod is kept at a constant temperature, and radiation takes place at the other end, or when one end is rendered impervious to heat, may be treated in the same way. Some results are given in §§ 3.10, 3.11.

† It is assumed throughout that $h > 0$. If $h = 0$, that is no flow of heat across $x = 0$ and $x = l$, the result (12) still holds [cf. 3.4 (6)] except that we have to add to it the term
$$\frac{1}{l}\int_0^l f(x) \, dx.$$

We stated above that the equation

$$\tan \alpha l = \frac{2\alpha h}{\alpha^2 - h^2}$$

cannot have a complex root of the form $a \pm ib$.

If this were possible, we would have two conjugate roots $a \pm ib$, and these would give the two expressions

$$X = \cos \alpha x + \frac{h}{\alpha} \sin \alpha x, \qquad X' = \cos \alpha' x + \frac{h}{\alpha'} \sin \alpha' x,$$

where

$$\alpha = a + ib \quad \text{and} \quad \alpha' = a - ib.$$

Now we have seen that for any two unequal roots of (5),

$$\int_0^l X_m X_n \, dx = 0,$$

and this applies also to X, X', so that

$$\int_0^l X X' \, dx = 0.$$

But dividing X into its real and imaginary parts, we have

$$X = R + iS \quad \text{and} \quad X' = R - iS,$$

so that we would have

$$\int_0^l (R^2 + S^2) \, dx = 0,$$

which is impossible.

Thus we see that (5) has only real roots.

3.10. The region $-l < x < l$. Radiation at the ends $x = \pm l$ into a medium at zero temperature. Initial temperature $f(x)$

It is often better to take the origin at the centre of the region, since any symmetry in the solution then shows up more clearly. Also it is often useful to take the thickness of the slab to be $2l$ as the results are then easy to compare with those for cylinders and spheres of diameter $2a$. For these reasons we restate the result 3.9 (12) in this notation. It is

$$v = \sum_{n=1}^{\infty} e^{-\kappa \alpha_n^2 t} \frac{c_n \cos \alpha_n x + d_n \sin \alpha_n x}{(\alpha_n^2 + h^2)l + h} \int_{-l}^{l} f(x)[c_n \cos \alpha_n x + d_n \sin \alpha_n x] \, dx, \tag{1}$$

where

$$c_n = h \sin \alpha_n l + \alpha_n \cos \alpha_n l, \tag{2}$$

$$d_n = h \cos \alpha_n l - \alpha_n \sin \alpha_n l, \tag{3}$$

and the α_n are the positive roots of

$$\tan 2\alpha l = \frac{2\alpha h}{\alpha^2 - h^2}. \tag{4}$$

Since (4) is equivalent to

$$(h \sin \alpha l + \alpha \cos \alpha l)(h \cos \alpha l - \alpha \sin \alpha l) = 0, \tag{5}$$

its positive roots α_n comprise the positive roots of the two equations

$$\alpha \tan \alpha l - h = 0, \tag{6}$$

$$\alpha \cot \alpha l + h = 0. \tag{7}$$

It follows from the results of § 3.9 that the roots† of (6) and (7) are all real and simple: this, of course, is easily proved directly. Some numerical values are given in Appendix IV.

If α_n is a root of (6), $d_n = 0$, and $c_n^2 = h^2 + \alpha_n^2$; while if α_n is a root of (7), $c_n = 0$, and $d_n^2 = h^2 + \alpha_n^2$.

If $f(x)$ is an even function of x, (1) reduces to

$$v = 2 \sum_{n=1}^{\infty} e^{-\kappa \alpha_n^2 t} \frac{(h^2 + \alpha_n^2) \cos \alpha_n x}{(\alpha_n^2 + h^2) l + h} \int_0^l f(x) \cos \alpha_n x \, dx, \tag{8}$$

where the α_n are the positive roots of (6). This is also the solution of the problem of *conduction of heat in the region* $0 < x < l$, *with no flow of heat over the boundary* $x = 0$, *radiation into a medium at zero at* $x = l$, *and initial temperature* $f(x)$; it could easily have been obtained directly by the method of § 3.9. In this case the expansion 3.9 (6) has become

$$f(x) = \sum_{n=1}^{\infty} \frac{2(h^2 + \alpha_n^2) \cos \alpha_n x}{(\alpha_n^2 + h^2) l + h} \int_0^l f(x) \cos \alpha_n x \, dx, \tag{9}$$

where the α_n are the positive roots of (6). In the special case $f(x) = 1$, (9) gives

$$\sum_{n=1}^{\infty} \frac{2h \cos \alpha_n x}{[(\alpha_n^2 + h^2) l + h] \cos \alpha_n l} = 1. \tag{10}$$

In the same way if $f(x)$ is an odd function of x, (1) reduces to

$$v = 2 \sum_{n=1}^{\infty} e^{-\kappa \alpha_n^2 t} \frac{(h^2 + \alpha_n^2) \sin \alpha_n x}{(h^2 + \alpha_n^2) l + h} \int_0^l f(x) \sin \alpha_n x \, dx, \tag{11}$$

where the α_n are the positive roots of (7). This is also the solution of the problem of *conduction of heat in the region* $0 < x < l$, *with* $x = 0$ *maintained at zero temperature for* $t > 0$, *radiation into a medium at zero temperature at* $x = l$, *and initial temperature* $f(x)$.

If $f(x)$ is neither even nor odd, the solution is given by (1), and involves the roots of both (6) and (7). The position is precisely analogous to that of the Fourier series for an arbitrary function $f(x)$ in $(-l, l)$ and its specialization into cosine or sine series if $f(x)$ is even or odd.

† The roots of (6) are all real if $h > 0$; those of (7) are all real if $lh > -1$, cf. § 9.4.

3.11. Special problems and numerical results for the slab with a radiation boundary condition

Many problems of practical importance involve this boundary condition and a great deal of numerical information in the form of tables and graphs is available about their solution. A number of results will be given here, and, to simplify numerical discussion, they will be stated in terms of the dimensionless parameters

$$L = lh, \quad T = \kappa t/l^2, \quad x/l. \tag{1}$$

Usually, consideration has been limited to important quantities such as centre, surface, or average temperature, which involve only two parameters L and T and so can be expressed as a family of curves. There are, in fact, eight functions† of L and T in terms of which many results of this sort may be expressed.

The first four of these are

$$f_1(L, T) = 1 - \sum_{n=1}^{\infty} \frac{2L}{L(L+1)+\alpha_n^2} e^{-\alpha_n^2 T}, \tag{2}$$

$$f_2(L, T) = 1 - \sum_{n=1}^{\infty} \frac{2L \sec \alpha_n}{L(L+1)+\alpha_n^2} e^{-\alpha_n^2 T}, \tag{3}$$

$$f_3(L, T) = 1 - \sum_{n=1}^{\infty} \frac{2L^2}{\alpha_n^2[L(L+1)+\alpha_n^2]} e^{-\alpha_n^2 T}, \tag{4}$$

$$f_4(L, T) = 1 - \sum_{n=1}^{\infty} \frac{4L^2 \sec \alpha_n}{(L+2)\alpha_n^2[L(L+1)+\alpha_n^2]} e^{-\alpha_n^2 T}, \tag{5}$$

where α_n, $n = 1, 2,...$ are the positive roots of

$$\alpha \tan \alpha = L, \tag{6}$$

which is 3.10 (6) in the present notation.

The other four functions, involving the roots of 3.10 (7), are

$$\phi_1(L, T) = 1 - \sum_{n=1}^{\infty} \frac{2(L+1)}{L(L+1)+\beta_n^2} e^{-\beta_n^2 T}, \tag{7}$$

$$\phi_2(L, T) = 1 + \sum_{n=1}^{\infty} \frac{2L(L+1)\sec \beta_n}{L(L+1)+\beta_n^2} e^{-\beta_n^2 T}, \tag{8}$$

$$\phi_3(L, T) = 1 - \sum_{n=1}^{\infty} \frac{6(L+1)^2}{\beta_n^2[L(L+1)+\beta_n^2]} e^{-\beta_n^2 T}, \tag{9}$$

† Graphs of these, plotted against $\tan^{-1}L$ for fixed values of T, with various applications are given by Jaeger and Clarke, *Phil. Mag.* (7) 38 (1947) 504–15. Values of $1-f_1$, $1-f_2$, $1-f_3$ are shown in Figs. 17–19, respectively.

$$\phi_4(L, T) = 1 + \sum_{n=1}^{\infty} \frac{12L(L+1)^2\sec\beta_n}{(L+3)\beta_n^2[L(L+1)+\beta_n^2]}\, e^{-\beta_n^2 T}, \tag{10}$$

where β_n, $n = 1, 2,...$, are the positive roots of

$$\beta\cot\beta + L = 0. \tag{11}$$

Some values of the roots of (6) and (11) are given in Appendix IV. It should be remarked that for large values of T, only the first terms of the series given above are of importance. In many cases this will be found to be true if $T > 0{\cdot}5$.

For small values of T, say $T < 0{\cdot}01$, the series become slowly convergent and alternative expressions may be deduced by the methods of § 12.5 IV. They do not lead to simple explicit formulae as in some of the cases discussed earlier.

A number of important special problems will now be discussed.

(i) *The region* $-l < x < l$ *with constant initial temperature* V *and radiation at its surface into medium at zero.*

Here, in the notation (1), (6), 3.10 (8) gives

$$\frac{v}{V} = \sum_{n=1}^{\infty} \frac{2L\cos(\alpha_n x/l)\sec\alpha_n}{L(L+1)+\alpha_n^2}\, e^{-\alpha_n^2 T}. \tag{12}$$

The surface temperature $v_s = V[1-f_1(L, T)]$, and the centre temperature $v_c = V[1-f_2(L, T)]$, where f_1 and f_2 are defined in (2) and (3). The average temperature in the slab $v_{av} = V[1-f_3(L, T)]$, and the quantity of heat lost by the slab (from both faces) up to time t is

$$Q = 2l\rho c[V-v_{av}] = 2lV\rho c\, f_3(L, T).$$

In Figs. 17, 18, and 19, values of v_s/V, v_c/V, and $1-Q/2Vl\rho c$ are shown plotted against $\log_{10} T$ for values 0·05, 0·1, 0·2, 0·5, 1, 2, 5, 10, ∞ of L.

Because of the importance of the present problems, a number of tables and charts of numerical values has been given. Gurney and Lurie† plot $\log(v/V)$ against T for fixed values of L and x/l; their curves are largely straight lines, corresponding to the case in which only the first term of (12) is of importance. Other authors have not attempted to cover all values of the three parameters, but study the quantities v_s, v_c, and Q or v_{av} defined above. Gröber‡ gives short tables and plots his results against

† *Ind. Eng. Chem.* **15** (1923) 1170. Their results are reproduced in McAdams, *Heat Transmission* (McGraw-Hill, edn. 2, 1942) Chap. 2.

‡ *Z. Ver. dtsch. Ing.* **69** (1925) 705. See also Gröber, Erk, and Grigull, *Wärmeübertragung* (Springer, 1955).

log L for fixed values of T, or against log T for fixed values of L. Schack†
has extended Gröber's results and replotted them against L with a linear
scale which changes abruptly at each power of ten. Ede‡ proposes a

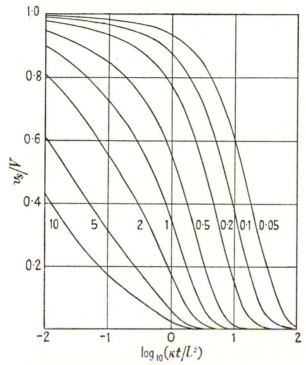

Fig. 17. The function $[1-f_1(L, T)]$ defined in (2). The figure shows the surface
temperature v_s of a slab of thickness $2l$, initially at temperature V, and cooling
by radiation into a medium at zero. The numbers on the curves are values of
$$L = lh.$$

plot of ln T against ln(L/K) for fixed values of v_c, v_s, and v_{av} and gives
sufficient numerical values for plotting such curves. Heisler§ gives a very
complete set of charts with a critical introduction. Newman gives
tables of results for the case of constant and parabolic initial distributions
of temperature.∥ Other numerical results, or solutions adapted for

† *Stahl u. Eisen*, **50** (1930) 1290. His curves are reproduced in Fishenden and Saunders,
Heat Transfer (Oxford, 1950); in Schack, Goldschmidt, and Partridge, *Industrial Heat
Transfer* (Wiley, 1933); and by Newman, *Ind. Eng. Chem.* **28** (1936) 545.
‡ *Phil. Mag.* (7) **36** (1945) 845–91.
§ *Trans. Amer. Soc. Mech. Engrs.* **69** (1947) 227–36. Some of his results are repro-
duced in Jakob, *Heat Transfer* (Wiley, 1949), §§ 13–19.
∥ *Trans. Amer. Inst. Chem. Engrs.* **27** (1931) 202; *Ind. Eng. Chem.* **28** (1936) 545.

computation, are given by Goldstein,† Pöschl,‡ Nistler,§ McKay,‖ and
Bachmann.††

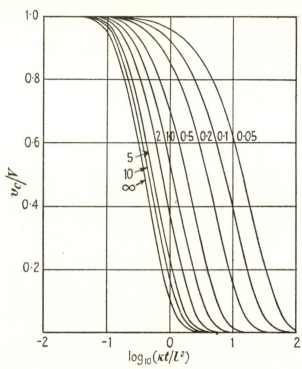

FIG. 18. The function $[1-f_2(L, T)]$ defined in (3). The figure shows the centre
temperature of a slab of thickness $2l$, initially at temperature V, and cooling by
radiation into a medium at zero. The numbers on the curves are values of
$L = lh$.

(ii) *The region* $-l < x < l$ *with radiation at* $x = \pm l$ *into a medium at zero
temperature. Initial temperature* $a - bx^2$.

$$v = \sum_{n=1}^{\infty} \frac{\{2La\alpha_n^2 - 2bl^2[\alpha_n^2(L+2) - 2L]\}\cos(\alpha_n x/l)}{\alpha_n^2[\alpha_n^2 + L^2 + L]\cos\alpha_n} e^{-\alpha_n^2 T}, \tag{13}$$

where the α_n are the roots of (6).

(iii) *The region* $-l < x < l$, *with zero initial temperature, heated by radiation
from a medium at* V.

As remarked in § 3.8 this is also the case of the slab heated by constant flux
at its surface combined with radiation into a medium at zero. The temperature
is just V minus the results of (i).

† *Z. angew. Math. Mech.* **12** (1932) 234; **14** (1934) 158.
‡ Ibid. **12** (1932) 280.   § Ibid. **17** (1937) 245.
‖ *Proc. Phys. Soc.* **42** (1930) 547.
†† *Tafeln über Abkühlungsvorgänge einfacher Körper* (Springer, 1938).

(iv) *The region* $0 < x < l$ *with zero initial temperature. Constant flux F_0 into the solid at $x = 0$. At $x = l$, radiation into a medium at zero.*[†]

$$v = \frac{lF_0}{KL}\left\{1+L\left(1-\frac{x}{l}\right) - \sum_{n=1}^{\infty} \frac{2L(\alpha_n^2+L^2)\cos(\alpha_n x/l)}{\alpha_n^2[L+L^2+\alpha_n^2]}\, e^{-\alpha_n^2 T}\right\}, \tag{14}$$

where the α_n are the positive roots of (6). The temperature at $x = 0$ is

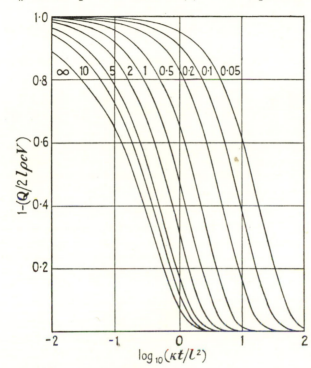

FIG. 19. The function $[1-f_3(L, T)]$ defined in (4). The figure shows the heat Q lost from unit area of a slab of thickness $2l$, initially at temperature V, and cooling by radiation into a medium at zero. The numbers on the curves are values of $L = lh$.

$(lF_0/KL)[f_1(L, T)+Lf_3(L, T)]$, and that at $x = l$ is $(lF_0/KL)f_2(L, T)$. The quantity of heat which crosses the plane $x = l$ per unit area from $t = 0$ to $t = t$ is $(F_0 l^2/\kappa)\{T-[(L+2)/2L]f_4(L, T)\}$.

(v) *The region* $0 < x < l$. *Zero initial temperature.* $x = 0$ *maintained at zero temperature for* $t > 0$. *At $x = l$, radiation to the solid from medium at constant temperature* V.

$$v = \frac{LVx}{l(1+L)} - 2LV \sum_{n=1}^{\infty} \frac{\sin(\beta_n x/l)}{(L+L^2+\beta_n^2)\sin\beta_n}\, e^{-\beta_n^2 T}, \tag{15}$$

where β_n, $n = 1, 2,...$ are the positive roots of (11). The temperature at $x = l$ is

† Newman and Green, *Trans. Electrochem. Soc.* **66** (1934) 345, give some numerical results.

$[LV/(L+1)]\phi_1(L, T)$, and the quantities of heat crossing unit area of the planes $x = 0$ and $x = l$ up to time t involve $\phi_4(L, T)$ and $\phi_3(L, T)$, respectively.

(vi) *The region $0 < x < l$ with zero initial temperature. The end $x = 0$ is kept at temperature V for $t > 0$. At $x = l$ there is radiation into a medium at zero temperature.*

$$\frac{v}{V} = \frac{1+L(1-x/l)}{1+L} - \sum_{n=1}^{\infty} \frac{2(\beta_n^2+L^2)\sin(\beta_n x/l)}{\beta_n[L+L^2+\beta_n^2]} e^{-\beta_n^2 T}, \tag{16}$$

where the β_n are the positive roots of (11). The temperature at $x = l$ is $[V/(1+L)]\phi_2(L, T)$.

(vii) *The region $0 < x < l$. Initial temperature $V_0 x/l$. $x = 0$ maintained at zero for $t > 0$. At $x = l$, radiation to the solid from a medium at V_1.*

$$v = \frac{V_0 x}{l} + \frac{(x/l)[L(V_1-V_0)-V_0]}{L+1} - 2[L(V_1-V_0)-V_0] \sum_{n=1}^{\infty} \frac{\sin(\beta_n x/l)}{[L^2+L+\beta_n^2]\sin\beta_n} e^{-\beta_n^2 T}, \tag{17}$$

where the β_n are the positive roots of (11). This illustrates the effect of changing the temperature outside one surface of a wall when steady conditions have been established. It is easy to show directly from the differential equation that the *change* in temperature is given by (15) with V replaced by

$$V_1-V_0-(V_0/L). \tag{18}$$

(viii) *The region $-l < x < l$ with zero initial temperature. At $x = l$ there is radiation to the solid from medium at V. At $x = -l$ there is radiation from the solid to medium at zero temperature, the surface conductance being the same at both surfaces.*

This result, which is better stated as above for the region $-l < x < l$, follows by combining (12) and (15). It is

$$v = \frac{LVx}{2l(1+L)} - LV \sum_{n=1}^{\infty} \frac{\sin(\beta_n x/l)}{(L^2+L+\beta_n^2)\sin\beta_n} e^{-\beta_n^2 T} +$$

$$+ \tfrac{1}{2}V - LV \sum_{n=1}^{\infty} \frac{\cos(\alpha_n x/l)}{(L^2+L+\alpha_n^2)\cos\alpha_n} e^{-\alpha_n^2 T}, \tag{19}$$

where α_n and β_n are the positive roots of (6) and (11), respectively.

(ix) *The region $0 < x < l$ with initial temperature $f(x)$. The boundary conditions*

$$\left. \begin{array}{l} k_1 \dfrac{\partial v}{\partial x} - h_1 v = 0, \quad x = 0, \\[2mm] k_2 \dfrac{\partial v}{\partial x} + h_2 v = 0, \quad x = l, \end{array} \right\} \tag{20}$$

where $k_1 \geqslant 0$, $h_1 \geqslant 0$, but they are not both to vanish, and k_2, h_2 satisfy the same conditions. These boundary conditions thus include the nine possible combinations of zero temperature, zero flux, or radiation into a medium at zero, at either face.

$$v = \sum_{n=1}^{\infty} Z_n(x)e^{-\kappa\beta_n^2 t} \int_0^l Z_n(x')f(x')\,dx', \tag{21}$$

where

$$Z_n(x) = \frac{[2(k_2^2\beta_n^2+h_2^2)]^{\frac{1}{2}}(k_1\beta_n\cos\beta_n x+h_1\sin\beta_n x)}{\{(k_1^2\beta_n^2+h_1^2)[l(k_2^2\beta_n^2+h_2^2)+k_2h_2]+k_1h_1(k_2^2\beta_n^2+h_2^2)\}^{\frac{1}{2}}}, \tag{22}$$

β_n, $n = 1, 2...$, are the positive roots of

$$(k_1 k_2 \beta^2 - h_1 h_2)\sin \beta l = \beta(k_1 h_2 + k_2 h_1)\cos \beta l, \qquad (23)$$

and if $h_1 = h_2 = 0$ an additional term

$$\frac{1}{l}\int_0^l f(x')\,dx' \qquad (24)$$

is to be added to (21).

This is the generalization of the result of § 3.9 to the case of different surface conductances at the two faces. The extension to the case of radiation into media at different temperatures is made as at the end of § 3.9.

3.12. The region $-l < x < l$ with zero initial temperature and radiation at the ends into medium at $\phi(t)$

It follows in the usual way from Duhamel's theorem, § 1.14, that

$$v = 2\kappa h \sum_{n=1}^{\infty} \frac{\alpha_n^2 e^{-\kappa \alpha_n^2 t}\cos \alpha_n x}{\{h + l(\alpha_n^2 + h^2)\}\cos \alpha_n l}\int_0^t e^{\kappa \alpha_n^2 \lambda}\phi(\lambda)\,d\lambda, \qquad (1)$$

where the α_n are the roots of $\qquad \alpha \tan \alpha l = h.$ $\qquad (2)$

If the medium is at V_0 for $0 < t < T$ and V_1 for $t > T$, the solution is

$$v = V_0\left\{1 - \sum_{n=1}^{\infty} e^{-\kappa \alpha_n^2 t}\frac{2h\cos \alpha_n x}{[(h^2 + \alpha_n^2)l + h]\cos l\alpha_n}\right\}, \quad 0 < t < T, \qquad (3)$$

$$v = V_1 - \sum_{n=1}^{\infty} e^{-\kappa \alpha_n^2 t}\frac{2h[V_0 + (V_1 - V_0)e^{\kappa \alpha_n^2 T}]\cos \alpha_n x}{[(h^2 + \alpha_n^2)l + h]\cos l\alpha_n}, \quad t > T. \qquad (4)$$

If the medium is at temperature $V\sin(\omega t + \epsilon)$, the solution is

$$v = \frac{hVM_0}{M_1}\sin(\omega t + \epsilon + \gamma_0 - \gamma_1) +$$

$$+ 2h\kappa V \sum_{n=1}^{\infty} \frac{\alpha_n^2(\omega \cos \epsilon - \kappa \alpha_n^2 \sin \epsilon)\cos \alpha_n x}{(\kappa^2 \alpha_n^4 + \omega^2)[(\alpha_n^2 + h^2)l + h]\cos \alpha_n l}e^{-\kappa \alpha_n^2 t}, \qquad (5)$$

where $\qquad M_0 e^{i\gamma_0} = \cosh \omega'x \cos \omega'x + i\sinh \omega'x \sin \omega'x, \qquad (6)$

$$M_1 e^{i\gamma_1} = \omega'\sinh \omega'l\cos \omega'l - \omega'\cosh \omega'l\sin \omega'l + h\cosh \omega'l\cos \omega'l +$$

$$+ i[\omega'\sinh \omega'l\cos \omega'l + \omega'\cosh \omega'l\sin \omega'l + h\sinh \omega'l\sin \omega'l], \qquad (7)$$

and $\qquad \omega' = \sqrt{(\omega/2\kappa)}. \qquad (8)$

If the medium is at temperature kt, the solution is

$$v = kt + \frac{k(hx^2 - l^2 h - 2l)}{2\kappa h} + \frac{2hk}{\kappa}\sum_{n=1}^{\infty} e^{-\kappa \alpha_n^2 t}\frac{\cos \alpha_n x}{\alpha_n^2[(h^2 + \alpha_n^2)l + h]\cos l\alpha_n}, \qquad (9)$$

where the α_n are the positive roots of (2).

3.13. The slab with one face in contact with a layer of perfect conductor or well-stirred fluid

If the slab is in perfect thermal contact at its face $x = l$ with mass M' per unit area of well-stirred fluid (or perfect conductor) of specific heat c', the boundary condition at $x = l$ is, as in 1.9 (14),

$$K\frac{\partial v}{\partial x} + M'c'\frac{\partial v}{\partial t} = Q, \tag{1}$$

where Q/M' is the external rate of supply (if any) of heat per unit mass of the fluid. It is assumed that there is no loss of heat from the fluid except to the slab. In the solutions of such problems the three dimensionless parameters $\kappa t/l^2$, lh, and x/l will occur, where here, and in (i) to (vi) below, we write

$$h = \rho c/M'c'. \tag{2}$$

The solution of a number of simple problems which may be expressed in terms of the roots of the equations

$$\alpha \tan \alpha l = h, \tag{3}$$

or

$$\beta \cot \beta l = -h, \tag{4}$$

are given below. Numerical values may be expressed in terms of the fundamental functions f and ϕ defined in 3.11 (1)–(10). In more complicated problems involving radiation at $x = 0$, or radiation from the fluid, or a contact resistance between the solid and the fluid, additional parameters appear, cf. (vii)–(ix).

(i) *The region $0 < x < l$ with zero initial temperature. At $x = l$ the solid is in contact with mass M' per unit area of well-stirred fluid of specific heat c' initially at zero temperature. No flow of heat at $x = 0$. Constant supply of heat Q to the fluid.* The solution is

$$v = \frac{Q}{M'c'(1+lh)}\Big[t + \frac{x^2}{2\kappa} - \frac{l^2(3+lh)}{6\kappa(1+lh)}\Big] - \frac{2Qh}{M'c\kappa}\sum_{n=1}^{\infty}\frac{e^{-\kappa\beta_n^2 t}\cos\beta_n x}{\beta_n^2\cos\beta_n l[l(\beta_n^2+h^2)+h]}, \tag{5}$$

where the β_n are the roots of (4).

(ii) *The problem of (i) except that $Q = 0$ and the initial temperature of the fluid is V.*

$$v = \frac{V}{1+lh} + \sum_{n=1}^{\infty}\frac{2hVe^{-\kappa\beta_n^2 t}\cos\beta_n x}{\cos\beta_n l[l(\beta_n^2+h^2)+h]}. \tag{6}$$

(iii) *The problem of (i) except that $x = 0$ is kept at zero temperature. Q constant.*

$$v = \frac{Ql}{K}\Big\{\frac{x}{l} - \sum_{n=1}^{\infty}\frac{2he^{-\kappa\alpha_n^2 t}\sin\alpha_n x}{l\alpha_n\cos\alpha_n l\{l(\alpha_n^2+h^2)+h\}}\Big\}, \tag{7}$$

where the α_n are the roots of (3).

(iv) *The problem of (i) with $Q = 0$, the initial temperature of the fluid V, and $x = 0$ kept at zero temperature.*

$$v = \sum_{n=1}^{\infty}\frac{2hVe^{-\kappa\alpha_n^2 t}\sin\alpha_n x}{\sin\alpha_n l[l(\alpha_n^2+h^2)+h]}, \tag{8}$$

where the α_n are the roots of (3).

(v) *The problem of (i) with $Q = 0$, the initial temperature of the fluid zero, but $x = 0$ kept at temperature V for $t > 0$.*

$$\frac{v}{V} = 1 - \sum_{n=1}^{\infty} \frac{2(\alpha_n^2 + h^2)e^{-\kappa \alpha_n^2 t}\sin \alpha_n x}{\alpha_n[l(\alpha_n^2 + h^2)+h]}, \tag{9}$$

where h is defined in (2) and the α_n are the roots of (3).

(vi) *The problem of (i) with $Q = 0$, the initial temperature of the fluid zero, but heat supplied at $x = 0$ at constant rate Q' per unit time per unit area.*

$$v = \frac{Q'l}{K(1+lh)}\left\{\frac{h\kappa t}{l} + \frac{l-x}{l} + \frac{h(l-x)^2}{2l} - \frac{lh(3+lh)}{6(1+lh)} - \right.$$
$$\left. - \sum_{n=1}^{\infty} \frac{2(1+lh)(h^2 + \beta_n^2)e^{-\kappa \beta_n^2 t}\cos \beta_n x}{l\beta_n^2\{l(\beta_n^2 + h^2)+h\}}\right\}, \tag{10}$$

where h is defined in (2) and the β_n are the roots of (4).

(vii) *The region $0 < x < l$ with initial temperature V. No loss of heat at $x = 0$. At $x = l$ contact with mass M' per unit area of well-stirred fluid of specific heat c' which loses heat by radiation at a rate H times its temperature. The initial temperature of the fluid zero.*

$$v = 2V \sum_{n=1}^{\infty} \frac{e^{-\kappa \alpha_n^2 t}(h - k\alpha_n^2)\cos \alpha_n x}{[l(h - k\alpha_n^2)^2 + \alpha_n^2(l+k)+h]\cos \alpha_n l}, \tag{11}$$

where $h = H/K$, $k = M'c'/\rho c$, and the α_n are the roots of

$$\alpha \tan \alpha l = h - k\alpha^2. \tag{12}$$

(viii) *The system of (vii) except that the initial temperatures of both solid and fluid are zero and heat is supplied to the fluid at the rate $f(t)/M'$ per unit time per unit mass.*

$$v = \frac{2\kappa}{K} \sum_{n=1}^{\infty} \frac{\alpha_n^2 \cos \alpha_n x e^{-\kappa \alpha_n^2 t}}{\cos \alpha_n l\{l(h - k\alpha_n^2)^2 + (l+k)\alpha_n^2 + h\}} \int_0^t e^{\kappa \alpha_n^2 \tau}f(\tau)\, d\tau, \tag{13}$$

where the α_n are the roots of (12).

(ix) *The region $0 < x < l$ with no flow of heat at $x = l$ and at $x = 0$ contact with mass M' per unit area of well-stirred fluid of specific heat c' to which heat is supplied at the constant rate Q/M' per unit mass per unit time. It is not assumed that the surface temperature of the solid is equal to the temperature of the fluid but that there is heat transfer between them at a rate H times their temperature difference†
(cf. 1.9 (16)).*

$$\frac{Kv}{Ql} = \frac{k}{1+k}\left\{T + \tfrac{1}{2}(1-x/l)^2 - \frac{3L+6+kL}{6L(1+k)} + \right.$$
$$\left. + 2L(1+k) \sum_{n=1}^{\infty} \frac{\alpha_n^2 - kL}{P_n \cos \alpha_n} e^{-\alpha_n^2 T}\cos \alpha_n(1-x/l)\right\}, \tag{14}$$

† For further problems of this type and some numerical results see Jaeger, *Proc. Camb. Phil. Soc.* **41** (1945) 43.

where $h = H/K$, $L = lh$, $k = lpc/M'c'$, $T = \kappa t/l^2$,

$$P_n = \alpha_n^6 + \alpha_n^4(L^2 + L - 2kL) + kL^2(1+k)\alpha_n^2,$$

and the α_n are the roots of $\tan \alpha = \dfrac{L\alpha}{\alpha^2 - kL}$.

3.14. The slab with heat produced within it

If heat is produced at the rate A per unit time per unit volume, the differential equation to be solved is

$$\frac{\partial^2 v}{\partial x^2} - \frac{1}{\kappa}\frac{\partial v}{\partial t} = -\frac{A}{K}. \tag{1}$$

The most important case is $A = A_0$, constant, for $t > 0$, which occurs in dielectric heating. Extensions to the case in which A is a function of position or time are easily made as in §§ 1.14, 12.6. As remarked in § 1.6, the general case in which A is a function of v is quite intractable. The case in which A is a linear function of v is treated in § 15.7. Here we shall use the method of § 1.14; alternatively, the Laplace transformation may be used.

(i) *The region* $-l < x < l$ *with zero surface temperature†, zero initial temperature, and heat production at the rate* A_0, *constant, for* $t > 0$.

Proceeding as in § 1.14 (I) we notice that

$$A_0(l^2 - x^2)/2K \tag{2}$$

satisfies all the conditions of the problem; it is, in fact, the solution for the case of steady temperature. We now substitute

$$v = \frac{A_0(l^2 - x^2)}{2K} + w \tag{3}$$

in (1) so that w has to satisfy

$$\frac{\partial^2 w}{\partial x^2} - \frac{1}{\kappa}\frac{\partial w}{\partial t} = 0, \qquad -l < x < l, \tag{4}$$

with $w = 0$, when $x = \pm l$, $t > 0$,   (5)

and $w = -A_0(l^2 - x^2)/2K$, when $t = 0$.   (6)

The solution of (4), (5), and (6) has been given in 3.3 (16) and, using this in (3), we get finally

$$v = \frac{A_0 l^2}{2K}\left\{1 - \frac{x^2}{l^2} - \frac{32}{\pi^3}\sum_{n=0}^{\infty}\frac{(-1)^n}{(2n+1)^3}\cos\frac{(2n+1)\pi x}{2l}\,e^{-\kappa(2n+1)^2\pi^2 t/4l^2}\right\}, \tag{7}$$

or, using 3.13 (17), the solution useful for small values of the time is

---

† The case of constant surface temperature, e.g. combined surface and dielectric heating, is treated by Nelson, *Brit. J. Appl. Phys.* **3** (1952) 79–86.

[cf. § 12.5 V for the corresponding result for heat production at the rate $A_0 t^{\frac{1}{2}n}$, where n may be -1, 0 or any positive integer]

$$v = \frac{A_0 t}{\rho c}\left\{1 - 4\sum_{n=0}^{\infty}(-1)^n\left[\text{i}^2\text{erfc}\,\frac{(2n+1)l-x}{2(\kappa t)^{\frac{1}{2}}} + \text{i}^2\text{erfc}\,\frac{(2n+1)l+x}{2(\kappa t)^{\frac{1}{2}}}\right]\right\}.$$
(8)

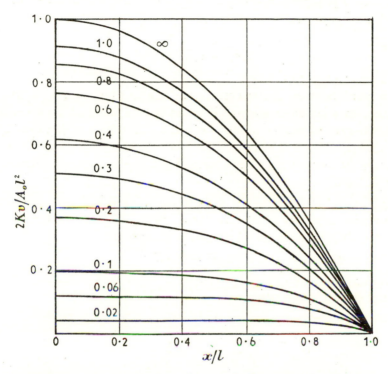

FIG. 20. Temperatures in the slab $-l < x < l$ with constant heat production at the rate A_0 per unit volume and zero surface temperature. The numbers on the curves are values of $\kappa t/l^2$.

Some values of v for various values of $\kappa t/l^2$ are shown in Fig. 20. It may be remarked that approximate results for the case of radiation at the surfaces into a medium at temperature zero may be obtained from Fig. 20 by the simple device of replacing l by $l+(K/H)$.

(ii) *If the rate of heat production, instead of being constant, is $A(t)$, it follows from* (7) and Duhamel's theorem, 1.14 (20), that

$$v = \frac{4}{\pi\rho c}\sum_{n=0}^{\infty}\frac{(-1)^n}{(2n+1)}\cos\frac{(2n+1)\pi x}{2l}\int_0^t A(\tau)e^{-\kappa(2n+1)^2\pi^2(t-\tau)/l^2}\,d\tau.$$
(9)

(iii) *If the rate of heat production is a function of position, $A(x)$,*

$$v = \frac{4l}{\pi^2 K} \sum_{n=1}^{\infty} \frac{1}{n^2} \{1 - e^{-\kappa n^2 \pi^2 t/4l^2}\} \cos \frac{n\pi x}{2l} \int_{-l}^{l} A(x') \cos \frac{n\pi x'}{2l} \, dx'. \tag{10}$$

An alternative form for the steady temperature is

$$v = \frac{1}{2Kl} \left\{ (x+l) \int_{x}^{l} d\eta \int_{-l}^{\eta} A(\xi) \, d\xi + (x-l) \int_{-l}^{x} d\eta \int_{-l}^{\eta} A(\xi) \, d\xi \right\}.$$

The case of heating by eddy currents may be treated in this way. In this case the rate of heat production is proportional to $(\cosh 2px - \cos 2px)$ where $p = 2\pi(\mu f \sigma)^{\frac{1}{2}}$, μ and σ are the permeability and electrical conductivity of the material, and f is the frequency.† For very high frequencies, the heat is almost all developed very close to the surface and the results of § 3.8 may be used.

(iv) *The region $-l < x < l$. Zero initial temperature. Radiation at $x = \pm l$ into a medium at zero temperature. The rate of heat production is A_0, constant, for $t > 0$.*

Here, in place of (3), the substitution is

$$v = \frac{A_0}{2Kh} (2l + hl^2 - hx^2) + w, \tag{11}$$

where the first term on the right-hand side is the solution for the case of steady temperature. Proceeding as in (i), but using 3.10 (8) gives

$$v = \frac{A_0}{2Kh} \left\{ 2l + hl^2 - hx^2 - 4h^2 \sum_{n=1}^{\infty} \frac{\cos \alpha_n x \, e^{-\kappa \alpha_n^2 t}}{\alpha_n^2 [l(\alpha_n^2 + h^2) + h] \cos l\alpha_n} \right\}, \tag{12}$$

where the α_n are the positive roots of $\alpha \tan \alpha l = h$. In the notation of 3.11 (4), (5), the temperatures at $x = 0$ and $x = l$ are $A_0 l^2 (L+2) f_4 [L, T]/(2KL)$ and $A_0 l^2 f_3 [L, T]/KL$, respectively.

(v) *The region $0 < x < l$. Zero initial temperature. $x = 0$ maintained at zero for $t > 0$. Radiation at $x = l$ into a medium at zero. The rate of heat production is A_0, constant, for $t > 0$.*

$$v = \frac{A_0}{2K} \left\{ \frac{(2+lh)lx}{1+lh} - x^2 \right\} + \frac{4A_0 h}{K} \sum_{n=1}^{\infty} \frac{\sin \alpha_n x (1 - \cos \alpha_n l)}{\alpha_n^2 \sin 2\alpha_n l [h(1+lh) + l\alpha_n^2]} e^{-\kappa \alpha_n^2 t}, \tag{13}$$

where the α_n are the positive roots of $\alpha \cot l\alpha + h = 0$. The temperature at $x = l$ can be expressed in terms of ϕ_3 and ϕ_4 of § 3.11.

(vi) *The slab $-l < x < l$ with zero initial and surface temperature. Heat production‡ at the rate $A_0 e^{-\lambda t}$ for $t > 0$.*

$$v = \frac{\kappa A_0}{\lambda K} \left\{ \frac{\cos x(\lambda/\kappa)^{\frac{1}{2}}}{\cos l(\lambda/\kappa)^{\frac{1}{2}}} - 1 \right\} e^{-\lambda t} + \frac{4\kappa A_0}{\pi \lambda K} \sum_{n=0}^{\infty} \frac{(-1)^n e^{-\kappa(2n+1)^2 \pi^2 t/4l^2} \cos(2n+1)\pi x/2l}{(2n+1)\{1 - [(2n+1)^2 \pi^2 \kappa/4\lambda l^2]\}}. \tag{14}$$

† Russell, *Alternating Currents* (Cambridge, 1914) Vol. I, p. 497.
‡ This result and the corresponding ones for radial flow, 7.9 (2) and 9.8 (14), are also of importance in connexion with the diffusion of radiogenic gases in geological materials, cf. Wasserburg in *Nuclear Geology*, ed. Faul (Wiley, 1954) § 9.5.

LINEAR FLOW OF HEAT IN THE ROD

4.1. Introductory

IN this chapter we consider problems on conduction of heat in rods of small cross-section. The rod is supposed to be so thin that the temperature at all points of the section may be taken to be the same.† The problem is thus one of linear flow in which the temperature is specified by the time and the distance x measured along the rod. If there is no loss of heat from the surface of the rod, the problems become identical with those discussed in Chapter III. The essential new feature of the problems of this chapter is that we assume each element of the surface of the rod to lose heat by radiation to a surrounding medium. Many of the older and some of the newer methods of determining thermal conductivity involve experimental arrangements of this type.

4.2. The differential equation satisfied by the temperature in a thin rod

We suppose the rod to have constant area of cross-section ω, perimeter p, conductivity K, density ρ, specific heat c, diffusivity κ, and surface conductance H.

Suppose the rod to lie along the axis of x, and consider the element of volume bounded by the sections at x and $x+dx$.

The rate at which heat flows into this element over the face at x is

$$-K\frac{\partial v}{\partial x}\omega.$$

Similarly, the rate at which heat flows across the face at $x+dx$ is

$$\left(-K\frac{\partial v}{\partial x}-K\frac{\partial^2 v}{\partial x^2}dx-\ldots\right)\omega.$$

Hence ultimately the rate at which the element gains heat by flow across these two faces is

$$\omega K\frac{\partial^2 v}{\partial x^2}dx.$$

---

† The general case in which the flow of heat in a body is predominantly in one direction, there being small losses in perpendicular directions, is discussed by Fox, *Phil. Mag.* (7) **18** (1934) 209–27. It appears that the average value of the temperature over planes perpendicular to the direction of flow satisfies an equation of type 4.2 (4) to a good approximation.

The rate at which heat is lost by radiation at the surface is

$$H(v-v_0)p \, dx,$$

where v_0 is the temperature of the medium into which the rod radiates. Also, the total rate of gain of heat in the element is ultimately

$$\omega c\rho \frac{\partial v}{\partial t} dx.$$

Thus we have
$$\frac{\partial v}{\partial t} = \frac{K}{c\rho} \frac{\partial^2 v}{\partial x^2} - \frac{Hp}{\rho c\omega}(v-v_0), \tag{1}$$

which becomes
$$\frac{\partial v}{\partial t} = \kappa \frac{\partial^2 v}{\partial x^2} - \nu(v-v_0) \tag{2}$$

on putting .
$$\frac{Hp}{c\rho\omega} = \nu \quad \text{and} \quad \frac{K}{\rho c} = \kappa. \tag{3}$$

When the surface of the rod is rendered impervious to heat, so that no radiation takes place, the equation for the temperature takes the form

$$\frac{\partial v}{\partial t} = \kappa \frac{\partial^2 v}{\partial x^2}$$

and the problems on the distribution of temperature in the rod are reduced to those of linear flow of Chapters II and III.

When radiation takes place into a medium at constant temperature, this may be taken as the zero of our scale, and the equation becomes

$$\frac{\partial v}{\partial t} = \kappa \frac{\partial^2 v}{\partial x^2} - \nu v. \tag{4}$$

This equation may be solved as in § 1.14 (III) by substituting†

$$v = ue^{-\nu t}, \tag{5}$$

which reduces it to
$$\frac{\partial u}{\partial t} = \kappa \frac{\partial^2 u}{\partial x^2}. \tag{6}$$

Thus the problem is reduced immediately to that of linear flow previously examined.

Ex. 1. *The semi-infinite rod* $x > 0$. *Zero initial temperature. The end* $x = 0$ *maintained at constant temperature* V_0 *for* $t > 0$.

Here the function u has to satisfy (6), to vanish for $t = 0$, and to have the value

$$V_0 e^{\nu t}, \quad x = 0, \quad t > 0.$$

The solution of this problem has been given in 2.5 (9) and is

$$u = \tfrac{1}{2}V_0 e^{\nu t}\left\{e^{-x\sqrt{(\nu/\kappa)}}\operatorname{erfc}\left[\frac{x}{2\sqrt{(\kappa t)}} - \sqrt{(\nu t)}\right] + e^{x\sqrt{(\nu/\kappa)}}\operatorname{erfc}\left[\frac{x}{2\sqrt{(\kappa t)}} + \sqrt{(\nu t)}\right]\right\}.$$

† Alternatively the Laplace transformation method of Chapter XII may be used.

Thus, finally,

$$v = \tfrac{1}{2}V_0\, e^{-x\sqrt{(\nu/\kappa)}}\mathrm{erfc}\left[\frac{x}{2\sqrt{(\kappa t)}} - \sqrt{(\nu t)}\right] + \tfrac{1}{2}V_0\, e^{x\sqrt{(\nu/\kappa)}}\mathrm{erfc}\left[\frac{x}{2\sqrt{(\kappa t)}} + \sqrt{(\nu t)}\right]. \qquad (7)$$

Ex. 2. *The finite rod* $-l < x < l$. *Zero initial temperature. The ends* $x = 0$ *and* $x = l$ *maintained at constant temperature* V_0 *for* $t > 0$.

Here the function u has to satisfy (6), to vanish for $t = 0$, and to take the values $V_0 e^{\nu t}$ at $x = \pm l$. Using the solution of this problem given in 3.5 (6) we obtain finally

$$v = V_0 \frac{\cosh x(\nu/\kappa)^{\frac{1}{2}}}{\cosh l(\nu/\kappa)^{\frac{1}{2}}} - \frac{4V_0}{\pi}\sum_{n=0}^{\infty}\frac{(-1)^n e^{-\nu t - [\kappa(2n+1)^2\pi^2 t/4l^2]}\cos(2n+1)\pi x/2l}{(2n+1)[1 + \{4\nu l^2/(2n+1)^2\pi^2\kappa\}]}. \qquad (8)$$

The first term is the steady state solution which will be obtained directly in § 4.5.

If the material of the rod is not homogeneous, or the thermal conductivity is a function of the temperature, it is clear that the temperature equation (1) must be replaced by

$$\frac{1}{\rho c}\frac{\partial}{\partial x}\left(K\frac{\partial v}{\partial x}\right) - \frac{Hp}{\rho c\omega}(v - v_0) = \frac{\partial v}{\partial t}. \qquad (9)$$

If the section of the rod is variable, ω and p will be functions of x, and we obtain in the same way, for a rod of constant conductivity K,

$$\frac{\kappa}{\omega}\frac{\partial}{\partial x}\left(\omega\frac{\partial v}{\partial x}\right) - \frac{H}{\rho c}\cdot\frac{p}{\omega}(v - v_0) = \frac{\partial v}{\partial t}. \qquad (10)$$

4.3. The semi-infinite rod. Steady temperatures. Forbes's method

For a semi-infinite rod with the end $x = 0$ maintained at constant temperature V_0, the solution of the equation of steady flow,

$$\kappa\frac{d^2v}{dx^2} - \nu v = 0,$$

is

$$v = V_0\, e^{-x(\nu/\kappa)^{\frac{1}{2}}}. \qquad (1)$$

This solution may be used for a bar of finite length l provided $l(\nu/\kappa)^{\frac{1}{2}}$ is large. If observations are made on bars of different metals with their surfaces treated so that they will all have the same values of ν, a comparison of temperatures† in different bars will give the ratios of their conductivities.‡ Such methods have the disadvantage that they only give relative conductivities, and also that it is difficult to ensure that different bars have the same values of ν.

† If α is the coefficient of linear expansion, the total expansion of the rod, $\alpha V_0(\kappa/\nu)^{\frac{1}{2}}$ provides another equation for comparison: cf. Todd, *J. Sci. Instr.* **4** (1927) 97.

‡ The methods of Ingenhausen, Despretz, and Wiedemann and Franz are based on this result, cf. Poynting and Thomson, *Text-book of Physics—Heat* (edn. 6) pp. 96 et seq.; Winkelmann, *Handbuch der Physik* (2. Aufl.) Bd. III, pp. 450 et seq.; Preston, *Theory of Heat* (edn. 3) §§ 296–9.

The classical experiments of Forbes[†] provide an interesting method for determining the conductivity of a single rod. Suppose one end of a semi-infinite rod is kept at constant temperature until the flow is steady. Then the rate of flow of heat along the bar at any point x, namely

$$-K\omega\frac{\partial v}{\partial x}, \tag{2}$$

where ω is the area of the rod, must be equal to the total rate of loss of heat from the region to the right of the point x. This is

$$p\int_x^\infty f(v)\,dx, \tag{3}$$

where p is the perimeter of the bar, and $f(v)$ is its rate of loss of heat from its surface at temperature v. Forbes made two separate sets of experiments: in the first of these he determined $\partial v/\partial x$ in (2) numerically from observations of the temperatures at a series of points along the bar; in the second he determined $pf(v)$ in (3) by heating the same bar to a constant temperature and allowing it to cool under similar conditions, its temperature being measured as a function of time, so that

$$pf(v) = -\omega\rho c\frac{\partial v}{\partial t}$$

could be found numerically as a function of v. This allowed the integral in (3) to be evaluated numerically, and equating (2) and (3) gave K. It may be observed that the method is quite general and does not assume that K or ν is independent of v, also it does not use any particular solution of the equation of conduction of heat. Forbes, in fact, measured the conductivity of iron as a function of temperature in this way.

4.4. The semi-infinite rod. Periodically varying temperature. Ångström's method[‡]

In the preceding article we have shown how the steady temperature of a long metal rod of small cross-section may be employed in obtaining the conductivity of the substance. The variable temperature has also been used, in the case in which one end of the bar is subjected to periodic variations of temperature which cause heat waves to travel down the bar. The conductivity is calculated from the march of these waves.

[†] *Trans. Roy. Soc. Edin.* **23** (1864) 133; Preston, loc. cit., § 301; Poynting and Thomson, loc. cit., p. 98; Winkelmann, loc. cit., p. 454. For the present status of Forbes's method see Griffiths, *Proc. Phys. Soc.* **41** (1928) 151.

[‡] *Ann. Physik, Lpz.* **114** (1861) 513; **123** (1864), 628; *Phil. Mag.* **25** (1863) 130; **26** (1863) 161; Preston, loc. cit., § 307; Bosanquet and Aris, *Brit. J. Appl. Phys.* **5** (1954) 252–5.

Ångström was the first to employ this method, and his work is of exceptional interest both from the neatness of the mathematical discussion and the novelty of his experimental method. Hagström† later discussed the same problem assuming that the conductivity and emissivity vary with the temperature; Neumann and Weber extended the method to the case of a short bar, both ends of which undergo periodic changes of temperature (see § 4.8).

Ångström employed long bars of small cross-section. The end $x = 0$ was subjected to periodic changes of temperature, being alternately heated by a current of steam and cooled by a current of cold water for equal intervals. When this has gone on for some time, the temperature in the bar will ultimately settle down to a periodic state, independent of the initial distribution. It is this periodic state which Ångström investigates and upon which his results depend. The bar is allowed to radiate‡ into a medium at a constant temperature, taken as the zero of the experiment. As before, it is supposed of such small cross-section that the temperature over the section may be taken as that at the centre, and of such length that the temperature at the farther end remains unaffected by the alterations at $x = 0$, so that in the mathematical treatment it is supposed unlimited in this direction.

We require a periodic solution of 4.2 (4) of the form

$$v = V(x)e^{in\omega t}.$$

Substituting this in 4.2 (4), V has to satisfy

$$\frac{d^2V}{dx^2} - \frac{v+in\omega}{\kappa}V = 0.$$

The solution of this which tends to zero as $x \to \infty$ is

$$V = V_0\exp(-q_n x - iq'_n x), \tag{1}$$

where

$$q_n + iq'_n = \left(\frac{v+in\omega}{\kappa}\right)^{\frac{1}{2}}, \tag{2}$$

so that

$$q_n = \{[v+(v^2+n^2\omega^2)^{\frac{1}{2}}]/2\kappa\}^{\frac{1}{2}}, \qquad q'_n = \{[-v+(v^2+n^2\omega^2)^{\frac{1}{2}}]/2\kappa\}^{\frac{1}{2}},$$

$$q_n q'_n = n\omega/2\kappa. \tag{3}$$

Thus

$$v = \sum_{n=0}^{\infty} A_n e^{-q_n x} \cos(n\omega t - q'_n x + \epsilon_n) \tag{4}$$

is a general periodic solution of (1) with period $2\pi/\omega$.

† Hagström, *Öfvers. Vetensk. Akad. Förh., Stockh.* **48** (1891) 45, 289, 381.
‡ If a guard ring is used, this correction can be eliminated.

In Ångström's experiments the period $2\pi/\omega$ was 24 minutes, and after a time long enough for steady conditions to be attained the temperatures at two points x_1 and x_2 were measured and analysed harmonically into the forms

$$\sum_{n=0}^{\infty} B_n \cos(n\omega t + \beta_n) \quad \text{and} \quad \sum_{n=0}^{\infty} C_n \cos(n\omega t + \gamma_n), \text{ respectively.}$$

Comparing these with (4) gives

$$B_1/C_1 = e^{q_1 l}, \qquad \beta_1 - \gamma_1 = l q_1', \tag{5}$$

where $l = x_2 - x_1$. Using these results in the last of (3) gives

$$\kappa = \frac{\omega l^2}{2(\beta_1 - \gamma_1)(\ln B_1 - \ln C_1)}. \tag{6}$$

The conductivity is thus determined independently of the surface conductance. By altering the nature of the surface of the bar so that ν changes, the values obtained for κ should not vary. Ångström made such changes, and his results confirmed the values given by his earlier experiments.

When κ is known, ν can be found at once.

If the temperature at $x = 0$ is 1 when $rT < t < (r+\tfrac{1}{2})T$ and zero when $(r+\tfrac{1}{2})T < t < (r+1)T$, $r = 0, 1, 2,...,$ the steady state temperature at the point x at time t is

$$\tfrac{1}{2}e^{-q_0 x} + \frac{2}{\pi}\sum_{n=0}^{\infty} \frac{1}{2n+1} e^{-x q_{2n+1}} \sin\{(2n+1)\omega t - x q'_{2n+1}\},$$

where q_n and q_n' are defined above and $\omega = 2\pi/T$.

If the temperature at $x = 0$ is

$$A_0 + A_1 \cos(\omega t + \epsilon_1),$$

only the first two terms of (4) are present and the temperature oscillations are propagated with velocity

$$v = 2\pi/T q_1'.$$

Using the value (3) of q_1' this gives

$$\nu = \frac{v^2}{4\kappa} - \frac{4\pi^2 \kappa}{T^2 v^2}.$$

If v_1 and v_2 are the velocities for periods T_1 and T_2, ν can be eliminated and we have

$$\kappa^2 = \frac{T_1^2 T_2^2 v_1^2 v_2^2 (v_1^2 - v_2^2)}{16\pi^2 (T_2^2 v_2^2 - T_1^2 v_1^2)}.$$

King[†] has used this arrangement, the end $x = 0$ of the wire being heated by a coil which carries current proportional to $\sin \frac{1}{2}\omega t$.

4.5. The finite rod. Ends at fixed temperatures. Steady temperature

When the surface is not impervious to heat and the temperature of the medium is taken as zero, the equation for the temperature is

$$\frac{\partial v}{\partial t} = \frac{K}{c\rho}\frac{\partial^2 v}{\partial x^2} - \frac{Hp}{c\rho\omega}v$$

with the notation of § 4.2 (3).

The observation of steady temperature in such a bar, when its ends are kept at constant temperatures V_1 and V_2, is one of the earliest methods of obtaining the relative values of the conductivities of different solids.

If we put $\mu^2 = Hp/K\omega$, we have the equations

$$\frac{d^2v}{dx^2} - \mu^2 v = 0 \quad (0 < x < l),$$

$$v = V_1, \quad \text{when } x = 0,$$

$$v = V_2, \quad \text{when } x = l,$$

and our solution is given by

$$\left. \begin{array}{l} v = Ae^{\mu x} + Be^{-\mu x} \\ V_1 = A + B \\ V_2 = Ae^{\mu l} + Be^{-\mu l} \end{array} \right\}.$$

where

and

Thus
$$v = \frac{V_1 \sinh \mu(l-x) + V_2 \sinh \mu x}{\sinh \mu l}. \tag{1}$$

Let the temperatures be v_1, v_2, and v_3 at the points x_1, x_2, and x_3, where

$$x_3 - x_2 = x_2 - x_1 = a.$$

Then
$$\frac{v_1 + v_3}{v_2} = 2\cosh \mu a = 2n, \quad \text{say.}$$

Hence
$$e^{\mu a} = n + \sqrt{(n^2 - 1)},$$

a result independent of V_1 and V_2. For two bars of the same perimeter, cross-section, and surface conductance, it follows that

$$\sqrt{\left(\frac{K_1}{K_2}\right)} = \frac{\mu_2}{\mu_1} = \frac{\log\{n_2 + \sqrt{(n_2^2 - 1)}\}}{\log\{n_1 + \sqrt{(n_1^2 - 1)}\}}. \tag{2}$$

This provides a method for comparing the thermal conductivities of two substances.[‡]

† *Phys. Rev.* (2) **6** (1915) 437.
‡ Jakob and Erk, *Z. Phys.* **35** (1926), 670; Preston, *Theory of Heat* (edn. 3) §§ 296–9.

In absolute measurements it is usual to supply heat to one end of the rod[†] at a known rate by means of a heating coil.

It follows from (1) that the rate of flow of heat F_0 into the rod at $x = 0$ is

$$F_0 = -K\omega\left[\frac{dv}{dx}\right]_{x=0} = \frac{K\omega(\mu V_1 \cosh \mu l - \mu V_2)}{\sinh \mu l}. \tag{3}$$

Also the rate of flow of heat F_l from the rod at $x = l$ is

$$F_l = -K\omega\left[\frac{dv}{dx}\right]_{x=l} = \frac{K\omega(\mu V_1 - \mu V_2 \cosh \mu l)}{\sinh \mu l}. \tag{4}$$

The lateral loss of heat from the rod between $x = 0$ and $x = l$ is thus

$$F_0 - F_l = K\omega\mu(V_1 + V_2)\tanh \tfrac{1}{2}\mu l. \tag{5}$$

In practice μ is usually small, so that the hyperbolic functions in (3), (4), (5) may be replaced by the first terms of their series expansions. This gives approximately

$$F_0 = \frac{K\omega}{l}(V_1 - V_2) + \tfrac{1}{6}K\omega\mu^2 l(2V_1 + V_2), \tag{6}$$

$$F_0 - F_l = \tfrac{1}{2}K\omega\mu^2 l(V_1 + V_2). \tag{7}$$

Two equations of type (6), obtained by changing F_0 or V_2 in the experiment, will give K and H. A generalization of (6) which allows for linear variation of thermal conductivity with temperature is given by Kannuluik and Laby, loc. cit. Equation (7) is useful for making corrections for loss of heat from the rod, for example between a heating coil and a thermocouple. Lees, loc. cit., gives a number of calculations of the corrections needed in apparatus of this type.

If there is no flow of heat from the end $x = l$ of the rod, we have in place of (1)

$$v = V_1 \frac{\cosh \mu(l-x)}{\cosh \mu l}.$$

The rate of flow of heat into the rod at $x = 0$ is

$$F_0 = -K\omega\left[\frac{dv}{dx}\right]_{x=0} = \mu K\omega V_1 \tanh \mu l.$$

Thus $$K = F_0^2/H p\omega V_1^2 \tanh^2\mu l,$$

this relation has been used[‡] for the determination of K.

† Lees, *Phil. Trans. Roy. Soc.* A, **208** (1908) 381; Kannuluik and Laby, *Proc. Roy. Soc.* A, **121** (1928) 640; Bridgman, *Proc. Amer. Acad. Arts Sci.* 57 (1922) 80; Barratt and Winter, *Phil. Mag.* (6) **49** (1925) 313.

‡ Barratt, *Proc. Phys. Soc.* **26** (1914) 347.

4.6. The rod of variable section. Cooling fins. Steady temperature

The differential equations of § 4.2 have applications in engineering to the theory of thin fins† attached to surfaces to assist their cooling by radiation or by forced convection. In all cases in this section the fins are supposed to be so thin that the temperature may be taken to be constant over their thickness; corresponding problems for thick fins are discussed in §§ 5.3, 8.3. Here the steady temperature only will be considered. Problems on variable temperature may either be solved directly by the Laplace transformation method of Chapter XII, or reduced by the substitution 4.2 (5).

We first consider from this point of view the problem already treated in § 4.5, and then proceed to other cases.

I. *A rectangular fin on a flat surface*

Taking the flat surface to be the plane $x = 0$, the fin consists of the region

$$-\tfrac{1}{2}D < z < \tfrac{1}{2}D, \quad -\infty < y < \infty, \quad 0 < x < l,$$

where D is the (small) thickness of the fin. We suppose the plane $x = 0$ to be at temperature V and there to be radiation at the surface of the fin into medium at zero temperature. Then, in the notation of § 4.2, unit length of the fin in the y-direction has the perimeter $p = 2$ and area $\omega = D$. Thus 4.2 (2) becomes

$$\frac{d^2v}{dx^2} - \mu^2 v = 0, \quad 0 < x < l, \tag{1}$$

where

$$\mu^2 = 2H/KD. \tag{2}$$

This has to be solved with

$$v = V, \quad x = 0, \tag{3}$$

and

$$\frac{dv}{dx} = 0, \quad x = l. \tag{4}$$

The simple boundary condition (4) implies that radiation from the end of the fin is negligible, and this is usually sufficiently accurate; if

---

† McAdams, *Heat Transmission* (McGraw-Hill, edn. 2) p. 232; Jakob, *Heat Transfer* (Wiley, 1949) p. 217; Harper and Brown, *Nat. Adv. Comm. Aeronaut. Tech. Rep.* No. 158 (1922); Binnie, *Phil. Mag.* (7) **2** (1926) 449; Schmidt, *Z. Ver. dtsch. Ing.* **70** (1926) 885, 947; Bueche and Schau, *Arch. Wärmew.* **17** (1936) 67; Gardner, *Trans. Amer. Soc. Mech. Engrs.* **67** (1945) 621; Jakob, ibid. **67** (1945) 629; Miles, *J. Appl. Phys.* **23** (1952) 372.

not,† (4) must be replaced by

$$K\frac{dv}{dx}+Hv = 0, \quad x = l, \tag{5}$$

which makes the formulae more clumsy and introduces no new features.

The solution of (1), (3), and (4) is

$$v = V\frac{\cosh\mu(l-x)}{\cosh\mu l}. \tag{6}$$

The most interesting quantity from the engineering point of view is the 'effectiveness' of the fin. This is defined as the ratio of the heat lost per unit time from the surface of the fin, to the heat which would be lost per unit time if the fin were removed and the area of its base radiated in the same way. This latter quantity is HDV, per unit length, and the effectiveness of the fin is

$$\frac{2}{HDV}\int_0^l \frac{HV\cosh\mu(l-x)\,dx}{\cosh\mu l} = \frac{2}{\mu D}\tanh\mu l. \tag{7}$$

II. *A tapering fin on a flat surface*

Suppose the sides of the fin taper at a small angle α, so that if the area of the fin at $x = 0$ is D, per unit length in the y-direction, it will be $D-2\alpha x$ at the point x. The perimeter, per unit length in the y-direction, will be $p = 2$, neglecting terms in α^2. Then 4.2 (10) becomes

$$\frac{d}{dx}\Big[(D-2\alpha x)\frac{dv}{dx}\Big]-\frac{2H}{K}v = 0, \quad 0 < x < l, \tag{8}$$

to be solved with the boundary conditions (3) and (4). Writing

$$\xi = (D-2\alpha x)^{\frac{1}{2}}, \tag{9}$$

equation (8) becomes

$$\frac{d^2v}{d\xi^2}+\frac{1}{\xi}\frac{dv}{d\xi}-\frac{2H}{K\alpha^2}v = 0, \quad D^{\frac{1}{2}} < \xi < (D-2\alpha l)^{\frac{1}{2}}. \tag{10}$$

The general solution of (10) is the linear combination of modified Bessel functions‡

$$AI_0[\xi(2H/K\alpha^2)^{\frac{1}{2}}]+BK_0[\xi(2H/K\alpha^2)^{\frac{1}{2}}]. \tag{11}$$

The constants A and B are found from the boundary conditions (3) and (4), which give

$$\left.\begin{aligned}AI_0(\xi_0)+BK_0(\xi_0) &= V\\ AI_1(\xi_1)-BK_1(\xi_1) &= 0\end{aligned}\right\}, \tag{12}$$

where for shortness we have written

$$\xi_0 = (2HD/K\alpha^2)^{\frac{1}{2}}, \qquad \xi_1 = [2H(D-2\alpha l)/K\alpha^2]^{\frac{1}{2}}. \tag{13}$$

Solving (12), the solution (11) is finally

$$v = V\frac{I_0[2H(D-2\alpha x)/K\alpha^2]^{\frac{1}{2}}K_1(\xi_1)+K_0[2H(D-2\alpha x)/K\alpha^2]^{\frac{1}{2}}I_1(\xi_1)}{I_0(\xi_0)K_1(\xi_1)+K_0(\xi_0)I_1(\xi_1)}. \tag{14}$$

† Cf. Jakob, *Phil. Mag.* (7) **28** (1939) 571.

‡ See Appendix III. The formulae $I_0'(z) = I_1(z)$, $K_0'(z) = -K_1(z)$ are used below.

III. *A thin fin round a cylindrical barrel of radius a. Constant thickness*

Here the plane of the fin is perpendicular to the axis of the cylinder, its thickness D parallel to the axis is small, its outer radius is b, and the flow of heat from the outer surface of radius b is neglected. Since the flow is purely radial we may use the differential equation 4.2 (10) taking for the area at radius r, $\omega = 2\pi r D$, and for the perimeter at radius r, $p = 4\pi r$.

Then 4.2 (10) gives

$$\frac{1}{r}\frac{d}{dr}\left(r\frac{dv}{dr}\right) - \frac{2H}{KD}v = 0, \quad a < r < b, \tag{15}$$

to be solved with

$$v = V, \quad r = a, \tag{16}$$

$$\frac{dv}{dr} = 0, \quad r = b. \tag{17}$$

The general solution of (15) is

$$v = AI_0(\mu r) + BK_0(\mu r), \tag{18}$$

where

$$\mu = (2H/KD)^{\frac{1}{2}}. \tag{19}$$

A and B are to be found from (16) and (17), and evaluating these we have finally

$$v = V\frac{I_0(\mu r)K_1(\mu b) + K_0(\mu r)I_1(\mu b)}{I_0(\mu a)K_1(\mu b) + K_0(\mu a)I_1(\mu b)}. \tag{20}$$

The effectiveness of the fin in this case is

$$\frac{2}{\mu D}\frac{I_1(\mu b)K_1(\mu a) - K_1(\mu b)I_1(\mu a)}{I_0(\mu a)K_1(\mu b) + K_0(\mu a)I_1(\mu b)}. \tag{21}$$

IV. *A thin fin round a cylinder of radius a. Variable thickness*

In this case a linearly tapering fin, as in II, does not lead to a solution in terms of tabulated functions. Two other laws of variation of thickness with radius do lead to simple solutions.

If the thickness, z, of the fin varies inversely as the radius, so that

$$z = D/r, \tag{22}$$

we have $\omega = 2\pi r z = 2\pi D$, and $p = 4\pi r$, so 4.2 (10) becomes

$$\frac{d^2v}{dr^2} - \mu^2 r v = 0, \quad a < r < b, \tag{23}$$

where μ is defined in (19). The general solution of (23) is

$$v = Ar^{\frac{1}{2}}I_{\frac{1}{3}}(\tfrac{2}{3}\mu r^{3/2}) + Br^{\frac{1}{2}}K_{\frac{1}{3}}(\tfrac{2}{3}\mu r^{3/2}), \tag{24}$$

and A and B can be found as before from the boundary conditions (16) and (17).

If the thickness, z, of the fin varies inversely as the square of the radius, so that

$$z = D/r^2, \tag{25}$$

ve have $\omega = 2\pi D/r$ and $p = 4\pi r$ so that 4.2 (10) becomes

$$\frac{1}{r}\frac{d}{dr}\left[\frac{1}{r}\frac{dv}{dr}\right] - \mu^2 v = 0, \quad a < r < b, \tag{26}$$

where μ is defined in (19). The solution of (26) which satisfies (16) and (17) is

$$v = V\frac{\cosh\frac{1}{2}\mu(b^2 - r^2)}{\cosh\frac{1}{2}\mu(b^2 - a^2)}. \tag{27}$$

4.7. The finite† rod with radiation at its surface. Variable temperature

For a rod of uniform cross-section the differential equation 4.2 (2) is

$$\frac{\partial v}{\partial t} = \kappa \frac{\partial^2 v}{\partial x^2} - \nu v, \tag{1}$$

where $\kappa = K/\rho c$, $\nu = Hp/\rho c\omega$, and the temperature v_0 of the medium into which the rod radiates is taken to be zero. As in 4.2 (5) this is solved by the substitution $v = ue^{-\nu t}$ so that u is to satisfy

$$\frac{\partial u}{\partial t} = \kappa \frac{\partial^2 u}{\partial x^2}. \tag{2}$$

The solution of a number of problems of interest can now be written down using results of Chapter III.

I. *The finite rod $0 < x < l$. Initial temperature $f(x)$. Ends at temperatures* $\phi_1(t)$ *and* $\phi_2(t)$

Here (2) has to be solved with

$$u = f(x), \; t = 0; \quad u = e^{\nu t}\phi_1(t), \; x = 0; \quad u = e^{\nu t}\phi_2(t), \; x = l. \tag{3}$$

Using 3.5 (2) we find

$$v = \frac{2}{l}\sum_{n=1}^{\infty} e^{-\nu t - (\kappa n^2 \pi^2 t / l^2)} \sin\frac{n\pi x}{l}\left\{\int_0^l f(x')\sin\frac{n\pi x'}{l}\,dx' + \right.$$

$$\left. +\frac{n\pi\kappa}{l}\int_0^t e^{(\kappa n^2 \pi^2 \lambda / l^2) + \nu\lambda}[\phi_1(\lambda) - (-1)^n \phi_2(\lambda)]\,d\lambda\right\}. \tag{4}$$

II. *The finite rod $0 < x < l$. Initial temperature $f(x)$. No flow of heat at $x = 0$.* $x = l$ *at temperature* $\phi_2(t)$

Using 3.5 (3) we find, writing $\beta_n = (2n+1)\pi/2l$,

$$v = \frac{2}{l}\sum_{n=0}^{\infty} e^{-\nu t - \kappa\beta_n^2 t}\cos\beta_n x\left\{\kappa\beta_n(-1)^n\int_0^t e^{(\kappa\beta_n^2 + \nu)\lambda}\phi_2(\lambda)\,d\lambda + \int_0^l f(x')\cos\beta_n x'\,dx'\right\}. \tag{5}$$

III. *The finite rod $0 < x < l$. Initial temperature $f(x)$. No flow of heat at the ends*

It follows from 3.4 (6) that

$$v = \frac{1}{l}e^{-\nu t}\int_0^l f(x')\,dx' + \frac{2}{l}\sum_{n=1}^{\infty} e^{-\nu t - (\kappa n^2 \pi^2 t / l^2)}\cos\frac{n\pi x}{l}\int_0^l f(x')\cos\frac{n\pi x'}{l}\,dx'. \tag{6}$$

IV. *The finite rod $-l < x < l$. Initial temperature V, constant. Radiation from the surface and ends into medium at zero*

It follows from 3.11 (12) that

$$\frac{v}{V} = \sum_{n=1}^{\infty} \frac{2h\cos\alpha_n x}{[(h^2 + \alpha_n^2)l + h]\cos\alpha_n l}e^{-(\kappa\alpha_n^2 + \nu)t}, \tag{7}$$

† The semi-infinite rod is considered by Lowan, *Quart. Appl. Math.* **4** (1946) 84–87.

where $\pm\alpha_n$, $n = 1, 2,...$, are the roots of

$$\alpha \tan \alpha l = h.$$

V. *The finite rod $0 < x < l$. Initial temperature $f(x)$. Radiation from the surface and ends into medium at zero*

Using 3.9 (12)

$$v = \sum_{n=1}^{\infty} e^{-(\nu+\kappa\alpha_n^2)t} A_n X_n, \tag{8}$$

where

$$X_n = \cos\alpha_n x + (h/\alpha_n)\sin\alpha_n x, \tag{9}$$

$$A_n = \frac{2\alpha_n^2}{(\alpha_n^2+h^2)l+2h} \int_0^l f(x)X_n \, dx, \tag{10}$$

and $\pm\alpha_n$, $n = 1, 2,...$, are the roots of

$$\tan \alpha l = \frac{2h\alpha}{\alpha^2 - h^2}. \tag{11}$$

Neumann† showed that this result may be used in determining the values of the thermal constants. His method requires the measurement of the temperatures v_0 and v_l at $x = 0$ and $x = l$.

It follows from (9) that $X_n = 1$ when $x = 0$. Also it follows from § 3.9 that $X_n^2 = 1$ when $x = l$; we proceed to determine the sign of X_n here.

Using (9) and (11) we have when $x = l$

$$X_n = \frac{\alpha_n^2+h^2}{2\alpha_n h} \sin\alpha_n l.$$

But

$$0 < \alpha_1 l < \pi < \alpha_2 l < 2\pi < \dots .$$

Therefore

$$X_n = (-1)^{n-1}, \quad \text{when } x = l.$$

Thus

$$\tfrac{1}{2}(v_0+v_l) = A_1 e^{-\beta_1 t} + A_3 e^{-\beta_3 t} + ..., \tag{12}$$

$$\tfrac{1}{2}(v_0-v_l) = A_2 e^{-\beta_2 t} + A_4 e^{-\beta_4 t} + ..., \tag{13}$$

where

$$\beta_n = \kappa\alpha_n^2 + \nu.$$

Since β_n increases with n, if t is chosen large enough we shall obtain a close approximation by using only the first term in each of these series. On this understanding

$$\tfrac{1}{2}(v_0+v_l) = A_1 e^{-\beta_1 t} \tag{14}$$

and

$$\tfrac{1}{2}(v_0-v_l) = A_2 e^{-\beta_2 t}. \tag{15}$$

In Neumann's experiment he first heated one end of the bar by a flame, and then allowed the bar to cool by radiation. After some time he began to take observations of $v_0 \pm v_l$ at equal intervals. These readings showed when the temperatures began to obey the law given above. By this means the constants β_1 and β_2 are found; and thus two equations are obtained from which the conductivity and emissivity may be obtained. However, as the values of α_1 and α_2 involve h, this calculation has to proceed by successive approximations and is somewhat complicated.

A simpler method is obtained by observing also the temperatures at the middle point of the bar. It follows as in § 3.10 that $\alpha_1, \alpha_3,...$ are the roots of

$$\alpha \tan \tfrac{1}{2}\alpha l = h \tag{16}$$

† *Ann. Chim. Phys.* (3) **66** (1862) 183. See also Glage, *Ann. Physik,* (4) **18** (1905) 904–40, who discusses the relationship between Ångström's method and the two methods of Neumann. It should be remarked that the methods of this section are completely independent of the initial distribution of temperature in the rod.

and that α_2, α_4,... are the roots of

$$\alpha \cot \tfrac{1}{2}\alpha l + h = 0. \qquad (17)$$

Also that, when $x = \tfrac{1}{2}l$,

$$X_n = \sec \tfrac{1}{2}l\alpha_n, \quad \text{when } n \text{ is odd},$$
$$= 0, \qquad \text{when } n \text{ is even}.$$

Thus
$$v_{\frac{1}{2}l} = A_1 e^{-\beta_1 t} \sec \tfrac{1}{2}\alpha_1 l + A_3 e^{-\beta_3 t} \sec \tfrac{1}{2}\alpha_3 l + \ldots . \qquad (18)$$

Thus, for large values of the time it follows from (14) and (18) that

$$\frac{v_0 + v_l}{v_{\frac{1}{2}l}} = 2 \cos \tfrac{1}{2}\alpha_1 l.$$

From this result we find α_1; then h follows from (16), and α_2 from (17).

Also
$$\beta_1 = \kappa\alpha_1^2 + \nu \quad \text{and} \quad \beta_2 = \kappa\alpha_2^2 + \nu$$

give κ and ν; so that the values of K and H may then be found.

4.8. The finite rod. The ends at periodically varying temperatures. Neumann's method

In his paper, 'Über das Wärmeleitungsvermögen von Eisen und Neusilber',† Weber describes a series of experiments which he conducted on a method suggested by Neumann in his lectures. The idea of this method is the same as that of Ångström, § 4.4, but in this case both ends of the rod are subjected to periodical changes of temperature. The end A of the rod AB is kept at temperature v_1, and B at v_2 for the interval $0 < t < T$. Then A is kept at v_2, and B at v_1 from $t = T$ to $t = 2T$; and this is repeated indefinitely. When this series of surface temperatures has gone on for a sufficient time, the influence of the initial temperature distribution will have disappeared and the resulting steady periodic oscillation may be studied by the method of § 3.6.

The temperature is given by 4.7 (4) with $f(x) = 0$, and

$$\phi_1(t) = v_1, \quad \text{when } 2rT < t < (2r+1)T,$$
$$\phi_1(t) = v_2, \quad \text{when } (2r+1)T < t < (2r+2)T,$$
$$\phi_2(t) = v_2, \quad \text{when } 2rT < t < (2r+1)T,$$
$$\phi_2(t) = v_1, \quad \text{when } (2r+1)T < t < (2r+2)T,$$

r being zero or any positive integer.

We call the intervals $2rT < t < (2r+1)T$ *even periods*, and the intervals $(2r+1)T < t < (2r+2)T$ *odd periods*. Then, as in § 3.6, we find that after the surface temperature oscillation has gone on for a long time, the temperature at time t after the commencement of one of the even periods is

$$v = \frac{4\kappa\pi}{l^2}(v_1 - v_2)\sum_{n=1}^{\infty}\frac{n}{p_{2n}}\sin\frac{2n\pi x}{l} + \frac{2\kappa\pi}{l^2}(v_1 + v_2)\sum_{n=0}^{\infty}\frac{2n+1}{p_{2n+1}}\sin\frac{(2n+1)\pi x}{l} -$$

$$-\frac{8\kappa\pi}{l^2}(v_1 - v_2)\sum_{n=1}^{\infty}\frac{n}{p_{2n}}\sin\frac{2n\pi x}{l}\frac{e^{-p_{2n}t}}{1+e^{-p_{2n}T}}, \quad (1)$$

† *Ann. Physik*, **146** (1872) 257.

and at time t from the beginning of one of the odd periods

$$v = -\frac{4\kappa\pi}{l^2}(v_1 - v_2)\sum_{n=1}^{\infty}\frac{n}{p_{2n}}\sin\frac{2n\pi x}{l} + \frac{2\kappa\pi}{l^2}(v_1 + v_2)\sum_{n=0}^{\infty}\frac{2n+1}{p_{2n+1}}\sin\frac{(2n+1)\pi x}{l} +$$

$$+ \frac{8\kappa\pi}{l^2}(v_1 - v_2)\sum_{n=1}^{\infty}\frac{n}{p_{2n}}\sin\frac{2n\pi x}{l}\frac{e^{-p_{2n}t}}{1+e^{-p_{2n}T}}, \quad (2)$$

where

$$p_n = \frac{\kappa n^2 \pi^2}{l^2} + \nu. \quad (3)$$

These results may be simplified further by using the Fourier series

$$U = \frac{\sinh\mu(l-x) - \sinh\mu x}{2\sinh\mu l} = \frac{4\kappa\pi}{l^2}\sum_{n=1}^{\infty}\frac{n}{p_{2n}}\sin\frac{2n\pi x}{l}, \quad (4)$$

$$V = \frac{\sinh\mu(l-x) + \sinh\mu x}{2\sinh\mu l} = \frac{2\kappa\pi}{l^2}\sum_{n=0}^{\infty}\frac{(2n+1)}{p_{2n+1}}\sin\frac{(2n+1)\pi x}{l}, \quad (5)$$

where p_n is defined in (3) and $\mu^2 = \nu/\kappa$.

Therefore for the even period we have

$$v = (v_1 - v_2)U + (v_1 + v_2)V - \frac{8\pi\kappa}{l^2}(v_1 - v_2)\sum_{n=1}^{\infty}\frac{n}{p_{2n}}\sin\frac{2n\pi x}{l}\frac{e^{-p_{2n}t}}{1+e^{-p_{2n}T}}, \quad (6)$$

and for the odd period

$$v = -(v_1 - v_2)U + (v_1 + v_2)V + \frac{8\pi\kappa}{l^2}(v_1 - v_2)\sum_{n=1}^{\infty}\frac{n}{p_{2n}}\sin\frac{2n\pi x}{l}\frac{e^{-p_{2n}t}}{1+e^{-p_{2n}T}}. \quad (7)$$

It follows from (6) and (7) that $v_{\frac{1}{2}l}$, the temperature at the middle point of the bar, has the constant value

$$v_{\frac{1}{2}l} = \frac{v_1 + v_2}{2\cosh\frac{1}{2}\mu l}. \quad (8)$$

This gives the ratio ν/κ. A simple way of obtaining another relation between the two unknown quantities is to take the difference of temperature at any instant between the points $x = \frac{1}{6}l$ and $x = \frac{2}{3}l$. For these points the terms in which n is a multiple of 2 or 3 disappear from the series in the expression for the difference of the temperatures, and this series is so rapidly convergent that we may neglect the term for $n = 5$ and those which follow. Thus, with this approximation, this temperature difference at time t after the beginning of one of the periods is

$$d = M - Ne^{-pt}, \quad (9)$$

where $p = \nu + (4\kappa\pi^2/l^2)$, and M and N do not involve t.

Let d_1, d_2, d_1', d_2' be the values of d for times $t_1, t_2, t_1 + \beta, t_2 + \beta$, then

$$\frac{d_1 - d_2}{d_1' - d_2'} = e^{\nu\beta}$$

and hence p can be found. Using this and the value of ν/κ derived from (8) the unknown ν and κ follow.

4.9. Problems on conduction of heat in a moving rod†

Suppose that the rod is of area ω and perimeter p, both assumed to be constants, and moves with velocity U in the direction of the x-axis. The rod is assumed to lose heat by radiation into a medium at temperature v_0, the surface conductance being H.

The differential equation may be obtained as in § 4.2, except that a term corresponding to convection of heat is to be added as in § 1.7 leading to an additional term $-(U/\kappa)\,\partial v/\partial x$ as in 1.7 (2), so that 4.2 (2) is replaced by

$$\frac{\partial^2 v}{\partial x^2} - \frac{U}{\kappa}\frac{\partial v}{\partial x} - \frac{\nu(v-v_0)}{\kappa} - \frac{1}{\kappa}\frac{\partial v}{\partial t} = 0, \tag{1}$$

where

$$\kappa = K/\rho c, \qquad \nu = Hp/\rho c\omega. \tag{2}$$

A number of transient problems involving (1) will be discussed in § 15.2. Here we discuss the case of steady temperature and radiation into a medium at zero, so that (1) becomes

$$\kappa\frac{d^2 v}{dx^2} - U\frac{dv}{dx} - \nu v = 0. \tag{3}$$

(i) *The semi-infinite rod moving with velocity U in the direction of its length. $x = 0$ maintained at constant temperature v_1*

The solution (valid for both signs of U) is

$$v = v_1 \exp\left\{\frac{U - (U^2 + 4\kappa\nu)^{\frac{1}{2}}}{2\kappa}\,x\right\}. \tag{4}$$

(ii) *The semi-infinite rod moving with velocity U in the direction of its length. $x = 0$ and $x = 2l$ maintained at constant temperatures v_1 and v_2 respectively*

The solution is

$$v = \frac{v_2\,e^{-U(2l-x)/2\kappa}\sinh \xi x + v_1\,e^{Ux/2\kappa}\sinh \xi(2l-x)}{\sinh 2\xi l}, \tag{5}$$

where

$$\xi = [(U^2/4\kappa^2) + (Hp/\omega K)]^{\frac{1}{2}}. \tag{6}$$

A number of experiments‡ has been made on mercury in which the liquid has moved with constant velocity along a tube connecting two reservoirs which are maintained at different temperatures. If v_l is the temperature at the mid-point $x = l$ of the tube, it follows from (5) that

$$2v_l \cosh \xi l = v_2\,e^{-lU/2\kappa} + v_1\,e^{lU/2\kappa}. \tag{7}$$

If v_1 and v_2 are adjusted so that $v_l = 0$, this reduces to

$$v_2/v_1 = -e^{-lU/\kappa}. \tag{8}$$

This equation does not involve ξ, and the diffusivity κ can be found immediately from it.

---

† Conduction of heat in a moving cylinder of finite diameter is discussed in § 7.11.

‡ Nettleton, *Proc. Phys. Soc.* **22** (1910) 278; ibid. **25** (1912) 44. In the latter paper the determination of the Thomson effect by this method is discussed; this involves treating the complete equation 4.10 (4) by the present method. The variable state is discussed by Somers, *Proc. Phys. Soc.* **25** (1912) 74, and extensively by Owen, *Proc. Lond. Math. Soc.* (2) **23** (1925) 238; see also § 15.2 below.

4.10. The equation of conduction in a thin wire heated by an electric current

The equation for the temperature in a thin wire along which an electric current of constant strength is flowing was given by Verdet† in 1872. For some time little use was made of this method of heating the metal though it has several obvious advantages. In the first place, electrical measurements can be made with such accuracy that it is possible to arrange the experiments so that the range of temperature in the wire is small; the error due to neglecting the variation of the electrical and thermal conductivities with temperature is thus reduced. Further, in studying the ratio of the electrical and thermal conductivities (the Wiedemann–Franz ratio‡) it is very desirable that both these quantities be determined in the same experiment.

We shall first give the equation of conduction and then show how the steady temperature and the variable temperature in such a wire have been used to determine the electrical and thermal conductivities of metals.

We suppose, as in § 4.2, that the wire is of area ω, perimeter p, and has surface conductance H. Let I be the current in the wire measured in amperes,§ and let σ be the electrical conductivity of the wire. Then the rate of generation of heat in length dx of the wire is

$$\frac{jI^2}{\omega\sigma}\,dx,$$

where $j = 0.239...$ is the number of calories in a joule. This is to be added to the other terms corresponding to rate of gain of heat in the calculation of § 4.2, so that 4.2 (2) is replaced by

$$\frac{\partial v}{\partial t} = \kappa\,\frac{\partial^2 v}{\partial x^2} - \nu(v-v_0) + \frac{jI^2}{\rho c\omega^2\sigma}, \tag{1}$$

where

$$\kappa = \frac{K}{\rho c}, \qquad \nu = \frac{Hp}{\rho c\omega}. \tag{2}$$

---

† Verdet, *Théorie mécanique de la chaleur*, T. II (1872) p. 197.

Alternating currents have also been used in this connexion. Several important papers may be referred to: Cranz, *Z. Math. Phys.* **34** (1889) 92; Ebeling, *Ann. Physik,* (4) **27** (1908) 391; Weinreich, *Z. Math. Phys.* **63** (1915) 1. The last-named memoir contains a valuable account of the literature on this subject, and the variation of the temperature over the section as well as along the length of the wire is taken into consideration. See also Fischer, *Z. tech. Phys.* **19** (1938) 25.

‡ *Ann. Physik,* **89** (1853) 497.

§ In this and the following sections, thermal quantities are measured in calories and c.g.s. units, and electrical quantities in practical units.

Several other effects which are of importance in connexion with experimental work can be taken into account without greatly complicating the mathematical theory.

(i) The electrical resistance of the conductor may be taken to vary linearly with temperature—this is a reasonable approximation over small ranges of temperature. If α is the temperature coefficient of resistance, we have only to replace $(jI^2/\rho c\omega^2\sigma)$ in (1) by

$$\frac{jI^2}{\rho c\omega^2\sigma_0}(1+\alpha v),\tag{3}$$

where σ_0 is the electrical conductivity at zero temperature.

(ii) Heat is supplied by the Thomson effect at a rate proportional to the current and to the temperature gradient, that is, a term

$$-\frac{Is}{\rho c\omega}\frac{\partial v}{\partial x},$$

where s is the Thomson coefficient, has to be added to the right-hand side of (1). This term involves the direction of the current and disappears if alternating current is used for heating the wire. The Thomson coefficient varies with temperature and may be of either sign. Usually, the Thomson heating is small, and the chief importance of this term is that several methods of measuring s are based on the differential equation (4).

(iii) If the rod moves in the direction of its length with velocity U as in § 4.9, a term

$$-U\frac{\partial v}{\partial x}$$

is to be added to the right-hand side of (1).

With these generalizations, (1) is replaced by†

$$\frac{\partial v}{\partial t} = \frac{K}{\rho c}\frac{\partial^2 v}{\partial x^2} - \left(U+\frac{sI}{\rho c\omega}\right)\frac{\partial v}{\partial x} - v\left(\frac{Hp}{\rho c\omega}-\frac{\alpha jI^2}{\rho c\omega^2\sigma_0}\right) + \frac{Hpv_0}{\rho c\omega} + \frac{jI^2}{\rho c\omega^2\sigma_0}.\tag{4}$$

This equation is of the form

$$a\frac{\partial^2 v}{\partial x^2} + b\frac{\partial v}{\partial x} + cv + d = \frac{\partial v}{\partial t},\tag{5}$$

where a, b, c, d are constants, and is linear in v. Equations of this type arise in many branches of mathematical physics, cf. § 1.13. They can always be solved by a simple modification of the classical methods for the case $b = c = d = 0$, or by the Laplace transformation method of Chapter XII which has the advantage that it treats all cases in the same way. Exact solution is possible also in a few cases in which a, b, c, or d are simple functions of x.

As a final extension, the variation of thermal conductivity with temperature must be considered. Over a restricted range of temperature the linear law

$$K = K_0(1+\beta v),\tag{6}$$

where β is the temperature coefficient of thermal conductivity and is usually negative, may be taken to hold. In this case (4) is replaced by

$$\frac{\partial v}{\partial t} = \frac{K_0}{\rho c}\frac{\partial}{\partial x}\left[(1+\beta v)\frac{\partial v}{\partial x}\right] - \left(U+\frac{sI}{\omega\rho c}\right)\frac{\partial v}{\partial x} - v\left(\frac{Hp}{\omega\rho c}-\frac{\alpha jI^2}{\rho c\omega^2\sigma_0}\right) + \frac{Hpv_0}{\rho c\omega} + \frac{jI^2}{\rho c\omega^2\sigma_0}.\tag{7}$$

† For the application of this equation in the steady state to the measurement of the Thomson effect see Nettleton, *Proc. Phys. Soc.* **25** (1912) 44; **29** (1916), 50; also § 4.9; also King, *Proc. Amer. Acad. Arts Sci.* **33** (1898) 353; Laws, *Phil. Mag.* (6) **7** (1904) 560; Berg, *Ann. Physik* (4) **32** (1910) 477.

This equation is not linear in v, and has been studied only in the steady state case $\partial v/\partial t = 0$. In this case, if, in addition, $U = 0$, c does not appear, so the fact that it, also, may vary with temperature need not be considered.

In § 4.11 a number of ways in which the differential equations of this section have been applied to the determination of thermal conductivity will be discussed. In § 4.14 the variable state will be considered.

4.11. The steady temperature. Determination of conductivity

I. *Kohlrausch's method of obtaining the ratio of the electrical and thermal conductivities*

Kohlrausch† has shown how the steady temperature may be employed in finding the ratio of the electrical and thermal conductivities.

The ends of the wire are kept at as nearly as possible equal temperatures. The surface is supposed rendered impervious to heat, and the current I is supposed to have been flowing long enough to allow the steady rate of temperature to have been reached.

In this case the equation of conduction becomes

$$K\frac{d^2v}{dx^2} + \frac{jI^2}{\sigma\omega^2} = 0.$$

Let u be the electric potential at the section x.

Then
$$I = -\omega\sigma\frac{du}{dx}.$$

But
$$\frac{dv}{dx} = \frac{dv}{du}\frac{du}{dx}.$$

Therefore
$$\frac{dv}{dx} = -\frac{I}{\omega\sigma}\frac{dv}{du}$$

and
$$\frac{d^2v}{dx^2} = \frac{I^2}{\omega^2\sigma^2}\frac{d^2v}{du^2}.$$

Therefore we have
$$\frac{K}{j\sigma}\frac{d^2v}{du^2} + 1 = 0. \tag{1}$$

Thus
$$\frac{K}{j\sigma}v = -\tfrac{1}{2}u^2 + Au + B, \tag{2}$$

where A and B are constants determined by the temperatures at the ends.

† Kohlrausch, *S.B. preuss. Akad. Wiss.* (1899) 714; *Ann. Physik*, (4) **1** (1900) 132. See also Czermak, *S.B. Akad. Wiss. Wien*, **103** (1894) 1107; Duncan, *Pap. Dep. Phys. McGill Univ.*, No. 11 (1900); and Weinreich, loc. cit., p. 4; Jaeger and Diesselhorst, *Wiss. Abh. phys.-tech. Reichsanst.* **3** (1900) 269; Meissner, *Ann. Physik*, (4) **47** (1915) 1001.

Let (u_1, v_1), (u_2, v_2), and (u_3, v_3) be the values of u and v at any three sections x_1, x_2, x_3 of the wire. It follows from (2) that

$$\frac{2K}{j\sigma}[v_1(u_2-u_3)+v_2(u_3-u_1)+v_3(u_1-u_2)] = (u_2-u_3)(u_3-u_1)(u_1-u_2).$$

(3)

When the temperatures at the ends of the wire are kept the same, the distribution of temperature in the wire will be symmetrical about its middle point. Let the points x_1 and x_3 be at equal distances from the middle point x_2 on either side.

Then
$$v_1 = v_3$$

and
$$u_1 - u_2 = u_2 - u_3.$$

Therefore we have from (3)

$$\frac{K}{j\sigma}(v_2-v_1) = \tfrac{1}{2}(u_1-u_2)^2,$$

and we have thus obtained a simple method of determining the value of the ratio K/σ of the thermal and electrical conductivities, involving only the reading of the difference of the temperatures and potentials at two points of the wire, when the current is so regulated that the temperature of the wire is steady.

II. *Callendar's and related methods*

We consider a wire $0 < x < 2l$, whose ends are kept at zero temperature, and which is surrounded by an enclosure at zero temperature. The wire is heated by alternating current so that the term in $\partial v/\partial x$ in 4.10 (4) due to the Thomson effect disappears. Then in the steady state 4.10 (4) becomes

$$\frac{d^2v}{dx^2} + a^2v = -k,$$

(4)

where
$$k = \frac{jI^2}{\omega^2\sigma_0 K}, \qquad a^2 = \alpha k - \frac{Hp}{\omega K}.$$

(5)

Equation (4) has to be solved with boundary conditions $v = 0$ when $x = 0$ and $x = 2l$. The solution is

$$\begin{aligned}
v &= \frac{k}{a^2}\left\{\frac{\cos a(l-x)}{\cos al} - 1\right\}, & \text{if } a^2 > 0 \\
v &= \tfrac{1}{2}kx(2l-x), & \text{if } a^2 = 0 \\
v &= \frac{k}{a'^2}\left\{1 - \frac{\cosh a'(l-x)}{\cosh a'l}\right\}, & \text{if } a^2 = -a'^2 < 0
\end{aligned}$$

(6)

The easiest quantity to measure accurately in these methods is the

change in electrical resistance of the wire caused by heating due to the
current. If σ_0 is the electrical conductivity at zero temperature, the resistance of the wire at zero temperature is

$$R_0 = 2l/\omega\sigma_0. \tag{7}$$

The resistance of the wire with the current I flowing and steady
conditions is

$$R = \frac{1}{\omega\sigma_0} \int_0^{2l} (1+\alpha v)\, dx, \tag{8}$$

and using the values (6) of v in this we find†

$$\frac{R-R_0}{R_0} = \frac{k\alpha}{a^2}\left(\frac{\tan al}{al} - 1\right), \quad \text{if } a^2 > 0, \tag{9}$$

$$= \tfrac{1}{3}\alpha k l^2 = \frac{\alpha j I^2 l R_0}{6K\omega}, \quad \text{if } a^2 = 0, \tag{10}$$

$$= \frac{k\alpha}{a'^2}\left(1 - \frac{\tanh a'l}{a'l}\right), \quad \text{if } a^2 = -a'^2 < 0. \tag{11}$$

In Callendar's method‡ the current is adjusted so that $a^2 = 0$, i.e.

$$\frac{\alpha j I^2 R_0}{2l} = Hp; \tag{12}$$

this implies that the coefficient of v in 4.10 (4) vanishes, that is, that the
loss of heat of the wire by radiation from its surface is exactly com-
pensated by the increase of heat supply caused by the increase of the
resistance of the wire with temperature. With this adjustment, K is
found from (10) in terms of known or measurable quantities. One way
of making the adjustment (12) is by measuring the temperature v_l at
the mid-point of the wire for a case in which $a^2 < 0$; for a long wire
the temperature will be nearly uniform in the central portion, that is
d^2v/dx^2 will be negligible, and (4) gives approximately

$$v_l = \frac{k}{a'^2} = \frac{jI^2}{\omega\sigma Hp - \alpha j I^2}. \tag{13}$$

Hp can be found from (13), and then I can be adjusted to satisfy (12).

Equations (9) and (11) are valid for any values of H, but their use is
restricted by the fact that both H and K depend on temperature.
H varies with temperature in a complicated manner,§ and all that can

---

† Kannuluik, *Proc. Roy. Soc.* A, **131** (1931) 320; Weber, *Ann. Physik*, **54** (1917) 165.
See also § 7.2 (VI) below.
‡ *Encyclopaedia Britannica*, 11th edn., article on Conduction of Heat.
§ Loss of heat by radiation *in vacuo* is proportional to $v^4 - v_0^4$ and only approximately
to $v - v_0$; for this case see § 4.12. The validity of an expression of type $H(v - v_0)$ for loss
of heat through gas surrounding the wire is discussed by Smoluchowski, cf. § 7.2 (VI).

be done in practice is to work at temperatures of the wire not very different from that of the surroundings.†

The variation of thermal conductivity with temperature may be studied as follows:‡ if 4.10 (7) is used with the condition (12), the temperature in the wire is given by

$$K_0 \frac{d}{dx}\left[(1+\beta v)\frac{dv}{dx}\right] = -\frac{jI^2}{\omega^2\sigma}. \tag{14}$$

The solution of (14), for which $v = 0$ when $x = 0$ and $x = 2l$, is

$$K_0 v(1+\tfrac{1}{2}\beta v) = jI^2 x(2l-x)/2\omega^2\sigma. \tag{15}$$

If v_l is the temperature at the mid-point of the wire, and

$$K_m = K_0(1+\tfrac{1}{2}\beta v_l) \tag{16}$$

is the mean thermal conductivity of the wire over the range $(0, v_l)$ of temperature, (15) gives

$$K_m = \frac{jI^2 l^2}{2\omega^2\sigma v_l}. \tag{17}$$

4.12. The case of high temperatures in a wire carrying electric current

The assumption of § 4.10 that the wire loses heat by radiation to its surroundings at a rate proportional to the temperature difference is only satisfactory if this temperature difference is small: if not, it must be replaced by the statement that if T is the absolute temperature of the wire, and T_0 that of its surroundings, the rate of loss of heat by black-body radiation from length dx of the wire is

$$H'p(T^4 - T_0^4)\, dx, \tag{1}$$

where H' is a constant.§

In this case, and in the steady state, 4.10 (1) must be replaced by

$$\frac{d^2T}{dx^2} - \frac{H'p}{K\omega}(T^4 - T_0^4) + \frac{jI^2}{K\omega^2\sigma} = 0. \tag{2}$$

† Knudsen, *Ann. Physik*, **34** (1911) 593, and Weber, loc. cit., derive formulae for the case in which H is reduced to a small correction by enclosing the wire in a high vacuum. Here the relevant formula is (9) and an approximate formula is obtained by using the series for tan al.

‡ O'Day, *Phys. Rev.* (2) **23** (1924) 245. Another method of adjusting the coefficients in 4.10 (7) so that this equation has a simple solution is also given.

§ In fact H' is the product of the Stefan–Boltzmann constant and a factor involving the emissivities of the wire and the enclosure surrounding it, cf. 1.9 (10) and McAdams, *Heat Transmission*, Chap. III. The formula (1) is the theoretical one for black-body radiation; a nearer approximation to the conditions in a heated filament is obtained by using a power law of the same type, $T^n - T_0^n$, where n is a constant determined experimentally, cf. Langmuir, *Phys. Rev.* (2) **7** (1916) 151; Langmuir and Taylor, ibid. **50** (1936) 68. In the latter paper variation of electrical resistance and thermal conductivity of the wire with temperature according to power laws is considered, and a full discussion of the various cases, such as long or short filaments, which arise in practice is given. A first integral of (2), corresponding to (4), can still be found if the electrical resistance and thermal conductivity of the wire vary with the temperature according to power laws.

We consider the temperature in a wire of length $2l$ whose ends, $x = 0$ and $x = 2l$, are maintained at absolute temperature T_0. It is sufficient to consider the region $0 < x < l$ with

$$\left.\begin{aligned} T &= T_0, & x &= 0 \\ \frac{dT}{dx} &= 0, & x &= l \end{aligned}\right\}. \tag{3}$$

A first integral of (2) is

$$\left(\frac{dT}{dx}\right)^2 - aT^5 + bT + C = 0, \tag{4}$$

where C is constant and

$$a = \frac{2H'p}{5K\omega}, \qquad b = \frac{2H'pT_0^4}{K\omega} + \frac{2jI^2}{K\omega^2\sigma}. \tag{5}$$

Writing T_l for the temperature at $x = l$, (3) and (4) give

$$C = aT_l^5 - bT_l.$$

Then from (4) with $T = T_0$ when $x = 0$, it follows that

$$x = \int_{T_0}^{T} \frac{dT}{\{a(T^5 - T_l^5) - b(T - T_l)\}^{\frac{1}{2}}}, \tag{6}$$

and the unknown T_l is found from the value of this for $x = l$, namely

$$l = \int_{T_0}^{T_l} \frac{dT}{\{a(T^5 - T_l^5) - b(T - T_l)\}^{\frac{1}{2}}}. \tag{7}$$

The heat loss from the ends of the wire is

$$2K\omega\left[\frac{dT}{dx}\right]_{x=0} = 2K\omega\{a(T_0^5 - T_l^5) - b(T_0 - T_l)\}^{\frac{1}{2}}. \tag{8}$$

It appears that the solution of this simple problem cannot be expressed in terms of elementary functions.†

For a semi-infinite wire $x > 0$, $d^2T/dx^2 \to 0$ as $x \to \infty$ and, using this in (2), it follows that $T \to T_m$ where

$$T_m^4 = T_0^4 + \frac{jI^2}{H'p\omega\sigma}. \tag{9}$$

Clearly, T_m is a good approximation for T_l for a long wire. Using this approximation, (8) becomes

$$\left\{\frac{8KjI^2(T_0^5 + 4T_m^5 - 5T_0\,T_m^4)}{5\sigma(T_m^4 - T_0^4)}\right\}^{\frac{1}{2}}, \tag{10}$$

and, if T_0 is small compared with T_m, this is approximately

$$\left(\frac{32KjI^2T_m}{5\sigma}\right)^{\frac{1}{2}}\left(1 - \frac{5T_0}{8T_m}\right). \tag{11}$$

† It has been discussed by Cox, *Phys. Rev.* **64** (1943) 241 and Baerwald, *Phil. Mag.* (7) **21** (1936) 641. Nagai, *J. Phys. Soc. Japan*, **11** (1956) 329–30, gives a graphical representation of the integral (7). Bush and Gould, *Phys. Rev.* **29** (1927) 337, give some numerical results obtained by using the differential analyser; they also consider the effects of electron emission and the variation of conductivity with temperature. Recently Jain and Krishnan have re-examined the whole question in a series of papers, *Proc. Roy. Soc.* A, **222** (1954) 167–80; A, **225** (1954) 1–7, 7–18, 19–32; A, **227** (1955) 141–54; A, **229** (1955) 439–45. They derive series solutions for (7) and (12), and discuss the case of temperature variation of thermal and electrical properties and comparison with experiment.

Using (9), (2) may be written

$$\frac{d^2T}{dx^2} + \frac{5a}{2}(T_m^4 - T^4) = 0.$$ (12)

An approximate solution for the variation of temperature near the centre of the wire may be obtained by putting

$$T' = T_m - T,$$ (13)

which reduces (12) to

$$\frac{d^2T'}{dx^2} - \frac{5a}{2}\{4T_m^3\,T' - 6T_m^2\,T'^2 + 4T_m\,T'^3 - T'^4\} = 0.$$ (14)

If T' is small, terms in T'^2, etc., in (14) can be neglected, and (14) becomes approximately

$$\frac{d^2T'}{dx^2} - 10aT_m^3\,T' = 0.$$ (15)

The solution of this which takes the value T_l' when $x = l$ is[†]

$$T' = T_l' \cosh\{(l-x)(10aT_m)^{\frac{1}{2}}\}.$$ (16)

4.13. Steady flow of heat in a composite wire

Problems on conduction of heat in composite wires are of some interest in connexion with the theory of thermocouples and current measuring devices. Here we consider the steady state only; the unsteady state is easily treated by the Laplace transformation method of Chapter XII.

We suppose the part of the wire $0 < x < a$ to have conductivity K_1, density ρ_1, specific heat c_1, diffusivity κ_1, temperature v_1, surface conductance H_1, area ω_1, perimeter p_1, $\mu_1^2 = H_1 p_1/K_1 \omega_1$, electrical conductivity σ_1. If electric current I amperes flows in the wire, we write $\beta_1 = jI^2/\omega_1^2 \sigma_1 K_1$, where j is the number of calories in a joule. The corresponding quantities for the part of the wire $a < x < b$ will be distinguished by the suffix 2.

If the wire has different areas in the two regions, we neglect any end effects at the junction $x = a$ so that the boundary conditions here are

$$v_1 = v_2,$$ (1)

$$\omega_1 K_1 \frac{dv_1}{dx} = \omega_2 K_2 \frac{dv_2}{dx}.$$ (2)

If the wire radiates into medium at temperature v_0, the differential equations to be satisfied are, as in 4.10 (1),

$$\frac{d^2v_1}{dx^2} - \mu_1^2(v_1 - v_0) + \beta_1 = 0, \quad 0 < x < a,$$ (3)

$$\frac{d^2v_2}{dx^2} - \mu_2^2(v_2 - v_0) + \beta_2 = 0, \quad a < x < b.$$ (4)

Some typical problems are considered briefly below.

(i) *The ends $x = 0$ and $x = b$ maintained at V_1 and V_2 respectively. Radiation into a medium at zero temperature. No current in the wire.*

[†] Expressions of this type are used by Worthing, *Phys. Rev.* **4** (1914) 523 ; *J. Franklin Inst.* **194** (1922) 597 ; Stead, *J. Instn. Elect. Engrs.* **58** (1920) 107 ; Prescott and Hincke, *Phys. Rev.* **31** (1928) 130. See also Jain and Krishnan, loc. cit.

Here solutions of (3) and (4) with $\beta_1 = \beta_2 = v_0 = 0$, which take the values V_1 and V_2 at $x = 0$ and $x = b$, are

$$v_1 = V_1 \cosh \mu_1 x + A \sinh \mu_1 x,$$
$$v_2 = V_2 \cosh \mu_2(b-x) + B \sinh \mu_2(b-x).$$

The boundary conditions (1) and (2) give A and B, and we obtain finally

$$v_1 = \frac{V_1}{\Delta}\{\mu_2 \omega_2 K_2 \cosh \mu_2(b-a)\sinh \mu_1(a-x) + \mu_1 \omega_1 K_1 \sinh \mu_2(b-a)\cosh \mu_1(a-x)\} +$$

$$+\frac{V_2}{\Delta}\mu_2 \omega_2 K_2 \sinh \mu_1 x, \quad (5)$$

where

$$\Delta = \mu_2 \omega_2 K_2 \sinh \mu_1 a \cosh \mu_2(b-a) + \mu_1 \omega_1 K_1 \cosh \mu_1 a \sinh \mu_2(b-a), \quad (6)$$

and there is a similar expression for v_2. The temperature at the junction $x = a$ is

$$\frac{1}{\Delta}\{V_1 \mu_1 \omega_1 K_1 \sinh \mu_2(b-a) + V_2 \mu_2 \omega_2 K_2 \sinh \mu_1 a\}. \quad (7)$$

(ii) *The ends at $x = 0$ and $x = b$ maintained at zero. Radiation into a medium at zero. Current I in the wire.*

The temperature at the junction $x = a$ is

$$\frac{1}{\Delta\mu_1\mu_2}\{\beta_2 K_2 \omega_2 \mu_1 \sinh \mu_1 a[\cosh \mu_2(b-a)-1]+$$

$$+\beta_1 K_1 \omega_1 \mu_2 \sinh \mu_2(b-a)[\cosh \mu_1 a -1]\}, \quad (8)$$

where Δ is defined in (6).

(iii) *The end $x = 0$ maintained at V_1. No flow at $x = b$. No current in the wire. Radiation into a medium at zero temperature.*

$$v_2 = [K_1 \omega_1 \mu_1 V_1 \cosh \mu_2(b-x)]/\Delta', \quad (9)$$

where

$$\Delta' = K_1 \omega_1 \mu_1 \cosh \mu_1 a \cosh \mu_2(b-a) + K_2 \omega_2 \mu_2 \sinh \mu_1 a \sinh \mu_2(b-a). \quad (10)$$

(iv) *The end $x = 0$ maintained at zero. No flow at $x = b$. Current I in the wire. Radiation into a medium at zero temperature.*

$$v_2 = \frac{\beta_2}{\mu_2^2} + \left\{\frac{K_1 \omega_1(\beta_1 \mu_2^2 - \beta_2 \mu_1^2)}{\mu_1 \mu_2^2 \Delta'}\cosh \mu_1 a - \frac{K_1 \omega_1 \beta_1}{\mu_1 \Delta'}\right\}\cosh \mu_2(b-x), \quad (11)$$

where Δ' is defined in (10).

4.14. Variable temperature in a wire carrying electric current

The variable temperature of a wire along which a constant electric current is flowing, while radiation takes place at the surface, has also been used in determining the thermal and electrical constants. The following investigation is due to Straneo.†

We have found the equation of conduction (4.10 (1)) in the form

$$\frac{\partial v}{\partial t} = \kappa \frac{\partial^2 v}{\partial x^2} - \nu(v-v_0)+a,$$

† Straneo, *R.C. Accad. Lincei* (5) **7**, Sem. ii (1898). See also: Schaufelberger, *Ann. Physik* (4) **7** (1902) 589; and Weinreich, loc. cit.; Fischer, *Z. tech. Phys.* **19** (1938) 105.

where $\qquad\qquad \kappa = \dfrac{K}{c\rho}, \quad \nu = \dfrac{Hp}{c\rho\omega}, \quad$ and $\quad a = \dfrac{jI^2}{c\rho\omega^2\sigma}.$

Suppose the temperature of the medium into which radiation takes place to be zero, and that

$$v = 0, \quad \text{when } t = 0 \quad (0 < x < l),$$

$$v = 0, \quad \text{when } x = 0 \text{ and } x = l,$$

are the initial and boundary conditions.

To integrate the equation of conduction, we proceed as usual to break up the problem into one of steady temperature and one of variable temperature.

Put $\qquad\qquad\qquad\qquad v = u + w,$

where u is independent of the time and satisfies the equations

$$\left.\begin{aligned} \kappa\dfrac{d^2u}{dx^2} - \nu u + a &= 0 \\ u = 0 \quad \text{at } x = 0 \text{ and } x = l & \end{aligned}\right\},$$

and w is a function of x and t which satisfies the equations

$$\left.\begin{aligned} \dfrac{\partial w}{\partial t} &= \kappa\dfrac{\partial^2 w}{\partial x^2} - \nu w \\ w = 0 \quad &\text{at } x = 0 \text{ and } x = l \\ w = -u \quad &\text{at } t = 0 \end{aligned}\right\}.$$

The value of u is obtained immediately in the form

$$u = b\left[1 - \frac{\sinh\mu x + \sinh\mu(l-x)}{\sinh\mu l}\right], \qquad\qquad (1)$$

where $\qquad\qquad \mu = \sqrt{(\nu/\kappa)} \quad \text{and} \quad b = a/\nu.$

This function may be expanded in the sine series

$$\sum_{n=1}^{\infty} \frac{4bl^2\mu^2}{\pi(2n-1)[(2n-1)^2\pi^2 + l^2\mu^2]} \sin\frac{(2n-1)\pi x}{l}.$$

With this value of u the solution of the equations for w follows immediately, and we have

$$w = -\frac{4b}{\pi}\sum_{n=1}^{\infty} \frac{l^2\mu^2}{(2n-1)\{(2n-1)^2\pi^2 + l^2\mu^2\}} \sin\frac{(2n-1)\pi x}{l} e^{-\nu t - t[\kappa(2n-1)^2\pi^2/l^2]}.$$

$$(2)$$

Therefore

$$v = b\left[1 - \frac{\sinh \mu x + \sinh \mu(l-x)}{\sinh \mu l}\right] -$$

$$- \frac{4b}{\pi} \sum_{n=1}^{\infty} \frac{l^2 \mu^2}{(2n-1)\{(2n-1)^2 \pi^2 + l^2 \mu^2\}} \sin \frac{(2n-1)\pi x}{l} e^{-\nu t - l[\kappa(2n-1)^2 \pi^2/l^2]}. \quad (3)$$

In applying this solution we note that the coefficients of the terms in the series for w diminish rapidly, and when $x = l/3$ or $2l/3$ the second term in w is zero. Hence to a close approximation the value of v at these points is given by

$$v = [u]_{\frac{1}{3}l} - \frac{2bl^2 \mu^2 \sqrt{3}}{\pi(\pi^2 + l^2 \mu^2)} e^{-\nu t - l(\kappa \pi^2/l^2)}. \quad (4)$$

Let v_1, v_2, v_3 be the temperatures at the point $x = \frac{1}{3}l$ at times t, $t+\tau$, $t+2\tau$, then from (4)

$$\ln \frac{v_2 - v_1}{v_3 - v_2} = \left(\frac{\kappa \pi^2}{l^2} + \nu\right)\tau, \quad (5)$$

and

$$v_2 - v_1 = \frac{2bl^2 \mu^2 \sqrt{3}}{\pi(\pi^2 + l^2 \mu^2)}\{e^{-(\nu + \kappa \pi^2/l^2)t_1} - e^{-(\nu + \kappa \pi^2/l^2)t_2}\}. \quad (6)$$

Also v_4, the value of the steady temperature at the mid-point of the wire, is

$$v_4 = b(1 - \mathrm{sech}\, \tfrac{1}{2}\mu l). \quad (7)$$

Equation (5) gives the value of $\nu + \kappa \pi^2/l^2$. Inserting this in (6), μ is found, and then b follows from (7). Using these results in the definitions of a, b, and μ, the values of κ, ν, and σ are determined.

In Straneo's original experimental work the wire was first heated by the current until the steady temperature was attained. If the current is then switched off at $t = 0$, the subsequent temperature is given by

$$v = \frac{4b}{\pi} \sum_{n=1}^{\infty} \frac{l^2 \mu^2}{(2n-1)[(2n-1)^2 \pi^2 + l^2 \mu^2]} \sin \frac{(2n-1)\pi x}{l} e^{-\nu t - \kappa l(2n-1)^2 \pi^2/l^2}.$$

The method of determining κ, ν, and σ described above, leading to (5), (6), and (7), still holds.

These results only apply to cases in which the temperature of the wire is not much greater than that of its surroundings. For high temperatures, where the wire loses heat by fourth-power radiation,† the equation of conduction of heat in the variable state corresponding to 4.12 (12) is

$$\frac{\partial^2 T}{\partial x^2} - \frac{1}{\kappa}\frac{\partial T}{\partial t} + \frac{5a}{2}(T_m^4 - T^4) = 0,$$

---

† For a full discussion see Jain and Krishnan, *Proc. Roy. Soc.* A, **227** (1955) 141–54, who discuss in particular the time-lag in attaining the steady state. See also Winter-gerst, *Z. angew. Phys.* **2** (1950) 167.

where T is the absolute temperature in the wire and T_m and a are defined in 4.12 (5), (9). One simple result may be noted: at the centre of a long wire where $\partial^2 T/\partial x^2$ is negligible, the temperature T at time t, for the case in which the current is switched on at $t = 0$ when the temperature is T_0, is given by

$$\exp\left\{-10a\kappa T_m^3 t + 2\tan^{-1}\frac{T}{T_m} - 2\tan^{-1}\frac{T_0}{T_m}\right\} = \frac{(T_m - T)(T_m + T_0)}{(T_m + T)(T_m - T_0)}. \tag{8}$$

4.15. Fourier's ring

One of the simplest and most suggestive problems in the conduction of heat, when the temperature depends only upon one coordinate and the time, is Fourier's problem of the ring. This problem is also of special interest, as it was the first to which Fourier applied his mathematical theory, and for which the results of his mathematical investigations were compared with the facts of experiment.†

The ring consists of a small cross-section twisted into a circle (or other closed curve). Then with the notation and assumptions of § 4.2 the differential equation for the temperature in the ring is 4.2 (4), that is

$$\frac{\partial v}{\partial t} = \kappa\frac{\partial^2 v}{\partial x^2} - \nu v. \tag{1}$$

We suppose the length of the ring to be $2l$, so that taking the origin at any convenient point we have to solve (1) in the region $-l \leqslant x \leqslant l$. Since the ring forms a closed curve we do not have boundary conditions at $x = \pm l$, but instead the condition that v is to be periodic with period $2l$ in x, that is

$$v(x, t) = v(x + 2nl, t), \quad n = 1, 2, \ldots . \tag{2}$$

I. *Initial temperature $f(x)$. No radiation*

We assume that $f(x)$ can be expanded in the Fourier series

$$f(x) = \sum_{n=0}^{\infty} a_n\cos\frac{n\pi x}{l} + \sum_{n=1}^{\infty} b_n\sin\frac{n\pi x}{l}. \tag{3}$$

Then
$$v = \sum_{n=0}^{\infty} a_n e^{-\kappa n^2\pi^2 t/l^2}\cos\frac{n\pi x}{l} + \sum_{n=1}^{\infty} b_n e^{-\kappa n^2\pi^2 t/l^2}\sin\frac{n\pi x}{l} \tag{4}$$

satisfies all the conditions of the problem. This may be verified‡ as in § 3.3.

The solution for the case of radiation follows on substituting $v = ue^{-\nu t}$ in (1).

II. *Steady temperature with $x = \pm l$ maintained at V*

Here the temperature distribution must be an even function of x which takes the value V at $x = \pm l$ and satisfies (1) with $\partial v/\partial t = 0$.

Thus the solution is
$$v = V\frac{\cosh\mu x}{\cosh\mu l}, \tag{5}$$

where
$$\mu^2 = \nu/\kappa. \tag{6}$$

This may be used for the comparison of the conductivities§ of two solids by a method similar to that leading to 4.5 (2).

† Fourier, *Théorie analytique de la chaleur*, Chaps. II and IV.

‡ It may also be verified that in this case v and $\partial v/\partial x$ are continuous at $x = \pm l$ for $t > 0$, as they should be since the ring forms a continuous curve. They need not be continuous there when $t = 0$.

§ Fourier, loc. cit., §§ 107–10.

III. *The ring with steady temperature given by* (5) *allowed to cool by radiation into medium at zero*

Putting $v = ue^{-\nu t}$ in (1), we have to find a solution of

$$\frac{\partial u}{\partial t} = \kappa \frac{\partial^2 u}{\partial x^2}, \quad -l \leqslant x \leqslant l, \tag{7}$$

which is periodic with period $2l$ and has initial value

$$V \frac{\cosh \mu x}{\cosh \mu l} = \frac{2\mu V \tanh \mu l}{l} \left[\frac{1}{2\mu^2} + \sum_{n=1}^{\infty} \frac{(-1)^n}{\mu^2 + (n^2\pi^2/l^2)} \cos \frac{n\pi x}{l} \right]. \tag{8}$$

Using (4), the solution is found to be

$$v = \frac{2V\mu}{l} \tanh \mu l \, e^{-\nu t} \left[\frac{1}{2\mu^2} + \sum_{n=1}^{\infty} \frac{(-1)^n}{\mu^2 + (n^2\pi^2/l^2)} \, e^{-\kappa n^2\pi^2 t/l^2} \cos \frac{n\pi x}{l} \right]. \tag{9}$$

(9) may be adapted for the determination† of κ and ν from observations of the temperature at $x = 0$ and $x = l$.

† Neumann, *Ann. Chim. Phys.* (3) **66** (1862) 183; *Phil. Mag.* (4) **25** (1863) 63.

V

FLOW OF HEAT IN A RECTANGLE

5.1. Introductory

IN the last three chapters we have been examining different cases of linear flow of heat. In these the temperature has been dependent only upon the time and upon one geometrical coordinate. Such problems may be referred to as one-dimensional. We proceed to the discussion of cases in which the flow of heat takes place in parallel planes. If these planes are taken parallel to the xy-plane, the temperature will depend only upon x and y, if it is a case of steady temperature, or upon x, y, and t, if we are dealing with variable temperature. We speak of these problems as two-dimensional.

The first problem in the conduction of heat discussed in detail by Fourier in his treatise is that of the Steady Temperature in the Infinite Solid bounded by the planes $x = \pm\frac{1}{2}\pi$, $y = 0$, and extending to infinity in the direction of y positive. The boundaries $x = \pm\frac{1}{2}\pi$ are kept at zero temperature, and the base $y = 0$ at temperature unity. His discussion led him to the expansion of unity in the interval $-\frac{1}{2}\pi < x < \frac{1}{2}\pi$ in the series

$$\frac{4}{\pi}\{\cos x - \tfrac{1}{3}\cos 3x + \tfrac{1}{5}\cos 5x - ...\},\dagger$$

and he then proceeded to consider the question of the development of an arbitrary function in trigonometrical series, and obtained the expansion now known as Fourier's series. He was thus able to give the distribution of temperature in this solid, when the base is kept at the temperature $v = f(x)$, $f(x)$ being an arbitrary function of x, while the faces $x = \pm\frac{1}{2}\pi$ are kept as before at zero.

5.2. The infinite rectangular solid. Steady temperature

Instead of taking Fourier's problem in the form which he adopted, we shall take the solid as bounded by the planes $x = 0$ and $x = l$, which are kept at zero temperature, and the plane $y = 0$, which is kept at the

---

† The series $\qquad \dfrac{4}{\pi}(\cos x - \tfrac{1}{3}\cos 3x + \tfrac{1}{5}\cos 5x - ...)$

may be obtained in the ordinary way as the cosine series for $f(x)$, when

$$f(x) = 1 \qquad (0 < x < \tfrac{1}{2}\pi)$$
$$f(x) = -1 \qquad (\tfrac{1}{2}\pi < x < \pi).$$

temperature $v = f(x)$. We assume that the function $f(x)$ is bounded and satisfies Dirichlet's conditions ($F.S.$, § 93) in $(0, l)$.

The equations for the temperature will thus be as follows:

$$\frac{\partial^2 v}{\partial x^2} + \frac{\partial^2 v}{\partial y^2} = 0, \quad 0 < x < l, y > 0, \tag{1}$$

$$v = 0, \quad \text{when } x = 0 \text{ and } x = l, \tag{2}$$

$$v = f(x), \quad \text{when } y = 0, 0 < x < l. \tag{3}$$

Also we have

$$\lim_{y \to \infty} v = 0.$$

Starting with the sine series for $f(x)$

$$\sum_{n=1}^{\infty} a_n \sin \frac{n\pi x}{l}, \tag{4}$$

where

$$a_n = \frac{2}{l} \int_0^l f(x') \sin \frac{n\pi x'}{l} \, dx', \tag{5}$$

let us examine the function v defined by the equation

$$v = \sum_{n=1}^{\infty} a_n e^{-n\pi y/l} \sin \frac{n\pi x}{l}. \tag{6}$$

Since $f(x)$ is bounded and integrable in $(0, l)$ it follows from (5) that

$$|a_n| < 2M, \quad \text{where } |f(x)| < M \text{ in } (0, l).$$

Also

$$\left| a_n \sin \frac{n\pi x}{l} e^{-n\pi y/l} \right| < 2M e^{-n\pi y_0/l}, \quad \text{when } y \geqslant y_0 > 0,$$

y_0 being an arbitrary positive number:

Now the series

$$\sum_{n=1}^{\infty} e^{-n\pi y_0/l}$$

is convergent and its terms are independent of x and y.

Thus the series (6), regarded as a function of x, is uniformly convergent for any interval of x, when $y > 0$; and, regarded as a function of y, it is uniformly convergent when $y \geqslant y_0 > 0$.

The same is true of the series obtained by term-by-term differentiation of (6) with respect to x and y in these intervals. Therefore

$$\frac{\partial^2 v}{\partial x^2} = -\frac{\pi^2}{l^2} \sum_{n=1}^{\infty} n^2 a_n e^{-n\pi y/l} \sin \frac{n\pi x}{l},$$

and

$$\frac{\partial^2 v}{\partial y^2} = \frac{\pi^2}{l^2} \sum_{n=1}^{\infty} n^2 a_n e^{-n\pi y/l} \sin \frac{n\pi x}{l}.$$

Hence
$$\frac{\partial^2 v}{\partial x^2} + \frac{\partial^2 v}{\partial y^2} = 0,$$

and (6) satisfies the differential equation (1) of the problem. Further, it satisfies the boundary conditions at the faces $x = 0$ and $x = l$, for, since (6) is uniformly convergent in the interval $0 \leqslant x \leqslant l$, and the sum of the series vanishes when $x = 0$ and $x = l$, the limit of v as x approaches these values is zero, y being positive.

We have assumed that $f(x)$ is bounded and satisfies Dirichlet's conditions in the interval $(0, l)$. It follows that the sine series (4) converges, and its sum is $f(x)$ at every point between 0 and l at which $f(x)$ is continuous, and $\frac{1}{2}\{f(x+0)+f(x-0)\}$ at all other points. It follows from $F.S.$, § 73, I,† that, if v is defined by the series (6),

$$\lim_{y \to 0} v = f(x) \quad \text{at a point of continuity}$$
$$= \tfrac{1}{2}\{f(x+0)+f(x-0)\} \quad \text{at all other points.}$$

Thus (6) is the solution of our problem. This may be written

$$v = \frac{2}{l} \int_0^l f(x') \sum_{n=1}^{\infty} \left(e^{-n\pi y/l} \sin\frac{n\pi x}{l} \sin\frac{n\pi x'}{l} \right) dx', \tag{7}$$

since the series under the integral is uniformly convergent.

If $f(x)$ is unity, the solution (6) becomes

$$v = \frac{2}{\pi} \sum_{n=1}^{\infty} \frac{1-\cos n\pi}{n} e^{-n\pi y/l} \sin\frac{n\pi x}{l} \tag{8}$$

$$= \frac{4}{\pi} \sum_{n=0}^{\infty} \frac{1}{(2n+1)} e^{-(2n+1)\pi y/l} \sin\frac{(2n+1)\pi x}{l} \tag{9}$$

$$= \frac{4}{\pi} \mathbf{I}\{e^{-\pi(y-ix)/l} + \tfrac{1}{3}e^{-3\pi(y-ix)/l} + \ldots\}$$

$$= \frac{2}{\pi} \mathbf{I}\left\{ \ln\frac{1+e^{\pi i(x+iy)/l}}{1-e^{\pi i(x+iy)/l}} \right\} \tag{10}$$

$$= \frac{2}{\pi} \mathbf{I}\left\{ \ln\frac{1-e^{-2\pi y/l}+2ie^{-\pi y/l}\sin(\pi x/l)}{1-2e^{-\pi y/l}\cos(\pi x/l)+e^{-2\pi y/l}} \right\}$$

$$= \frac{2}{\pi} \tan^{-1}\frac{\sin(\pi x/l)}{\sinh(\pi y/l)}. \tag{11}$$

The conjugate function‡ to (11), that is, the real part of the logarithm in (10), is

$$\frac{1}{\pi}\ln\left(\frac{1+2e^{-\pi y/l}\cos(\pi x/l)+e^{-2\pi y/l}}{1-2e^{-\pi y/l}\cos(\pi x/l)+e^{-2\pi y/l}}\right) = \frac{1}{\pi}\ln\left(\frac{\cosh(\pi y/l)+\cos(\pi x/l)}{\cosh(\pi y/l)-\cos(\pi x/l)}\right). \tag{12}$$

† The argument of $F.S.$, § 73, II, also applies to this case.
‡ For the definition and use of conjugate functions in this connexion see §§ 16.6–9. **I** is written for 'the imaginary part of'.

It follows that the lines of flow are given by

$$\frac{\cosh(\pi y/l) + \cos(\pi x/l)}{\cosh(\pi y/l) - \cos(\pi x/l)} = \text{const.},$$

these being orthogonal to the isothermals†

$$\frac{\sin(\pi x/l)}{\sinh(\pi y/l)} = \text{const.}$$

The corresponding problem, but with radiation at one or both of the surfaces $x = 0$ and $x = l$, is treated as in § 5.3.

We now consider the problem of *steady temperature in the strip* $0 < x < l$, $y > 0$, *with boundary conditions*

$$v = f(y), \quad x = 0, \qquad y > 0, \tag{13}$$

$$v = 0, \qquad x = l, \qquad y > 0, \tag{14}$$

$$v = 0, \qquad 0 < x < l, \; y = 0. \tag{15}$$

Here we use Fourier's sine integral, 2.3 (8), in place of the sine series which was used above. We notice that, whatever ξ may be,

$$\sin \xi y \sinh \xi(l-x)$$

satisfies the differential equation (1) and the boundary conditions (14) and (15). Thus

$$\int_0^\infty \sin \xi y \sinh \xi(l-x) F(\xi) \, d\xi, \tag{16}$$

where $F(\xi)$ is an arbitrary function, will also satisfy them. If, in addition, (16) is to satisfy (13), we must have

$$f(y) = \int_0^\infty \sin \xi y \sinh \xi l F(\xi) \, d\xi,$$

and thus, by 2.3 (9)

$$\sinh \xi l \, F(\xi) = \frac{2}{\pi} \int_0^\infty f(y') \sin \xi y' \, dy'.$$

Using this value of $F(\xi)$, (16) gives for the solution of our problem

$$v = \frac{2}{\pi} \int_0^\infty \sin \xi y \, \frac{\sinh \xi(l-x)}{\sinh \xi l} \, d\xi \int_0^\infty f(y') \sin \xi y' \, dy'. \tag{17}$$

If we assume that the orders of integration may be interchanged, this becomes

$$v = \frac{2}{\pi} \int_0^\infty f(y') \, dy' \int_0^\infty \frac{\sinh \xi(l-x)}{\sinh \xi l} \sin \xi y \sin \xi y' \, d\xi$$

$$= \frac{1}{2l} \sin \frac{\pi x}{l} \int_0^\infty f(y') \left\{ \frac{1}{\cos[\pi(l-x)/l] + \cosh[\pi(y-y')/l]} - \right.$$

$$\left. - \frac{1}{\cos[\pi(l-x)/l] + \cosh[\pi(y+y')/l]} \right\} dy'. \tag{18}$$

Clearly problems with radiation at the surfaces $y = 0$ or $x = l$ could be treated in the same way (cf. § 5.3). But the above analysis leading to (18) is to be regarded

† Fourier, loc. cit., § 205.

as formal and suggestive only: not only have we interchanged the orders of integration in (17), but the use of 2.3 (8) implies very severe restrictions on $f(y)$; these, in fact, are not needed, and a fuller discussion† shows that (18) is valid provided $|f(y)|$ is of exponential type, i.e. $|f(y)| < Ke^{c|y|}$, where K and c are positive constants.

Two other results of importance may be derived formally in a similar manner using Fourier's integral theorem.

For *the doubly infinite strip* $0 < x < l$, $-\infty < y < \infty$ *with* $x = 0$ *maintained at temperature* $f(y)$ *and* $x = l$ *at zero*, the steady temperature is given by

$$\frac{1}{2l} \sin \frac{\pi x}{l} \int_{-\infty}^{\infty} \frac{f(y')\,dy'}{\cos[\pi(l-x)/l] + \cosh[\pi(y-y')/l]}. \tag{19}$$

In *the half-plane*‡ $x > 0$, $-\infty < y < \infty$ *with* $x = 0$ *maintained at temperature* $f(y)$, the steady temperature is

$$\frac{x}{\pi} \int_{-\infty}^{\infty} \frac{f(y')\,dy'}{x^2 + (y-y')^2}. \tag{20}$$

5.3. Steady temperature in the rectangle $0 < x < a$, $0 < y < b$

Here the differential equation to be solved is

$$\frac{\partial^2 v}{\partial x^2} + \frac{\partial^2 v}{\partial y^2} = 0, \quad 0 < x < a, \; 0 < y < b, \tag{1}$$

with boundary conditions depending on the problem considered.

I. $y = 0$ *kept at* $f(x)$, *the other surfaces at zero*

Here the boundary conditions are

$$v = f(x), \quad y = 0, \quad 0 \le x < a, \tag{2}$$

$$v = 0, \quad y = b, \quad 0 < x < a, \tag{3}$$

$$v = 0, \quad x = 0, \quad 0 < y < b, \tag{4}$$

$$v = 0, \quad x = a, \quad 0 < y < b. \tag{5}$$

As in § 5.2 we start with the sine series for $f(x)$

$$f(x) = \sum_{n=1}^{\infty} a_n \sin \frac{n\pi x}{a}, \tag{6}$$

where

$$a_n = \frac{2}{a} \int_0^a f(x') \sin \frac{n\pi x'}{a}\,dx'. \tag{7}$$

Now, for any n, a term of type

$$\sin \frac{n\pi x}{a} \sinh \frac{(b-y)n\pi}{a} \tag{8}$$

---

† Titchmarsh, *Fourier Integrals* (Oxford, 1937) § 10.11.
‡ Steady temperatures in the half-plane, $x > 0$, $-\infty < y < \infty$ and the rectangular corner $x > 0$, $y > 0$, are discussed by Karush, *J. Appl. Phys.* **23** (1952) 492–4.

satisfies (1), (3), (4), and (5). Thus we are led to consider the expression

$$v = \sum_{n=1}^{\infty} a_n \sin \frac{n\pi x}{a} \sinh \frac{(b-y)n\pi}{a} \operatorname{cosech} \frac{n\pi b}{a} \qquad (9)$$

with a_n given by (7). A complete verification that (9) satisfies (1)–(5) follows† as in § 5.2.

If $f(x) = V$, constant, (9) becomes

$$\frac{4V}{\pi} \sum_{n=0}^{\infty} \frac{1}{(2n+1)} \sin \frac{(2n+1)\pi x}{a} \sinh \frac{(b-y)(2n+1)\pi}{a} \operatorname{cosech} \frac{(2n+1)\pi b}{a}. \qquad (10)$$

If some of the other sides of the rectangle are also kept at prescribed temperatures the result can be obtained by combining solutions of type (9).

II. *$y = 0$ kept at $f(x)$. No flow of heat over $y = b$ and $x = 0$. Radiation into a medium at zero temperature over $x = a$*

Here the boundary conditions are

$$v = f(x), \quad y = 0, \quad 0 < x < a, \qquad (11)$$

$$\frac{\partial v}{\partial y} = 0, \qquad y = b, \quad 0 < x < a, \qquad (12)$$

$$\frac{\partial v}{\partial x} = 0, \qquad x = 0, \quad 0 < y < b, \qquad (13)$$

$$\frac{\partial v}{\partial x} + hv = 0, \qquad x = a, \quad 0 < y < b. \qquad (14)$$

The expression $\qquad \cos \alpha x \cosh \alpha (b-y)$

satisfies (1), (13), and (12) for any α. It also satisfies (14) if α is a root of

$$\alpha \tan \alpha a = h. \qquad (15)$$

The roots α_n of this equation, $n = 1, 2,...,$ have been discussed in § 3.10. If we now assume that $f(x)$ can be expanded in the form 3.10 (9), we have as the solution of the problem

$$v = \sum_{n=1}^{\infty} \frac{2(h^2 + \alpha_n^2) \cos \alpha_n x \cosh \alpha_n (b-y)}{[(\alpha_n^2 + h^2)a + h]\cosh \alpha_n b} \int_0^a f(x) \cos \alpha_n x \, dx, \qquad (16)$$

where the α_n are the positive roots of (15).

† We need the result $|\sinh[n\pi(b-y)/a]\operatorname{cosech}[n\pi b/a]| < 2e^{-n\pi y/a}$, if $n >$ some fixed n_0, and then the argument is precisely that of § 5.2.

If $f(x) = V$, constant, (16) becomes

$$v = 2hV \sum_{n=1}^{\infty} \frac{\cos \alpha_n x \cosh \alpha_n(b-y)}{[(\alpha_n^2+h^2)a+h]\cos \alpha_n a \cosh \alpha_n b}. \tag{17}$$

By symmetry this is also the solution for the region $-a < x < a$, $0 < y < 2b$ with $y = 0$ and $y = 2b$ kept at temperature V, and with radiation at $x = -a$ and at $x = a$.

III. $y = 0$ *kept at* $f(x)$. *No flow of heat over* $x = 0$. *Radiation at* $x = a$ *and at* $y = b$ *into a medium at zero temperature*

Here the boundary conditions are (11), (13), (14) and

$$\frac{\partial v}{\partial y}+hv = 0, \quad y = b, \quad 0 < x < a. \tag{18}$$

The expression $\cos \alpha x\{\alpha \cosh \alpha(b-y)+h \sinh \alpha(b-y)\}$

satisfies (1), (13), and (18) for all α. Proceeding as in II the solution is found to be

$$v = 2 \sum_{n=1}^{\infty} \frac{(h^2+\alpha_n^2)\cos \alpha_n x\{\alpha_n \cosh \alpha_n(b-y)+h \sinh \alpha_n(b-y)\}}{[(\alpha_n^2+h^2)a+h]\{\alpha_n \cosh \alpha_n b+h \sinh \alpha_n b\}} \int_0^a f(x)\cos \alpha_n x \, dx, \tag{19}$$

where the α_n are the positive roots of (15).

If $f(x) = V$, constant, this becomes

$$v = 2hV \sum_{n=1}^{\infty} \frac{\cos \alpha_n x\{\alpha_n \cosh \alpha_n(b-y)+h \sinh \alpha_n(b-y)\}}{[(\alpha_n^2+h^2)a+h]\{\alpha_n \cosh \alpha_n b+h \sinh \alpha_n b\}\cos \alpha_n a}, \tag{20}$$

and this is the solution for the region $-a < x < a$, $0 < y < b$ with $y = 0$ kept at constant temperature V, and with radiation into medium at zero over the other faces. This solution has been used in the study of the temperature distribution in cooling fins of finite thickness.† The corresponding problem for a thin fin has been treated in § 4.6 and that of a fin in the form of a rectangular parallelepiped is discussed in § 6.2.

IV. $y = 0$ *kept at* $f(x)$. $y = b$ *kept at zero temperature. No flow of heat over* $x = 0$. *At* $x = a$ *radiation into a medium at zero temperature*

The solution is

$$v = 2 \sum_{n=1}^{\infty} \frac{(h^2+\alpha_n^2)\cos \alpha_n x \sinh \alpha_n(b-y)}{[(\alpha_n^2+h^2)a+h]\sinh \alpha_n b} \int_0^a f(x)\cos \alpha_n x \, dx, \tag{21}$$

where the α_n are the positive roots of (15).

If $f(x) = V$, constant, this becomes

$$v = 2hV \sum_{n=1}^{\infty} \frac{\cos \alpha_n x \sinh \alpha_n(b-y)}{[(\alpha_n^2+h^2)a+h]\cos \alpha_n a \sinh \alpha_n b}, \tag{22}$$

and this is the solution for the region $-a < x < a$, $0 < y < b$ with $y = 0$ kept at V, $y = b$ kept at zero, and radiation from the faces $x = \pm a$.

† Harper and Brown, *Nat. Adv. Comm. Aero. Rep*, No. 158 (1923); Avrami and Little, *J. App. Phys.* **13** (1942) 255.

If $y = 0$ is kept at V_1 and $y = b$ at V_2, and there is radiation at the other two faces into medium at zero,

$$v = 2h \sum_{n=1}^{\infty} \frac{[V_1 \sinh \alpha_n(b-y) + V_2 \sinh \alpha_n y]\cos \alpha_n x}{[(\alpha_n^2 + h^2)a + h]\cos \alpha_n a \sinh \alpha_n b}. \tag{23}$$

V. *The case of different conductivities in the x- and y-directions*

This case, which arises frequently in engineering practice, may be treated by a trivial extension of the method above. Suppose the conductivity in the x-direction is k^2 times that in the y-direction, where k^2 is a constant. Then (1) is replaced by

$$k^2 \frac{\partial^2 v}{\partial x^2} + \frac{\partial^2 v}{\partial y^2} = 0. \tag{24}$$

Considering the problem of **I** above, a term of type

$$\sin \frac{n\pi x}{a} \sinh \frac{k(b-y)n\pi}{a} \tag{25}$$

satisfies (24) and replaces (8). Then the solution of this problem corresponding to (9) is

$$v = \sum_{n=1}^{\infty} a_n \sin \frac{n\pi x}{a} \sinh \frac{k(b-y)n\pi}{a} \operatorname{cosech} \frac{kn\pi b}{a}. \tag{26}$$

5.4. The thin rectangular plate with radiation from its surface

We suppose the plate to lie in the XY-plane, and its thickness D in the direction of Z to be so small that the temperature may be taken to be constant over it. Let H be the surface conductance of the material, K its thermal conductivity, ρ its density, and c its specific heat, then as in § 4.2 the differential equation satisfied by the temperature in the plate is found to be

$$\frac{\partial^2 v}{\partial x^2} + \frac{\partial^2 v}{\partial y^2} - \frac{\rho c}{K}\frac{\partial v}{\partial t} - \frac{2H}{KD}(v - v_0) = 0, \tag{1}$$

where v_0 is the temperature of the surrounding medium.

In the steady state for rectangular boundaries this equation may be solved by the methods of §§ 5.2, 5.3. As an example† we consider the *steady temperature in the region* $0 < x < a$, $0 < y < b$, *which radiates from its surface into medium at zero, for the case in which the edge* $y = 0$ *is kept at temperature* $f(x)$ *and the other edges at zero.*

Here, writing
$$k^2 = 2H/KD, \tag{2}$$

(1) becomes
$$\frac{\partial^2 v}{\partial x^2} + \frac{\partial^2 v}{\partial y^2} - k^2 v = 0, \quad 0 < x < a, \quad 0 < y < b, \tag{3}$$

with
$$v = f(x), \quad y = 0, \quad 0 < x < a, \tag{4}$$
$$v = 0, \quad y = b, \quad 0 < x < a, \tag{5}$$
$$v = 0, \quad x = 0, \quad 0 < y < b, \tag{6}$$
$$v = 0, \quad x = a, \quad 0 < y < b. \tag{7}$$

The expression
$$\operatorname{s n} \frac{n\pi x}{a} \sinh(b-y)\left[k^2 + \frac{n^2\pi^2}{a^2}\right]^{\frac{1}{2}}, \quad n = 1, 2, \dots \tag{8}$$

† For other problems and methods see Malkin, *J. Franklin Inst.* **232** (1941) 129.

satisfies (3), (5), (6), and (7). If we assume as in § 5.3 that $f(x)$ can be expanded in the series

$$f(x) = \frac{2}{a} \sum_{n=1}^{\infty} \sin\frac{n\pi x}{a} \int_0^a f(x)\sin\frac{n\pi x}{a}\, dx, \tag{9}$$

the solution is given by

$$v = \frac{2}{a} \sum_{n=1}^{\infty} \frac{\sin(n\pi x/a)\sinh(b-y)(k^2+n^2\pi^2/a^2)^{\frac{1}{2}}}{\sinh b(k^2+n^2\pi^2/a^2)^{\frac{1}{2}}} \int_0^a f(x)\sin\frac{n\pi x}{a}\, dx. \tag{10}$$

5.5. Steady temperature in the rectangle $-a < x < a$, $-b < y < b$ with heat generation

Problems of this type are of considerable technical importance† since many types of electrical windings may be regarded approximately as cylinders of rectangular cross-section. Here we shall illustrate the application of the methods of § 5.3—the Green's function may also be used as in § 16.2. In many problems of this type the conductivities in the x- and y-directions are different, and we shall take them here to be K and (K/k^2) where k is a constant.

I. *Heat production at a constant rate A_0 in the rectangle, with heat transfer into a medium at zero through surface resistance R per unit area over the surfaces $x = \pm a$, and R' over the surfaces $y = \pm b$*

The differential equation is

$$\frac{\partial^2 v}{\partial x^2} + \frac{1}{k^2}\frac{\partial^2 v}{\partial y^2} = -\frac{A_0}{K}. \tag{1}$$

The expression

$$aRA_0 + \frac{A_0(a^2-x^2)}{2K} + \sum_{n=1}^{\infty} a_n \cos\alpha_n x \cosh ky\alpha_n \tag{2}$$

satisfies (1) and the boundary conditions at $x = \pm a$ if α_n, $n = 1, 2,...$, are the positive roots of

$$\alpha \tan a\alpha = 1/RK. \tag{3}$$

The boundary conditions at $y = \pm b$ require

$$\sum_{n=1}^{\infty} a_n \cos\alpha_n x[k^{-1}KR'\alpha_n \sinh kb\alpha_n + \cosh kb\alpha_n] = -aRA_0 - \frac{A_0(a^2-x^2)}{2K}. \tag{4}$$

† Cockroft, *Proc. Camb. Phil. Soc.* **22** (1925) 759–72. Higgins, *Elect. Engng.* **64** (1945) 190–4, discusses the more general case of heat transfer with different surface resistances, and into media at different temperatures, at all four surfaces. The rises of temperature in practice are sufficiently large for the variations of heat production and thermal conductivity with temperature to be of importance. The present method is applicable to heat production $A_0(1+\alpha v)$. The variation of thermal conductivity with temperature may be taken into account by using the variable Θ of 1.6 (10) in place of v. See also Jakob, *Trans. Amer. Soc. Mech. Engrs.* **65** (1943) 593–602. Buchholz, *Z. angew. Math. Mech.* **14** (1934) 285–94, discusses a rather more complicated boundary condition. Toroidal coils of rectangular cross-section are discussed by Higgins, *J. Franklin Inst.* **240** (1945) 97–112.

a_n is then found from **3.10** (9) and we get finally

$$v = aRA_0 + \frac{A_0(a^2 - x^2)}{2K} -$$

$$-\frac{4A_0}{K}\sum_{n=1}^{\infty} \frac{\sin a\alpha_n \cos x\alpha_n \cosh ky\alpha_n}{\alpha_n^2[2a\alpha_n + \sin 2a\alpha_n][k^{-1}KR'\alpha_n \sinh kb\alpha_n + \cosh kb\alpha_n]}. \quad (5)$$

II. *Heat production at a constant rate A_0 in the rectangle. The surfaces $x = \pm a$, $y = \pm b$ kept at zero temperature*

$$v = \frac{A_0(a^2 - x^2)}{2K} - \frac{16A_0 a^2}{K\pi^3} \sum_{n=0}^{\infty} \frac{(-1)^n \cos[(2n+1)\pi x/2a]\cosh[(2n+1)\pi ky/2a]}{(2n+1)^3\cosh[(2n+1)\pi kb/2a]}. \quad (6)$$

5.6. Variable state. Product solutions

It was remarked in § 1.15 that for certain types of initial and boundary conditions, the solution of problems in several space variables can be written down as the product of solutions of one-variable problems. In the two-dimensional case the initial temperature has to be expressible as a product $f(x)f(y)$, and the boundary conditions have to be zero temperature, zero flux, or radiation into a medium at zero.

The case of the greatest practical importance is that of constant initial temperature, and only this will be considered here. In all cases solutions for zero initial temperature with unit temperature, or radiation into medium at unit temperature, at the surface, may be obtained by subtracting the solutions given below from unity.

I. *The rectangular corner $x > 0$, $y > 0$, with unit initial temperature*

The solution for the region $x > 0$ with unit initial temperature and zero surface temperature was found in 2.4 (3) to be

$$\phi(x) = \operatorname{erf} \frac{x}{2\sqrt{(\kappa t)}}. \quad (1)$$

Thus the solution for the rectangular corner $x > 0$, $y > 0$ with unit initial temperature and zero surface temperature is

$$v = \phi(x)\phi(y) = \operatorname{erf} \frac{x}{2\sqrt{(\kappa t)}} \operatorname{erf} \frac{y}{2\sqrt{(\kappa t)}}. \quad (2)$$

In Fig. 21 isothermals corresponding to the values $0, 0\cdot1, 0\cdot2, \ldots, 0\cdot9$ of v are plotted against $x/2\sqrt{(\kappa t)}$ and $y/2\sqrt{(\kappa t)}$.

For a body bounded by planes meeting at right angles, the solution (2) will hold approximately near any external corner.

The flux of heat at the point $(0, y)$ of the surface is

$$-\frac{K}{\sqrt{(\pi\kappa t)}} \operatorname{erf} \frac{y}{2\sqrt{(\kappa t)}},$$

which is less than that for the semi-infinite solid $x > 0$ by the amount

$$\frac{K}{\sqrt{(\pi\kappa t)}}\operatorname{erfc}\frac{y}{2\sqrt{(\kappa t)}}.$$

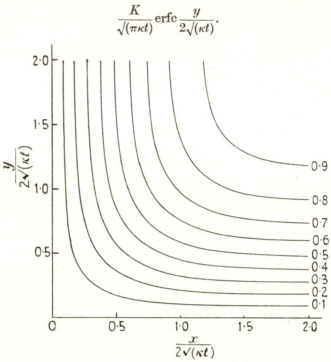

FIG. 21. Isothermals 0·1, 0·2,..., 0·9 for a rectangular corner, initially at unit temperature, with its surface kept at zero temperature.

Thus the rate of loss of heat per unit time, per unit length perpendicular to the bounding planes, from the rectangular corner is less than that from the semi-infinite solid by

$$\frac{2K}{\sqrt{(\pi\kappa t)}}\int\limits_{0}^{\infty}\operatorname{erfc}\frac{y}{2\sqrt{(\kappa t)}}\,dy = \frac{4K}{\pi}, \tag{3}$$

where the integral has been evaluated by the use of Appendix II (15).

If the region $x > 0$, $y > 0$ has unit initial temperature and radiates at its surfaces into medium at zero temperature, the temperature is

$$v = \phi(h,x)\phi(h,y), \tag{4}$$

where $\phi(h,x)$ is now the quantity defined in 2.7 (1), namely

$$\phi(h,x) = \operatorname{erf}\frac{x}{2\sqrt{(\kappa t)}} + e^{hx+h^2\kappa t}\operatorname{erfc}\left\{\frac{x}{2\sqrt{(\kappa t)}} + h\sqrt{(\kappa t)}\right\}. \tag{5}$$

II. *The semi-infinite rectangle $-l < x < l$, $y > 0$, with unit initial temperature*

If the faces $x = \pm l$ are kept at zero, the relevant one-dimensional solution is that of 3.3 (8)

$$\psi(x,l) = \frac{4}{\pi} \sum_{n=0}^{\infty} \frac{(-1)^n}{(2n+1)} e^{-\kappa(2n+1)^2\pi^2 t/4l^2} \cos \frac{(2n+1)\pi x}{2l}, \qquad (6)$$

while, if there is radiation at these faces, the relevant solution is that of 3.11 (12)

$$\psi(x,l,h) = \sum_{n=1}^{\infty} \frac{2h \cos \alpha_n x}{[(h^2+\alpha_n^2)l+h]\cos \alpha_n l} e^{-\kappa \alpha_n^2 t}, \qquad (7)$$

where the α_n are the positive roots of

$$\alpha \tan \alpha l = h.$$

If all surfaces of the region are kept at zero, the solution is

$$v = \phi(y)\psi(x,l). \qquad (8)$$

If there is radiation into a medium at zero at all surfaces,

$$v = \phi(h,y)\psi(x,l,h). \qquad (9)$$

If there is radiation at $y = 0$ into a medium at zero, and the faces $x = \pm l$ are kept at zero,

$$v = \phi(h,y)\psi(x,l), \qquad (10)$$

and so on.

III. *The finite rectangle $-l < x < l$, $-b < y < b$ with unit initial temperature*

Here the solutions of the corresponding one-variable problems are defined in (6) and (7).

If all surfaces of the region are kept at zero, the solution is

$$v = \psi(x,l)\psi(y,b). \qquad (11)$$

Values of the function $\psi(x,l)$ may be read off from Fig. 11, and the isothermals for any value of the time are easily constructed. In Figs. 22 and 23 the isothermals $v = 0\cdot1, 0\cdot2,..., 0\cdot9$ for a square, and for a rectangle in which $b = 2l$, are shown for $\kappa t/l^2 = 0\cdot08$.

Similarly, if there is radiation at the surfaces into medium at zero, the solution is

$$v = \psi(x,l,h)\psi(y,b,h). \qquad (12)$$

As remarked in § 1.15, if the solid has different diffusivities in the x- and y-directions and different surface conductances at different faces, the result may still be written down in the same way.

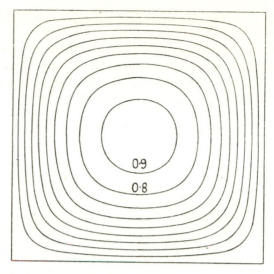

Fig. 22. Isothermals 0·1, 0·2,..., 0·9 for a square of side $2l$ with unit initial and zero surface temperature: $\kappa t/l^2 = 0·08$.

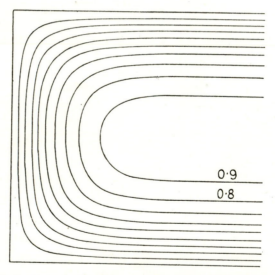

Fig. 23. Isothermals 0·1, 0·2,..., 0·9 for a rectangle of sides $2l$, $4l$, with unit initial and zero surface temperature: $\kappa t/l^2 = 0·08$.

5.7. Variable state. Arbitrary initial and surface conditions

The general problem can, as in § 1.14, be reduced to a problem of steady temperature and one of prescribed initial temperature and zero surface temperature (or radiation into medium at zero). If this initial temperature is a product of functions of x and y, the solution may be expressed as above as a product: if not, the theory of double Fourier series must be used and the solution appears as a double series. The method is discussed in detail for the rectangular parallelepiped in §§ 6.3, 6.6. Alternatively† the Green's function, §§ 14.4, 14.5, may be used. A general survey of the methods available is given in § 6.1.

† Lowan, *Phil. Mag.* (7) **24** (1937) 410, gives a number of solutions using the Laplace transformation.

VI

THE FLOW OF HEAT IN A RECTANGULAR PARALLELEPIPED

6.1. Introductory

THE three-dimensional problems which have been studied most are those of conduction of heat in regions bounded by surfaces of the Rectangular Cartesian, Cylindrical, and Spherical Polar coordinate systems. For Radial Flow in cylinders and spheres the solution involves only one space variable and the time, and is considered in Chapters VII and IX; in this chapter, and Chapter VIII, we consider problems on regions such as the rectangular parallelepiped and the finite cylinder in which two or more space variables are involved. As the results can be obtained in several different ways, it is desirable at this stage to mention the methods which will be used and the relationships between them.

(i) The simplest and most important problems are those whose solutions can be expressed by § 1.15 as a product of solutions of one-variable problems.

The fundamental case is that of unit initial temperature and zero surface temperature (or radiation into medium at zero temperature); from this the solution for zero initial temperature and unit surface temperature follows, and hence, by Duhamel's theorem, that for surface temperature $\phi(t)$.

(ii) Arbitrary initial and surface temperatures can be discussed by the use of multiple Fourier series or their generalizations.

(iii) The use of the Green's function, Chapter XIV, also gives a complete solution of the general problem of arbitrary initial and surface temperature. For the simple cases referred to in (i) it gives the same result after some reduction. It also gives immediately solutions for heat production in the solid at a rate which is a given function of position and time.

(iv) The direct application of the Laplace transformation method, § 15.11, is particularly useful when some of the boundary surfaces are maintained at temperatures which are simple functions of the time: constant, kt, $\sin \omega t$, etc. In such cases the results obtained by (ii) and (iii) (and (i), unless the surface temperature is constant), in the form of integrals, only give the simplest form of the solution after certain series have been summed.

In this chapter we apply the first two of these methods to problems on the rectangular parallelepiped. Further problems on this region are discussed by the other methods in §§ 14.5, 15.11.

Conduction of heat in solids of simple geometrical forms such as the rectangular parallelepiped and the finite cylinder is of considerable importance since such solids occur so commonly in practice, as in tinned foods, cased fruit, etc. Also many of the older methods of determining the thermal conductivities of poor conductors used those materials in the form of cubes, spheres, and finite cylinders, since the 'bar' methods discussed earlier are unsuitable because the amount of heat lost by radiation from the surface of a badly conducting bar may be large compared with that conducted along the bar.

6.2. Steady temperature

In this section we consider problems in which certain faces of the solid are maintained at constant temperature, while others are at zero temperature or radiate into a medium at zero. More complicated cases in which the surface temperatures are prescribed functions of position can be dealt with, as in § 6.4, by using the theory of double Fourier series.

I. *The solid $0 < x < a$, $0 < y < b$, $0 < z < c$. Surface temperature v_1, constant, on $x = 0$; v_2, constant, on $x = a$; the other faces at zero*

Here the differential equation for the temperature is

$$\frac{\partial^2 v}{\partial x^2} + \frac{\partial^2 v}{\partial y^2} + \frac{\partial^2 v}{\partial z^2} = 0. \tag{1}$$

It is clear that the expression

$$\frac{v_1 \sinh l(a-x) + v_2 \sinh lx}{\sinh la} \sin \frac{m\pi y}{b} \sin \frac{n\pi z}{c} \tag{2}$$

satisfies (1) provided $\quad l^2 = \left(\frac{m^2}{b^2} + \frac{n^2}{c^2}\right)\pi^2. \tag{3}$

Also, if m and n are integers, it vanishes when $y = 0$ and $y = b$, and when $z = 0$ and $z = c$. The expression

$$v = \sum_{m=1}^{\infty} \sum_{n=1}^{\infty} A_{m,n} \frac{v_1 \sinh l(a-x) + v_2 \sinh lx}{\sinh la} \sin \frac{m\pi y}{b} \sin \frac{n\pi z}{c} \tag{4}$$

has the same properties, and also has the value v_1 when $x = 0$, and the value v_2 when $x = a$, provided that

$$\sum_{m=1}^{\infty} \sum_{n=1}^{\infty} A_{m,n} \sin\frac{m\pi y}{b} \sin\frac{n\pi z}{c} = 1.\dagger \tag{5}$$

Now‡ expanding unity in a sine series in $(0, b)$ gives

$$1 = \frac{4}{\pi} \sum_{p=0}^{\infty} \frac{\sin[(2p+1)\pi y/b]}{2p+1}. \tag{6}$$

And similarly $\qquad 1 = \frac{4}{\pi} \sum_{q=0}^{\infty} \frac{\sin[(2q+1)\pi z/c]}{2q+1}. \tag{7}$

Multiplying (6) and (7) gives

$$1 = \frac{16}{\pi^2} \sum_{p=0}^{\infty} \sum_{q=0}^{\infty} \frac{\sin[(2p+1)\pi y/b]\sin[(2q+1)\pi z/c]}{(2p+1)(2q+1)}. \tag{8}$$

Comparing (8) with (5) we see that $A_{m,n}$ must be zero unless m and n are both odd, and in that case it is equal to $16/\pi^2 mn$. Thus finally from (4)

$$v = \frac{16}{\pi^2} \sum_{p=0}^{\infty} \sum_{q=0}^{\infty} \frac{[v_1 \sinh l(a-x)+v_2 \sinh lx]\sin[(2p+1)\pi y/b]\sin[(2q+1)\pi z/c]}{(2p+1)(2q+1)\sinh la}, \tag{9}$$

where $\qquad l^2 = \dfrac{(2p+1)^2\pi^2}{b^2} + \dfrac{(2q+1)^2\pi^2}{c^2}. \tag{10}$

II. *Steady temperature in the solid $0 < x < a$, $-b < y < b$, $-c < z < c$ when the faces $x = 0$ and $x = a$ are maintained at temperatures v_1 and v_2 respectively, and there is radiation at the other faces into medium at zero*

Here the surface conditions are

$$v = v_1 \quad \text{when } x = 0, \quad \text{and} \quad v = v_2 \quad \text{when } x = a; \tag{11}$$

$$\frac{\partial v}{\partial y} - hv = 0, \quad \text{when } y = -b; \qquad \frac{\partial v}{\partial y} + hv = 0, \quad \text{when } y = b; \tag{12}$$

$$\frac{\partial v}{\partial z} - hv = 0, \quad \text{when } z = -c; \qquad \frac{\partial v}{\partial z} + hv = 0, \quad \text{when } z = c. \tag{13}$$

† For a rigorous treatment of this question it would be necessary to justify the term-by-term differentiation of the double series for v. The same remark applies to the other problems discussed in this chapter. Cf. § 3.3 above, and Moore, 'On convergence factors in double series and the double Fourier's series', *Trans. Amer. Math. Soc.* **14** (1913) 73, also *Bull. Amer. Math. Soc.* **25** (1919) 274.

‡ This artifice avoids the use of double Fourier series in this case. They have to be used if the surface temperatures are arbitrary functions of y and z, cf. § 6,3 below. The product series (8) is known to be convergent, cf. Hardy, *Proc. Lond. Math. Soc.* (2) **6** (1908) 410.

The expression $\dfrac{v_1\sinh l(a-x)+v_2\sinh lx}{\sinh la}\cos\alpha_r y\cos\beta_s z$ (14)

satisfies the equation of conduction, provided

$$l^2 = \alpha_r^2+\beta_s^2.$$ (15)

Also it satisfies the surface conditions (12) and (13), provided α_r is a root of

$$\alpha\tan\alpha b = h,$$ (16)

and provided β_s is a root of $\beta\tan\beta c = h.$ (17)

Thus the solution of the problem is given by

$$\sum_{r=1}^{\infty}\sum_{s=1}^{\infty} A_{rs}\frac{v_1\sinh l(a-x)+v_2\sinh lx}{\sinh la}\cos\alpha_r y\cos\beta_s z,$$ (18)

provided the constants A_{rs} are chosen so that

$$\sum_{r=1}^{\infty}\sum_{s=1}^{\infty} A_{rs}\cos\alpha_r y\cos\beta_s z = 1.$$ (19)

Now by 3.10 (10), if α_r and β_s are positive roots of (16) and (17) respectively,

$$2h\sum_{r=1}^{\infty}\frac{\cos\alpha_r y}{[(\alpha_r^2+h^2)b+h]\cos\alpha_r b} = 1,$$ (20)

and $2h\displaystyle\sum_{s=1}^{\infty}\frac{\cos\beta_s z}{[(\beta_s^2+h^2)c+h]\cos\beta_s c} = 1.$ (21)

Multiplying these we have

$$\sum_{r=1}^{\infty}\sum_{s=1}^{\infty}\frac{4h^2\cos\alpha_r y\cos\beta_s z}{\cos\alpha_r b\cos\beta_s c[(\alpha_r^2+h^2)b+h][(\beta_s^2+h^2)c+h]} = 1.$$ (22)

As before A_{rs} in (18) is found by comparing (19) and (22), and (18) becomes finally

$$v = \sum_{r=1}^{\infty}\sum_{s=1}^{\infty}\frac{4h^2[v_1\sinh l(a-x)+v_2\sinh lx]\cos\alpha_r y\cos\beta_s z}{\cos\alpha_r b\cos\beta_s c[(\alpha_r^2+h^2)b+h][(\beta_s^2+h^2)c+h]\sinh la},$$ (23)

where l is defined in (15).

This solution is not very suitable for numerical calculation or for determination of thermal conductivity.†

III. *Steady temperature in the solid $0 < x < a$, $-b < y < b$, $-c < z < c$, with $x = 0$ maintained at temperature V, constant, and radiation at the other faces into medium at zero*

Here the boundary conditions are (12) and (13), and

$$v = V, \quad x = 0,$$ (24)

$$\frac{\partial v}{\partial x}+hv = 0, \quad x = a.$$ (25)

† Chivers, *Phil. Mag.* (7) **6** (1928) 305.

The expression corresponding to (14) is, in this case,

$$V \frac{h \sinh l(a-x) + l \cosh l(a-x)}{h \sinh al + l \cosh al} \cos \alpha_r y \cos \beta_s z, \tag{26}$$

where l is given by (15), and α_r and β_s are positive roots of (16) and (17). Proceeding as in II we have finally

$$v = \sum_{r=1}^{\infty} \sum_{s=1}^{\infty} \frac{4h^2 V[h \sinh l(a-x) + l \cosh l(a-x)] \cos \alpha_r y \cos \beta_s z}{[h \sinh al + l \cosh al][(\alpha_r^2 + h^2)b + h][(\beta_s^2 + h^2)c + h] \cos \alpha_r b \cos \beta_s c}. \tag{27}$$

This is a generalization to the case of a short cooling fin of the two-dimensional result of 5.3 (20).

IV. Problems on the infinite rectangular solid are treated in the same way, cf. also § 5.2. As an example consider *the solid $x > 0$, $-b < y < b$, $-c < z < c$, with the surface $x = 0$ at V, and radiation at the other surfaces into medium at zero*

Here the expression corresponding to (14) is

$$Ve^{-lx} \cos \alpha_r y \cos \beta_s z, \tag{28}$$

and we get finally

$$v = 4h^2 V \sum_{r=1}^{\infty} \sum_{s=1}^{\infty} \frac{e^{-lx} \cos \alpha_r y \cos \beta_s z}{[(\alpha_r^2 + h^2)b + h][(\beta_s^2 + h^2)c + h] \cos \alpha_r b \cos \beta_s c}, \tag{29}$$

where α_r, β_s, and l are defined in (16), (17), and (15) respectively.

6.3. Double and multiple Fourier series

The usual formal discussion (*F.S.*, § 90) of the Fourier series of a function $f(x)$ of one variable, defined in $(-l, l)$, assumes that $f(x)$ can be expanded in the series

$$f(x) = a_0 + \sum_{n=1}^{\infty} \left(a_n \cos \frac{n\pi x}{l} + b_n \sin \frac{n\pi x}{l} \right), \tag{1}$$

and derives the coefficients a_0, a_n, b_n by multiplying (1) by unity, $\cos(n\pi x/l)$, and $\sin(n\pi x/l)$, respectively, and integrating the resulting series term by term.

In this way, using the results

$$\int_{-l}^{l} \sin \frac{m\pi x}{l} \sin \frac{n\pi x}{l} \, dx = 0, \qquad \int_{-l}^{l} \cos \frac{m\pi x}{l} \cos \frac{n\pi x}{l} \, dx = 0, \quad m \neq n, \tag{2}$$

$$\int_{-l}^{l} \sin \frac{m\pi x}{l} \cos \frac{n\pi x}{l} \, dx = 0, \tag{3}$$

$$\int_{-l}^{l} \left[\sin \frac{m\pi x}{l} \right]^2 dx = \int_{-l}^{l} \left[\cos \frac{m\pi x}{l} \right]^2 dx = l, \quad m = 1, 2, ..., \tag{4}$$

it is found that

$$a_0 = \frac{1}{2l} \int_{-l}^{l} f(x) \, dx; \qquad a_n = \frac{1}{l} \int_{-l}^{l} f(x') \cos \frac{n\pi x'}{l} \, dx';$$

$$b_n = \frac{1}{l} \int_{-l}^{l} f(x') \sin \frac{n\pi x'}{l} \, dx'. \tag{5}$$

This formal theory is then made rigorous by a careful discussion of the series on the right-hand side of (1) with coefficients given by (5), and it is found that, if $f(x)$ satisfies certain conditions (e.g. Dirichlet's conditions, *F.S.*, § 93), this series is convergent, and its sum is $f(x)$ at every point of continuity of $f(x)$, and $\frac{1}{2}\{f(x+0)+f(x-0)\}$ at all other points.

The formal theory of double and multiple Fourier series follows precisely the same lines. Suppose we have a function $f(x,y)$ defined in the rectangle $-a < x < a$, $-b < y < b$. Corresponding to the two sets of 'orthogonal functions', $\sin(n\pi x/l)$, $n = 1, 2,...$, and $\cos(m\pi x/l)$, $m = 0, 1,...$, such that the integral from $-l$ to l of the product of any two different functions vanishes, we have in the two-variable case the four sets

$$
\left.
\begin{aligned}
&\sin\frac{m\pi x}{a}\sin\frac{n\pi y}{b}; && m = 1, 2,...; && n = 1, 2,... \\[2mm]
&\sin\frac{m\pi x}{a}\cos\frac{n\pi y}{b}; && m = 1, 2,...; && n = 0, 1,... \\[2mm]
&\cos\frac{m\pi x}{a}\sin\frac{n\pi y}{b}; && m = 0, 1,...; && n = 1, 2,... \\[2mm]
&\cos\frac{m\pi x}{a}\cos\frac{n\pi y}{b}; && m = 0, 1,...; && n = 0, 1,...
\end{aligned}
\right\}, \tag{6}
$$

with the properties corresponding to (2), (3), and (4). That is, the only products of such functions whose integrals over the rectangle do not vanish are the following:

$$
\left.
\begin{aligned}
\int_{-a}^{a}\int_{-b}^{b}\left[\sin\frac{m\pi x}{a}\sin\frac{n\pi y}{b}\right]^2 dx\,dy &= \int_{-a}^{a}\int_{-b}^{b}\left[\sin\frac{m\pi x}{a}\cos\frac{n\pi y}{b}\right]^2 dx\,dy = ab \\[2mm]
\int_{-a}^{a}\int_{-b}^{b}\left[\cos\frac{m\pi x}{a}\sin\frac{n\pi y}{b}\right]^2 dx\,dy &= \int_{-a}^{a}\int_{-b}^{b}\left[\cos\frac{m\pi x}{a}\cos\frac{n\pi y}{b}\right]^2 dx\,dy = ab
\end{aligned}
\right\}, \tag{7}
$$

for $m \geqslant 1, n \geqslant 1$. If $n = 0$ in the second, $m = 0$ in the third, or either $m = 0$ or $n = 0$ in the last of these integrals, the result is doubled; while if $m = n = 0$ in the last integral its value is $4ab$.

The expansion of $f(x,y)$ analogous to (1) is

$$
f(x,y) = \sum_{m=1}^{\infty}\sum_{n=1}^{\infty} A_{m,n}\sin\frac{m\pi x}{a}\sin\frac{n\pi y}{b} + \sum_{m=1}^{\infty}\sum_{n=0}^{\infty} A'_{m,n}\sin\frac{m\pi x}{a}\cos\frac{n\pi y}{b} +
$$

$$
+ \sum_{m=0}^{\infty}\sum_{n=1}^{\infty} B'_{m,n}\cos\frac{m\pi x}{a}\sin\frac{n\pi y}{b} + \sum_{m=0}^{\infty}\sum_{n=0}^{\infty} B_{m,n}\cos\frac{m\pi x}{a}\cos\frac{n\pi y}{b}. \tag{8}
$$

To find the coefficients we multiply both sides of (8) by one of the functions (6) and integrate from $-a$ to a with respect to x, and from $-b$ to b with respect to y. Then, using (7) and the fact that all the other double integrals of products of the functions (6) vanish, we find

$$A_{m,n} = \frac{1}{ab} \int_{-a}^{a} \int_{-b}^{b} f(x,y)\sin\frac{m\pi x}{a}\sin\frac{n\pi y}{b}\,dxdy;$$

$$A'_{m,n} = \frac{1}{ab} \int_{-a}^{a} \int_{-b}^{b} f(x,y)\sin\frac{m\pi x}{a}\cos\frac{n\pi y}{b}\,dxdy;$$

$$B'_{m,n} = \frac{1}{ab} \int_{-a}^{a} \int_{-b}^{b} f(x,y)\cos\frac{m\pi x}{a}\sin\frac{n\pi y}{b}\,dxdy;$$

$$B_{m,n} = \frac{1}{ab} \int_{-a}^{a} \int_{-b}^{b} f(x,y)\cos\frac{m\pi x}{a}\cos\frac{n\pi y}{b}\,dxdy, \tag{9}$$

except that $A'_{m,0}$, $B'_{0,n}$, $B_{0,n}$, $B_{m,0}$ are one-half, and $B_{0,0}$ one-quarter, of the above values.

We shall usually be concerned with the case in which $f(x,y)$ is an odd function of both x and y, so that all the $A'_{m,n}$, $B'_{m,n}$, and $B_{m,n}$ vanish, and it follows that

$$f(x,y) = \sum_{m=1}^{\infty}\,' \sum_{n=1}^{\infty} A_{m,n}\sin\frac{m\pi x}{a}\sin\frac{n\pi y}{b}, \tag{10}$$

where

$$A_{m,n} = \frac{4}{ab} \int_{0}^{a}\int_{0}^{b} f(x',y')\sin\frac{m\pi x'}{a}\sin\frac{n\pi y'}{b}\,dx'dy', \tag{11}$$

which is the analogue of the Fourier sine series.

Similarly if $f(x,y,z)$ is defined in $-a < x < a$, $-b < y < b$, $-c < z < c$, and is an odd function of x, y, and z, we have the triple sine series

$$f(x,y,z) = \sum_{l=1}^{\infty}\sum_{m=1}^{\infty}\sum_{n=1}^{\infty} A_{l,m,n}\sin\frac{l\pi x}{a}\sin\frac{m\pi y}{b}\sin\frac{n\pi z}{c}, \tag{12}$$

where

$$A_{l,m,n} = \frac{8}{abc} \int_{0}^{a}\int_{0}^{b}\int_{0}^{c} f(x',y',z')\sin\frac{l\pi x'}{a}\sin\frac{m\pi y'}{b}\sin\frac{n\pi z'}{c}\,dx'dy'dz'. \tag{13}$$

If $f(x,y,z)$ is an even function of x, y, and z, we find a cosine series in the same way.

Alternatively, problems involving multiple Fourier series may be looked at from the following point of view which will also be found useful in Chapter VIII, where combinations of Fourier and Fourier–Bessel series occur. As an example we consider the case in which $f(x,y)$ is an odd function of both x and y in $-a < x < a$, $-b < y < b$. For any fixed y in $-b < y < b$, $f(x,y)$ is an odd function of x, and may thus be expanded in the sine series

$$f(x,y) = \sum_{m=1}^{\infty} a_m(y)\sin\frac{m\pi x}{a}, \tag{14}$$

where the coefficients

$$a_m(y) = \frac{2}{a} \int_0^a f(x', y) \sin \frac{m\pi x'}{a} \, dx' \tag{15}$$

are now functions of y. These are odd functions of y in $-b < y < b$, and thus may be expanded in the series

$$a_m(y) = \sum_{n=1}^{\infty} A_{m,n} \sin \frac{n\pi y}{b}, \tag{16}$$

where

$$A_{m,n} = \frac{2}{b} \int_0^b a_m(y') \sin \frac{n\pi y'}{b} \, dy' = \frac{4}{ab} \int_0^b \sin \frac{n\pi y'}{b} \, dy' \int_0^a f(x', y') \sin \frac{m\pi x'}{a} \, dx'. \tag{17}$$

Thus, finally, $\qquad f(x, y) = \sum_{m=1}^{\infty} \sum_{n=1}^{\infty} A_{m,n} \sin \frac{m\pi x}{a} \sin \frac{n\pi y}{b}$,

with $A_{m,n}$ given by (17), in agreement with (10) and (11). All the other cases may be treated similarly. The same argument may be applied to the expansion of a function of several variables in terms of the functions of §§ 3.9, 3.10.

The solution of the problems of § 6.2 in which the surface temperatures are functions of position on the surface may now be written down.

Taking the problem of § 6.2 I, we consider *steady temperature in the solid* $0 < x < a$, $0 < y < b$, $0 < z < c$ *with* $x = 0$ *maintained at temperature* $f(y, z)$ *and the other faces at zero.*

Here $\qquad \dfrac{\sinh l(a-x)}{\sinh al} \sin \dfrac{m\pi y}{b} \sin \dfrac{n\pi z}{c} \tag{18}$

satisfies the differential equation of the problem if

$$l^2 = \left(\frac{m^2}{b^2} + \frac{n^2}{c^2}\right)\pi^2, \tag{19}$$

and vanishes on all surfaces of the solid except $x = 0$.

Then, if $f(y, z)$ can be expanded in a sine series (10), the solution of our problem is

$$v = \sum_{m=1}^{\infty} \sum_{n=1}^{\infty} A_{m,n} \frac{\sinh l(a-x)}{\sinh al} \sin \frac{m\pi y}{b} \sin \frac{n\pi z}{c}, \tag{20}$$

where $\qquad A_{m,n} = \dfrac{4}{bc} \int_0^b \int_0^c f(y', z') \sin \dfrac{m\pi y'}{b} \sin \dfrac{n\pi z'}{c} \, dy'dz'. \tag{21}$

6.4. Variable temperature. Product solutions

As in § 5.6 the solutions of a number of important problems can be written down by using the method of § 1.15 and known solutions of one-dimensional problems. Here we give results for unit initial temperature and zero surface temperature (or radiation into a medium at zero). Solutions for zero initial temperature and unit surface temperature (or radiation into medium at unity) are obtained by subtracting the results given below from unity. Solutions for arbitrary surface temperatures then follow by Duhamel's theorem, § 1.14. If the solid is anisotropic with thermal axes parallel to the coordinate planes, and if the surface conductances at the faces are different the method is still applicable.

I. *The region $x > 0$, $y > 0$, $z > 0$, with unit initial temperature and zero surface temperature*

$$v = \operatorname{erf} \frac{x}{2\sqrt{(\kappa t)}} \operatorname{erf} \frac{y}{2\sqrt{(\kappa t)}} \operatorname{erf} \frac{z}{2\sqrt{(\kappa t)}}. \tag{1}$$

II. *The same region with unit initial temperature and radiation at the surface into medium at zero*

$$v = \phi(h, x)\phi(h, y)\phi(h, z), \tag{2}$$

where $\phi(h, x)$ is defined in 5.6 (5).

III. *The region $-a < x < a$, $-b < y < b$, $z > 0$, with unit initial temperature and zero surface temperature*

$$v = \psi(x, a)\psi(y, b)\operatorname{erf} \frac{x}{2\sqrt{(\kappa t)}}, \tag{3}$$

where

$$\psi(x, a) = \frac{4}{\pi} \sum_{n=0}^{\infty} \frac{(-1)^n}{(2n+1)} e^{-\kappa(2n+1)^2\pi^2 t/4a^2} \cos \frac{(2n+1)\pi x}{2a}. \tag{4}$$

IV. *The region $-a < x < a$, $-b < y < b$, $-c < z < c$, with unit initial temperature and zero surface temperature*

$$v = \psi(x, a)\psi(y, b)\psi(z, c) \tag{5}$$

$$= \frac{64}{\pi^3} \sum_{l=0}^{\infty} \sum_{m=0}^{\infty} \sum_{n=0}^{\infty} \frac{(-1)^{l+m+n}}{(2l+1)(2m+1)(2n+1)} \times$$

$$\times \cos \frac{(2l+1)\pi x}{2a} \cos \frac{(2m+1)\pi y}{2b} \cos \frac{(2n+1)\pi z}{2c} e^{-\alpha_{l,m,n}t}, \tag{6}$$

where

$$\alpha_{l,m,n} = \frac{\kappa\pi^2}{4}\left[\frac{(2l+1)^2}{a^2} + \frac{(2m+1)^2}{b^2} + \frac{(2n+1)^2}{c^2}\right]. \tag{7}$$

V. *For the region $-a < x < a$, $-b < y < b$, $-c < z < c$, with unit initial temperature and radiation at the surface into medium at zero*[†]

$$v = \psi(x, a, h)\psi(y, b, h)\psi(z, c, h), \tag{8}$$

where

$$\psi(x, a, h) = \sum_{n=1}^{\infty} \frac{2h \cos \alpha_n x}{[(h^2 + \alpha_n^2)a + h]\cos \alpha_n a} e^{-\kappa\alpha_n^2 t}, \tag{9}$$

[†] Berger, *Z. angew. Math. Mech.* **8** (1928) 479, gives figures of some isothermals.

and the α_n are the positive roots of

$$\alpha \tan \alpha a = h. \tag{10}$$

VI. *The region* $-a < x < a,\ -b < y < b,\ -c < z < c,$ *with zero initial temperature and surface temperature* $\phi(t)$

If $\phi(t) = V$, constant, the solution, which follows from (6) and (7), is

$$v = V - \frac{64V}{\pi^3} \sum_{l=0}^{\infty} \sum_{m=0}^{\infty} \sum_{n=0}^{\infty} \frac{(-1)^{l+m+n}}{(2l+1)(2m+1)(2n+1)} \times$$

$$\times \cos\frac{(2l+1)\pi x}{2a} \cos\frac{(2m+1)\pi y}{2b} \cos\frac{(2n+1)\pi z}{2c} e^{-\alpha_{l,m,n}t}. \tag{11}$$

For surface temperature $\phi(t)$, by Duhamel's theorem,

$$v = \frac{64}{\pi^3} \sum_{l=0}^{\infty} \sum_{m=0}^{\infty} \sum_{n=0}^{\infty} \frac{\alpha_{l,m,n}(-1)^{l+m+n}}{(2l+1)(2m+1)(2n+1)} \times$$

$$\times \cos\frac{(2l+1)\pi x}{2a} \cos\frac{(2m+1)\pi y}{2b} \cos\frac{(2n+1)\pi z}{2c} e^{-\alpha_{l,m,n}t} \int_0^t e^{\alpha_{l,m,n}\lambda}\phi(\lambda)\, d\lambda. \tag{12}$$

If $\phi(t) = kt$, this gives†

$$v = kt - \frac{64k}{\pi^3} \sum_{l=0}^{\infty} \sum_{m=0}^{\infty} \sum_{n=0}^{\infty} \frac{(-1)^{l+m+n}}{(2l+1)(2m+1)(2n+1)\alpha_{l,m,n}} (1 - e^{-\alpha_{l,m,n}t}) \times$$

$$\times \cos\frac{(2l+1)\pi x}{2a} \cos\frac{(2m+1)\pi y}{2b} \cos\frac{(2n+1)\pi z}{2c}, \tag{13}$$

where we have used the value of (11) with $t = 0$ to reduce one of the series. An expression for the case of harmonic surface temperature can be written down from (12), but the final form is not a very useful one. These problems are again discussed in § 15.11.

6.5. Application to the determination of thermal conductivity and the extrapolation of cooling curves

The solutions 6.4 (6) and 6.4 (8) for the rectangular parallelepiped with unit initial temperature and its surface either at zero temperature or radiating into a medium at zero provide an extremely simple method for determining the diffusivity of poor conductors.

For large values of the time, the series 6.4 (6) converges very rapidly and the temperature is very nearly given by its first term. Thus, if the logarithm of the temperature at any point in the solid be plotted against

† Williamson and Adams, *Phys. Rev.* (2) **14** (1919) 99, give a number of results of this type.

the time, this curve will ultimately become a straight line of slope

$$-\frac{\kappa\pi^2}{4}\left(\frac{1}{a^2}+\frac{1}{b^2}+\frac{1}{c^2}\right). \tag{1}$$

If the temperature at the point $(\frac{1}{3}a, \frac{1}{3}b, \frac{1}{3}c)$ is measured, the terms in 6.4 (6) with l, m, and n unity all disappear, and the first term of 6.4 (6) is in this case a particularly good approximation.

If the solid loses heat by radiation into a medium at zero temperature, both κ and h have to be determined. Treating, for simplicity, the case of a cube, $a = b = c$, the temperature at any point for large values of the time, i.e. when the curve of $\ln v$ against t is linear, is given by

$$v = \frac{8h^3 \cos \alpha_1 x \cos \alpha_1 y \cos \alpha_1 z}{[(h^2+\alpha_1^2)a+h]^3 \cos^3\alpha_1 a}e^{-3\kappa\alpha_1^2 t}. \tag{2}$$

The ratio of the temperature at the centre to that at any other convenient point gives α_1. h is then found from 6.4 (10), and $\kappa\alpha_1^2$ from the slope of the curve of $\ln v$ against t. The initial temperature of the solid, provided it is constant, need not be known: it can be calculated from the values of α_1, h, κ, and a measurement of the temperature (or of the intercept on the axis of t of the asymptote of the curve of $\ln v$ against t). This forms a check if the initial temperature is known, also it makes possible the calculation of the initial temperature of a solid·which has been cooling by radiation for some time.

The same method may be used for other geometrically simple bodies such as finite cylinders, § 8.5, or spheres, § 9.5. An alternative method of reducing the observations is discussed in connexion with the sphere.

6.6. Variable temperature. Triple Fourier series

If the initial temperature in the parallelepiped is not the product of functions of x, y, and z, or if the boundary conditions are not of the type discussed in § 1.15, the method of § 6.4 is not available. Since the problem of the solid with prescribed initial and surface temperatures can be reduced as in § 1.14 to a problem of steady temperature, already discussed in § 6.2, and a problem with zero surface temperature, we need only consider here *the region* $0 < x < a, 0 < y < b, 0 < z < c,$ *with zero surface temperature and initial temperature* $f(x, y, z)$.

We assume that $f(x, y, z)$ can be expanded in the triple sine series 6.3 (12),

$$f(x,y,z) = \sum_{l=1}^{\infty} \sum_{m=1}^{\infty} \sum_{n=1}^{\infty} A_{l,m,n} \sin\frac{l\pi x}{a} \sin\frac{m\pi y}{b} \sin\frac{n\pi z}{c}, \tag{1}$$

where $\quad A_{l,m,n} = \frac{8}{abc} \int_0^a\int_0^b\int_0^c f(x',y',z')\sin\frac{l\pi x'}{a} \sin\frac{m\pi y'}{b} \sin\frac{n\pi z'}{c}\ dx'dy'dz'. \tag{2}$

Now
$$e^{-\kappa t \alpha_{l,m,n}^2} \sin\frac{l\pi x}{a} \sin\frac{m\pi y}{b} \sin\frac{n\pi z}{c}, \tag{3}$$

where
$$\alpha_{l,m,n}^2 = \pi^2\left(\frac{l^2}{a^2} + \frac{m^2}{b^2} + \frac{n^2}{c^2}\right), \tag{4}$$

satisfies the differential equation and vanishes on the boundary surfaces.

Thus we are led to

$$v = \sum_{l=1}^{\infty} \sum_{m=1}^{\infty} \sum_{n=1}^{\infty} e^{-\kappa \alpha_{l,m,n}^2 t} A_{l,m,n} \sin\frac{l\pi x}{a} \sin\frac{m\pi y}{b} \sin\frac{n\pi z}{c} \tag{5}$$

as the solution of our problem.

This and the corresponding result for the radiation boundary condition† can be obtained by the use of the Green's function, cf. § 14.5, where the problem of the rectangular parallelepiped with heat generated within it is also discussed.

† The finite Fourier transform may also be used, cf. § 17.5.

VII

THE FLOW OF HEAT IN AN INFINITE CIRCULAR CYLINDER

7.1. Introductory

WE have seen in § 1.8 that the equation of conduction, when expressed in cylindrical coordinates, becomes

$$\frac{1}{r}\frac{\partial}{\partial r}\left(r\frac{\partial v}{\partial r}\right)+\frac{1}{r^2}\frac{\partial^2 v}{\partial \theta^2}+\frac{\partial^2 v}{\partial z^2}=\frac{1}{\kappa}\frac{\partial v}{\partial t}, \tag{1}$$

or

$$\frac{\partial v}{\partial t}=\kappa\left(\frac{\partial^2 v}{\partial r^2}+\frac{1}{r}\frac{\partial v}{\partial r}+\frac{1}{r^2}\frac{\partial^2 v}{\partial \theta^2}+\frac{\partial^2 v}{\partial z^2}\right). \tag{2}$$

If a circular cylinder whose axis coincides with the axis of z is heated, and the initial and boundary conditions are independent of the coordinates θ and z, the temperature will be a function of r and t only, and this equation reduces to

$$\frac{\partial v}{\partial t}=\kappa\left(\frac{\partial^2 v}{\partial r^2}+\frac{1}{r}\frac{\partial v}{\partial r}\right). \tag{3}$$

In this case the flow of heat takes place in planes perpendicular to the axis, and the lines of flow are radial.

When the initial and boundary conditions do not contain z, the flow of heat again takes place in planes perpendicular to the axis, and the equation of conduction reduces to

$$\frac{\partial v}{\partial t}=\kappa\left(\frac{\partial^2 v}{\partial r^2}+\frac{1}{r}\frac{\partial v}{\partial r}+\frac{1}{r^2}\frac{\partial^2 v}{\partial \theta^2}\right). \tag{4}$$

Again, when the initial and boundary conditions do not contain θ, the flow of heat takes place in planes through the axis, and the equation of conduction becomes

$$\frac{\partial v}{\partial t}=\kappa\left(\frac{\partial^2 v}{\partial r^2}+\frac{1}{r}\frac{\partial v}{\partial r}+\frac{\partial^2 v}{\partial z^2}\right). \tag{5}$$

In this chapter solutions will be given for problems on the cylinder and hollow cylinder with various boundary conditions. These solutions will always be in 'Fourier–Bessel' form: solutions useful for small values of $\kappa t/a^2$ are rather more difficult to find and are deferred to Chapter XIII, also, since they do not take simple closed forms, they will not be quoted here. Problems on composite cylindrical regions and on the region bounded internally by a circular cylinder are discussed in Chapter XIII.

7.2. Steady temperature. Radial flow

If the solid is a hollow cylinder, whose inner and outer radii are a and b, the equation 7.1 (1) for the temperature becomes

$$\frac{d}{dr}\left(r\frac{dv}{dr}\right) = 0, \quad a < r < b. \tag{1}$$

The general solution of this is

$$v = A + B\ln r, \tag{2}$$

where A and B are constants to be determined from the boundary conditions at $r = a$ and $r = b$.

I. *$r = a$ kept at temperature v_1 and $r = b$ at temperature v_2*

Here
$$v = \frac{v_1\ln(b/r) + v_2\ln(r/a)}{\ln(b/a)}. \tag{3}$$

The rate of flow of heat per unit length is

$$-2\pi r K\frac{dv}{dr} = \frac{2\pi K(v_1 - v_2)}{\ln(b/a)}. \tag{4}$$

II. *$r = a$ kept at temperature v_1. At $r = b$ there is radiation into a medium at v_2, the boundary condition there being*

$$\frac{dv}{dr} + h(v - v_2) = 0, \quad r = b.$$

Here
$$v = \frac{v_1[1 + hb\ln(b/r)] + hbv_2\ln(r/a)}{1 + hb\ln(b/a)}. \tag{5}$$

The outward rate of flow of heat per unit length of the cylinder is

$$2\pi K(v_1 - v_2)\frac{hb}{1 + hb\ln(b/a)}. \tag{6}$$

If $ah > 1$ the expression (6) decreases steadily as b increases from a, but if $ah < 1$ it has a maximum at $b = 1/h$. This implies that, in certain circumstances, it is possible to increase the heat loss from a pipe by surrounding it with insulating material.[†]

III. *Heat supplied at a constant rate F_0 per unit length of the inner cylinder*

Here, since it follows from (1) that $r(dv/dr)$ is constant, the flow of heat over any cylinder is independent of its radius, and

$$F_0 = -2\pi r K\frac{dv}{dr}, \quad a < r < b. \tag{7}$$

† Cf. Porter and Martin, *Phil. Mag.* **20** (1910) 511.

Then, if v_1 and v_2 are the temperatures at radii r_1 and r_2 respectively, integrating (7) we have

$$2\pi K(v_1 - v_2) = F_0 \ln(r_2/r_1). \tag{8}$$

This relation is independent of how the heat is supplied, and of the boundary conditions at the cylindrical surfaces.† If the heat is supplied by a wire along the axis of the cylinder, of resistance R ohms per unit length and carrying current I amps., we have

$$F_0 = jI^2R,$$

where j is the number of calories in a joule.

If the thermal conductivity K depends on the temperature, a relation of type (8) still holds. For (7) is still true, and if we introduce

$$K_m = \frac{1}{v_2 - v_1} \int_{v_1}^{v_2} K \, dv, \tag{9}$$

the mean conductivity over the range of temperature from v_1 to v_2, integrating (7) gives

$$2\pi K_m(v_1 - v_2) = F_0 \ln(r_2/r_1). \tag{10}$$

IV. *The composite hollow cylinder of n regions (a_1, a_2), (a_2, a_3),...,*
   (a_n, a_{n+1}) of conductivities $K_1,...,$ K_n

If v_1, v_2,..., v_{n+1} are the temperatures at a_1, a_2,..., a_{n+1}, repeated application of (4) shows that the rate of flow of heat per unit length of the system, F, is

$$F = \frac{2\pi K_1(v_1 - v_2)}{\ln(a_2/a_1)} = ... = \frac{2\pi K_n(v_n - v_{n+1})}{\ln(a_{n+1}/a_n)}. \tag{11}$$

Therefore

$$v_1 - v_{n+1} = \frac{F}{2\pi} \sum_{r=1}^{n} \frac{\ln(a_{r+1}/a_r)}{K_r}. \tag{12}$$

If, in addition, there are contact resistances R_1, R_2,..., R_n, R_{n+1} per unit area over the surfaces a_1, a_2,..., a_n, a_{n+1}, and v_0 and v_{n+2} are the temperatures inside and outside the composite cylinder,

$$v_0 - v_{n+2} = \frac{F}{2\pi} \left\{ \sum_{r=1}^{n} \frac{\ln(a_{r+1}/a_r)}{K_r} + \sum_{r=1}^{n+1} \frac{R_r}{a_r} \right\}. \tag{13}$$

(6), above, is a simple special case of this.

† It has been much used for determinations of thermal conductivity. Cf. Lamb and Wilson, *Proc. Roy. Soc.* A, **65** (1899) 285; Niven, ibid. **76** (1905) 34; Poole, *Phil. Mag.* **24** (1912) 45; **27** (1914) 58; Bridgman, *Proc. Amer. Acad. Arts Sci.* **57** (1922) 80.

V. *Heat production in a cylinder*

If the rate of heat generation is A_0, constant, 1.6 (7) becomes for steady radial flow

$$\frac{1}{r}\frac{d}{dr}\left(r\frac{dv}{dr}\right)+\frac{A_0}{K}=0. \tag{14}$$

The general solution of (14) is

$$v = A+B\ln r-(A_0 r^2/4K). \tag{15}$$

For a solid cylinder,† the term in $\ln r$ is inadmissible and (15) becomes

$$v = v_0-(A_0 r^2/4K), \tag{16}$$

where v_0 is the temperature on the axis of the cylinder. If the surface $r = a$ of the cylinder is maintained at zero temperature,

$$v = A_0(a^2-r^2)/4K, \tag{17}$$

while if there is radiation at $r = a$ into a medium at zero temperature through surface conductance H,

$$v = \frac{aA_0}{2H}+\frac{A_0(a^2-r^2)}{4K}. \tag{18}$$

For a hollow cylindrical wire of internal and external radii a and b, if v_1 and v_2 are the internal and external surface temperatures, (15) gives

$$v_1-v_2 = B\ln(a/b)-(a^2-b^2)A_0/4K. \tag{19}$$

The constant B is determined from the boundary conditions. If, as is usually the case, there is no flow of heat at the inner surface, we have

$$\left[\frac{\partial v}{\partial r}\right]_{r=a} = \frac{B}{a}-\frac{aA_0}{2K} = 0. \tag{20}$$

From (19) and (20) it follows that

$$K = A_0\{b^2-a^2-2a^2\ln(b/a)\}/4(v_1-v_2). \tag{21}$$

(16) and (21) have been used for the determination of thermal conductivity.‡

For an insulated wire, the insulation, of thermal conductivity K_1, fills the region $a < r < b$ surrounding the wire $0 < r < a$, of thermal

---

† The simplest case is that of the heating of a wire by electric current of density I amp/cm² flowing in it; here $A_0 = jI^2/\sigma$, where σ is the electrical conductivity of the wire and j is the number of calories in a joule. Steady temperature in a wire heated by alternating current (in which case A_0 is a function of r because of skin-effect) or by induction are discussed by Strutt, *Phil. Mag.* (7) **5** (1928) 904–14. The present problem is also that of the heating of a long cylindrical magnet coil, cf. Emmerich, *J. Appl. Phys.* **21** (1950) 75–80. The cylinder heated by induction is discussed by Thorn and Simpson, *J. Appl. Phys.* **24** (1953) 297–9.

‡ Angell, *Phys. Rev.* (1) **33** (1911) 421; Worthing, ibid. (2) **4** (1914) 536; Langmuir, ibid. (2) **7** (1916) 151; Powell and Schofield, *Proc. Phys. Soc.* **51** (1939) 153.

conductivity K, in which heat is generated at the rate A_0 per unit time per unit volume. If there is radiation at $r = b$, that is,

$$\frac{\partial v}{\partial r} + hv = 0, \quad r = b,$$

the solution, obtained from (2) and (16), is

$$v = \frac{A_0}{4K}\left\{a^2 - r^2 + \frac{2Ka^2}{hbK_1} + \frac{2Ka^2}{K_1}\ln\frac{b}{a}\right\}, \quad 0 \leqslant r < a, \tag{22}$$

$$v = \frac{a^2 A_0}{2K_1}\left\{\frac{1}{hb} + \ln\frac{b}{r}\right\}, \quad a < r < b. \tag{23}$$

If linear variation of the resistance of the wire with temperature is allowed for, (14) is replaced by

$$\frac{d^2 v}{dr^2} + \frac{1}{r}\frac{dv}{dr} + \beta^2 v = -\frac{A_0}{K}, \tag{24}$$

where

$$\beta^2 = \alpha A_0 / K, \tag{25}$$

and α is the temperature coefficient of resistance.

The solution of (24) which is finite when $r = 0$ is

$$v = A J_0(\beta r) - (1/\alpha). \tag{26}$$

If the surface $r = a$ is maintained at zero temperature, the temperature at r is given by[†]

$$v = \frac{1}{\alpha}\left\{\frac{J_0(\beta r)}{J_0(\beta a)} - 1\right\}, \tag{27}$$

where $J_0(z)$ is defined in Appendix III.

For the hollow cylinder $a < r < b$, with $r = b$ maintained at zero and no flow at $r = a$, the temperature is

$$v = \frac{1}{\alpha}\left\{\frac{J_0(\beta r)Y_1(\beta a) - Y_0(\beta r)J_1(\beta a)}{J_0(\beta b)Y_1(\beta a) - Y_0(\beta b)J_1(\beta a)} - 1\right\}. \tag{28}$$

VI. *Hot wire methods*[‡]

Various combinations of (4) and the results of § 4.11 II have been used for measurement[§] of thermal conductivity of gases and poor conductors. As in § 4.11 II a wire of radius a and length $2l$ is heated by electric current and the change in its electric resistance is measured. The annular region $a < r < b$, $0 < x < 2l$ surrounding the wire is filled

[†] If α is negative, as for graphite, $J_0(\beta r)$ in (27) is to be replaced by $I_0[r(-\alpha A_0/K)^{\frac{1}{2}}]$. Numerical values of (27) are given by Jakob, *Trans. Amer. Soc. Mech. Engrs.* **65** (1943) 593–605, for the case of positive α, and ibid. **70** (1948) 25–30, for negative α.

[‡] Schleiermacher, *Wied. Ann.* **34** (1888) 623; **36** (1889) 346.

[§] Kannuluik and Martin, *Proc. Roy. Soc.* A, **141** (1933) 144; **144** (1934) 496; Gregory and Archer, ibid. **110** (1926) 91; *Phil. Mag.* (7) **3** (1927) 931; (7) **15** (1933) 301. The variable state is discussed by Fischer, *Ann. Physik*, (5) **34** (1939) 669; see also § 13.4.

with the substance whose conductivity K_1 is to be measured, and the outside surfaces $r = b$, and $z = 0$ and $z = 2l$, are maintained at zero temperature.

If v is the temperature at the point x of the wire (assumed constant over its cross-section as in § 4.10) and the flow in the region $a < r < b$ is assumed to be radial,† and it is assumed that there is no temperature discontinuity‡ at the surface of separation $r = a$ between the wire and the surrounding medium, (4) gives for the rate of loss of heat from the wire at the point x

$$\frac{2\pi K_1}{\ln(b/a)}\,v. \tag{29}$$

The differential equation for the temperature in the wire is then 4.10 (4) with (29) in place of Hpv, and it is studied as in § 4.11 II.

7.3. Steady periodic temperature in circular cylinders§

As in § 3.7 we seek solutions of period $2\pi/\omega$ containing a time-factor $e^{i\omega t}$ which will be understood to multiply all temperatures and fluxes and will be omitted throughout. Then 7.1 (3) becomes

$$\frac{d^2v}{dr^2}+\frac{1}{r}\frac{dv}{dt}-\frac{i\omega}{\kappa}v = 0. \tag{1}$$

The general solution of (1) is [cf. Appendix III (6)]

$$v = PI_0(kri^{\frac12})+QK_0(kri^{\frac12}), \tag{2}$$

where
$$k = (\omega/\kappa)^{\frac12}. \tag{3}$$

The flux f is given by

$$f = -K\frac{\partial v}{\partial r} = -Kki^{\frac12}PI_1(kri^{\frac12})+Kki^{\frac12}QK_1(kri^{\frac12}). \tag{4}$$

The Bessel functions of complex argument in (2) and (4) are expressed in terms of the tabulated ber, bei, ker, and kei functions by the relations‖

$$i^nI_n(zi^{\frac12}) = \operatorname{ber}_n z+i\operatorname{bei}_n z, \tag{5}$$

$$i^{-n}K_n(zi^{\frac12}) = \operatorname{ker}_n z+i\operatorname{kei}_n z, \tag{6}$$

and so may be regarded as being known numerically for any values of r and k.

Suppose, now, that v_1 and F_1 are the values of the temperature and flux at $r = a_1$,

† The complete two-variable problem in which this assumption is not made is treated in § 8.3 IX. Cf. Kannuluik and Martin, loc. cit.

‡ As remarked in § 1.9 G, a temperature discontinuity is always to be expected at a surface of separation of two solids unless their contact is extremely intimate. For contact between a solid and a gas it has been shown by Smoluchowski, *Ann. Physik*, **64** (1898) 101; **35** (1911) 983, that there is a temperature discontinuity at the surface whose magnitude depends on the temperature gradient there. See also Gregory, *Proc. Roy. Soc.* A, **149** (1935) 35. If there is a temperature discontinuity at the surface, a modification of (29) has to be made to allow for this.

§ Dahl, *Trans. Amer. Soc. Mech. Engrs.* **46** (1924) 161–208; Awbery, *Phil. Mag.* (7) **28** (1939) 447; van Gorcum, *Appl. Sci. Res.* A, **2** (1951) 272–80; Vodicka, ibid. A, **5** (1955) 115–20, 268–72, 327–37.

‖ Cf. McLachlan, *Bessel Functions for Engineers* (Oxford, 1934) Chap. 8.

and that v_1' and f_1' are their values at $r = a_2$, then, solving (2) and (4) with $r = a_1$ gives P and Q in terms of v_1 and f_1, and using these values of P and Q in (2) and (4) with $r = a_2$ gives v_1' and f_1' in terms of v_1 and f_1 in the form

$$\begin{pmatrix} v_1' \\ f_1' \end{pmatrix} = \begin{pmatrix} A & B \\ C & D \end{pmatrix} \begin{pmatrix} v_1 \\ f_1 \end{pmatrix}, \qquad (7)$$

where

$$A = a_1 k i^{\frac{1}{2}} [I_0(ka_2 i^{\frac{1}{2}}) K_1(ka_1 i^{\frac{1}{2}}) + K_0(ka_2 i^{\frac{1}{2}}) I_1(ka_1 i^{\frac{1}{2}})], \qquad (8)$$

$$B = (a_1/K) [I_0(ka_1 i^{\frac{1}{2}}) K_0(ka_2 i^{\frac{1}{2}}) - K_0(ka_1 i^{\frac{1}{2}}) I_0(ka_2 i^{\frac{1}{2}})], \qquad (9)$$

$$C = K k^2 a_1 [I_1(ka_2 i^{\frac{1}{2}}) K_1(ka_1 i^{\frac{1}{2}}) - K_1(ka_2 i^{\frac{1}{2}}) I_1(ka_1 i^{\frac{1}{2}})], \qquad (10)$$

$$D = a_1 k i^{\frac{1}{2}} [I_0(ka_1 i^{\frac{1}{2}}) K_1(ka_2 i^{\frac{1}{2}}) + K_0(ka_1 i^{\frac{1}{2}}) I_1(ka_2 i^{\frac{1}{2}})], \qquad (11)$$

and

$$AD - BC = a_1/a_2. \qquad (12)$$

In the reduction, Appendix III (22) is used. The matrix notation (7) has been discussed in § 3.7. Because of the relation (12), v_1 and f_1 may be expressed in terms of v_1' and f_1' in the form

$$\begin{pmatrix} v_1 \\ f_1 \end{pmatrix} = \frac{a_2}{a_1} \begin{pmatrix} D & -B \\ -C & A \end{pmatrix} \begin{pmatrix} v_1' \\ f_1' \end{pmatrix}. \qquad (13)$$

Just as in § 3.7, results for composite cylindrical regions may be written down immediately by matrix multiplication. For example, if $a_1 < r < a_2$ is of one material with values A_1, B_1, C_1, D_1 given by (8)–(12), and $a_2 < r < a_3$ is of another with A_2, B_2, C_2, D_2, and if there is thermal resistance R per unit area between the two materials at a_2,

$$\begin{pmatrix} v_2' \\ f_2' \end{pmatrix} = \begin{pmatrix} A_2 & B_2 \\ C_2 & D_2 \end{pmatrix} \begin{pmatrix} 1 & -R \\ 0 & 1 \end{pmatrix} \begin{pmatrix} A_1 & B_1 \\ C_1 & D_1 \end{pmatrix} \begin{pmatrix} v_1 \\ f_1 \end{pmatrix}. \qquad (14)$$

As remarked in § 3.7, the general expressions rapidly become hopelessly complicated, but for any definite set of conditions the A, B, C, D can be found numerically from tables, and the multiplication of the square matrices in (14) or its equivalent is easily carried out numerically. The boundary conditions at the inner and outer surfaces provide two additional conditions on the temperatures and fluxes at these surfaces, and thus all four quantities can be determined.

For the region $0 \leqslant r < a$, we must have $Q = 0$ in (2) since $K_0(kri^{\frac{1}{2}}) \to \infty$ as $r \to 0$. It follows that at $r = a$,

$$f = -\frac{K k i^{\frac{1}{2}} I_1(kai^{\frac{1}{2}})}{I_0(kai^{\frac{1}{2}})} v. \qquad (15)$$

For the infinite region $r > b$, we must have $P = 0$ in (2) since $I_0(kri^{\frac{1}{2}}) \to \infty$ as $r \to \infty$. Therefore, in this case at $r = b$,

$$f = \frac{K k i^{\frac{1}{2}} K_1(kbi^{\frac{1}{2}})}{K_0(kbi^{\frac{1}{2}})} v. \qquad (16)$$

7.4. Infinite cylinder. Radial flow. Variable temperature

Let the initial temperature be given by $v = f(r)$ and let the surface $r = a$ be kept at a constant temperature, which may be taken as zero.†

† If the constant surface temperature is v_0, we may reduce this to the case of zero temperature by putting $v = v_0 + w$.

The equations for v are as follows:

$$\frac{\partial v}{\partial t} = \kappa \left(\frac{\partial^2 v}{\partial r^2} + \frac{1}{r} \frac{\partial v}{\partial r} \right), \quad 0 < r < a.$$

$$v = 0, \quad \text{when } r = a,$$

and
$$v = f(r), \quad \text{when } t = 0.$$

If we put $v = e^{-\kappa \alpha^2 t} u$, where u is a function of r only, then we must have

$$\frac{d^2 u}{dr^2} + \frac{1}{r} \frac{du}{dr} + \alpha^2 u = 0,$$

which is Bessel's equation of order zero.

As the solution of the second kind is infinite at $r = 0$, the particular integral of the temperature equation suitable for our problem is

$$v = A J_0(\alpha r) e^{-\kappa \alpha^2 t},$$

where $J_0(x)$ is the Bessel function of order zero of the first kind.†

To satisfy the boundary condition α must be a root of

$$J_0(a\alpha) = 0.$$

It is known‡ that this equation has no complex roots or repeated roots, and that it has an infinite number of real positive roots

$$\alpha_1, \quad \alpha_2, \quad \alpha_3, \quad \dots .$$

Also to each positive root α there corresponds a negative root $-\alpha$. The first few roots are given in Appendix IV, Table III, under $C = \infty$.

If $f(r)$ can be expanded in the series

$$A_1 J_0(\alpha_1 r) + A_2 J_0(\alpha_2 r) + \dots, \tag{1}$$

the conditions of the problem will be satisfied by

$$v = \sum_{n=1}^{\infty} A_n J_0(\alpha_n r) e^{-\kappa \alpha_n^2 t}. \tag{2}$$

Assuming the possibility of the expansion§ and that the series can

---

† A brief account of Bessel functions is given in Appendix III. The standard works are Watson, *Theory of Bessel Functions* (Cambridge, edn. 2, 1944), and Gray and Mathews, *Treatise on Bessel Functions* (Macmillan, edn. 2, 1922). These will be referred to as W.B.F. and G. and M. respectively.

‡ W.B.F., §§ 15.21, 15.25.

§ For a discussion of the possibility of expanding an arbitrary function in a series of Bessel functions see Watson, loc. cit., Chap. 18; Hobson, *Proc. Lond. Math. Soc.* (2) **7** (1909) 359–88; Moore, *Trans. Amer. Math. Soc.* **10** (1909) 391–435; **12** (1911) 181–206; **21** (1920) 107–56; Young, *Proc. Lond. Math. Soc.* (2) **18** (1920) 163–200. The subject is also treated in Dini, *Serie di Fourier* (1880), pp. 246–69, and Ford, *Studies in Divergent Series and Summability* (1916), Chap. V.

If $f(r)$ is such that it can be expanded in the series (1) it is still necessary to verify that (2) does satisfy the differential equation and initial and boundary conditions, the position being precisely analogous to that for linear flow in § 3.3. For discussion on

be integrated term by term, we can obtain the values of the coefficients by the help of the definite integrals discussed in the next section.

The 'Fourier–Bessel' series (1) is the expansion of $f(r)$ suited to the problem of the cylinder with zero surface temperature. If, instead, there had been radiation at the surface into medium at zero temperature, the boundary condition would have been

$$\frac{\partial v}{\partial r} + hv = 0, \quad r = a,$$

and for $J_0(\alpha r)$ to satisfy this, α must be a root of

$$\alpha J_0'(\alpha a) + h J_0(\alpha a) = 0. \tag{3}$$

In this case we assume that $f(r)$ can be expanded in the form (1) where now the α_n are the roots of (3). These expansions (Dini series) and the corresponding expansions in terms of Bessel functions of order n are discussed in Watson (loc. cit., Chap. 18).

In the next section we derive fundamental definite integrals which will evaluate the coefficients in all the series expansions we need, and in the subsequent section various problems on conduction of heat in the cylinder will be solved, assuming always the possibility of the expansions, and that they may be integrated term by term. All the solutions can be obtained by the use of the Laplace transformation as in Chapters XIII, XIV.

7.5. The integrals†

$$\int_0^a r J_n(\alpha r) J_n(\beta r) \, dr \quad \text{and} \quad \int_0^a r [J_n(\alpha r)]^2 \, dr$$

Putting $u = J_n(\alpha r)$ and $v = J_n(\beta r)$, we have from Bessel's equation,

$$\frac{1}{r} \frac{d}{dr}\left(r \frac{du}{dr}\right) + \left(\alpha^2 - \frac{n^2}{r^2}\right)u = 0,$$

$$\frac{1}{r} \frac{d}{dr}\left(r \frac{dv}{dr}\right) + \left(\beta^2 - \frac{n^2}{r^2}\right)v = 0.$$

Thus

$$(\beta^2 - \alpha^2) \int_0^a r u v \, dr = \int_0^a \left[v \frac{d}{dr}\left(r \frac{du}{dr}\right) - u \frac{d}{dr}\left(r \frac{dv}{dr}\right)\right] dr = a\left[v \frac{du}{dr} - u \frac{dv}{dr}\right]_{r=a},$$

these points reference may be made to Moore's papers cited above. The same remarks apply to all the expansions in series of Bessel functions in this chapter and the next. The verification process may also be carried out along the lines used for the Laplace transformation in Appendix I.

† The convergence of the integrals when $r = 0$ requires that the real part of n shall be greater than -1. In the applications of these integrals in the text we shall be dealing with n real and not less than zero.

and this vanishes when

$$\alpha J_n(\beta a)J'_n(\alpha a)-\beta J_n(\alpha a)J'_n(\beta a) = 0,$$

where

$$J'_n(\alpha a) = \left[\frac{d}{dr}J_n(r)\right]_{r=\alpha a}.$$

Thus when α and β are two different positive roots of

(i)
$$J_n(\alpha a) = 0,$$

or (ii)
$$J'_n(\alpha a) = 0,$$

or (iii)
$$\alpha J'_n(\alpha a)+hJ_n(\alpha a) = 0,$$

we have
$$\int_0^a rJ_n(\alpha r)J_n(\beta r)\, dr = 0. \tag{1}$$

Again, since
$$\frac{1}{r}\frac{d}{dr}\left(r\frac{du}{dr}\right)+\left(\alpha^2-\frac{n^2}{r^2}\right)u = 0,$$

$$2r\frac{du}{dr}\frac{d}{dr}\left(r\frac{du}{dr}\right)+2\left(\alpha^2-\frac{n^2}{r^2}\right)r^2u\frac{du}{dr} = 0.$$

Therefore
$$\frac{d}{dr}\left(r\frac{du}{dr}\right)^2+\alpha^2r^2\frac{du^2}{dr}-n^2\frac{du^2}{dr} = 0,$$

and
$$\alpha^2\int_0^a r^2\frac{du^2}{dr}\, dr+\left[\left(r\frac{du}{dr}\right)^2-n^2u^2\right]_0^a = 0.$$

Integrate by parts and it follows that

$$2\alpha^2\int_0^a ru^2dr = \left[r^2\left(\frac{du}{dr}\right)^2+(\alpha^2r^2-n^2)u^2\right]_0^a.$$

Therefore, using Appendix III (1) at the lower limit,

$$\int_0^a r\{J_n(\alpha r)\}^2\, dr = \frac{1}{2\alpha^2}[a^2\alpha^2\{J'_n(\alpha a)\}^2+(a^2\alpha^2-n^2)\{J_n(\alpha a)\}^2]$$

$$= \frac{a^2}{2}\left[\{J'_n(\alpha a)\}^2+\left(1-\frac{n^2}{a^2\alpha^2}\right)\{J_n(a\alpha)\}^2\right].$$

Thus (i) when α is a root† of $J_n(\alpha a) = 0$,

$$\int_0^a r\{J_n(\alpha r)\}^2\, dr = \frac{a^2}{2}\{J'_n(\alpha a)\}^2; \tag{2}$$

† It is known that the roots of this equation and the others in (ii) and (iii) are all real and not repeated. Cf. *W.B.F.*, §§ 15.23, 15.25.

(ii) when α is a root of $J'_n(\alpha a) = 0$,

$$\int_0^a r\{J_n(\alpha r)\}^2 \, dr = \frac{a^2}{2}\left(1 - \frac{n^2}{a^2\alpha^2}\right)\{J_n(\alpha a)\}^2; \qquad (3)$$

and (iii) when α is a root of $\alpha J'_n(\alpha a) + h J_n(\alpha a) = 0$,

$$\int_0^a r\{J_n(\alpha r)\}^2 \, dr = \frac{1}{2\alpha^2}\{a^2 h^2 + (a^2\alpha^2 - n^2)\}\{J_n(\alpha a)\}^2. \qquad (4)$$

Finally we remark that from the recurrence formula (Appendix III (17))

$$\frac{d}{dz}\{z^{n+1}J_{n+1}(z)\} = z^{n+1}J_n(z)$$

it follows immediately that for *any* α, and $n > -1$,

$$\int_0^r r^{n+1}J_n(\alpha r) \, dr = \frac{1}{\alpha} r^{n+1}J_{n+1}(\alpha r). \qquad (5)$$

7.6. The infinite cylinder. Surface at $\phi(t)$. Initial temperature $f(r)$

Treating first the case of *initial temperature $f(r)$ and zero surface temperature* we take

$$f(r) = A_1 J_0(\alpha_1 r) + A_2 J_0(\alpha_2 r) + ..., \qquad (1)$$

where $\alpha_1, \alpha_2, ...$ are the positive roots of

$$J_0(\alpha a) = 0. \qquad (2)$$

Multiplying both sides of (1) by $rJ_0(\alpha_n r)$, integrating from 0 to a, and using the results 7.5 (1), (2), namely

$$\int_0^a rJ_0(\alpha_m r) J_0(\alpha_n r) \, dr = 0, \quad m \neq n,$$

$$\int_0^a r[J_0(\alpha_n r)]^2 \, dr = \tfrac{1}{2}a^2[J'_0(a\alpha_n)]^2 = \tfrac{1}{2}a^2 J_1^2(a\alpha_n),$$

where the result $J'_0(z) = -J_1(z)$ has been used, we find

$$A_n = \frac{2}{a^2 J_1^2(a\alpha_n)} \int_0^a rf(r)J_0(r\alpha_n) \, dr. \qquad (3)$$

Therefore $\quad v = \dfrac{2}{a^2} \displaystyle\sum_{n=1}^{\infty} e^{-\kappa\alpha_n^2 t} \dfrac{J_0(r\alpha_n)}{J_1^2(a\alpha_n)} \int_0^a rf(r)J_0(r\alpha_n) \, dr. \qquad (4)$

For the case of constant initial temperature, $f(r) = V$, the integral in (4) is evaluated by 7.5 (5), and we get

$$v = \frac{2V}{a} \sum_{n=1}^{\infty} e^{-\kappa \alpha_n^2 t} \frac{J_0(r\alpha_n)}{\alpha_n J_1(a\alpha_n)}. \tag{5}$$

Using 7.5 (5) again, *the average temperature of the cylinder* is found to be

$$\frac{2}{a^2} \int_0^a rv \, dr = \frac{4V}{a^2} \sum_{n=1}^{\infty} \frac{1}{\alpha_n^2} e^{-\kappa \alpha_n^2 t}. \tag{6}$$

If the initial distribution of temperature is parabolic, say

$$f(r) = V_0 - kr^2,$$

we find, integrating by parts and using 7.5 (5),

$$\int_0^a r^3 J_0(\alpha r) \, dr = \frac{a^3}{\alpha} J_1(a\alpha) - \frac{2a^2}{\alpha^2} J_2(\alpha a).$$

Thus, using this result in (4),

$$v = \frac{2}{a} \sum_{n=1}^{\infty} e^{-\kappa \alpha_n^2 t} \frac{J_0(r\alpha_n)}{\alpha_n^2 J_1^2(a\alpha_n)} \{\alpha_n(V_0 - ka^2)J_1(a\alpha_n) + 2kaJ_2(a\alpha_n)\}. \tag{7}$$

If the cylinder has *zero initial temperature and its surface is maintained at V, constant, for $t > 0$*, the solution, obtained by subtracting (5) from V, is

$$v = V - \frac{2V}{a} \sum_{n=1}^{\infty} e^{-\kappa \alpha_n^2 t} \frac{J_0(r\alpha_n)}{\alpha_n J_1(a\alpha_n)}. \tag{8}$$

In numerical work it is a little more convenient to use dimensionless variables. Thus we write

$$a\alpha_n = \beta_n \quad \text{and} \quad \kappa t/a^2 = T, \tag{9}$$

and (8) becomes†

$$\frac{v}{V} = 1 - 2 \sum_{n=1}^{\infty} e^{-\beta_n^2 T} \frac{J_0(r\beta_n/a)}{\beta_n J_1(\beta_n)}, \tag{10}$$

where now $\pm\beta_n$, $n = 1, 2, \ldots$, are the roots of

$$J_0(\beta) = 0. \tag{11}$$

The first few roots of this equation are given in Appendix IV, Table III, under $C = \infty$; the first fifty roots, with the corresponding values of $J_1(\beta_n)$, are given in *G. and M.*

---

† The quantities $J_0(r\beta_n/a)$ occurring in (10) are tabulated by Goodwin and Staton, *Quart. J. Mech. Appl. Math.* **1** (1948) 220–4, to 5D for $n = 1, 2, \ldots, 10$ and

$$(r/a) = 0.00(0.01)1.00.$$

In Fig. 24 values of v/V calculated from (10) for various values of T are plotted against r/a. The curves are seen to be very similar to the corresponding ones for the slab, Fig. 11, and in fact[†] it is possible to find a value $T_1 = \kappa t_1/a^2$ for a slab of thickness $2a$, such that the temperature distribution in the cylinder at time t is very nearly the same as

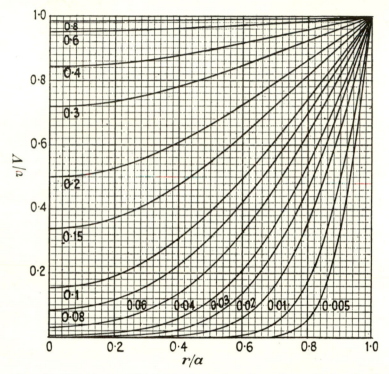

FIG. 24. Temperature distribution at various times in a cylinder of radius a with zero initial temperature and surface temperature V. The numbers on the curves are the values of $\kappa t/a^2$.

that in the slab at time t_1. The temperature on the axis of the cylinder,[‡] and the average temperature of the cylinder, have already been plotted in Fig. 12 as functions of T. The series (10) converges quite rapidly except for small values of T. Alternative solutions suitable for numerical work in this case are given in § 13.3.

If the initial temperature of the cylinder is zero and its surface tempera-ture is $\phi(t)$, *the solution, obtained from (8) and Duhamel's theorem,*

[†] Jaeger, *Proc. Phys. Soc.* **56** (1944) 197; Macey, ibid. **54** (1942) 128. The same remark applies to other boundary conditions.

[‡] Numerical values for the temperature on the axis are given by Olson and Schultz, *Ind. Eng. Chem.* **34** (1942) 874–7.

§ 1.14, is

$$v = \frac{2\kappa}{a} \sum_{n=1}^{\infty} e^{-\kappa\alpha_n^2 t} \frac{\alpha_n J_0(r\alpha_n)}{J_1(a\alpha_n)} \int_0^t e^{\kappa\alpha_n^2 \lambda}\phi(\lambda)\, d\lambda. \tag{12}$$

If the initial temperature is zero and the surface temperature is kt,

$$v = k\left(t - \frac{a^2 - r^2}{4\kappa}\right) + \frac{2k}{a\kappa} \sum_{n=1}^{\infty} e^{-\kappa\alpha_n^2 t} \frac{J_0(r\alpha_n)}{\alpha_n^3 J_1(a\alpha_n)}. \tag{13}$$

If the initial temperature is zero and the surface temperature is $V\sin(\omega t + \epsilon)$,

$$= V\mathbf{R}\left\{\frac{I_0[r(i\omega/\kappa)^{\frac{1}{2}}]}{iI_0[a(i\omega/\kappa)^{\frac{1}{2}}]}e^{i(\omega t + \epsilon)}\right\} +$$

$$+ \frac{2\kappa V}{a} \sum_{n=1}^{\infty} e^{-\kappa\alpha_n^2 t} \frac{\alpha_n(\omega\cos\epsilon - \kappa\alpha_n^2\sin\epsilon)J_0(r\alpha_n)}{(\kappa^2\alpha_n^4 + \omega^2)J_1(a\alpha_n)}. \tag{14}$$

The first term of (14) is the steady periodic part of the solution and the second the transient part. Writing†

$$I_0(zi^{\frac{1}{2}}) = \operatorname{ber} z + i\operatorname{bei} z = M_0(z)e^{i\theta_0(z)}, \tag{15}$$

and $\omega' = (\omega/\kappa)^{\frac{1}{2}}$, the steady periodic part becomes

$$\frac{VM_0(\omega'r)}{M_0(\omega'a)}\sin\{\omega t + \epsilon + \theta_0(\omega'r) - \theta_0(\omega'a)\}. \tag{16}$$

7.7. The infinite cylinder. Radiation at the surface

In this section we consider the circular cylinder with initial temperature $f(r)$ and radiation at its surface. We assume

$$f(r) = A_1 J_0(\alpha_1 r) + A_2 J_0(\alpha_2 r) + \ldots, \tag{1}$$

where $\pm\alpha_1, \pm\alpha_2, \ldots$ are the roots of

$$\alpha J_0'(a\alpha) + hJ_0(a\alpha) = 0\,; \tag{2}$$

these are all real and simple (*W.B.F.*, §§ 15.23, 15.25).

Then, since by 7.5 (4)

$$\int_0^a r[J_0(\alpha_n r)]^2\, dr = \frac{a^2}{2\alpha_n^2}(h^2 + \alpha_n^2)J_0^2(a\alpha_n),$$

we have

$$v = \frac{2}{a^2} \sum_{n=1}^{\infty} e^{-\kappa\alpha_n^2 t} \frac{\alpha_n^2 J_0(\alpha_n r)}{(h^2 + \alpha_n^2)J_0^2(\alpha_n a)} \int_0^a rf(r)J_0(r\alpha_n)\, dr. \tag{3}$$

† For the notation and numerical values of $M_0(z)$ and $\theta_0(z)$ see McLachlan, *Bessel Functions for Engineers* (Oxford, 1934) p. 182.

For the case in which $f(r) = V$, constant, this becomes, by 7.5 (5),

$$v = \frac{2V}{a} \sum_{n=1}^{\infty} e^{-\kappa \alpha_n^2 t} \frac{\alpha_n J_0(r\alpha_n)J_1(a\alpha_n)}{(h^2+\alpha_n^2)J_0^2(a\alpha_n)}$$

$$= \frac{2hV}{a} \sum_{n=1}^{\infty} e^{-\kappa \alpha_n^2 t} \frac{J_0(r\alpha_n)}{(h^2+\alpha_n^2)J_0(a\alpha_n)}. \tag{4}$$

Writing $a\alpha_n = \beta_n$, $\kappa t/a^2 = T$, $ah = A$, $\qquad$ (5)

to introduce dimensionless quantities, this may be written

$$\frac{v}{V} = \sum_{n=1}^{\infty} e^{-\beta_n^2 T} \frac{2A J_0(r\beta_n/a)}{(\beta_n^2+A^2)J_0(\beta_n)}, \tag{6}$$

where $\pm\beta_n$, $n = 1, 2,...$, are the roots of

$$\beta J_1(\beta) = A J_0(\beta). \tag{7}$$

A table of values of these roots is given in Appendix IV.

If the initial temperature of the cylinder is zero and there is radiation at its surface into a medium at V,

$$\frac{v}{V} = 1 - 2A \sum_{n=1}^{\infty} e^{-\beta_n^2 T} \frac{J_0(r\beta_n/a)}{(\beta_n^2+A^2)J_0(\beta_n)}. \tag{8}$$

Numerical values of (6) and (8) have been given by many authors.[†]

The average temperature, v_{av}, of the cylinder is

$$\frac{v_{\mathrm{av}}}{V} = 1 - 4A \sum_{n=1}^{\infty} e^{-\beta_n^2 T} \frac{J_1(\beta_n)}{\beta_n(\beta_n^2+A^2)J_0(\beta_n)}, \tag{9}$$

and the heat content per unit length of the cylinder is $\pi a^2 \rho c v_{\mathrm{av}}$. *If the initial temperature of the cylinder is zero and there is radiation at its surface into a medium at kt,*

$$v = kt - \frac{ka^2}{4\kappa}\left(1 - \frac{r^2}{a^2} + \frac{2}{A}\right) + \frac{2Aka^2}{\kappa} \sum_{n=1}^{\infty} e^{-\beta_n^2 T} \frac{J_0(r\beta_n/a)}{\beta_n^2(A^2+\beta_n^2)J_0(\beta_n)}, \tag{10}$$

where A, T, β_n are defined in (5) and (7).

If the initial temperature of the cylinder is zero and there is radiation at its surface into a medium at V sin (ωt+ε),

$$v = \mathbf{R}\left\{ \frac{hV I_0[r(i\omega/\kappa)^{\frac{1}{2}}]e^{i(\omega t+\epsilon)}}{i\{(i\omega/\kappa)^{\frac{1}{2}}I_1[a(i\omega/\kappa)^{\frac{1}{2}}] + hI_0[a(i\omega/\kappa)^{\frac{1}{2}}]\}} \right\} +$$

$$+ 2\kappa A V \sum_{n=1}^{\infty} e^{-\beta_n^2 T} \frac{\beta_n^2(a^2\omega \cos\epsilon - \kappa\beta_n^2 \sin\epsilon)J_0(r\beta_n/a)}{(\kappa^2\beta_n^4+\omega^2 a^4)(\beta_n^2+A^2)J_0(\beta_n)}. \tag{11}$$

† Gurney and Lurie, *Ind. Eng. Chem.* **15** (1923) 1170; Schack, *Stahl u. Eisen,* **50** (1930) 1290; Newman, *Trans. Amer. Inst. Chem. Engrs.* **27** (1931) 203; *Ind. Eng. Chem.* **28** (1936) 545–8; Heisler, *Trans. Amer. Soc. Mech. Engrs.* **69** (1947) 227–36. An application to the radiation correction in calorimetry is made by Vasileff, *J. Appl. Phys.* **23** (1952) 979–83.

Using the notation 7.6 (15), the steady part of (11) may be written in the form

$$V\eta_0 \sin(\omega t + \epsilon - \epsilon_0),$$

where ϵ_0 and η_0 are complicated expressions involving ber and bei functions.[†]

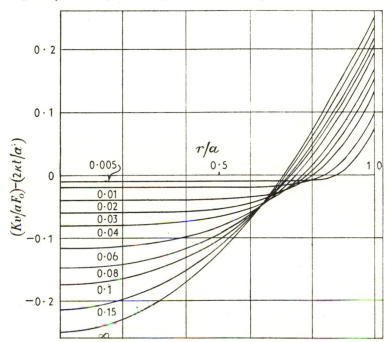

FIG. 25. The temperature in a circular cylinder with constant flux at the surface. The numbers on the curves are the values of $\kappa t/a^2$.

7.8. The infinite cylinder. Constant flux at the surface[‡]

I. *Zero initial temperature. Constant flux F_0 into the cylinder*

$$v = \frac{2F_0 \kappa t}{Ka} + \frac{F_0 a}{K}\left\{\frac{r^2}{2a^2} - \frac{1}{4} - 2\sum_{n=1}^{\infty} e^{-\kappa\alpha_n^2 t/a^2} \frac{J_0(r\alpha_n/a)}{\alpha_n^2 J_0(\alpha_n)}\right\}, \qquad (1)$$

where α_n, $n = 1, 2, ...$, are the positive roots of

$$J_1(\alpha) = 0. \qquad (2)$$

Some values of $v - (2F_0 \kappa t/Ka)$ are given in Fig. 25. For a proof of (1) see § 13.2 III.

[†] Numerical values of ϵ_0 and η_0 are plotted by Gröber, *Z. Ver. dtsch. Ing.* **70** (1926) 1266.

[‡] The problem arises in induction heating and in connexion with the drying of clay. Cf. Macey, *Proc. Phys. Soc.* **52** (1940) 625; **54** (1942) 128; Awbery, ibid. **55** (1943) 202; Jaeger, ibid. **56** (1944) 197; Newman and Church, *J. Appl. Mech.* **2** (1935) A–96.

II. *Initial temperature $f(r)$. Zero surface flux*

$$v = \frac{2}{a^2} \int_0^a r'f(r')\,dr' + \frac{2}{a^2} \sum_{n=1}^{\infty} e^{-\kappa\alpha_n^2 t/a^2} \frac{J_0(r\alpha_n/a)}{J_0^2(\alpha_n)} \int_0^a r'f(r')J_0(r'\alpha_n/a)\,dr', \qquad (3)$$

where the α_n are the roots of (2).

This is the case $h = 0$ of the problem of § 7.7, but it must be noticed that putting $h = 0$ in 7.7 (3) does not give (3), the first term being absent. It was remarked in § 1.9 that in boundary conditions of type

$$\frac{\partial v}{\partial r} + hv = 0, \qquad r = a,$$

h is always supposed to be positive, and that putting $h = 0$ in results for this boundary condition often does not give the correct result for the problem in which the boundary is impervious to heat. Physically, in the present problem, the distinction is that if h is positive the final temperature of the cylinder is zero, no matter how small h may be, while if $h = 0$, so that the surface is impervious to heat, the final temperature of the cylinder is not zero but the average of the initial temperature. Mathematically, a constant first term, analogous to the constant first term of a Fourier cosine series, must be added to the expansion.† The same effect appears in connexion with the sphere with no flow of heat at its surface, § 9.7.

7.9. The infinite cylinder with heat generated within it‡

I. *Zero initial and surface temperature. Heat production at the rate A_0, constant, per unit time per unit volume for $t > 0$*

$$v = \frac{A_0(a^2 - r^2)}{4K} - \frac{2A_0}{aK} \sum_{n=1}^{\infty} e^{-\kappa\alpha_n^2 t} \frac{J_0(r\alpha_n)}{\alpha_n^3 J_1(a\alpha_n)}, \qquad (1)$$

where the α_n are the positive roots of $J_0(a\alpha) = 0$. Some values of the temperature are given in Fig. 26. For a proof of (1) see § 13.2 VI.

II. *Zero initial and surface temperature. Heat production at the rate $A_0 e^{-\lambda t}$ per unit time per unit volume for $t > 0$*

$$v = \frac{\kappa A_0}{K\lambda} e^{-\lambda t} \left\{ \frac{J_0[r(\lambda/\kappa)^{\frac{1}{2}}]}{J_0[a(\lambda/\kappa)^{\frac{1}{2}}]} - 1 \right\} - \frac{2A_0\kappa}{aK} \sum_{n=1}^{\infty} \frac{e^{-\kappa\alpha_n^2 t}J_0(r\alpha_n)}{\alpha_n(\kappa\alpha_n^2 - \lambda)J_1(a\alpha_n)}, \qquad (2)$$

where the α_n are the positive roots of $J_0(a\alpha) = 0$.

---

† This difficulty does not appear with the Laplace transformation method of Chapter XIII, since there the terms which appear in the solution are determined by the singularities of a function of a complex variable. See also *W.B.F.*, § 18.3. In *C.H.*, p. 118, the constant term is omitted and thus the solution given there is incorrect, the first term of (3) being absent.

‡ This problem, and the corresponding one for the hollow cylinder, is discussed with biological applications by Thews, *Acta Biotheoretica*, A **10** (1953) 105–38.

III. *Zero initial temperature. Heat production A_0, constant, for $t > 0$. Radiation at $r = a$ into a medium at zero*

$$v = \frac{A_0 a^2}{4K}\left(1 - \frac{r^2}{a^2} + \frac{2}{ah}\right) - \frac{2hA_0}{aK}\sum_{n=1}^{\infty} e^{-\kappa \alpha_n^2 t}\frac{J_0(r\alpha_n)}{\alpha_n^2(h^2 + \alpha_n^2)J_0(a\alpha_n)}, \quad (3)$$

where the α_n are the positive roots of $\alpha J_1(a\alpha) = hJ_0(a\alpha)$.

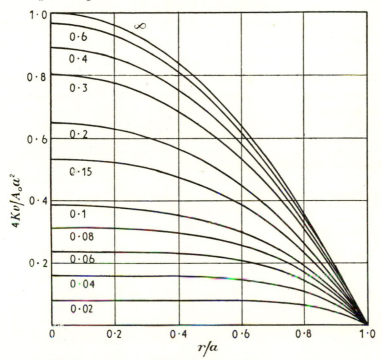

FIG. 26. Temperature in a circular cylinder with a constant rate of heat generation and zero surface temperature. The numbers on the curves are the values
of $\kappa t/a^2$.

7.10. The infinite hollow cylinder. Radial flow

I. *Infinite hollow cylinder. Surfaces $r = a$ and $r = b$ kept at zero.
Initial temperature $f(r)$*

In this case we have

$$\frac{\partial v}{\partial t} = \kappa\left(\frac{\partial^2 v}{\partial r^2} + \frac{1}{r}\frac{\partial v}{\partial r}\right), \quad a < r < b. \quad (1)$$

If we put $v = ue^{-\kappa\alpha^2 t}$, where u depends on r only, the equation for u is

$$\frac{d^2u}{dr^2} + \frac{1}{r}\frac{du}{dr} + \alpha^2 u = 0,$$

or Bessel's equation of order zero.

As the range of r does not extend to the origin, Bessel functions of the second kind are not excluded. We consider the function

$$U_0(\alpha r) = J_0(\alpha r)Y_0(\alpha b) - J_0(\alpha b)Y_0(\alpha r). \tag{2}$$

This vanishes when $r = b$, and it vanishes also when $r = a$ provided α is a root of

$$J_0(\alpha a)Y_0(\alpha b) - J_0(\alpha b)Y_0(\alpha a) = 0. \tag{3}$$

It is known† that the roots of (3) are all real and simple, and that to every positive root α there corresponds a negative root $-\alpha$. Some values of these roots are given in Appendix IV, Table IV. We first require integrals for the function $U_0(\alpha r)$ corresponding to those of § 7.5. Repeating the analysis of that section with $u = U_0(\alpha r)$ and $v = U_0(\beta r)$, it follows that

$$\int_a^b rU_0(\alpha r)U_0(\beta r)\, dr = 0, \tag{4}$$

if α and β are two different roots of (3).

And also

$$\int_a^b rU_0^2(\alpha r)\, dr = \frac{1}{2\alpha^2}\left[\left(r\frac{dU_0}{dr}\right)^2\right]_a^b. \tag{5}$$

Now, using Appendix III (20),

$$\left[r\frac{dU_0}{dr}\right]_{r=b} = \alpha b[J_0'(\alpha b)Y_0(\alpha b) - Y_0'(\alpha b)J_0(\alpha b)] = -\frac{2}{\pi}. \tag{6}$$

Also

$$\left[r\frac{dU_0}{dr}\right]_{r=a} = \alpha a[J_0'(\alpha a)Y_0(\alpha b) - Y_0'(\alpha a)J_0(\alpha b)]. \tag{7}$$

But, when α is a root of (3),

$$\frac{J_0(a\alpha)}{J_0(b\alpha)} = \frac{Y_0(a\alpha)}{Y_0(b\alpha)} = \rho, \quad \text{say.}$$

Thus (7) becomes

$$\left[r\frac{dU_0}{dr}\right]_{r=a} = \frac{\alpha a}{\rho}[J_0'(\alpha a)Y_0(\alpha a) - Y_0'(\alpha a)J_0(\alpha a)] = -\frac{2}{\pi\rho}. \tag{8}$$

Therefore we have from (5)

$$\int_a^b rU_0^2(\alpha r)\, dr = \frac{2\{J_0^2(\alpha a) - J_0^2(\alpha b)\}}{\pi^2\alpha^2 J_0^2(\alpha a)}. \tag{9}$$

Two other integrals which are also needed may be derived in much the same way from the differential equation, using (6) and (8). These

† Gray and Mathews, *Treatise on Bessel Functions* (edn. 2, 1922) p. 82, Theorem X. They also give formulae for calculating the large roots of this equation. Values of the roots for $b/a = 1\cdot2$, $1\cdot5$, $2\cdot0$ are given by Kalähne (*Z. Math. Phys.* **54** (1907) 55), also in Jahnke–Emde, *Funktionen Tafeln* (Teubner, edn. 3, 1933). See also Lowan and Hillman, *J. Math. Phys.* **22** (1943) 208.

are, if α is a root of (3),

$$\int_a^b rU_0(\alpha r)\,dr = -\frac{1}{\alpha^2}\left[r\frac{dU_0}{dr}\right]_a^b = \frac{2\{J_0(a\alpha)-J_0(b\alpha)\}}{\pi\alpha^2 J_0(a\alpha)}, \tag{10}$$

$$\int_a^b rU_0(\alpha r)\ln r\,dr = -\frac{1}{\alpha^2}\left[r\ln r\frac{dU_0}{dr}\right]_a^b = \frac{2\{J_0(a\alpha)\ln b - J_0(b\alpha)\ln a\}}{\pi\alpha^2 J_0(a\alpha)}. \tag{11}$$

Assuming, now, that the initial temperature $f(r)$ can be expanded in the series
$$f(r) = A_1 U_0(\alpha_1 r) + A_2 U_0(\alpha_2 r) + ...,$$
and that the series can be integrated term by term, we have from (4) and (9)

$$A_n = \frac{\pi^2\alpha_n^2}{2}\frac{J_0^2(a\alpha_n)}{J_0^2(a\alpha_n)-J_0^2(b\alpha_n)}\int_a^b rf(r)U_0(r\alpha_n)\,dr.$$

Thus we are led to the solution of our problem in the form

$$v = \frac{\pi^2}{2}\sum_{n=1}^{\infty}\frac{\alpha_n^2 J_0^2(a\alpha_n)}{J_0^2(a\alpha_n)-J_0^2(b\alpha_n)}e^{-\kappa\alpha_n^2 t}U_0(r\alpha_n)\int_a^b rf(r)U_0(r\alpha_n)\,dr, \tag{12}$$

the summation being taken over the positive roots of (3).

For the case of constant initial temperature $f(r) = V$, constant, we have, using (8),

$$v = V\pi\sum_{n=1}^{\infty}e^{-\kappa\alpha_n^2 t}\frac{J_0(a\alpha_n)U_0(r\alpha_n)}{J_0(a\alpha_n)+J_0(b\alpha_n)}. \tag{13}$$

II. Another case of practical importance is that of *initial temperature* $f(r)$, *the surfaces* $r = a$ *and* $r = b$ *being kept at constant temperatures* v_1 *and* v_2 *for* $t > 0$.

Here, as in § 1.14, we write
$$v = u+w,$$
where
$$u = \frac{v_1\ln(b/r)+v_2\ln(r/a)}{\ln(b/a)} \tag{14}$$

is by 7.2 (3) the temperature in steady flow between $r = a$ at v_1 and $r = b$ at v_2, and w is to be given by (12) with $f(r)$ replaced by $f(r)-u$. Using the result (11) the solution is found to be

$$v = \frac{\pi^2}{2}\sum_{n=1}^{\infty}\frac{\alpha_n^2 J_0^2(a\alpha_n)}{J_0^2(a\alpha_n)-J_0^2(b\alpha_n)}e^{-\kappa\alpha_n^2 t}U_0(r\alpha_n)\int_a^b rf(r)U_0(r\alpha_n)\,dr-$$

$$-\pi\sum_{n=1}^{\infty}\frac{\{v_2 J_0(a\alpha_n)-v_1 J_0(b\alpha_n)\}J_0(a\alpha_n)U_0(r\alpha_n)}{J_0^2(a\alpha_n)-J_0^2(b\alpha_n)}e^{-\kappa\alpha_n^2 t}+$$

$$+\frac{v_1\ln(b/r)+v_2\ln(r/a)}{\ln(b/a)}. \tag{15}$$

A number of other results for the hollow cylinder, derived by the methods above, is given by Muskat.† Problems with radiation or other boundary conditions at the surfaces may be treated as above using appropriate generalizations of the cylinder function $U_0(\alpha r)$; these, however, are probably more satisfactorily discussed by the Laplace transformation method and are treated in § 13.4.

7.11. The infinite cylinder. Steady temperature. General case

I. *Steady temperature in the cylinder* $0 \leqslant r < a$. *Surface at* $F(\theta, z)$

We expand $F(\theta, z)$ in the Fourier series

$$\sum_{n=0}^{\infty} \{\phi_n(z)\cos n\theta + \psi_n(z)\sin n\theta\}. \tag{1}$$

Then we take the Fourier integrals for $\phi_n(z)$ and $\psi_n(z)$

$$\left. \begin{array}{l} \phi_n(z) = \dfrac{1}{\pi} \displaystyle\int_0^\infty d\alpha \int_{-\infty}^\infty \phi_n(\beta)\cos\alpha(\beta - z)\, d\beta, \\[3mm] \psi_n(z) = \dfrac{1}{\pi} \displaystyle\int_0^\infty d\alpha \int_{-\infty}^\infty \psi_n(\beta)\cos\alpha(\beta - z)\, d\beta. \end{array} \right\} \tag{2}$$

Then, since
$$I_n(\alpha r)\cos\alpha(\beta - z)\begin{smallmatrix}\cos\\\sin\end{smallmatrix} n\theta,$$

where $I_n(z)$ is defined in Appendix III, is a solution of Laplace's equation in cylindrical coordinates which is finite when $r = 0$, the solution of our problem is

$$v = \frac{1}{\pi} \sum_{n=0}^{\infty} \int_0^\infty \frac{I_n(\alpha r)}{I_n(\alpha a)}\, d\alpha \int_{-\infty}^\infty \{\phi_n(\beta)\cos n\theta + \psi_n(\beta)\sin n\theta\}\cos\alpha(\beta - z)\, d\beta$$

$$= \frac{1}{\pi} \sum_{n=0}^{\infty} \int_{-\infty}^\infty \{\phi_n(\beta)\cos n\theta + \psi_n(\beta)\sin n\theta\}\, d\beta \int_0^\infty \frac{I_n(\alpha r)}{I_n(\alpha a)}\cos\alpha(\beta - z)\, d\alpha, \tag{3}$$

assuming that the orders of integration can be interchanged.

To evaluate the inner integral in (3) consider

$$\int \frac{J_n(\alpha r)}{J_n(\alpha a)} e^{-\alpha(\beta - z)}\, d\alpha \tag{4}$$

taken round a contour consisting, if $\beta > z$, of the imaginary axis and a large semicircle in the right-hand half-plane. The integrand has poles at α_s, $s = 1, 2,...$, the positive roots of
$$J_n(a\alpha) = 0.$$

Evaluating the residues at these poles we obtain finally

$$\int_0^\infty \cos\alpha(\beta - z)\frac{I_n(\alpha r)}{I_n(\alpha a)}\, d\alpha = -\frac{\pi}{a} \sum_{s=1}^{\infty} e^{-\alpha_s|\beta - z|} \frac{J_n(r\alpha_s)}{J_n'(a\alpha_s)}. \tag{5}$$

† *Flow of Homogeneous Fluids* (McGraw-Hill, 1937) Chap. X; also *J. Appl. Phys.* **5** (1934) 71.

Thus (3) gives

$$v = -\frac{1}{a}\sum_{n=0}^{\infty}\sum_{s=1}^{\infty}\frac{J_n(r\alpha_s)}{J_n'(a\alpha_s)}\int_{-\infty}^{\infty}\{\phi_n(\beta)\cos n\theta + \psi_n(\beta)\sin n\theta\}e^{-\alpha_s|\beta-z|}\,d\beta. \qquad (6)$$

If the surface temperature is a function of z only, say f(z), this becomes

$$v = \frac{1}{a}\sum_{s=1}^{\infty}\frac{J_0(r\alpha_s)}{J_1(a\alpha_s)}\int_0^{\infty}e^{-u\alpha_s}\{f(z+u)+f(z-u)\}\,du, \qquad (7)$$

where now the α_s are the positive roots of

$$-J_0(a\alpha) = 0.$$

For example if
$$f(z) = 1, \quad z > 0,$$
$$f(z) = 0, \quad z < 0,$$

(7) gives

$$v = 1 - \frac{1}{a}\sum_{s=1}^{\infty}\frac{J_0(r\alpha_s)}{\alpha_s J_1(a\alpha_s)}e^{-\alpha_s z}, \quad z > 0,$$
$$= \frac{1}{a}\sum_{s=1}^{\infty}\frac{J_0(r\alpha_s)}{\alpha_s J_1(a\alpha_s)}e^{\alpha_s z}, \qquad z < 0, \qquad (8)$$

where we have used the result

$$\frac{2}{a}\sum_{s=1}^{\infty}\frac{J_0(r\alpha_s)}{\alpha_s J_1(a\alpha_s)} = 1, \qquad (9)$$

which may be obtained by putting $t = 0$ in 7.6 (5).

(8) may also be obtained directly by the method used below.

II. *Steady temperature in the cylinder $0 \leqslant r < a$ which moves with velocity U in the direction of its length. The surface temperature unity for $z < 0$, and zero for $z > 0$*

Here, as in § 1.7, the differential equation is

$$\frac{\partial^2 v}{\partial r^2} + \frac{1}{r}\frac{\partial v}{\partial r} + \frac{\partial^2 v}{\partial z^2} - \frac{U}{\kappa}\frac{\partial v}{\partial z} = 0. \qquad (10)$$

If we seek a solution of (10) of type

$$J_0(\alpha r)u(z),$$

u has to satisfy
$$\frac{d^2 u}{dz^2} - \frac{U}{\kappa}\frac{du}{dz} - \alpha^2 u = 0.$$

Thus, if $\pm\alpha_s$ are the roots of
$$J_0(a\alpha) = 0,$$

$$v = 1 + \sum_{s=1}^{\infty}a_s J_0(r\alpha_s)e^{z\{k+\sqrt{(k^2+\alpha_s^2)}\}}, \quad z < 0,$$
$$v = \sum_{s=1}^{\infty}b_s J_0(r\alpha_s)e^{z\{k-\sqrt{(k^2+\alpha_s^2)}\}}, \qquad z > 0, \qquad (11)$$

where k is written for $U/2\kappa$, satisfy the differential equation and boundary conditions. The values of a_s and b_s are to be found from the fact that v and $\partial v/\partial z$ are to be continuous at $z = 0$. Using the result (9), these continuity conditions give

$$a_s + \frac{2}{a\alpha_s J_1(a\alpha_s)} = b_s,$$

$$[k + \sqrt{(k^2 + \alpha_s^2)}]a_s = [k - \sqrt{(k^2 + \alpha_s^2)}]b_s.$$

Hence, finally,

$$v = 1 + \sum_{n=1}^{\infty} \frac{[k - \sqrt{(k^2 + \alpha_s^2)}]J_0(r\alpha_s)}{a\alpha_s J_1(a\alpha_s)\sqrt{(k^2 + \alpha_s^2)}} e^{z\{k + \sqrt{(k^2 + \alpha_s^2)}\}}, \quad z < 0,$$

$$v = \sum_{n=1}^{\infty} \frac{J_0(r\alpha_s)[k + \sqrt{(k^2 + \alpha_s^2)}]}{a\alpha_s J_1(a\alpha_s)\sqrt{(k^2 + \alpha_s^2)}} e^{z\{k - \sqrt{(k^2 + \alpha_s^2)}\}}, \qquad z > 0.$$

Wilson[†] has discussed a number of problems of this type, both by this method and by the use of moving sources of heat (cf. Chapter X).

7.12. The infinite cylinder. Variable temperature. General case

In this section a number of problems on the infinite cylinder in which flow is not radial are discussed using the preceding methods and the integrals of § 7.5.

I. *Surface $r = a$ at zero temperature. Initial temperature $v = f(r, \theta)$*

In this case the equation of conduction becomes

$$\frac{\partial v}{\partial t} = \kappa\left(\frac{\partial^2 v}{\partial r^2} + \frac{1}{r}\frac{\partial v}{\partial r} + \frac{1}{r^2}\frac{\partial^2 v}{\partial \theta^2}\right),$$

and the expression $e^{-\kappa \alpha^2 t}J_n(\alpha r)(A_n \cos n\theta + B_n \sin n\theta)$ satisfies this equation, n being taken integral as the temperature is periodic in θ with period 2π.

Now take the Fourier series for $f(r, \theta)$, namely,

$$f(r, \theta) = \sum_{n=0}^{\infty} (a_n \cos n\theta + b_n \sin n\theta),$$

where

$$a_n = \frac{1}{\pi}\int_{-\pi}^{\pi} f(r, \theta)\cos n\theta\, d\theta \quad (n \geqslant 1),$$

$$b_n = \frac{1}{\pi}\int_{-\pi}^{\pi} f(r, \theta)\sin n\theta\, d\theta,$$

and

$$a_0 = \frac{1}{2\pi}\int_{-\pi}^{\pi} f(r, \theta)\, d\theta.$$

These coefficients are functions of r. Expand them in the series of Bessel functions of the nth order, e.g.

$$a_n = \sum_{s=1}^{\infty} A_{n,s} J_n(\alpha_s r),$$

where $\alpha_1, \alpha_2,..., \alpha_s,...$ are the positive roots of

$$J_n(\alpha a) = 0;$$

these are tabulated for $n = 0$ to 5 in *W.B.F.*

† *Proc. Camb. Phil. Soc.* **12** (1904) 406. See also Owen, *Proc. Lond. Math. Soc.* (2) **23** (1925) 238.

Then we have

$$A_{0,s} = \frac{1}{\pi a^2 \{J_0'(\alpha_s a)\}^2} \int_0^a \int_{-\pi}^\pi f(r,\theta) J_0(\alpha_s r) r \, dr d\theta,$$

$$A_{n,s} = \frac{2}{\pi a^2 \{J_n'(\alpha_s a)\}^2} \int_0^a \int_{-\pi}^\pi f(r,\theta) \cos n\theta J_n(\alpha_s r) r \, dr d\theta,$$

and

$$B_{n,s} = \frac{2}{\pi a^2 \{J_n'(\alpha_s a)\}^2} \int_0^a \int_{-\pi}^\pi f(r,\theta) \sin n\theta J_n(\alpha_s r) r \, dr d\theta.$$

Thus we obtain our solution in the form

$$v = \sum_{s=1}^\infty \sum_{n=0}^\infty (A_{n,s} \cos n\theta + B_{n,s} \sin n\theta) J_n(\alpha_s r) e^{-\kappa \alpha_s^2 t}. \tag{1}$$

II. *Radiation at the surface $r = a$ into a medium at zero temperature. Initial temperature $v = f(r, \theta)$*

In this case we take the Fourier series for $f(r,\theta)$, namely,

$$f(r,\theta) = \sum_{n=0}^\infty (a_n \cos n\theta + b_n \sin n\theta),$$

as in I.

The coefficients are functions of r. Expand them in the series of Bessel functions of the nth order, e.g.

$$a_n = \sum_{s=1}^\infty A_{n,s} J_n(\alpha_s r),$$

where $\alpha_1, \alpha_2, \dots$ are the positive roots of the equation

$$\alpha J_n'(\alpha a) + h J_n(\alpha a) = 0.$$

Then we have

$$A_{0,s} = \frac{\alpha_s^2}{\pi a^2 (\alpha_s^2 + h^2) \{J_0(\alpha_s a)\}^2} \int_0^a \int_{-\pi}^\pi f(r,\theta) J_0(\alpha_s r) r \, dr d\theta,$$

$$A_{n,s} = \frac{2\alpha_s^2}{\pi a^2 (\alpha_s^2 + h^2 - n^2/a^2) \{J_n(\alpha_s a)\}^2} \int_0^a \int_{-\pi}^\pi f(r,\theta) \cos n\theta J_n(\alpha_s r) r \, dr d\theta,$$

$$B_{n,s} = \frac{2\alpha_s^2}{\pi a^2 (\alpha_s^2 + h^2 - n^2/u^2) \{J_n(\alpha_s a)\}^2} \int_0^a \int_{-\pi}^\pi f(r,\theta) \sin n\theta J_n(\alpha_s r) r \, dr d\theta,$$

and

$$v = \sum_{s=1}^\infty \sum_{n=0}^\infty (A_{n,s} \cos n\theta + B_{n,s} \sin n\theta) J_n(\alpha_s r) e^{-\kappa \alpha_s^2 t}. \tag{2}$$

III. *Surface $r = a$ at zero. Initial temperature $v = f(r, \theta, z)$*

In this case we have

$$\frac{\partial v}{\partial t} = \kappa \left(\frac{\partial^2 v}{\partial r^2} + \frac{1}{r} \frac{\partial v}{\partial r} + \frac{1}{r^2} \frac{\partial^2 v}{\partial \theta^2} + \frac{\partial^2 v}{\partial z^2} \right),$$

and

$$\cdot \; e^{-\kappa(\alpha^2 + \mu^2)t} J_n(\mu r) \frac{\cos}{\sin} n\theta \frac{\cos}{\sin} \alpha z$$

is a particular integral.

Now expand $f(r, \theta, z)$ in the Fourier series

$$\sum_{n=0}^{\infty} (a_n \cos n\theta + b_n \sin n\theta).$$

The coefficients a_n and b_n are functions of r and z, denoted by $F_n(r, z)$ and $G_n(r, z)$.

Expand these functions in the series of Bessel functions given by the positive roots of
$$J_n(\mu a) = 0,$$

and let
$$F_n(r, z) = \sum_{\mu} \phi_n(z) J_n(\mu r),$$

$$G_n(r, z) = \sum_{\mu} \psi_n(z) J_n(\mu r).$$

Finally, take the Fourier integrals for $\phi_n(z)$ and $\psi_n(z)$, namely,

$$\phi_n(z) = \frac{1}{\pi} \int_0^{\infty} d\alpha \int_{-\infty}^{\infty} \phi_n(\beta) \cos \alpha(\beta - z) \, d\beta,$$

$$\psi_n(z) = \frac{1}{\pi} \int_0^{\infty} d\alpha \int_{-\infty}^{\infty} \psi_n(\beta) \cos \alpha(\beta - z) \, d\beta.$$

Thus we get our solution in the form

$$v = \frac{1}{\pi} \sum_{\mu} \sum_{n=0}^{\infty} \int_0^{\infty} \int_{-\infty}^{\infty} e^{-\kappa(\alpha^2 + \mu^2)t} J_n(\mu r)[\phi_n(\beta) \cos n\theta + \psi_n(\beta) \sin n\theta] \cos \alpha(\beta - z) \, d\alpha d\beta,$$

$$\tag{3}$$

the summation with regard to μ being over the positive roots of

$$J_n(\mu a) = 0.$$

Interchanging the orders of integration in (3) we obtain finally

$$v = \frac{1}{2\sqrt{(\pi \kappa t)}} \sum_{n=0}^{\infty} \sum_{\mu} e^{-\kappa \mu^2 t} J_n(\mu r) \int_{-\infty}^{\infty} e^{-(\beta - z)^2/4\kappa t} \{\phi_n(\beta) \cos n\theta + \psi_n(\beta) \sin n\theta\} \, d\beta. \quad (4)$$

IV. *Surface* $r = a$ *at* $v = F(\theta, z)$. *Initial temperature* $v = f(r, \theta, z)$

As shown in § 1.14 we reduce this to case III and § 7.11, case I, by putting $v = u + w$.

V. *Infinite cylinder. The surface* $r = a$ *and the planes* $\theta = 0$ *and* $\theta = \theta_0$ *kept at zero. Initial temperature* $f(r, \theta)$

In this case we have

$$\frac{\partial v}{\partial t} = \kappa \left(\frac{\partial^2 v}{\partial r^2} + \frac{1}{r} \frac{\partial v}{\partial r} + \frac{1}{r^2} \frac{\partial^2 v}{\partial \theta^2} \right),$$

and
$$J_{m\pi/\theta_0}(\alpha r) \sin \frac{m\pi\theta}{\theta_0} e^{-\kappa \alpha^2 t}$$

is a particular integral of this equation.

Also the conditions at $r = a$, $\theta = 0$, and $\theta = \theta_0$ are satisfied, provided that m is a positive integer and α is a root of

$$J_{m\pi/\theta_0}(\alpha a) = 0.$$

Expand $f(r, \theta)$ in the sine series

$$\sum_1^\infty a_m \sin \frac{m\pi\theta}{\theta_0},$$

the coefficient a_m being a function of r, say $F_m(r)$.

Then expand $F_m(r)$ in the series of Bessel functions given by the positive roots of

$$J_{m\pi/\theta_0}(\alpha a) = 0.$$

In this way we are brought to the solution of our problem in the form

$$v = \sum_\alpha \sum_{m=1}^\infty A_{\alpha,m} e^{-\kappa\alpha^2 t} \sin \frac{m\pi\theta}{\theta_0} J_{m\pi/\theta_0}(\alpha r), \qquad (5)$$

where, using 7.5 (2),

$$A_{\alpha,m} = \frac{4}{a^2\theta_0\{J'_{m\pi/\theta_0}(\alpha a)\}^2} \int_0^a \int_0^{\theta_0} f(r, \theta)\sin \frac{m\pi\theta}{\theta_0} J_{m\pi/\theta_0}(\alpha r)r \, dr d\theta,$$

the summation in α being over the positive roots of

$$J_{m\pi/\theta_0}(\alpha a) = 0.$$

The solution for the wedge given by $\theta = 0$ and $\theta = \theta_0$ can be deduced from the above by letting $a \to \infty$.

If the initial temperature is V, constant,† (5) becomes

$$v = \frac{8}{\pi a^2} \sum_{n=0}^\infty \frac{1}{(2n+1)} \sin s\theta \sum_{m=1}^\infty e^{-\kappa\alpha_m^2 t} \frac{J_s(r\alpha_m)}{[J'_s(a\alpha_m)]^2} \int_0^a r J_s(r\alpha_m) \, dr, \qquad (6)$$

where $s = (2n+1)\pi/\theta_0$, and $\pm\alpha_m$, $m = 1, 2,...,$ are the roots of

$$J_s(a\alpha) = 0. \qquad (7)$$

† Some numerical results for this case are given by Jaeger, *Phil. Mag.* (7) **33** (1942) 527. The semi-infinite cylinder with this cross-section is discussed by Craggs, ibid. **36** (1945) 220.

THE FLOW OF HEAT IN REGIONS BOUNDED BY SURFACES OF THE CYLINDRICAL COORDINATE SYSTEM

8.1. Introductory

IN this chapter we discuss a number of problems on the flow of heat in regions bounded by surfaces of the cylindrical coordinate system, such as finite and semi-infinite cylinders, finite hollow cylinders, etc. The methods used will be those of the preceding chapters. Problems of the same type which involve regions bounded internally by a circular cylinder can be treated in the same way by using the solutions of § 13.5.

Further problems on cylindrical regions are solved in §§ 14.10–14.15 by the use of the Green's function, and in § 15.11 by the Laplace transformation.

8.2. The steady temperature in an infinite or semi-infinite medium due to heat supply over a circular area

Suppose heat is supplied over the circular area $0 \leqslant r < a$ in the plane $z = 0$ at a rate depending on r only.

The differential equation for the temperature,

$$\frac{\partial^2 v}{\partial r^2} + \frac{1}{r}\frac{\partial v}{\partial r} + \frac{\partial^2 v}{\partial z^2} = 0, \tag{1}$$

is satisfied by $\quad e^{-\lambda|z|}J_0(\lambda r)$

for any λ. Thus

$$\int_0^\infty e^{-\lambda|z|}J_0(\lambda r)f(\lambda)\,d\lambda \tag{2}$$

will be a solution of the problem, if $f(\lambda)$ can be chosen to satisfy the prescribed conditions in the plane $z = 0$. This can be done by using Neumann's integral theorem (*W.B.F.*, Chap. XIV), but the two most interesting cases follow from the well-known integrals involving Bessel functions†

$$\int_0^\infty J_0(\lambda r)\sin \lambda a\, \frac{d\lambda}{\lambda} = \begin{cases} \sin^{-1}(a/r), & r > a, \\ \tfrac{1}{2}\pi, & r \leqslant a; \end{cases} \tag{3}$$

$$\int_0^\infty J_0(\lambda r)J_1(\lambda a)\,d\lambda = \begin{cases} 0, & r > a, \\ 1/(2a), & r = a, \\ 1/a, & r < a. \end{cases} \tag{4}$$

† *W.B.F.*, § 13.42; *G. and M.*, pp. 141, 78.

Now consider the problem in which *the circular disc*, $0 \leqslant r < a$, $z = 0$, *is maintained at temperature V*. Here we choose $f(\lambda)$ in (2) to be $[(2V/\pi)\sin \lambda a]/\lambda$, so that

$$v = \frac{2V}{\pi} \int_0^\infty e^{-\lambda|z|} J_0(\lambda r) \sin \lambda a \frac{d\lambda}{\lambda} \tag{5}$$

is the solution of our problem. By (3) this has the value V for $z = 0$ and $r \leqslant a$. This is also the solution of the problem of the region $z > 0$ with the region $0 \leqslant r < a$ of the plane $z = 0$ maintained at temperature V, and with no flow of heat over the rest of the plane $z = 0$.

If the infinite solid has *heat supplied at constant rate Q per unit area per unit time over a circular area of radius a in the plane $z = 0$*, for example by a flat circular heating element, the condition to be satisfied in the plane $z = 0$ is

$$- 2K \left[\frac{\partial v}{\partial z} \right]_{z=+0} = Q, \quad 0 < r < a, \atop = 0, \qquad r > a. \tag{6}$$

Thus, using (2) and (4), the solution is

$$v = \frac{Qa}{2K} \int_0^\infty e^{-\lambda|z|} J_0(\lambda r) J_1(\lambda a) \frac{d\lambda}{\lambda}. \tag{7}$$

These results may be applied to the important problem of the semi-infinite solid $z > 0$ with various conditions at its surface.† For problems of this type in the variable state see § 10.5.

I. *The region $z > 0$ with constant temperature, V, over $0 < r < a$ and no flow of heat over $r > a$*

By (5),

$$v = \frac{2V}{\pi} \int_0^\infty e^{-\lambda z} J_0(\lambda r) \sin \lambda a \frac{d\lambda}{\lambda} \tag{8}$$

$$= \frac{2V}{\pi} \sin^{-1} \left\{ \frac{2a}{[(r-a)^2 + z^2]^{\frac{1}{2}} + [(r+a)^2 + z^2]^{\frac{1}{2}}} \right\}. \tag{9}$$

The rate of flow of heat, F, over the circle $0 < r < a$, is

$$F = -2\pi K \int_0^a \left[\frac{\partial v}{\partial z} \right]_{z=0} r\, dr = 4KVa \int_0^\infty J_1(\lambda a) \sin \lambda a \frac{d\lambda}{\lambda} = 4KVa, \tag{10}$$

† For other problems of this type see Lowan, *Phil. Mag.* (7) **29** (1940) 93; Thomas, *Quart. J. Mech. Appl. Math.* **10** (1957) 482.

using $W.B.F.$, § 13.42 (2) to evaluate the integral. The value $R = V/F$ may be regarded as the thermal resistance to steady flow from a circle of radius a into a half space. By (10) it is

$$R = V/F = 1/(4Ka). \tag{11}$$

II. *The regions $z > 0$ and $z < 0$ are of conductivities K_1 and K_2, respectively, and their temperatures at great distances from the origin are 0 and V respectively. Steady flow takes place through a circle of radius a in the plane $z = 0$, the remainder of the plane being impervious to heat*

The temperatures v_1 and v_2 in the regions $z > 0$ and $z < 0$ are found as in I to be

$$v_1 = \frac{2K_2 V}{\pi(K_1+K_2)} \int_0^\infty e^{-\lambda z} J_0(\lambda r)\sin \lambda a \frac{d\lambda}{\lambda}, \tag{12}$$

$$v_2 = V - \frac{2K_1 V}{\pi(K_1+K_2)} \int_0^\infty e^{\lambda z} J_0(\lambda r)\sin \lambda a \frac{d\lambda}{\lambda}, \tag{13}$$

which can be further simplified as in (8) and (9). The thermal resistance is $(K_1+K_2)/4aK_1 K_2$.

III. *The region $z > 0$ with constant flux Q over the circular area $r < a$ and zero flux over $r > a$*

By (7) the temperature is

$$v = \frac{Qa}{K} \int_0^\infty e^{-\lambda z} J_0(\lambda r) J_1(\lambda a) \frac{d\lambda}{\lambda}. \tag{14}$$

The average temperature v_{av} over $0 < r < a$ is

$$v_{av} = \frac{2Q}{K} \int_0^\infty J_1^2(\lambda a) \frac{d\lambda}{\lambda^2} = \frac{8Qa}{3\pi K}, \tag{15}$$

using $W.B.F.$, § 13.33 (1). Since the rate of flow of heat over the circle of radius a is $F = \pi a^2 Q$, (15) gives

$$v_{av}/F = 8/3\pi^2 Ka. \tag{16}$$

The variable state, and the corresponding problem for heating over a strip, are discussed in § 10.5.

IV. *The region $z > 0$ with constant flux Q over the circle $0 < r < a$ and zero temperature† over $r > a$*

$$v = \frac{2Q}{K\pi} \int_0^\infty e^{-\lambda z} J_0(\lambda r)\{\sin \lambda a - \lambda a \cos \lambda a\}\frac{d\lambda}{\lambda^2}. \tag{17}$$

V. *Constriction resistance*

When heat or electricity flows over a circle of radius a into a half-space, it is frequently important to know the steady thermal or electrical resistance to flow into the half-space.‡ The simplest approximation which is frequently made for small circles is to replace the circle by a hemisphere of radius a (that is, effectively, to regard the material within the hemisphere as a perfect conductor) so that flow is radial. In this case the thermal resistance R, defined as V/F where V is the contact temperature and F the flow across it, is, by 9.2 (14),

$$R = V/F = 1/(2\pi K a). \tag{18}$$

The accurate value (11) for the case in which the circle is maintained at constant temperature is $(\frac{1}{2}\pi)$ times this, indicating the importance of the region near the origin.

In practical problems, heat or electricity is usually supplied to the half-space through a wire, and in this case the boundary condition of constant temperature assumed in I is not completely accurate, in fact, constant flux as in II may be regarded as having equal status, so that comparison of the factor $0 \cdot 25$ of (11) and $8/3\pi^2 = 0 \cdot 270...$ of (16) may be regarded as giving a measure of the uncertainty in the resistance.

8.3. Steady temperature in finite and semi-infinite cylinders

In this section we shall give a number of simple results on steady flow in finite and semi-infinite cylinders. By superposition of these, the results of a great many more problems may be written down. For example, by superposing the results of III and IV below, the problem of the finite cylinder with prescribed temperatures on all its surfaces may be solved; by putting $h = 0$ in results for the radiation boundary condition, various problems involving no flow of heat over some boundaries

---

† Karush and Young, *J. Appl. Phys.* **23** (1952) 1191–3. It should be noticed that the boundary conditions in the plane $z = 0$ are 'mixed', that is, prescribed flux over part of the plane and prescribed temperature over the remainder. For such problems the theory of 'dual' integral equations is available, cf. Titchmarsh, *Theory of Fourier Integrals* (Oxford, 1937) § 11.16. The corresponding problem with an infinite strip $|x| < a$ in place of a circle is discussed by Tranter, *J. Appl. Phys.* **24** (1953) 369.

‡ For the case of flow into a bounded region see *G. and M.*, Chap. XII.

are solved; by taking $f(z)$ symmetrical about $\frac{1}{2}l$ in IV and V, two additional problems on the cylinder with no flow over one plane surface are solved; and by taking $f(z)$ antisymmetrical about $\frac{1}{2}l$ in V, a result is obtained for the cylinder with zero temperature over one plane face and radiation over the other. Taken together, the results given contain all possible combinations of boundary conditions for the finite circular cylinder. In the same way, many additional results for the finite hollow cylinder may be derived from those given. Also, solutions for the important cases in which the conductivities are different in the r- and z-directions, as well as those for problems in which the surface conductances are different over different surfaces, may be written down with equal ease though they are a little more complicated. Finally, it should be mentioned that in many special problems involving radiation, the results can often be expressed more simply in terms of the roots of 3.10 (6) and (7) than in terms of those of 3.9 (5) as has been done here.

I. *The finite cylinder* $0 \leqslant r < a$, $0 < z < l$. $z = 0$ *kept at prescribed temperature* $f(r)$, *and* $z = l$ *at zero temperature. Radiation at* $r = a$ *into a medium at zero*

Here v has to satisfy

$$\frac{\partial^2 v}{\partial r^2} + \frac{1}{r}\frac{\partial v}{\partial r} + \frac{\partial^2 v}{\partial z^2} = 0, \quad 0 \leqslant r < a, 0 < z < l, \tag{1}$$

with

$$v = 0, \quad z = l, \quad 0 \leqslant r < a, \tag{2}$$

$$v = f(r), \quad z = 0, \quad 0 \leqslant r < a, \tag{3}$$

$$\frac{\partial v}{\partial r} + hv = 0, \quad 0 < z < l, \quad r = a. \tag{4}$$

The function†

$$J_0(\alpha r)\sinh \alpha(l-z) \tag{5}$$

satisfies (1) and (2) for all values of α. It also satisfies (4) if α is a root of

$$\alpha J_0'(\alpha a) + h J_0(\alpha a) = 0. \tag{6}$$

As in § 7.7, we assume that $f(r)$ can be expanded in the series

$$f(r) = \sum_{n=1}^{\infty} A_n J_0(r\alpha_n), \tag{7}$$

---

† (1) is satisfied *either* by products of Bessel functions with exponential functions, *or* by products of modified Bessel functions with trigonometric functions. We choose the former solution when the temperature is prescribed as an arbitrary function on a plane boundary, and the latter when it is prescribed on a circular boundary.

where the α_n are the positive roots of (6), and

$$A_n = \frac{2\alpha_n^2}{a^2(h^2+\alpha_n^2)J_0^2(a\alpha_n)} \int_0^a rf(r)J_0(r\alpha_n)\, dr. \tag{8}$$

Then, with this value of A_n, the solution of our problem is

$$v = \sum_{n=1}^{\infty} A_n \frac{J_0(r\alpha_n)\sinh(l-z)\alpha_n}{\sinh l\alpha_n}. \tag{9}$$

If $f(r) = V$, constant, using 7.5 (5) in (8), we find

$$\frac{v}{V} = \sum_{n=1}^{\infty} \frac{2hJ_0(r\alpha_n)\sinh(l-z)\alpha_n}{a(h^2+\alpha_n^2)J_0(a\alpha_n)\sinh l\alpha_n}. \tag{10}$$

The steady temperature in this case has been used for the measurement of thermal conductivity.†

The flow of heat into the cylinder over the surface $z = 0$ is, from (10),

$$-2\pi K \int_0^a r\left[\frac{\partial v}{\partial z}\right]_{z=0} dr = 4\pi Kh^2 V \sum_{n=1}^{\infty} \frac{\coth l\alpha_n}{\alpha_n(h^2+\alpha_n^2)}. \tag{11}$$

Similarly the flow out of the cylinder over the face $z = l$ is

$$4\pi Kh^2 V \sum_{n=1}^{\infty} \frac{\operatorname{cosech} l\alpha_n}{\alpha_n(h^2+\alpha_n^2)}. \tag{12}$$

These results are the generalization of those for the thin wire discussed in § 4.5. They can be used to discuss the effect of finite thickness of the wire.

II. *The finite cylinder $0 \leqslant r < a$, $0 < z < l$. $z = 0$ kept at prescribed temperature $f(r)$. Radiation at the other surfaces into medium at zero*

$$v = \sum_{n=1}^{\infty} A_n J_0(r\alpha_n) \frac{\alpha_n \cosh \alpha_n(l-z)+h \sinh \alpha_n(l-z)}{\alpha_n \cosh \alpha_n l+h \sinh \alpha_n l}, \tag{13}$$

where the α_n and A_n are defined in (6) and (8). Some numerical results are given by Nancarrow (loc. cit.).

III. *The finite cylinder $0 \leqslant r < a$, $0 < z < l$. $z = 0$ kept at prescribed temperature $f(r)$, the other surfaces at zero*

$$v = \frac{2}{a^2} \sum_{n=1}^{\infty} \frac{J_0(r\alpha_n)\sinh(l-z)\alpha_n}{J_1^2(a\alpha_n)\sinh l\alpha_n} \int_0^a rf(r)J_0(r\alpha_n)\, dr, \tag{14}$$

where the α_n are the positive roots of

$$J_0(a\alpha_n) = 0. \tag{15}$$

† Nancarrow, *Proc. Phys. Soc.* **45** (1933) 447.

IV. *The finite cylinder $0 \leqslant r < a$, $0 < z < l$. $r = a$ kept at prescribed temperature $f(z)$, the other surfaces at zero*

$$v = \frac{2}{l} \sum_{n=1}^{\infty} \frac{I_0(n\pi r/l)}{I_0(n\pi a/l)} \sin\frac{n\pi z}{l} \int_0^l f(z')\sin\frac{n\pi z'}{l}\, dz'. \tag{16}$$

V. *The finite cylinder $0 \leqslant r < a$, $0 < z < l$. $r = a$ kept at prescribed temperature $f(z)$, radiation into medium at zero at the other surfaces*

$$v = 2 \sum_{n=1}^{\infty} \frac{I_0(\alpha_n r)\{\alpha_n \cos\alpha_n z + h\sin\alpha_n z\}}{I_0(\alpha_n a)\{(\alpha_n^2 + h^2)l + 2h\}} \int_0^l f(z')\{\alpha_n\cos\alpha_n z' + h\sin\alpha_n z'\}\, dz', \tag{17}$$

where the α_n are the positive roots (cf. § 3.9) of

$$\tan\alpha l = \frac{2\alpha h}{\alpha^2 - h^2}. \tag{18}$$

VI. *The finite cylinder $0 \leqslant r < a$, $-l < z < l$. No flow of heat over the ends $z = \pm l$. Heat supplied at rate Q through a strip of the curved surface of length b at one end, and abstracted in the same way at the other end. No flow of heat over the remainder of the curved surface*†

Here the boundary condition at $r = a$ is

$$-K\frac{\partial v}{\partial r} = f(z), \quad r = a, \quad -l < z < l,$$

where
$$\begin{aligned}
f(z) &= -Q, & l > z > l-b, \\
&= 0, & l-b > z > -l+b, \\
&= Q, & -l+b > z > -l.
\end{aligned}$$

Here $f(z)$ is an odd function and may be expanded in the series

$$f(z) = -\frac{4Q}{\pi} \sum_{n=0}^{\infty} \frac{(-1)^n}{(2n+1)} \sin\frac{(2n+1)\pi b}{2l} \sin\frac{(2n+1)\pi z}{2l}.$$

Therefore

$$v = \frac{8Ql}{K\pi^2} \sum_{n=0}^{\infty} \frac{(-1)^n}{(2n+1)^2} \frac{I_0[(2n+1)\pi r/2l]}{I_1[(2n+1)\pi a/2l]} \sin\frac{(2n+1)\pi b}{2l} \sin\frac{(2n+1)\pi z}{2l}. \tag{19}$$

VII. *The finite hollow cylinder $a < r < b$, $0 < z < l$. $r = a$ maintained at $f(z)$, and the other faces at zero*

$$v = \frac{2}{l} \sum_{n=1}^{\infty} \frac{F_0(n\pi r/l;\, n\pi b/l)}{F_1(n\pi a/l;\, n\pi b/l)} \sin\frac{n\pi z}{l} \int_0^l f(z')\sin\frac{n\pi z'}{l}\, dz', \tag{20}$$

where
$$F_0(x;\, y) = I_0(x)K_0(y) - K_0(x)I_0(y). \tag{21}$$

---

† This is the problem of a finite bar heated over the portion of width b of its surface at one end by a heating coil, and cooled at the opposite end. See Lees, *Phil. Trans. Roy. Soc.* A, **208** (1908) 381.

VIII. *The finite hollow cylinder $a < r < b$, $0 < z < l$. Flux of heat into the solid at $r = a$, a prescribed function $f(z)$. The other surfaces kept at zero temperature*

$$v = -\frac{2}{K\pi} \sum_{n=1}^{\infty} \frac{F_0(n\pi r/l;\, n\pi b/l)}{nF_1(n\pi a/l;\, n\pi b/l)} \sin\frac{n\pi z}{l} \int_0^l f(z')\sin\frac{n\pi z'}{l}\, dz', \tag{22}$$

where F_0 is defined in (21), and

$$F_1(x;\, y) = I_1(x)K_0(y) + K_1(x)I_0(y). \tag{23}$$

If† $f(z) = Q$,   when $L < z < l - L$,

     $= 0$,   when $0 < z < L$ and $l - L < z < l$,

(22) becomes

$$-\frac{4Ql}{K\pi^2} \sum_{m=0}^{\infty} \frac{F_0[(2m+1)\pi r/l;\, (2m+1)\pi b/l]}{(2m+1)^2 F_1[(2m+1)\pi a/l;\, (2m+1)\pi b/l]} \cos\frac{(2m+1)\pi L}{l} \cdot \sin\frac{(2m+1)\pi z}{l}.$$

$$\tag{24}$$

IX. *The region $-l < z < l$, $a < r < b$ contains medium of conductivity K_1. The surfaces $r = b$ and $z = \pm l$ are maintained at zero. The region $r < a$ is occupied by a wire of conductivity K, which is heated by electric current.‡*

We assume that the temperature of the wire is constant over its cross-section, and that there is no temperature discontinuity between the wire and its surroundings at $r = a$.

If v is the temperature in the hollow cylinder, we have to solve (1) in this region with $v = 0$ on $r = b$ and on the surfaces $z = \pm l$. The solution of these will be

$$v = \sum_{n=0}^{\infty} a_n \cos\frac{(2n+1)\pi z}{2l} F_0\left[\frac{(2n+1)\pi r}{2l};\, \frac{(2n+1)\pi b}{2l}\right], \tag{25}$$

where F_0 is defined in (21), and the a_n are constants to be determined from the boundary condition at $r = a$. This is derived from the equation 4.10 (4) for temperature in the wire, with $U = s = 0$, and with the quantity $Hp(v-v_0)$, which in 4.10 (4) represented the loss of heat from the wire to its surroundings, replaced by

$$-2\pi a K_1\left[\frac{\partial v}{\partial r}\right]_{r=a}.$$

This gives for the boundary condition at $r = a$

$$K\frac{\partial^2 v}{\partial z^2} + \frac{2K_1}{a}\frac{\partial v}{\partial r} + \frac{\alpha j I^2}{\omega^2\sigma_0} v + \frac{jI^2}{\omega^2\sigma_0} = 0, \quad r = a, \quad -l < z < l. \tag{26}$$

---

† This is a rough approximation to the arrangement in a tube furnace.

‡ This is the complete theory of the experimental method discussed approximately in § 7.2 VI.

Substituting (25) in (26) gives

$$\sum_{n=0}^{\infty} a_n \cos\frac{(2n+1)\pi z}{2l}\left\{\left[\frac{\alpha j I^2}{\omega^2\sigma_0}-\frac{K(2n+1)^2\pi^2}{4l^2}\right]F_0\left[\frac{(2n+1)\pi a}{2l};\frac{(2n+1)\pi b}{2l}\right]+\right.$$

$$\left.+\frac{K_1(2n+1)\pi}{al}F_1\left[\frac{(2n+1)\pi a}{2l};\frac{(2n+1)\pi b}{2l}\right]\right\}$$

$$=-\frac{jI^2}{\omega^2\sigma_0}$$

$$=-\frac{4jI^2}{\omega^2\sigma_0\pi}\sum_{n=0}^{\infty}\frac{(-1)^n}{(2n+1)}\cos\frac{(2n+1)\pi z}{2l},\tag{27}$$

where F_0 and F_1 are defined in (21) and (23). Equation (27) gives a_n for all n, and the solution (25) is complete.

X. *The finite hollow cylinder $a < r < b$, $0 < z < l$. The surface $z = 0$ kept at $f(r)$, the other surfaces at zero*

$$v = \frac{\pi^2}{2}\sum_{n=1}^{\infty}\frac{\alpha_n^2 J_0^2(a\alpha_n)U_0(r\alpha_n)\sinh(l-z)\alpha_n}{[J_0^2(a\alpha_n)-J_0^2(b\alpha_n)]\sinh l\alpha_n}\int_a^b rf(r)U_0(r\alpha_n)\,dr,\tag{28}$$

where $U_0(\alpha r)$ is defined in 7.10 (2), and the α_n are the positive roots of $U_0(\alpha a) = 0$.

XI. *The finite hollow cylinder $a < r < b$, $0 < z < l$. $r = a$ kept at temperature $f(r)$. Radiation at the other surfaces into medium at zero*†

$$v = 2\sum_{n=1}^{\infty}\frac{(\alpha_n\cos\alpha_n z+h\sin\alpha_n z)\phi(r;n)}{[(\alpha_n^2+h^2)l+2h]\phi(a;n)}\int_0^l f(z)(\alpha_n\cos\alpha_n z+h\sin\alpha_n z)\,dz,\tag{29}$$

where the α_n are the positive roots of (18), and

$$\phi(r;n) = I_0(r\alpha_n)[\alpha_n K_1(b\alpha_n)-hK_0(b\alpha_n)]+K_0(r\alpha_n)[\alpha_n I_1(b\alpha_n)+hI_0(b\alpha_n)].\tag{30}$$

XII. *The semi-infinite cylinder $0 \leqslant r < a$, $z > 0$. $z = 0$ kept at prescribed temperature $f(r)$, and $r = a$ at zero temperature*

$$v = \frac{2}{a^2}\sum_{n=1}^{\infty}\frac{J_0(r\alpha_n)}{J_1^2(a\alpha_n)}e^{-\alpha_n z}\int_0^a rf(r)J_0(r\alpha_n)\,dr,\tag{31}$$

where the α_n are the positive roots of (15).

XIII. *The semi-infinite cylinder $0 \leqslant r < a$, $z > 0$. $r = a$ kept at prescribed temperature $f(z)$, and $z = 0$ at zero temperature*

$$v = \frac{1}{a}\sum_{n=1}^{\infty}\frac{J_0(r\alpha_n)}{J_1(a\alpha_n)}\int_0^{\infty} f(\beta)\{e^{-\alpha_n|\beta-z|}-e^{-\alpha_n(\beta+z)}\}\,d\beta,\tag{32}$$

where the α_n are the positive roots of (15).

† This is the problem of steady flow of heat in a cylindrical cooling fin, cf. § 4.6.

XIV. *The semi-infinite cylinder $0 \leqslant r < a$, $z > 0$. $z = 0$ kept at prescribed temperature $f(r)$, and at $r = a$ radiation into medium at zero*

$$v = \frac{2}{a^2} \sum_{n=1}^{\infty} \frac{\alpha_n^2 J_0(r\alpha_n)}{(h^2+\alpha_n^2)J_0^2(a\alpha_n)} \, e^{-\alpha_n z} \int_0^a rf(r)J_0(r\alpha_n) \, dr, \tag{33}$$

where the α_n are the positive roots of (6).

When a is small the roots $\alpha_1, \alpha_2,...$ increase rapidly, and we may take for v the first term in this expansion. Further, if $\alpha_1^2 a^2$ may be neglected, we have

$$J_0(a\alpha_1) = 1 \quad \text{and} \quad J_0'(a\alpha_1) = -\tfrac{1}{2}a\alpha_1.$$

Therefore, from (6) $\alpha_1 = \sqrt{(2h/a)}$.

It will be noticed that this requires ah to be small, and to this approximation

$$v = Ve^{-z\sqrt{(2h/a)}},$$

where V is the average value of the temperature over the end. This agrees with the solution of § 4.3.

XV. *The semi-infinite cylinder $0 \leqslant r < a$, $z > 0$. $r = a$ kept at prescribed temperature $f(z)$, and at $z = 0$ there is radiation into medium at zero temperature*

$$v = \frac{1}{a} \sum_{n=1}^{\infty} \frac{J_0(r\alpha_n)}{J_1(a\alpha_n)} \int_0^{\infty} \left\{ e^{-\alpha_n|z-z'|} + \frac{\alpha_n - h}{\alpha_n + h} e^{-\alpha_n(z+z')} \right\} f(z') \, dz', \tag{34}$$

where the α_n are the positive roots of (15).

XVI. *The finite cylinder $0 \leqslant r < b$, $0 < z < l$. Constant flux F over the circle $0 \leqslant r < a$, $z = 0$ and constant flux F over the circle $0 \leqslant r < a$, $z = l$. No flow over the remainder of the surface*

The difference between the average temperatures of the two regions over which heat is supplied and removed is

$$\frac{Fl}{K}\left\{1 - \frac{16}{\pi^2} \sum \frac{I_1(n\pi a/l)}{n^2 I_1(n\pi b/l)}\left[I_1\left(\frac{n\pi b}{l}\right)K_1\left(\frac{n\pi a}{l}\right) - K_1\left(\frac{n\pi b}{l}\right)I_1\left(\frac{n\pi a}{l}\right)\right]\right\}, \tag{35}$$

where the summation is over odd values of n.

XVII. *The finite cylinder $0 < z < l$, $0 \leqslant r < a$ with heat production† at the constant rate A_0 per unit time per unit volume. Zero surface temperature*

The equation to be solved is

$$\frac{\partial^2 v}{\partial r^2} + \frac{1}{r}\frac{\partial v}{\partial r} + \frac{\partial^2 v}{\partial z^2} = -\frac{A_0}{K}. \tag{36}$$

A particular integral of this which vanishes when $z = 0$ and $z = l$ is

$$A_0 z(l-z)/2K, \tag{37}$$

and the complete solution is obtained by adding to this the solution of (1) which has the value $-A_0 z(l-z)/2K$ when $r = a$, and is zero on $z = 0$ and $z = l$. This

† Problems involving heat production occur in the heating of the windings of magnet coils, cf. Emmerich, *J. Appl. Phys.* **21** (1950) 75–80, Szalay, *Frequenz*, **7** (1953) 81–84, and in some biological problems, cf. Thews, *Acta Biotheoretica*, A **10** (1953) 105–38.

is given by (16), and the final result is

$$v = -\frac{A_0 z(l-z)}{2K} - \frac{4l^2 A_0}{K\pi^3} \sum_{n=0}^{\infty} \frac{I_0[(2n+1)\pi r/l]}{(2n+1)^3 I_0[(2n+1)\pi a/l]} \sin\frac{(2n+1)\pi z}{l}. \qquad (38)$$

XVIII. *The finite hollow cylinder $0 < z < l$, $a < r < b$ with no flow over $r = a$, zero temperature on $z = 0$ and $z = l$, and temperature $f(z)$ on $r = b$*

$$v = \frac{2}{l} \sum_{n=1}^{\infty} \frac{F_1(n\pi a/l; n\pi r/l)}{F_1(n\pi a/l; n\pi b/l)} \sin\frac{n\pi z}{l} \int_0^l f(z')\sin\frac{n\pi z'}{l} dz', \qquad (39)$$

where $F_1(x, y)$ is defined in (23).

XIX. *The finite hollow cylinder $0 < z < l$, $a < r < b$ with heat production at a constant rate A_0 per unit time per unit volume, no flow over $r = a$, zero temperature over the other surfaces*

Using (37) and (39) gives

$$v = \frac{A_0 z(l-z)}{2K} - \frac{4l^2 A_0}{K\pi^3} \sum_{n=0}^{\infty} \frac{F_1[(2n+1)\pi a/l; (2n+1)\pi r/l]}{F_1[(2n+1)\pi a/l; (2n+1)\pi b/l]} \frac{\sin(2n+1)\pi z/l}{(2n+1)^3}. \qquad (40)$$

XX. *The finite hollow cylinder $0 < z < l$, $a < r < b$ with heat production at the constant rate A_0, zero temperature over $z = 0$, $z = l$, and $r = b$, and water-cooling over $r = a$*

Problems of this type may readily be treated in the following manner. Suppose that the surface $r = a$ is in contact with fluid of specific heat c' at temperature V, and that mass M of this fluid is removed per unit time and replaced with the same amount of fluid at zero temperature, the fluid being kept well stirred. Then the rate of removal of heat by the fluid is $Mc'V$ and this must be equal to f_a, the rate of flow of heat over the surface $r = a$, $0 < z < l$. Writing down this latter, using (37) and (20), gives an equation for V.

XXI. *The finite cylinder $0 \leqslant r < a$, $0 < z < l$ with zero surface temperature and heat production at the rate $A_0(1+\beta v)$*

In this case the equation to be solved is

$$\frac{\partial^2 v}{\partial r^2} + \frac{1}{r}\frac{\partial v}{\partial r} + \frac{\partial^2 v}{\partial z^2} + \frac{A_0 \beta}{K} v = -\frac{A_0}{K}.$$

A particular integral which vanishes when $z = 0$ and $z = l$ is

$$\frac{1}{\beta}\left\{\frac{\cos\gamma(\tfrac{1}{2}l-z)}{\cos\tfrac{1}{2}\gamma l} - 1\right\},$$

where $\gamma = (A_0\beta/K)^{\frac{1}{2}}$.

Using this as before, the complete solution is found to be

$$v = \frac{1}{\beta}\left\{\frac{\cos\gamma(\tfrac{1}{2}l-z)}{\cos\tfrac{1}{2}\gamma l} - 1\right\} - \frac{4A_0 l^2}{\pi K} \sum_{n=0}^{\infty} \frac{I_0(r\gamma_n/l)\sin(2n+1)\pi z/l}{I_0(a\gamma_n/l)(2n+1)\gamma_n^2}, \qquad (41)$$

where $\gamma_n^2 = (2n+1)^2\pi^2 - \gamma^2 l^2$. There are restrictions on β of the type discussed in § 15.7.

XXII. *Other problems*

Weber† has given a number of solutions of problems on steady flow of electricity (or heat) in regions bounded by cylinders and planes perpendicular to their axis. In many of these electric current flows in and out through small electrodes on the surface. Problems of this type are not of great interest in conduction of heat and a full account of them is given in Gray and Mathews (loc. cit.).

8.4. Variable state. Product solutions

As in Chapters V and VI the solutions of many important problems can be written down as in § 1.15 as products of the solutions of simple problems. These were the functions $\phi(x)$ and $\phi(x,h)$, 5.6 (1) and (5), appropriate to the semi-infinite solid with prescribed surface temperature and radiation respectively, and the functions $\psi(x,l)$ and $\psi(x,l;h)$, 5.6 (6) and (7), appropriate to the slab $-l < x < l$ with prescribed surface temperature and radiation respectively. To these must now be added the functions

$$\chi(r,a) = \frac{2}{a} \sum_{n=1}^{\infty} e^{-\kappa \alpha_n^2 t} \frac{J_0(r\alpha_n)}{\alpha_n J_1(a\alpha_n)}, \tag{1}$$

where the α_n are the positive roots of

$$J_0(a\alpha) = 0, \tag{2}$$

and

$$\chi(r,a,h) = \frac{2h}{a} \sum_{s=1}^{\infty} e^{-\kappa \alpha_s^2 t} \frac{J_0(r\alpha_s)}{(h^2+\alpha_s^2)J_0(a\alpha_s)}, \tag{3}$$

where the α_s are the positive roots of

$$\alpha J_0'(a\alpha)+h J_0(a\alpha) = 0. \tag{4}$$

The expressions (1) and (3) were found in 7.6 (5), 7.7 (4) to be the solutions of the problems of the infinite cylinder with unit initial temperature, and with its surface kept at zero, and radiating into medium at zero, respectively.

   I. *The finite cylinder $-l < z < l$, $0 \leqslant r < a$. Unit initial temperature; zero surface temperature*

$$v = \psi(z,l)\chi(r,a). \tag{5}$$

Numerical values of the functions ψ and χ have been given in §§ 3.4, 7.6. In Figs. 27 and 28 the isothermals $v = 0.1, 0.2, 0.3,...$ are shown for $\kappa t/a^2 = 0.08$ for axial planes in the cylinders for which $l = a$ and $l = 2a$. These give an idea of the importance of end effects.

If the cylinder has different conductivities in the radial and axial

---

† *Crelle*, **75** (1873); **76** (1873). Most of the material of the first of these papers is given in Gray and Mathews, *Treatise on Bessel Functions* (edn. 2, 1922) Chap. XII. See also Weinberg, *Bull. Math. Biophysics*, **3** (1941) 39.

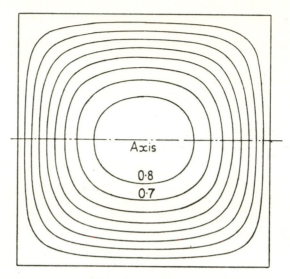

Fig. 27. Isothermals 0·1, 0·2,..., 0·8 for a cylinder of length and diameter $2a$, with unit initial and zero surface temperature: $\kappa t/a^2 = 0.08$.

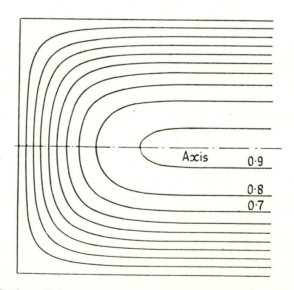

Fig. 28. Isothermals 0·1, 0·2,..., 0·9 for a cylinder of length $4a$ and diameter $2a$, with unit initial and zero surface temperature: $\kappa t/a^2 = 0.08$.

directions, as in a log of wood, the solution is still the product (5) of the same functions, each with its appropriate value of κ.

II. *The finite cylinder* $-l < z < l,\ 0 \leqslant r < a.$ *Zero initial temperature*

If the surface temperature is unity the solution is

$$v = 1 - \psi(z, l)\chi(r, a)$$

$$= 1 - \frac{8}{\pi a} \sum_{n=0}^{\infty} \sum_{m=1}^{\infty} \frac{(-1)^n J_0(r\alpha_m)}{(2n+1)\alpha_m J_1(a\alpha_m)} \cos \frac{(2n+1)\pi x}{2l} e^{-\kappa l\{\alpha_m^2 + (2n+1)^2\pi^2/4l^2\}},$$

(6)

where the α_m are the positive roots of (2). The result for surface temperature $\phi(t)$ follows from Duhamel's theorem.

III. *The finite cylinder* $-l < z < l,\ 0 \leqslant r < a.$ *Unit initial temperature. Radiation at the surface into medium at zero*

$$v = \psi(z, l, h)\chi(r, a, h).$$

(7)

Numerical values of ψ and χ are given in the literature (for references see §§ 3.11, 7.7). Some numerical values are given by Berger.[†]

IV. *The finite cylinder* $-l < z < l,\ 0 \leqslant r < a.$ *Unit initial temperature. Radiation at* $z = \pm l$ *into medium at zero.* $r = a$ *kept at zero for* $t > 0$

$$v = \psi(z, l, h)\chi(r, a).$$

(8)

V. *The semi-infinite cylinder* $z > 0,\ 0 \leqslant r < a,$ *with unit initial temperature*

If there is radiation at all surfaces into medium at zero

$$v = \phi(z, h)\chi(r, a, h).$$

(9)

If all surfaces are kept at zero

$$v = \phi(z)\chi(r, a).$$

(10)

For $\kappa t/a^2 = 0{\cdot}08$, the isothermals near the end $z = 0$ are in this case indistinguishable from those of Fig. 28.

If $z = 0$ is kept at zero, and there is radiation at $r = a$ into medium at zero,

$$v = \phi(z)\chi(r, a, h).$$

(11)

If $r = a$ is kept at zero, and there is radiation at $z = 0$ into medium at zero,

$$v = \phi(z, h)\chi(r, a).$$

(12)

† *Z. angew. Math. Mech.* **11** (1931) 45.

Clearly, the solutions of many other problems of the same type can be written down immediately; in particular, problems on hollow cylinders.

In Chapter XIII solutions for temperature in regions bounded internally by cylinders will be given, and then solutions of the present type for regions such as $r > a$, $z > 0$, or $r > a$, $-l < z < l$, can be written down.

8.5. Determination of conductivity from cylinders

A number of methods, similar to those of § 6.5, has been used for determining the conductivity of material in the form of a finite cylinder.

For the finite cylinder $0 \leqslant r < a$, $-l < z < l$, with unit initial temperature,† radiating at its surface into medium at zero, the first term of the solution of 8.4 (7) is

$$\frac{8h \sin \lambda_1 l \cos \lambda_1 z J_0(\alpha_1 r)}{a(h^2 + \alpha_1^2)[2l\lambda_1 + \sin 2l\lambda_1] J_0(a\alpha_1)} e^{-\kappa t(\lambda_1^2 + \alpha_1^2)}, \tag{1}$$

where α_1 is the smallest root of

$$\alpha J_1(a\alpha) = h J_0(a\alpha) \tag{2}$$

and λ_1 is the smallest root of $\qquad \lambda \tan \lambda l = h.$ $\qquad\qquad$ (3)

If the cylinder with unit initial temperature has its curved surface kept at zero and radiation at its ends into medium at zero,‡ the first term of the solution 8.4 (8) is

$$\frac{8 \sin \lambda_1 l \cos \lambda_1 z J_0(\mu_1 r)}{a\mu_1 J_1(a\mu_1)[2l\lambda_1 + \sin 2\lambda_1 l]} e^{-\kappa t(\lambda_1^2 + \mu_1^2)}, \tag{4}$$

where λ_1 is the smallest root of (3), and μ_1 is the smallest root of

$$J_0(a\mu) = 0. \tag{5}$$

If the cylinder has unit initial temperature, and for $t > 0$ the surface $z = -l$ is kept at zero temperature and the other surfaces radiate into medium at zero,§ the first term of the solution is

$$\frac{8h(1 - \cos 2\lambda_1 l)\sin \lambda_1(z+l) J_0(\alpha_1 r)}{a(h^2 + \alpha_1^2)[4\lambda_1 l - \sin 4\lambda_1 l] J_0(a\alpha_1)} e^{-\kappa t(\lambda_1^2 + \alpha_1^2)}, \tag{6}$$

where α_1 is the smallest root of (2), and λ_1 is the smallest root of

$$\lambda \cot 2\lambda l + h = 0. \tag{7}$$

Any of these results (1), (4), (6), or others of the same type may be reduced by any of the methods of §§ 6.5, 9.5 so as to give values of h, κ, and, if desired, the initial temperature of the cylinder.

8.6. The finite cylinder $-l < z < l$, $0 \leqslant r < a$. Initial temperature $f(r, \theta, z)$

If the initial temperature or the boundary conditions are such that the method of § 1.15 is not available, we use combinations of Fourier and Fourier–Bessel series.

---

† Weber, *Ann. Physik*, N.F. **10** (1880) 103.
‡ Weber, *S.B. preuss. Akad. Wiss.* (1880) 457.
§ Beglinger, *Verh. Ver. GewFleiss., Berl.* **75** (1896); Hall, *Phys. Rev.* **10** (1900) 277.

Alternatively the Green's function, Chapter XIV, may be used, or the Laplace transformation applied directly† as in Chapter XV.

I. *Surface at zero temperature*

The equations for the temperature are

$$\frac{\partial v}{\partial t} = \kappa \nabla^2 v \quad (0 \leqslant r < a, -l < z < l), \tag{1}$$

$$v = f(r, \theta, z), \quad \text{when } t = 0, \tag{2}$$

and

$$v = 0, \quad \text{when } r = a \text{ and } z = \pm l. \tag{3}$$

The expression $e^{-\kappa(\mu^2 + m^2\pi^2/4l^2)t} J_n(\mu r) \begin{matrix} \cos \\ \sin \end{matrix} n\theta \sin \dfrac{m\pi(z+l)}{2l}$

satisfies (1) and (3), if m is any integer and μ is a root of

$$J_n(\mu a) = 0.$$

Now expand $f(r, \theta, z)$ in the Fourier series

$$\sum_{n=0}^{\infty} (a_n \cos n\theta + b_n \sin n\theta),$$

a_n and b_n being functions of r and z, denoted by $F_n(r, z)$ and $G_n(r, z)$. Then expand $F_n(r, z)$ and $G_n(r, z)$ in the series of Bessel functions given by the positive roots of $J_n(\mu a) = 0$, and take the sine series, whose terms are the sines of multiples of $\pi(z+l)/2l$, for the coefficients in this series.

In this way we find the solution in the form

$$v = \sum_{\mu} \sum_{m=1}^{\infty} \sum_{n=0}^{\infty} e^{-\kappa(\mu^2 + m^2\pi^2/4l^2)t} J_n(\mu r) \sin \frac{m\pi(z+l)}{2l} (A_{\mu,m,n} \cos n\theta + B_{\mu,m,n} \sin n\theta),$$

where

$$A_{\mu,m,n} = \frac{2}{\pi a^2 l \{J_n'(\mu a)\}^2} \int_0^a r J_n(\mu r)\, dr \int_{-l}^l \sin \frac{m\pi(z+l)}{2l}\, dz \int_{-\pi}^{\pi} \cos n\theta f(r, \theta, z)\, d\theta,$$

and a similar expression holds for $B_{\mu,m,n}$.‡

II. *Other cases*

The solutions of other problems on the finite cylinder may be derived by the method used above and in § 7.12. The case of prescribed surface temperature is discussed in *C.H.*, § 60; that of radiation at the surface in ibid., § 61, also Heine, loc. cit., Bd. II, § 83. The region bounded by the planes $z = \pm l$, the planes $\theta = 0$ and $\theta = \theta_0$, and the surface $r = a$ is discussed in *C.H.*, § 62.

8.7. The semi-infinite cylinder

Problems on the semi-infinite cylinder with arbitrary initial or surface temperature may be discussed along the lines of §§ 7.12, 8.6. The solution for arbitrary initial temperature and zero surface temperature is given in *C.H.*, § 59. Some results for the semi-infinite hollow cylinder are given by Tranter (*Phil. Mag.* (7) **35** (1944) 102) and Lowan (*Quart. Appl. Math.* **2** (1945) 348).

† The variable state for the systems I, II, XIII, XIV of § 8.3 is discussed in this way in § 15.11.

‡ Cf. Heine, *Handbuch der Kugelfunctionen* (2. Aufl.), Bd. II, § 81; when $n = 0$, the expression is to be halved.

THE FLOW OF HEAT IN A SPHERE AND CONE

9.1. Introductory

W$_E$ have seen in § 1.8 that the equation of conduction, when expressed in spherical polar coordinates, becomes

$$\frac{\partial v}{\partial t} = \kappa \left\{ \frac{1}{r^2} \frac{\partial}{\partial r} \left(r^2 \frac{\partial v}{\partial r} \right) + \frac{1}{r^2 \sin \theta} \frac{\partial}{\partial \theta} \left(\sin \theta \frac{\partial v}{\partial \theta} \right) + \frac{1}{r^2 \sin^2 \theta} \frac{\partial^2 v}{\partial \phi^2} \right\}. \tag{1}$$

In the case of Flow of Heat in the Sphere, when the initial and surface conditions are such that the isothermal surfaces are concentric spheres, and the temperature thus depends only upon the coordinates r and t, this equation becomes

$$\frac{\partial v}{\partial t} = \kappa \left(\frac{\partial^2 v}{\partial r^2} + \frac{2}{r} \frac{\partial v}{\partial r} \right). \tag{2}$$

On putting $u = vr$, we have

$$\frac{\partial u}{\partial t} = \kappa \frac{\partial^2 u}{\partial r^2}. \tag{3}$$

In this chapter, a number of important problems on the sphere, the hollow sphere, and the region bounded internally by a sphere will be solved by the classical method involving the change of variable (3). For completeness, a number of results which are better derived by the methods of Chapters XIII and XIV will be stated without proof. Topics which are deferred to § 13.9 are composite spheres, spherical or infinite regions containing a spherical core of perfect conductor, and heat production in an infinite medium.

9.2. Steady temperature. Radial flow

In this case the differential equation is

$$\frac{d}{dr} \left(r^2 \frac{dv}{dr} \right) = 0. \tag{1}$$

The general solution of (1) is

$$v = \frac{A}{r} + B, \tag{2}$$

where A and B are constants to be determined from the boundary conditions.

I. *The hollow sphere $a < r < b$. $r = a$ at v_1, and $r = b$ at v_2*

$$v = \frac{av_1(b-r)+bv_2(r-a)}{r(b-a)}. \tag{3}$$

II. *The hollow sphere $a < r < b$. $r = a$ at v_1, and at $r = b$ radiation into medium at v_2*

$$v = \frac{av_1[hb^2+r(1-hb)]+hb^2 v_2(r-a)}{r[hb^2+a(1-hb)]}. \tag{4}$$

III. *The hollow sphere $a < r < b$. At $r = a$ radiation from medium at v_1, and at $r = b$ radiation into medium at v_2*

If the boundary conditions are

$$\frac{\partial v}{\partial r}+h_1(v_1-v) = 0, \quad r = a; \qquad \frac{\partial v}{\partial r}+h_2(v-v_2) = 0, \quad r = b,$$

the solution is

$$v = \frac{v_1 a^2 h_1[b^2 h_2-r(bh_2-1)]+v_2 b^2 h_2[r(ah_1+1)-a^2 h_1]}{r[b^2 h_2(ah_1+1)-a^2 h_1(bh_2-1)]}. \tag{5}$$

IV. *Constant flux $Q_0/4\pi a^2$ at the inner surface $r = a$ of the hollow sphere $a < r < b$*

Since

$$-4\pi r^2 K \frac{dv}{dr}$$

is the rate of flow of heat over any spherical surface of radius r, and by (1) this is constant, we have

$$Q_0 = -4\pi r^2 K \frac{dv}{dr}, \quad a < r < b. \tag{6}$$

If v_1 and v_2 are the temperatures at $r = a$ and $r = b$ respectively, it follows on integrating again that

$$Q_0 = \frac{4\pi K(v_1-v_2)ab}{b-a}. \tag{7}$$

Arrangements in which heat is produced within a hollow sphere have been used† for the measurement of thermal conductivity. The use of a sphere has the advantage of eliminating end effects, but introduces other difficulties.

If the conductivity K is a function of the temperature, (6) is still true, and integrating gives

$$Q_0\left(\frac{1}{a}-\frac{1}{b}\right) = 4\pi \int_{v_2}^{v_1} K\, dv = 4\pi K_m(v_1-v_2), \tag{8}$$

† Laws, Bishop, and McJunkin, *Proc. Amer. Acad. Arts Sci.* **41** (1906) 457; Müller, *Ann. Phys. und Chem.* **60** (1897) 82; Green, *Proc. Phys. Soc.* **44** (1932) 295.

where K_m is the mean conductivity over the range of temperature from a to b. Thus (7) remains true with K replaced by K_m.

V. *The composite hollow sphere of n regions (a_1, a_2), (a_2, a_3),..., (a_n, a_{n+1}) of conductivities K_1,..., K_n*

If $v_1, v_2,..., v_{n+1}$ are the temperatures at $a_1, a_2,..., a_{n+1}$, repeated application of (7) gives

$$Q_0 = \frac{4\pi K_1(v_1-v_2)a_1 a_2}{a_2-a_1} = ... = \frac{4\pi K_n(v_n-v_{n+1})a_{n+1} a_n}{a_{n+1}-a_n}.$$

Therefore

$$v_1-v_{n+1} = \frac{Q_0}{4\pi} \sum_{r=1}^{n} \frac{1}{K_r}\left(\frac{1}{a_r} - \frac{1}{a_{r+1}}\right). \tag{9}$$

If, in addition, there are contact resistances $R_1, R_2,..., R_{n+1}$ per unit area over the surfaces $a_1, a_2,..., a_{n+1}$, and v_0 and v_{n+2} are the temperatures inside and outside the composite sphere,

$$v_0-v_{n+2} = \frac{Q_0}{4\pi}\left\{ \sum_{r=1}^{n} \frac{1}{K_r}\left(\frac{1}{a_r} - \frac{1}{a_{r+1}}\right) + \sum_{r=1}^{n+1} \frac{R_r}{a_r^2}\right\}. \tag{10}$$

VI. *The solid sphere $0 \leqslant r < a$ with heat produced in it at a constant rate A_0 per unit time per unit volume*

Here the differential equation, 1.6 (7), becomes

$$\frac{1}{r^2} \frac{d}{dr}\left(r^2 \frac{dv}{dr}\right) = -\frac{A_0}{K}, \tag{11}$$

and we require a solution which is finite at $r = 0$. If the surface temperature is zero, this is

$$v = \frac{A_0}{6K}(a^2-r^2). \tag{12}$$

If there is radiation at $r = a$ into medium at zero, the solution is

$$v = \frac{A_0}{6hK}\{h(a^2-r^2)+2a\}. \tag{13}$$

VII. *The region $r > a$ with temperature V_0 at $r = a$*

$$v = aV_0/r, \quad -4\pi a^2 K\left[\frac{\partial v}{\partial r}\right]_{r=a} = 4\pi KaV_0. \tag{14}$$

VIII. *The region $0 \leqslant r < a$ has conductivity K_0 and heat production at the constant rate A_0 per unit time per unit volume. The region $r > a$ has conductivity K and no heat production. Contact resistance R per unit area over the surface $r = a$*

$$v = A_0\{a^2-r^2+2RaK_0+2a^2(K_0/K)\}/6K_0, \quad 0 \leqslant r < a,$$
$$v = A_0 a^3/3Kr, \quad r > a.$$

9.3. The sphere $0 \leqslant r < a$ with initial temperature $f(r)$, and surface temperature $\phi(t)$

As remarked in § 9.1, we substitute

$$u = vr, \tag{1}$$

and the equations for u are

$$\frac{\partial u}{\partial t} = \kappa \frac{\partial^2 u}{\partial r^2}, \quad 0 \leqslant r < a, \tag{2}$$

with

$$u = 0, \qquad \text{when } r = 0,$$

$$u = a\phi(t), \quad \text{when } r = a,$$

$$u = rf(r), \quad \text{when } t = 0.$$

These are the equations of flow of heat in a slab of thickness a, with its ends $r = 0$ and $r = a$ kept at zero and $a\phi(t)$ respectively, and with initial temperature $rf(r)$. The solution of this problem has been given in 3.5 (2). Using this we have finally

$$v = \frac{2}{ar} \sum_{n=1}^{\infty} e^{-\kappa n^2 \pi^2 t/a^2} \sin \frac{n\pi r}{a} \times$$

$$\times \left\{ \int_0^a r' f(r') \sin \frac{n\pi r'}{a} \, dr' - n\pi\kappa(-1)^n \int_0^t e^{\kappa n^2 \pi^2 \lambda/a^2} \phi(\lambda) \, d\lambda \right\}. \tag{3}$$

Results for some important special cases are given below. In most of them, alternative solutions which are useful for small values of the parameter $\kappa t/a^2$ are also given.

I. *Zero initial temperature. Surface temperature V, constant*

$$v = V + \frac{2aV}{\pi r} \sum_{n=1}^{\infty} \frac{(-1)^n}{n} \sin \frac{n\pi r}{a} e^{-\kappa n^2 \pi^2 t/a^2} \tag{4}$$

$$= \frac{aV}{r} \sum_{n=0}^{\infty} \left\{ \operatorname{erfc} \frac{(2n+1)a - r}{2(\kappa t)^{\frac{1}{2}}} - \operatorname{erfc} \frac{(2n+1)a + r}{2(\kappa t)^{\frac{1}{2}}} \right\}. \tag{5}$$

The temperature v_c at the centre, given by the limit as $r \to 0$ of (4) or (5), is

$$v_c = V + 2V \sum_{n=1}^{\infty} (-1)^n e^{-\kappa n^2 \pi^2 t/a^2} \tag{6}$$

$$= \frac{aV}{(\pi \kappa t)^{\frac{1}{2}}} \sum_{n=0}^{\infty} e^{-(2n+1)^2 a^2/4\kappa t}. \tag{7}$$

The average temperature v_{av} of the sphere at any time† is

$$v_{av} = V - \frac{6V}{\pi^2} \sum_{n=1}^{\infty} \frac{1}{n^2} e^{-\kappa n^2 \pi^2 t/a^2} \tag{8}$$

$$= \frac{6V(\kappa t)^{\frac{1}{2}}}{a\pi^{\frac{1}{2}}} - \frac{3\kappa Vt}{a^2} + \frac{12V(\kappa t)^{\frac{1}{2}}}{a} \sum_{n=1}^{\infty} \text{ierfc} \frac{na}{(\kappa t)^{\frac{1}{2}}}. \tag{9}$$

The heat content of the sphere at any time is $4\pi a^3 \rho c v_{av}/3$.

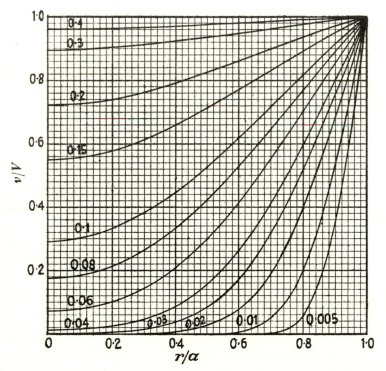

FIG. 29. Temperature distribution at various times in a sphere of radius a with zero initial temperature and surface temperature V. The numbers on the curves are the values of $\kappa t/a^2$.

In Fig. 29 values of v/V calculated from (4) for various values of $T = \kappa t/a^2$ are plotted as a function of r/a. These may be compared with the corresponding curves of Fig. 24 for the cylinder and Fig. 11 for the slab. Values of the centre and average temperatures, calculated from (6) and (8), have been given in Fig. 12.

† Evans. *Phil. Mag.* (7) **22** (1936) 833–7, remarks that by observing the weight of steam condensed at any time on a sphere in a Joly steam calorimeter both κ and ρc can be determined.

Equation (8) has an application to the theory of time-lag in thermometers.† Suppose a mercury-in-glass thermometer with a spherical bulb at zero temperature is plunged at $t = 0$ into a medium at V. If we neglect the effects of motion in the mercury, and the thermal resistance of the glass which contains it, the increase of volume of the mercury, that is the reading of the thermometer, is proportional to the average temperature (8). Similarly (11) gives the reading of a thermometer in a medium whose temperature increases linearly, for example, in a descending aircraft; in this case it appears that there may be a permanent lag. The effect of the glass container may be taken into account by using the result 9.4 (10) of the next section; expansion of the glass must also be allowed for.

II. *Zero initial temperature. Surface temperature* kt

$$v = k\left[t - \frac{a^2 - r^2}{6\kappa}\right] - \frac{2ka^3}{\kappa\pi^3 r} \sum_{n=1}^{\infty} \frac{(-1)^n}{n^3} e^{-\kappa n^2\pi^2 t/a^2} \sin\frac{n\pi r}{a}. \tag{10}$$

The average temperature is

$$k\left\{t - \frac{a^2}{15\kappa}\right\} + \frac{6ka^2}{\kappa\pi^4} \sum_{n=1}^{\infty} \frac{1}{n^4} e^{-\kappa n^2\pi^2 t/a^2}. \tag{11}$$

III. *Zero initial temperature. Surface temperature* $\sin(\omega t + \epsilon)$

$$v = \frac{aA}{r}\sin(\omega t + \epsilon + \phi) + \frac{2a\kappa\pi}{r} \sum_{n=1}^{\infty} \frac{(-1)^n n(\kappa n^2\pi^2 \sin\epsilon - \omega a^2 \cos\epsilon)}{(\kappa^2 n^4\pi^4 + \omega^2 a^4)} e^{-\kappa n^2\pi^2 t/a^2} \sin\frac{n\pi r}{a}, \tag{12}$$

where

$$A = \left|\frac{\sinh \omega' r(1+i)}{\sinh \omega' a(1+i)}\right| = \left\{\frac{\cosh 2\omega' r - \cos 2\omega' r}{\cosh 2\omega' a - \cos 2\omega' a}\right\}^{\frac{1}{2}}, \qquad \phi = \arg\left\{\frac{\sinh \omega' r(1+i)}{\sinh \omega' a(1+i)}\right\},$$

and $\omega' = \sqrt{(\omega/2\kappa)}$. The result follows from 3.6 (5).

IV. *Constant initial temperature* V. *Zero surface temperature*
Results are obtained by subtracting (4) to (9) from V.

V. *Initial temperature* $V(a-r)/a$. *Zero surface temperature*

$$v = \frac{8aV}{r\pi^3} \sum_{n=0}^{\infty} \frac{1}{(2n+1)^3} e^{-\kappa(2n+1)^2\pi^2 t/a^2} \sin\frac{(2n+1)\pi r}{a} \tag{13}$$

$$= \frac{V(a-r)}{a} - \frac{2V\kappa t}{ar} + \frac{8V\kappa t}{ar} \sum_{n=0}^{\infty} (-1)^n\left\{i^2\text{erfc}\frac{na+r}{2(\kappa t)^{\frac{1}{2}}} + i^2\text{erfc}\frac{(n+1)a-r}{2(\kappa t)^{\frac{1}{2}}}\right\}. \tag{14}$$

† Bromwich, *Phil. Mag.* (6) **37** (1919) 407–19; McLeod, ibid. (6) **37** (1919) 134–44; Jeffreys, *Operational Methods in Mathematical Physics* (edn. 2, 1931) § 6.4. Evans, *Proc. Phys. Soc.* **59** (1947) 242–56, gives an approximate discussion of time-lags in more complicated systems.

VI. *Initial temperature $V(a^2-r^2)/a^2$. Zero surface temperature*

$$v = \frac{12aV}{r\pi^3} \sum_{n=1}^{\infty} \frac{(-1)^{n-1}}{n^3} \sin \frac{n\pi r}{a} e^{-\kappa n^2 \pi^2 t/a^2} \tag{15}$$

$$= \frac{V(a^2-r^2)}{a^2} - \frac{6V\kappa t}{a^2} + \frac{24V\kappa t}{ra} \sum_{n=0}^{\infty} \left\{ i^2\mathrm{erfc}\, \frac{(2n+1)a-r}{2(\kappa t)^{\frac{1}{2}}} - i^2\mathrm{erfc}\, \frac{(2n+1)a+r}{2(\kappa t)^{\frac{1}{2}}} \right\}. \tag{16}$$

VII. *Initial temperature $(V/r)\sin(\pi r/a)$. Zero surface temperature*

$$v = \frac{V}{r} \sin \frac{\pi r}{a} e^{-\kappa\pi^2 t/a^2}. \tag{17}$$

VIII. *Initial temperature $V \exp[\alpha(r-a)]$*

$$v = \frac{2Va\pi}{r} \sum_{n=1}^{\infty} \frac{n}{(n^2\pi^2+\alpha^2 a^2)^2} e^{-\kappa n^2\pi^2 t/a^2} \sin \frac{n\pi r}{a} \times$$
$$\times \{(-1)^{n+1}[\alpha^2 a^2 - 2\alpha a + n^2\pi^2] - 2\alpha a e^{-\alpha a}\}. \tag{18}$$

IX. *Initial temperature V, constant, in $0 < r < b$, zero in $b < r < a$. Zero surface temperature*

$$v = \frac{2V}{r} \sum_{n=1}^{\infty} \left\{ \frac{a}{n^2\pi^2} \sin \frac{n\pi b}{a} - \frac{b}{n\pi} \cos \frac{n\pi b}{a} \right\} \sin \frac{n\pi r}{a} e^{-\kappa n^2\pi^2 t/a^2}. \tag{19}$$

Alternatively, for $0 < r < b$,

$$v = V - \frac{bV}{2r} \sum_{n=0}^{\infty} \left\{ \mathrm{erfc}\, \frac{2na+b-r}{2(\kappa t)^{\frac{1}{2}}} - \mathrm{erfc}\, \frac{2na+b+r}{2(\kappa t)^{\frac{1}{2}}} + \mathrm{erfc}\, \frac{(2n+2)a-r-b}{2(\kappa t)^{\frac{1}{2}}} - \right.$$
$$\left. - \mathrm{erfc}\, \frac{(2n+2)a+r-b}{2(\kappa t)^{\frac{1}{2}}} \right\} - \frac{(\kappa t)^{\frac{1}{2}}V}{r} \sum_{n=0}^{\infty} \left\{ \mathrm{ierfc}\, \frac{2na+b-r}{2(\kappa t)^{\frac{1}{2}}} - \mathrm{ierfc}\, \frac{2na+b+r}{2(\kappa t)^{\frac{1}{2}}} - \right.$$
$$\left. - \mathrm{ierfc}\, \frac{(2n+2)a-b-r}{2(\kappa t)^{\frac{1}{2}}} + \mathrm{ierfc}\, \frac{(2n+2)a+r-b}{2(\kappa t)^{\frac{1}{2}}} \right\},$$

and for $b < r < a$,

$$v = \frac{bV}{2r} \sum_{n=0}^{\infty} \left\{ \mathrm{erfc}\, \frac{2na+r-b}{2(\kappa t)^{\frac{1}{2}}} + \mathrm{erfc}\, \frac{2na+b+r}{2(\kappa t)^{\frac{1}{2}}} - \mathrm{erfc}\, \frac{(2n+2)a-r-b}{2(\kappa t)^{\frac{1}{2}}} - \right.$$
$$\left. - \mathrm{erfc}\, \frac{(2n+2)a-r+b}{2(\kappa t)^{\frac{1}{2}}} \right\} - \frac{(\kappa t)^{\frac{1}{2}}V}{r} \sum_{n=0}^{\infty} \left\{ \mathrm{ierfc}\, \frac{2na+r-b}{2(\kappa t)^{\frac{1}{2}}} - \mathrm{ierfc}\, \frac{2na+r+b}{2(\kappa t)^{\frac{1}{2}}} - \right.$$
$$\left. - \mathrm{ierfc}\, \frac{(2n+2)a-b-r}{2(\kappa t)^{\frac{1}{2}}} + \mathrm{ierfc}\, \frac{(2n+2)a-r+b}{2(\kappa t)^{\frac{1}{2}}} \right\}. \tag{20}$$

By combining this result with that for constant initial temperature V in $0 < r < a$, the solution for an initial temperature which is zero in $0 < r < b$ and constant in $b < r < a$ is obtained. From these results, again, that for an initial temperature which is constant in $b < r < c$ and zero in $0 < r < b$ and $c < r < a$ follows.

X. *Initial temperature $f(r)$. Zero surface temperature*

Fourier's result has been given in (3). The solution useful for small values of $\kappa t/a^2$ is

$$v = \frac{1}{2r(\pi\kappa t)^{\frac{1}{2}}} \sum_{n=-\infty}^{\infty} \int_0^a r'f(r')\left\{\exp\left[-\frac{(2na+r'-r)^2}{4\kappa t}\right] - \exp\left[-\frac{(2na+r'+r)^2}{4\kappa t}\right]\right\} dr'.$$

(21)

XI. *Zero surface temperature. Initial temperature*†

$$f(r) = b_0 + br + cr^2 + dr^3 + \dots . \tag{22}$$

$$v = \frac{2}{ar} \sum_{n=1}^{\infty} \sin\frac{n\pi r}{a}\left\{\frac{b_0 a^2}{n\pi}(-1)^{n+1} + \frac{ba^3}{n^3\pi^3}[(n^2\pi^2-2)(-1)^{n+1}-2] + \right.$$

$$\left. + \frac{ca^4}{n^4\pi^4}(n^3\pi^3-6n\pi)(-1)^{n+1} + \frac{da^5}{n^5\pi^5}[24-(n^4\pi^4-12n^2\pi^2+24)(-1)^n]+\dots\right\} \times$$

$$\times e^{-\kappa n^2\pi^2 t/a^2}. \quad (23)$$

9.4. The sphere $0 \leqslant r < a$. Initial temperature $f(r)$. Radiation at the surface

If the sphere radiates into medium at zero the equations for v are

$$\frac{\partial v}{\partial t} = \kappa\left(\frac{\partial^2 v}{\partial r^2} + \frac{2}{r}\frac{\partial v}{\partial r}\right), \quad 0 \leqslant r < a, \tag{1}$$

$$\frac{\partial v}{\partial r} + hv = 0, \quad \text{when } r = a, \tag{2}$$

and
$$v = f(r), \quad \text{when } t = 0. \tag{3}$$

Putting $u = vr$, we have

$$\frac{\partial u}{\partial t} = \kappa\frac{\partial^2 u}{\partial r^2}, \quad 0 < r < a, \tag{4}$$

$$u = 0, \quad \text{when } r = 0, \tag{5}$$

$$\frac{\partial u}{\partial r} + \left(h - \frac{1}{a}\right)u = 0, \quad \text{when } r = a, \tag{6}$$

and
$$u = rf(r), \quad \text{when } t = 0. \tag{7}$$

The problem is thus reduced to that of linear flow of heat in a slab, one end being kept at zero temperature, while at the other end radiation takes place into a medium at zero. This problem has been solved in 3.10 (11), and in this we have only to replace l by a, x by r, and h by $(ah-1)/a$. Thus the solution of (1)–(3) is‡

$$v = \frac{2}{ar} \sum_{n=1}^{\infty} e^{-\kappa\alpha_n^2 t} \frac{a^2\alpha_n^2 + (ah-1)^2}{a^2\alpha_n^2 + ah(ah-1)} \sin\alpha_n r \int_0^a r'f(r')\sin\alpha_n r'\, dr', \quad (8)$$

---

† van Orstrand, *Geophysics*, **5** (1940) 57–79. He considers two more powers in the series (22), and gives numerical values for the coefficients of the first forty terms of the series (23).

‡ This solution is easily obtained directly, cf. *C.H.*, § 65. See also § 14.7 II.

where $\pm\alpha_n$, $n = 1, 2,...$ are the roots of

$$a\alpha \cot a\alpha + ah - 1 = 0. \tag{9}$$

The equation (9) is simply the equation 3.10 (7) which has already been discussed and whose roots are tabulated in Appendix IV, except that the parameter ah, which was always positive in § 3.11, is replaced by $ah-1$ which may be negative. Provided $h > 0$, i.e. $ah-1 > -1$, the remarks of §§ 3.10, 3.11 hold, and the roots of (9) are all real.†

If the initial temperature $f(r)$ is V, constant, (8) becomes‡

$$v = \frac{2hV}{r} \sum_{n=1}^{\infty} e^{-\kappa\alpha_n^2 t} \frac{a^2\alpha_n^2 + (ah-1)^2}{\alpha_n^2[a^2\alpha_n^2 + ah(ah-1)]} \sin a\alpha_n \sin r\alpha_n. \tag{10}$$

If the sphere has zero initial temperature and is heated by radiation from medium at temperature kt, the solution is

$$v = k\left\{t + \frac{r^2ah - a^2(2+ah)}{6\kappa ah}\right\} + \frac{2a^2hk}{\kappa r} \sum_{n=1}^{\infty} \frac{\sin r\alpha_n}{\alpha_n^2\{a^2\alpha_n^2 + ah(ah-1)\}\sin a\alpha_n} e^{-\kappa\alpha_n^2 t}, \tag{11}$$

where the α_n are the positive roots of (9).

If the sphere has zero initial temperature and is heated by radiation from medium at temperature $V\sin(\omega t + \epsilon)$, the temperature is

$$v = \frac{2ah\kappa V}{r} \sum_{n=1}^{\infty} \frac{\alpha_n(\kappa\alpha_n^2 \sin\epsilon - \omega\cos\epsilon)(ah-1)\sin r\alpha_n}{(\kappa^2\alpha_n^4 + \omega^2)[a^2\alpha_n^2 + ah(ah-1)]\cos a\alpha_n} e^{-\kappa\alpha_n^2 t} +$$
$$+ \frac{a^2hVA_1}{rA_2} \sin(\omega t + \epsilon + \phi_1 - \phi_2), \tag{12}$$

where
$$A_1 e^{i\phi_1} = \sinh\omega'r\cos\omega'r + i\cosh\omega'r\sin\omega'r,$$
$$A_2 e^{i\phi_2} = a\omega'(1+i)\cosh a\omega'(1+i) + (ah-1)\sinh a\omega'(1+i),$$
$$\omega' = \sqrt{(\omega/2\kappa)},$$

and the α_n are the positive roots of (9).

9.5. Application to the determination of the conductivities of poor conductors

The expression we have just obtained for the temperature in a sphere cooling by radiation at the surface converges so rapidly that when a sufficient time has passed the terms after the first may be neglected.

† If $h < 0$ there is a pair of imaginary roots, but this case is, as always, excluded on physical grounds. If $h = 0$, that is, no flow of heat at the surface, (9) has a zero root, and a term

$$\frac{3}{a^3} \int_0^a r^2 f(r)\, dr$$

has to be added to the value of (8) with $h = 0$. Cf. 3.4 (6), 7.8 (3).

‡ Surface and centre temperatures for this case are plotted against ah for various values of $\kappa t/a^2$ by Schack, *Stahl u. Eisen*, **50** (1930) 1290. See also Heisler, *Trans. Amer. Soc. Mech. Engrs.* **69** (1947) 227–36.

This gives an expression suitable for numerical calculation, and it has been applied in different experiments where the initial temperature of the sphere is constant.

For example, a ball of the material to be tested is immersed in a bath at a constant temperature V for a sufficient time to allow the whole ball to acquire the temperature of the bath. It is then removed and allowed to cool by radiation in a medium at constant temperature. After cooling has gone on for a certain time, observations of the temperature are taken. In one set of experiments these readings are for the temperature at the centre and the surface. In another set of experiments the temperature at the centre alone is required.

With the notation of § 9.4 and to our approximation

$$v = \frac{2hV[a^2\alpha_1^2+(ah-1)^2]\sin a\alpha_1}{r\alpha_1^2[a^2\alpha_1^2+ah(ah-1)]} e^{-\kappa\alpha_1^2 t}\sin r\alpha_1. \tag{1}$$

Hence if v_a is the temperature at $r = a$ at time t, and v_0 is the temperature at $r = 0$ at time t,

$$\frac{v_a}{v_0} = \frac{\sin a\alpha_1}{a\alpha_1}. \tag{2}$$

This gives α_1, remembering that $0 < a\alpha_1 < \pi$. Also

$$\frac{(v_a)_{t=t_1}}{(v_a)_{t=t_2}} = e^{\kappa\alpha_1^2(t_2-t_1)} \tag{3}$$

gives $\kappa\alpha_1^2$. Thus κ is known[†] and h is found from 9.4 (9).

Ayrton and Perry used the second method in determining the conductivity of stone.[‡]

Using 9.4 (9), we can write the centre temperature obtained from (1) in the form

$$2V\frac{\sin a\alpha_1-a\alpha_1\cos a\alpha_1}{a\alpha_1-\sin a\alpha_1\cos a\alpha_1} e^{-\kappa\alpha_1^2 t} = Ne^{-nt}. \tag{4}$$

The value of n is obtained by two observations of the temperature, and, n being known, the value of N may be found. Also a table of the values of the expression

$$\frac{\sin x-x\cos x}{x-\sin x\cos x}$$

will give α_1 from the known value of N. But $n = \kappa\alpha_1^2$, and thus κ is determined.

† Weber, *Vjschr. naturf. Ges. Zürich*, **23** (1878) 209.
‡ *Phil. Mag.* (5) **5** (1878) 241.

9.6. The sphere in contact with a well-stirred fluid

We suppose the sphere to have initial temperature $f(r)$, and at $r = a$ to be in contact with mass M' of well-stirred fluid of specific heat c', which is initially at zero temperature. If there is no loss of heat from the fluid, the boundary condition at $r = a$ is

$$-4\pi a^2 K \frac{\partial v}{\partial r} = M'c' \frac{\partial v}{\partial t}, \quad r = a, \quad t > 0, \tag{1}$$

on the assumption that for $t > 0$ the temperature of the fluid is equal to the surface temperature of the sphere.

The solution is†

$$v = \frac{3}{(k+1)a^3} \int_0^a r^2 f(r) \, dr + \frac{2}{ar} \sum_{n=1}^\infty e^{-\kappa\alpha_n^2 t} \frac{k^2 a^4 \alpha_n^4 + 3(2k+3)a^2\alpha_n^2 + 9}{k^2 a^4 \alpha_n^4 + 9(k+1)a^2\alpha_n^2} \times$$

$$\times \sin r\alpha_n \int_0^a rf(r)\sin \alpha_n r \, dr, \tag{2}$$

where the $\pm\alpha_n$, $n = 1, 2,...$ are the roots of

$$\tan a\alpha = \frac{3a\alpha}{3 + ka^2\alpha^2}, \tag{3}$$

and

$$k = \frac{3M'c'}{4\pi a^3 \rho c} \tag{4}$$

is the ratio of the heat capacity of the fluid to that of the sphere.

If the initial temperature of the sphere is V, constant, (2) becomes

$$v = \frac{V}{k+1} - \frac{2kaV}{3r} \sum_{n=1}^\infty e^{-\kappa\alpha_n^2 t} \frac{k^2 a^4 \alpha_n^4 + 3(2k+3)a^2\alpha_n^2 + 9}{k^2 a^4 \alpha_n^4 + 9(k+1)a^2\alpha_n^2} \sin r\alpha_n \sin a\alpha_n. \tag{5}$$

And the temperature of the fluid is

$$v_a = \frac{V}{k+1} - 6kV \sum_{n=1}^\infty e^{-\kappa\alpha_n^2 t} \frac{1}{k^2 a^2 \alpha_n^2 + 9(k+1)}. \tag{6}$$

In Fig. 30 values of $(k+1)v_a/V$, that is, the ratio of the temperature of the fluid to its final temperature, are plotted against $T = \kappa t/a^2$ for the values 0·5, 1, 2, 4, 10, ∞ of k. They show the way in which the temperature of fluid in a calorimeter rises when a spherical solid is introduced: theoretically they can be used to determine both κ and ρc.

† Peddie, *Proc. Edin. Math. Soc.* **19** (1901) 34; Dougall, ibid. **19** (1901) 50. Peddie uses a modification of the Fourier method (the functions $\sin r\alpha_n$ are not orthogonal), and Dougall contour integration. The solution is easily obtained by the Laplace transformation. Paterson, *Proc. Phys. Soc.* **59** (1947) 50–58, gives more extensive numerical results than those in Fig. 30.

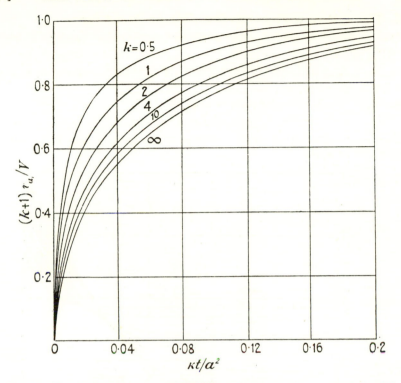

Fig. 30. The rise in temperature of fluid in a calorimeter when a spherical solid is introduced.

If the fluid loses heat to its surroundings† by radiation into medium at zero temperature, (1) is replaced by

$$-4\pi a^2 K \frac{\partial v}{\partial r} = M'c' \frac{\partial v}{\partial t} + Hv, \quad r = a, \tag{7}$$

and the temperature of the fluid, for constant initial temperature V of the sphere, is given by

$$6V \sum_{n=1}^{\infty} \frac{(3H' - ka^2\alpha_n^2)}{k^2a^4\alpha_n^4 + 3a^2\alpha_n^2(3 + 3k - 2kH') - 9H'(1 - H')} e^{-\kappa\alpha_n^2 t}, \tag{8}$$

where $H' = (H/4\pi aK)$, k is defined in (4), and $\pm\alpha_n$, $n = 1, 2,...$, are the roots of

$$\tan a\alpha = \frac{3a\alpha}{3(1 - H') + ka^2\alpha^2}. \tag{9}$$

† The boundary conditions 1.9 (14)–(16) which give a better approximation to conditions in a calorimeter may also be used. Hoare, *Phil. Mag.* (7) **29** (1940) 52, discusses the case of a perfectly conducting solid.

9.7. The sphere with prescribed flux at its surface

For the case of zero initial temperature and constant flux F_0 into the sphere, the temperature v is

$$v = \frac{3F_0 t}{\rho ca} + \frac{F_0(5r^2 - 3a^2)}{10Ka} - \frac{2F_0 a^2}{Kr} \sum_{n=1}^{\infty} \frac{\sin(r\alpha_n/a)}{\alpha_n^2 \sin \alpha_n} e^{-\kappa \alpha_n^2 t/a^2}, \tag{1}$$

where α_n, $n = 1, 2,...$, are the positive roots of

$$\tan \alpha = \alpha. \tag{2}$$

Numerical values of $(K/aF_0)[v - (3F_0 t/\rho ca)]$ are given in Fig. 31 for various values of $T = \kappa t/a^2$.

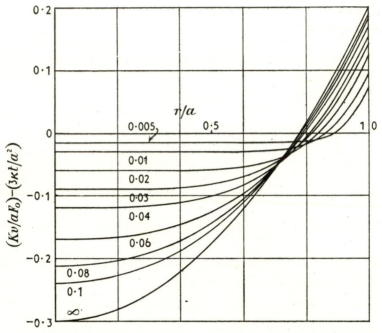

FIG. 31. Temperature in a sphere due to constant surface flux F_0. The numbers on the curves are the values of $\kappa t/a^2$.

9.8. The sphere $0 \leqslant r < a$ with heat generation

If heat is produced at the rate A_0 per unit time per unit volume, the differential equation 1.6 (7) becomes

$$\frac{1}{\kappa} \frac{\partial v}{\partial t} = \frac{1}{r^2} \frac{\partial}{\partial r}\left(r^2 \frac{\partial v}{\partial r}\right) + \frac{A_0}{K}. \tag{1}$$

Substituting $u = vr$, as usual, gives

$$\frac{1}{\kappa} \frac{\partial u}{\partial t} = \frac{\partial^2 u}{\partial r^2} + \frac{A_0 r}{K}. \tag{2}$$

A number of special cases will now be considered.

I. *Heat generation at a constant rate A_0. Zero initial and surface temperatures*

Putting
$$u = w - \frac{A_0 r^3}{6K},$$

gives as the equation for w the equation of linear flow in a slab

$$\frac{1}{\kappa} \frac{\partial w}{\partial t} = \frac{\partial^2 w}{\partial r^2}, \quad 0 < r < a, \tag{3}$$

with $w = 0$ at $r = 0$.

If the initial and surface temperatures of the sphere are zero, (3) has to be solved with

$$w = \frac{A_0 r^3}{6K}, \quad \text{when } t = 0, \, 0 < r < a, \tag{4}$$

and
$$w = \frac{A_0 a^3}{6K}, \quad \text{when } r = a, \, t > 0. \tag{5}$$

The value of w is obtained from 3.4 (1), and we get finally

$$v = \frac{A_0}{6K}(a^2 - r^2) + \frac{2A_0 a^3}{K\pi^3 r} \sum_{n=1}^{\infty} \frac{(-1)^n}{n^3} \sin \frac{n\pi r}{a} e^{-\kappa n^2 \pi^2 t / a^2}. \tag{6}$$

The alternative form of solution, useful for small values of $\kappa t / a^2$, is

$$v = \frac{\kappa A_0 t}{K} - \frac{4\kappa A_0 a t}{Kr} \sum_{n=0}^{\infty} \left\{ \text{i}^2 \text{erfc} \frac{(2n+1)a - r}{2(\kappa t)^{\frac{1}{2}}} - \text{i}^2 \text{erfc} \frac{(2n+1)a + r}{2(\kappa t)^{\frac{1}{2}}} \right\}. \tag{7}$$

Some values of the temperature distribution for various values of $\kappa t / a^2$ are shown in Fig. 32.

II. *Heat generation at the rate $A_0(a-r)/a$. Zero initial and surface temperatures*

$$v = \frac{A_0(a^3 - 2ar^2 + r^3)}{12aK} - \frac{8a^3 A_0}{Kr\pi^5} \sum_{n=0}^{\infty} \frac{1}{(2n+1)^5} e^{-\kappa(2n+1)^2 \pi^2 t / a^2} \sin \frac{(2n+1)\pi r}{a} \tag{8}$$

$$= \frac{\kappa A_0(a-r)t}{Ka} - \frac{A_0 \kappa^2 t^2}{Kar} + \frac{32A_0 \kappa^2 t^2}{Kar} \sum_{n=0}^{\infty} (-1)^n \left\{ \text{i}^4 \text{erfc} \frac{na+r}{2(\kappa t)^{\frac{1}{2}}} + \text{i}^4 \text{erfc} \frac{(n+1)a - r}{2(\kappa t)^{\frac{1}{2}}} \right\}. \tag{9}$$

III. *Heat generation at the rate $A_0(a^2 - r^2)/a^2$. Zero initial and surface temperatures*

$$v = \frac{A_0(a^2 - r^2)(7a^2 - 3r^2)}{60Ka^2} - \frac{12A_0 a^3}{rK\pi^5} \sum_{n=1}^{\infty} \frac{(-1)^{n-1}}{n^5} \sin \frac{n\pi r}{a} e^{-\kappa n^2 \pi^2 t / a^2} \tag{10}$$

$$= \frac{A_0(a^2 - r^2)\kappa t}{Ka^2} - \frac{3A_0 \kappa^2 t^2}{Ka^2} + \frac{96A_0 \kappa^2 t^2}{Kar} \sum_{n=0}^{\infty} \left\{ \text{i}^4 \text{erfc} \frac{(2n+1)a - r}{2(\kappa t)^{\frac{1}{2}}} - \text{i}^4 \text{erfc} \frac{(2n+1)a + r}{2(\kappa t)^{\frac{1}{2}}} \right\}. \tag{11}$$

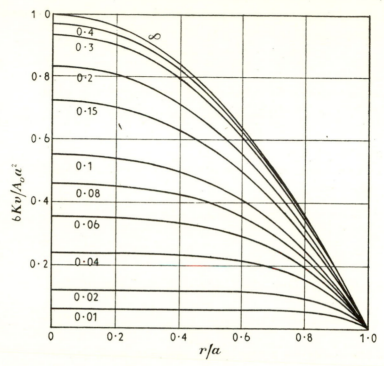

FIG. 32. Temperature distribution in the sphere $0 \leqslant r < a$ with zero surface temperature and constant heat production A_0 for $t > 0$. The numbers on the curves are values of $\kappa t/a^2$.

IV. *Heat generation at the rate $(A_0/r)\sin(\pi r/a)$. Zero initial and surface temperatures*†

$$v = \frac{A_0 a^2}{\pi^2 r K}(1 - e^{-\kappa \pi^2 t/a^2})\sin\frac{\pi r}{a}. \tag{12}$$

V. *Heat generation‡ at the rate $A_0 e^{\alpha(r-a)}$. Zero initial and surface temperatures*

$$v = \frac{A_0}{K\alpha^2}\Big\{1 - \frac{2}{a\alpha} - \Big(1 - \frac{2}{r\alpha}\Big)e^{\alpha(r-a)} - \frac{2(a-r)}{\alpha ar}e^{-\alpha a}\Big\} -$$

$$-\frac{2A_0 a^3}{\pi K r}\sum_{n=1}^{\infty}\frac{1}{n(n^2\pi^2 + \alpha^2 a^2)^2}\sin\frac{n\pi r}{a}e^{-\kappa n^2\pi^2 t/a^2} \times$$

$$\times [(-1)^n(2\alpha a - n^2\pi^2 - \alpha^2 a^2) - 2\alpha a e^{-\alpha a}]. \tag{13}$$

† Slichter, *Bull. Geol. Soc. Amer.* **52** (1941) 561–600, indicates the value of simple results of this type when the variation with r of the heat production is not well known. He also discusses constant heat production, and heat production which is constant within a layer and zero outside it.

‡ The problem is discussed by Lowan, *Phys. Rev.* (2) **44** (1933) 769–75, who obtains a different form for the solution. van Orstrand, *Geophysics*, **5** (1940) 57–79, gives numerical results for this case. Lowan, *Amer. J. Math.* **57** (1935) 174–82, extends the result to take into account the effect of contraction.

VI. *Heat generation at the rate $A_0 e^{-\lambda t}$. Zero initial and surface temperatures*

$$v = \frac{\kappa A_0}{K\lambda} e^{-\lambda t}\left\{\frac{a \sin r(\lambda/\kappa)^{\frac{1}{2}}}{r \sin a(\lambda/\kappa)^{\frac{1}{2}}} - 1\right\} + \frac{2a^3 A_0}{r\pi^3 K}\sum_{n=1}^{\infty}\frac{(-1)^n}{n(n^2 - \lambda a^2/\kappa\pi^2)}\sin\frac{n\pi r}{a}e^{-\kappa n^2\pi^2 t/a^2}, \quad (14)$$

$$= \frac{\kappa A_0}{K\lambda}(1 - e^{-\lambda t}) -$$

$$-\frac{\kappa A_0 a}{K\lambda r}\sum_{n=0}^{\infty}\left\{\text{erfc}\frac{(2n+1)a-r}{2(\kappa t)^{\frac{1}{2}}} - e^{-\lambda t}\mathbf{R}\left[e^{i[(2n+1)a-r](\lambda/\kappa)^{\frac{1}{2}}}\text{erfc}\left\{\frac{(2n+1)a-r}{2(\kappa t)^{\frac{1}{2}}} - i(\lambda t)^{\frac{1}{2}}\right\}\right]\right\} +$$

$$+\frac{\kappa A_0 a}{K\lambda r}\sum_{n=0}^{\infty}\left\{\text{erfc}\frac{(2n+1)a+r}{2(\kappa t)^{\frac{1}{2}}} - e^{-\lambda t}\mathbf{R}\left[e^{i[(2n+1)a+r](\lambda/\kappa)^{\frac{1}{2}}}\text{erfc}\left\{\frac{(2n+1)a+r}{2(\kappa t)^{\frac{1}{2}}} - i(\lambda t)^{\frac{1}{2}}\right\}\right]\right\}. \quad (15)$$

VII. *Heat generation† at the rate $A_0 e^{-\lambda t}$ in $b < r < a$, zero in $0 < r < b$. Zero initial and surface temperatures*

If $b < r < a$,

$$v = \frac{\kappa A_0 a}{K\lambda r}e^{-\lambda t}\left\{\frac{\sin r(\lambda/\kappa)^{\frac{1}{2}}}{\sin a(\lambda/\kappa)^{\frac{1}{2}}} - \frac{r}{a}\right\} +$$

$$+\frac{\kappa A_0 a}{K\lambda r}e^{-\lambda t}\frac{\sin(a-r)(\lambda/\kappa)^{\frac{1}{2}}}{\sin a(\lambda/\kappa)^{\frac{1}{2}}}\left\{\frac{b\cos b(\lambda/\kappa)^{\frac{1}{2}}}{a} - \frac{\sin b(\lambda/\kappa)^{\frac{1}{2}}}{a(\lambda/\kappa)^{\frac{1}{2}}}\right\} +$$

$$+\frac{2a^3 A_0}{\pi r K}\sum_{n=1}^{\infty}\frac{1}{(n^2\pi^2 - a^2\lambda/\kappa)}\left\{\frac{(-1)^n}{n} + \frac{1}{\pi n^2}\sin\frac{n\pi b}{a} - \frac{b}{an}\cos\frac{n\pi b}{a}\right\}\sin\frac{n\pi r}{a}e^{-\kappa n^2\pi^2 t/a^2}. \quad (16)$$

An expression of type (15) can also be derived in the usual way.

VIII. *The region $0 < r < a$ with heat production $f(r)$. Zero initial and surface temperatures*

$$v = -\frac{2a}{\pi^2 r K}\sum_{n=1}^{\infty}\frac{1}{n^2}e^{-\kappa n^2\pi^2 t/a^2}\sin\frac{n\pi r}{a}\int_0^a r'\sin\frac{n\pi r'}{a}f(r')\,dr' +$$

$$+\frac{1}{K}\left\{\frac{1}{r}\int_0^r r'^2 f(r')\,dr' + \int_r^a r'f(r')\,dr' - \frac{1}{a}\int_0^a r'^2 f(r')\,dr'\right\} \quad (17)$$

$$= \frac{(\kappa t)^{\frac{1}{2}}}{Kr}\sum_{n=0}^{\infty}\int_0^a r'f(r')\,dr'\left\{\text{ierfc}\frac{2na+r'-r}{2(\kappa t)^{\frac{1}{2}}} - \text{ierfc}\frac{2na+r'+r}{2(\kappa t)^{\frac{1}{2}}} - \right.$$

$$\left. - \text{ierfc}\frac{(2n+2)a-r'-r}{2(\kappa t)^{\frac{1}{2}}} + \text{ierfc}\frac{(2n+2)a-r'+r}{2(\kappa t)^{\frac{1}{2}}}\right\}. \quad (18)$$

IX. *Zero initial temperature. Constant heat generation A_0. Radiation at the surface into a medium at zero*

$$v = \frac{A_0}{6hK}\{h(a^2 - r^2) + 2a\} - \frac{2ha^2 A_0}{rK}\sum_{n=1}^{\infty}\frac{\sin r\alpha_n e^{-\kappa\alpha_n^2 t}}{\alpha_n^2[a^2\alpha_n^2 + ah(ah-1)]\sin a\alpha_n}, \quad (19)$$

where $\pm\alpha_n$, $n = 1, 2,...$, are the roots of

$$a\alpha\cot a\alpha = 1 - ah. \quad (20)$$

† Allen, *Amer. Math. Monthly*, **63** (1956) 315–23.

$$\tan a\alpha = \frac{a\alpha}{1 - ah}$$

$$\tan R\lambda = \frac{R\lambda}{1 - Rh}$$

X. *If the initial temperature*† *is*

$$\frac{A_0(a^2-r^2)}{6K}$$

(which is the steady temperature for zero surface temperature), and at $r = a$ for $t > 0$ there is radiation into a medium at V,

$$v = \frac{(a^2-r^2)A_0}{6K}+\left(V+\frac{aA_0}{3hK}\right)\left[1-\frac{2a^2h}{r}\sum_{n=1}^{\infty}\frac{\sin r\alpha_n}{\{a^2\alpha_n^2+ah(ah-1)\}\sin a\alpha_n}e^{-\kappa\alpha_n^2 t}\right],$$

where the α_n are the positive roots of (20).

9.9. The hollow sphere‡ $a < r < b$

If the initial temperature of the sphere is $f(r)$, and the surfaces $r = a$ and $r = b$ are kept at v_1 and v_2, constants, for $t > 0$, the solution,§ obtained from 3.4 (1), is

$$v = \frac{av_1}{r}+\frac{(bv_2-av_1)(r-a)}{r(b-a)}+$$

$$+\frac{2}{r\pi}\sum_{n=1}^{\infty}\frac{bv_2\cos n\pi-av_1}{n}\sin\frac{n\pi(r-a)}{(b-a)}e^{-\kappa n^2\pi^2 t/(b-a)^2}+$$

$$+\frac{2}{r(b-a)}\sum_{n=1}^{\infty}\sin\frac{n\pi(r-a)}{b-a}e^{-\kappa n^2\pi^2 t/(b-a)^2}\int_a^b r'f(r')\sin\frac{n\pi(r'-a)}{b-a}\,dr'. \quad (1)$$

If the surfaces are at temperatures $\phi_1(t)$ and $\phi_2(t)$, the result follows in the same way from 3.5 (2).

If there is radiation at $r = a$ and $r = b$ into medium at zero, so that the boundary conditions there are

$$k_1\frac{\partial v}{\partial r}-h_1 v = 0, \quad r = a,$$

$$k_2\frac{\partial v}{\partial r}+h_2 v = 0, \quad r = b,$$

the solution is
$$v = \frac{2}{r}\sum_{n=1}^{\infty}e^{-\kappa\alpha_n^2 t}R_n(r)\int_a^b r'R_n(r')f(r')\,dr', \quad (2)$$

where
$$G = ah_1+k_1, \quad H = bh_2-k_2,$$

$$R_n(r) = \frac{(H^2+b^2k_2^2\alpha_n^2)^{\frac{1}{2}}\{G\sin(r-a)\alpha_n+ak_1\alpha_n\cos(r-a)\alpha_n\}}{\{(b-a)(a^2k_1^2\alpha_n^2+G^2)(b^2k_2^2\alpha_n^2+H^2)+(Hak_1+Gbk_2)(GH+abk_1k_2\alpha_n^2)\}^{\frac{1}{2}}},$$

and $\pm\alpha_n$, $n = 1, 2,...$, are the roots of

$$(GH-abk_1k_2\alpha^2)\sin(b-a)\alpha+\alpha(ak_1 H+bk_2 G)\cos(b-a)\alpha = 0.$$

† Awbery, *Phil. Mag.* (7) **4** (1927) 629.
‡ A number of results with practical applications is given by Barrer, *Phil. Mag.* (7) **35** (1944) 802.
§ These results, of course, apply also to radial flow in a conical region. For a slightly tapering tube, see Talbot and Kitchener, *Brit. J. Appl. Phys.* **7** (1956) 96–97.

In the above k_1, k_2, h_1, h_2 are to be positive or zero, and (2) thus gives the solution for the various cases in which there is zero temperature, radiation, or no flow of heat at the surfaces. If $h_1 = h_2 = 0$, i.e. no flow of heat at either surface, a term

$$\frac{3}{(b^3-a^3)} \int_a^b r^2 f(r)\, dr$$

is to be added to (2).

The extension to the case in which there is radiation at $r = a$ and $r = b$ into media at different temperatures is made as at the end of § 3.9.

If the region $a < r < b$ is initially at zero temperature, and for $t > 0$ there is constant flux F_0 at $r = a$ while $r = b$ is maintained at zero,

$$v = \frac{a^2(b-r)F_0}{Kbr} + \frac{2a^2 F_0}{Kr} \sum_{n=1}^{\infty} \frac{(1+a^2\alpha_n^2)^{\frac{1}{2}} \sin \alpha_n(b-r)}{\alpha_n[b+a^2(b-a)\alpha_n^2]} e^{-\kappa \alpha_n^2 t}, \tag{3}$$

where α_n, $n = 1, 2,...$ are the positive roots of

$$\tan(b-a)\alpha + a\alpha = 0.$$

9.10. The region bounded internally by the sphere $r = a$

We suppose the initial temperature to be $f(r)$, and the surface temperature $\phi(t)$.

Substituting $u = vr$ as usual, we have to solve 9.3 (2) in $r > a$ with $u = a\phi(t)$ at $r = a$, and $u = rf(r)$ initially. The solution which follows from 2.4 (1), 2.5 (1) is

$$v = \frac{1}{2r\sqrt{(\pi\kappa t)}} \int_a^{\infty} r'f(r')\{e^{-(r-r')^2/4\kappa t} - e^{-(r+r'-2a)^2/4\kappa t}\}\, dr' +$$

$$+ \frac{2a}{r\sqrt{\pi}} \int_{(r-a)/2\sqrt{(\kappa t)}}^{\infty} \phi\left[t - \frac{(r-a)^2}{4\kappa\mu^2}\right] e^{-\mu^2}\, d\mu. \tag{1}$$

In particular, for zero initial temperature, and surface temperature V, constant, the solution is

$$v = \frac{aV}{r} \operatorname{erfc} \frac{r-a}{2\sqrt{(\kappa t)}}. \tag{2}$$

If the initial temperature is zero, and at $r = a$ there is radiation from a medium at temperature V, so that the boundary condition for v is

$$\frac{\partial v}{\partial r} = h(v-V), \quad r = a,$$

that for $u = rv$ is

$$\frac{\partial u}{\partial r} = u\left(h + \frac{1}{a}\right) - hVa, \quad r = a.$$

Thus, writing $h' = h+(1/a)$, it follows from 2.7 (5) that

$$v = \frac{ha^2V}{r(ah+1)}\left\{\text{erfc}\,\frac{r-a}{2\sqrt{(\kappa t)}} - e^{h'(r-a)+h'^2\kappa t}\,\text{erfc}\left[\frac{r-a}{2\sqrt{(\kappa t)}}+h'\sqrt{(\kappa t)}\right]\right\}. \qquad (3)$$

The solution for the region $r > a$, initially at V and cooling by radiation at $r = a$ into medium at zero, is obtained by subtracting (3) from V. The solution for any initial temperature follows from 14.7 (15).

If the initial temperature is zero and there is constant flux† F_0 *at* $r = a$, the solution is

$$v = \frac{a^2F_0}{Kr}\left\{\text{erfc}\,\frac{r-a}{2(\kappa t)^{\frac{1}{2}}} - \exp\left[\frac{r-a}{a}+\frac{\kappa t}{a^2}\right]\text{erfc}\left[\frac{r-a}{2(\kappa t)^{\frac{1}{2}}}+\frac{(\kappa t)^{\frac{1}{2}}}{a}\right]\right\}. \qquad (4)$$

9.11. The sphere. Surface $r = a$ at zero temperature. Initial temperature $f(r, \theta, \phi)$

In this case the equations for v are as follows:

$$\frac{\partial v}{\partial t} = \kappa\left\{\frac{\partial^2 v}{\partial r^2}+\frac{2}{r}\frac{\partial v}{\partial r}+\frac{1}{r^2\sin\theta}\frac{\partial}{\partial\theta}\left(\sin\theta\,\frac{\partial v}{\partial\theta}\right)+\frac{1}{r^2\sin^2\theta}\frac{\partial^2 v}{\partial\phi^2}\right\}, \qquad (1)$$

$$v = f(r, \theta, \phi), \quad \text{when } t = 0, \qquad (2)$$

$$v = 0, \qquad\qquad \text{when } r = a. \qquad (3)$$

Put $v = e^{-\kappa\alpha^2 t}u$, where u is a function of r, θ, and ϕ only, and write $\mu = \cos\theta$.

Then we have from (1),

$$\frac{\partial^2 u}{\partial r^2}+\frac{2}{r}\frac{\partial u}{\partial r}+\frac{1}{r^2}\frac{\partial}{\partial\mu}\left\{(1-\mu^2)\,\frac{\partial u}{\partial\mu}\right\}+\frac{1}{r^2(1-\mu^2)}\frac{\partial^2 u}{\partial\phi^2}+\alpha^2 u = 0. \qquad (4)$$

Now the zonal harmonic $P_n(\mu)$, when n is a positive integer, is the coefficient of h^n in the expansion of $(1-2\mu h+h^2)^{-\frac{1}{2}}$, and it satisfies Legendre's equation,

$$\frac{d}{d\mu}\left\{(1-\mu^2)\,\frac{dP_n}{d\mu}\right\}+n(n+1)P_n = 0.$$

Also the associated Legendre function‡

$$P_n^m(\mu) = (1-\mu^2)^{\frac{1}{2}m}\frac{d^m}{d\mu^m}P_n(\mu) \qquad (5)$$

---

† Holm, *J. Appl. Phys.* **19** (1948) 361–6; Ingersoll, Adler, Plass, and Ingersoll, *Heat. Pip. Air Condit.* **22** (1950) 113–22; von Bertele, *Brit. J. Appl. Phys.* **3** (1952) 127–32.

‡ Bateman, *Partial Differential Equations of Mathematical Physics* (Cambridge, 1932) Chap. VI; Whittaker and Watson, *Modern Analysis* (Cambridge, edn. 3, 1920) Chap. XV; MacRobert, *Spherical Harmonics* (Methuen, 1927); Byerly, *Fourier's Series and Spherical Cylindrical and Ellipsoidal Harmonics* (Boston, 1893) p. 196; Hobson, *Spherical and Ellipsoidal Harmonics* (Cambridge, 1931) Chap. III.

satisfies $\dfrac{d}{d\mu}\left\{(1-\mu^2)\dfrac{dP_n^m}{d\mu}\right\}+\left\{n(n+1)-\dfrac{m^2}{1-\mu^2}\right\}P_n^m=0.$ (6)

It follows that the expression†

$$R_n(r)P_n^m(\mu)\,{\cos\atop\sin}\,m\phi \qquad (7)$$

will satisfy (4) provided that $R_n(r)$ is a function of r only, and

$$\dfrac{d^2R_n}{dr^2}+\dfrac{2}{r}\dfrac{dR_n}{dr}+\left[\alpha^2-\dfrac{n(n+1)}{r^2}\right]R_n=0. \qquad (8)$$

This leads us to‡ $R_n=(\alpha r)^{-\frac{1}{2}}J_{n+\frac{1}{2}}(\alpha r),$

the solution $J_{-(n+\frac{1}{2})}(\alpha r)$ being inadmissible as it would make R_n tend to infinity as $r\to 0$.

We are thus brought to the following solution of (1):

$$e^{-\kappa\alpha^2 t}(\alpha r)^{-\frac{1}{2}}J_{n+\frac{1}{2}}(\alpha r)P_n^m(\mu)\,{\cos\atop\sin}\,m\phi, \qquad (9)$$

m and n being positive integers.

The condition at the surface is satisfied by (9), if α is a root of

$$J_{n+\frac{1}{2}}(a\alpha)=0. \qquad (10)$$

If, as before, we assume that $f(r,\theta,\phi)$ can be expanded in a series whose terms are of the form

$$(ar)^{-\frac{1}{2}}J_{n+\frac{1}{2}}(\alpha r)P_n^m(\mu)\,{\cos\atop\sin}\,m\phi,$$

and that this series can be integrated term by term, we can find the coefficients in the expansion.

For, let

$$f(r,\theta,\phi)=\sum_{n=0}^{\infty}\sum_{\alpha}\sum_{m=0}^{n}(\alpha r)^{-\frac{1}{2}}J_{n+\frac{1}{2}}(\alpha r)P_n^m(\mu)\{A_{n,m,\alpha}\cos m\phi+B_{n,m,\alpha}\sin m\phi\},$$

the summation in α being over the positive roots of (10).

Then we have

$$\int_0^{2\pi} f(r,\theta,\phi)\cos m\phi\,d\phi = \pi\sum_{n=0}^{\infty}\sum_{\alpha}(\alpha r)^{-\frac{1}{2}}J_{n+\frac{1}{2}}(\alpha r)P_n^m(\mu)A_{n,m,\alpha}.$$

Also we know that§

$$\int_{-1}^{1}[P_n^m(\mu)]^2\,d\mu = \dfrac{2(n+m)!}{(2n+1)(n-m)!} \qquad (11)$$

---

† The second solution of (6) has an infinity at $\theta=\pi$, and is thus inadmissible.
‡ These functions can be expressed in terms of trigonometric functions, *W.B.F.*, § 3.4. Numerical values are given in *W.B.F.*
§ Bateman, loc. cit., § 6.28; Whittaker and Watson, loc. cit., § 15.51; Byerly, loc. cit., § 106.

and
$$\int_{-1}^{1} P_n^m(\mu)P_{n'}^m(\mu)\, d\mu = 0, \quad n \neq n'. \tag{12}$$

Therefore

$$\int_{-1}^{1} P_n^m(\mu)\, d\mu \int_{0}^{2\pi} f(r,\theta,\phi)\cos m\phi\, d\phi$$

$$= \frac{2\pi(n+m)!}{(2n+1)(n-m)!} \sum_{\alpha} A_{n,m,\alpha}(\alpha r)^{-\frac{1}{2}}J_{n+\frac{1}{2}}(\alpha r).$$

Finally, from 7.5 (2),

$$\int_{0}^{a} r^{\frac{3}{2}}J_{n+\frac{1}{2}}(\alpha r)\, dr \int_{-1}^{1} P_n^m(\mu)\, d\mu \int_{0}^{2\pi} f(r,\theta,\phi)\cos m\phi\, d\phi$$

$$= \frac{\pi a^2 \alpha^{-\frac{1}{2}}[J'_{n+\frac{1}{2}}(a\alpha)]^2(n+m)!}{(2n+1)(n-m)!} A_{n,m,\alpha}. \tag{13}$$

In these results π must be replaced by 2π when $m = 0$.

Also $B_{n,m,\alpha}$ can be found in the same way.

Thus we are led to the solution of our problem in the form

$$v = \sum_{n=0}^{\infty} \sum_{\alpha} \sum_{m=0}^{n} e^{-\kappa\alpha^2 t}(\alpha r)^{-\frac{1}{2}}J_{n+\frac{1}{2}}(\alpha r)P_n^m(\mu)\{A_{n,m,\alpha}\cos m\phi + B_{n,m,\alpha}\sin m\phi\},$$

the constants $A_{n,m,\alpha}$ and $B_{n,m,\alpha}$ being determined as above, the summation in α being over the positive roots of the equation $J_{n+\frac{1}{2}}(a\alpha) = 0$. If there is radiation or no flow of heat at the surface we proceed in the same way, using the obvious modifications of (10).

9.12. The sphere with its surface $r = a$ at temperature $F(\theta, \phi)$

As in § 9.11 we expand $F(\theta,\phi)$ in the form

$$F(\theta,\phi) = \sum_{n=0}^{\infty} \sum_{m=0}^{n} P_n^m(\mu)\{A_{n,m}\cos m\phi + B_{n,m}\sin m\phi\}, \tag{1}$$

where

$$A_{n,m} = \frac{(2n+1)}{2\pi}\frac{(n-m)!}{(n+m)!}\int_{-1}^{1} P_n^m(\mu)\, d\mu \int_{0}^{2\pi} F(\theta,\phi)\cos m\phi\, d\phi, \tag{2}$$

$$B_{n,m} = \frac{(2n+1)}{2\pi}\frac{(n-m)!}{(n+m)!}\int_{-1}^{1} P_n^m(\mu)\, d\mu \int_{0}^{2\pi} F(\theta,\phi)\sin m\phi\, d\phi, \tag{3}$$

and if $m = 0$, π is to be replaced by 2π in the above.

For the steady temperature we need a solution of

$$\nabla^2 v = 0$$

which takes the value (1) at $r = a$. This is†

$$v = \sum_{n=0}^{\infty} \sum_{m=0}^{n} \left(\frac{r}{a}\right)^n P_n^m(\mu)\{A_{n,m}\cos m\phi + B_{n,m}\sin m\phi\}. \tag{4}$$

† The term involving r has to satisfy 9.11 (8) with $\alpha = 0$. Thus it must be r^n or r^{-n-1}, and the latter is inadmissible in a region including the origin.

If the surface temperature is a function of θ only, say

$$\sum_{n=0}^{\infty} A_n P_n(\mu), \tag{5}$$

the steady temperature is $$\sum_{n=0}^{\infty} A_n\left(\frac{r}{a}\right)^n P_n(\mu). \tag{6}$$

The temperature in the interior of the Earth's crust due to the decrease of mean temperature from the tropics to the poles will be of this form.

For the sphere with initial temperature $f(r,\theta,\phi)$ and surface temperature $F(\theta,\phi)$ we put as in § 1.14

$$v = u+w,$$

where u is the solution of the steady temperature problem above, and w is the solution of the problem of § 9.11 with initial temperature $f(r,\theta,\phi)-u$.

9.13. The part of the sphere cut out by the cone $\theta = \theta_0$. Surface temperature zero. Initial temperature $f(r, \theta, \phi)$

In this case the differential equation for v is

$$\frac{1}{\kappa}\frac{\partial v}{\partial t} = \frac{\partial^2 v}{\partial r^2}+\frac{2}{r}\frac{\partial v}{\partial r}+\frac{1}{r^2}\frac{\partial}{\partial\mu}\left\{(1-\mu^2)\frac{\partial v}{\partial\mu}\right\}+\frac{1}{r^2(1-\mu^2)}\frac{\partial^2 v}{\partial\phi^2}, \tag{1}$$

where $\mu = \cos\theta$. This has to be solved with

$$v = 0, \qquad\qquad \text{when } r = a, \tag{2}$$

$$v = 0, \qquad\qquad \text{when } \theta = \theta_0, \tag{3}$$

and $$v = f(r,\theta,\phi), \quad \text{when } t = 0.$$

Proceeding as in § 9.11 we find that (1) is satisfied by

$$e^{-\kappa\alpha^2 t}(\alpha r)^{-\frac{1}{2}}J_{n+\frac{1}{2}}(\alpha r)P_n^{-m}(\mu)\frac{\cos}{\sin}m\phi, \tag{4}$$

where $P_n^{-m}(\mu)$ is the generalized Legendre function† defined by

$$P_n^{-m}(\mu) = \frac{1}{\Gamma(1+m)}\left(\frac{1-\mu}{1+\mu}\right)^{\frac{1}{2}m} F\{-n, n+1; 1+m; \tfrac{1}{2}(1-\mu)\}, \tag{5}$$

where in (5), F denotes the Gaussian hypergeometric function. In (4), m has to be zero or a positive integer, and n must be greater‡ than $-\frac{1}{2}$ but it need not be integral. The function (5) with non-integral n has a singularity at $\mu = -1$, so it would not have been admissible in the problem of § 9.11 on the whole sphere.

If we write $\mu_0 = \cos\theta_0$, the condition (3) requires

$$P_n^{-m}(\mu_0) = 0, \tag{6}$$

so that n must be a root greater than $-\frac{1}{2}$ of this equation.§

Finally the condition (2) requires α to be a positive root of

$$J_{n+\frac{1}{2}}(\alpha a) = 0. \tag{7}$$

With this choice of n and α the surface conditions are satisfied by (4) and there is no infinity within the solid.

† Cf. Barnes, *Quart. J. Math.* **39** (1908) 97, or Bateman, loc. cit.
‡ Since it follows from (5) that $P_{-n-1}^{-m}(\mu) = P_n^{-m}(\mu)$, this function is symmetrical about $n = -\frac{1}{2}$.
§ For numerical values see Pal, *Bull. Calcutta Math. Soc.* **9** (1917) 85; **10** (1918) 187.

If, as before, we assume that the function $f(r, \theta, \phi)$ can be expanded in the series

$$f(r,\theta,\phi) = \sum_{m=0}^{\infty} \sum_{\alpha} \sum_{n} (\alpha r)^{-\frac{1}{2}} J_{n+\frac{1}{2}}(\alpha r) P_n^{-m}(\mu) \{A_{m,\alpha,n} \cos m\phi + B_{m,\alpha,n} \sin m\phi\}, \quad (8)$$

and that this series can be integrated term by term, we can find the coefficients in the expansion as in § 9.11. The only change is that now we need in place of 9.11 (11), (12), the result that if m is any positive number or zero, and n, n' are two different roots greater than $-\frac{1}{2}$ of $P_n^{-m}(\mu_0) = 0$, then

$$\int_{\mu_0}^{1} P_n^{-m}(\mu) P_{n'}^{-m}(\mu) \, d\mu = 0, \quad (9)$$

$$\int_{\mu_0}^{1} [P_n^{-m}(\mu)]^2 \, d\mu = -\frac{(1-\mu_0^2)}{2n+1} \frac{d}{dn} P_n^{-m}(\mu_0) \frac{d}{d\mu_0} P_n^{-m}(\mu_0). \quad (10)$$

These are proved at the end of this section. Using them we find as in § 9.11

$$-A_{m,\alpha,n} \frac{\pi a^2 \alpha^{-\frac{1}{2}}(1-\mu_0^2)}{2(2n+1)} \frac{d}{dn} P_n^{-m}(\mu_0) \frac{d}{d\mu_0} P_n^{-m}(\mu_0)[J'_{n+\frac{1}{2}}(a\alpha)]^2$$

$$= \int_0^a r^{\frac{3}{2}} J_{n+\frac{1}{2}}(\alpha r) \, dr \int_{\mu_0}^1 P_n^{-m}(\mu) \, d\mu \int_0^{2\pi} f(r,\theta,\phi) \cos m\phi \, d\phi. \quad (11)$$

In these results π must be replaced by 2π, when $m = 0$. Further $B_{m,\alpha,n}$ can be found in the same way. Thus we are brought to the solution of our problem in the form

$$v = \sum_{m=0}^{\infty} \sum_{\alpha} \sum_{n} e^{-\kappa \alpha^2 t} (\alpha r)^{-\frac{1}{2}} J_{n+\frac{1}{2}}(\alpha r) P_n^{-m}(\mu) \{A_{m,\alpha,n} \cos m\phi + B_{m,\alpha,n} \sin m\phi\}, \quad (12)$$

the constants being determined as above.

If the solid consists of the part of the sphere $r = a$ cut out by the cone $\theta = \theta_0$, and the planes $\phi = 0$, $\phi = \phi_0$, the surfaces being at zero temperature, we expand $f(r, \theta, \phi)$ in the series

$$\sum_{m=1}^{\infty} \sum_{\alpha} \sum_{n} A_{\alpha,m,n}(\alpha r)^{-\frac{1}{2}} J_{n+\frac{1}{2}}(\alpha r) P_n^{-m\pi/\phi_0}(\mu) \sin \frac{m\pi\phi}{\phi_0},$$

on the same lines as before. The same procedure applies to the solids bounded by other surfaces of the spherical polar coordinate system.

The solution for the cone $\theta = \theta_0$ may be obtained as the limit of (12) as $a \to \infty$. It will be derived in § 14.17.

It remains to prove the results (9) and (10) used above.

Let

$$u = P_n^{-m}(\mu) \quad \text{and} \quad u' = P_{n'}^{-m}(\mu).$$

Then we have

$$\frac{d}{d\mu}\left((1-\mu^2)\frac{du}{d\mu}\right) + \left(n(n+1) - \frac{m^2}{1-\mu^2}\right)u = 0,$$

$$\frac{d}{d\mu}\left((1-\mu^2)\frac{du'}{d\mu}\right) + \left(n'(n'+1) - \frac{m^2}{1-\mu^2}\right)u' = 0.$$

Therefore

$$(n'-n)(n'+n+1)\int_{\mu_0}^1 uu' \, d\mu = \int_{\mu_0}^1 \left\{u' \frac{d}{d\mu}\left((1-\mu^2)\frac{du}{d\mu}\right) - u\frac{d}{d\mu}\left((1-\mu^2)\frac{du'}{d\mu}\right)\right\} d\mu$$

$$= \left[(1-\mu^2)\left\{u'\frac{du}{d\mu} - u\frac{du'}{d\mu}\right\}\right]_{\mu_0}^1.$$

It follows that, when n, n' are two different roots of $P_n^{-m}(\mu_0) = 0$ greater than $-\frac{1}{2}$,

$$\int_{\mu_0}^{1} P_n^{-m}(\mu) P_{n'}^{-m}(\mu) \, d\mu = 0.$$

Also

$$\int_{\mu_0}^{1} \{P_n^{-m}(\mu)\}^2 \, d\mu = -\frac{(1-\mu_0^2)}{2n+1} \lim_{n' \to n} \frac{1}{n'-n} \left[P_{n'}^{-m}(\mu_0) \frac{d}{d\mu_0} P_n^{-m}(\mu_0) \right]$$

$$= -\frac{(1-\mu_0^2)}{2n+1} \frac{d}{dn} P_n^{-m}(\mu_0) \frac{d}{d\mu_0} P_n^{-m}(\mu_0),$$

when

$$P_n^{-m}(\mu_0) = 0.$$

9.14. Temperatures within the Earth

A knowledge of the temperature within the Earth is essential for the understanding of many geophysical phenomena, for example, the Earth's magnetic field, the plastic behaviour of the Earth's material, and questions such as the origin of vulcanicity and tectonic movements. But while other quantities, such as the pressure, density, and elastic properties, are sufficiently well known for differences of only a few per cent to be a matter of argument, estimates of the temperature vary greatly. So far, no adequate method of determining temperature from observation has appeared, so that, at present, estimates of the internal temperature have to be made by means of theoretical calculations based on the known value of the surface heat flux together with an assumed distribution of radioactive materials and an assumed initial distribution of temperature. These calculations have been made on the assumption of constant diffusivity,[†] though results for a two-layer Earth go back to Heaviside (cf. § 12.8).

With regard to the initial temperature, two possibilities are now entertained: (i) an initially cold earth formed by accretion, and (ii) a hot earth which was initially gaseous and subsequently condensed to a liquid. The 'cold' earth would have a constant distribution of radioactivity and initial temperature, and would heat, possibly to its melting-point.[‡] If it did melt, there would be a redistribution of radioactive matter and subsequent conditions would be very similar to the initially hot earth. For the initially hot earth, consideration begins at a time when it was wholly liquid and cooling rapidly by radiation from its surface: heat transfer in the liquid interior would be by convection, and

---

[†] It appears that, while K may vary considerably with depth, there are reasons for supposing that κ may be reasonably constant (at about 0·007 in the mantle), cf. Jacobs *Encycl. of Phys.*, Vol. 47 (Geophysics) p. 391.

[‡] Urry, *Trans. Amer. Geophys. Union*, **30** (1949) 171; Birch, *J. Geophys. Res.* **56** (1951) 107.

the temperature gradient[†] would be the adiabatic gradient which is of the order of 0·2° C/km. Solidification would commence at the point where the temperature first falls to the melting-point: since the rate of increase of the melting-point with depth (due to increasing pressure) is of the order of 2° C/km, the melting-point will first be reached at some point in the interior, possibly at the core-mantle boundary.[‡] Solidification will then proceed outwards to the surface. Thus the initial temperature for the present purpose is the melting-point-depth curve for which various theoretical formulae are available.[§]

When an estimate of the initial temperature and the distribution of radioactivity has been made, calculation of subsequent temperatures is essentially a matter of integrating the solution 14.7 (7) for the spherical surface source. The fact that some of the radioactive elements decay substantially over the times involved produces an additional complication which was ignored by early workers. Numerical calculations based on various models of the Earth's interior have been made by Urry, loc. cit., Jacobs, loc. cit., and others.

It should be noticed that assuming $\kappa = 0·007$, the Earth's radius $a = 6·4 \times 10^8$ cm, $t = 1·26 \times 10^{17}$ sec (4×10^9 years), the parameter $\kappa t/a^2 = 0·0022$. The present case is thus an ideal one for the use of the 'small time' solutions of § 12.5 as pointed out by Heaviside and Jeffreys. Despite this, many writers have used the conventional slowly convergent series of which a great many terms frequently have to be used. A number of results is given in §§ 9.3, 9.8.

[†] For a general account see Jeffreys, *The Earth* (Cambridge, edn. 3, 1952) § 10.01; Bullard in *The Earth as a Planet*, ed. Kuiper (Univ. of Chicago Press, 1954); Jacobs, loc. cit.

[‡] Adams, *J. Wash. Acad. Sci.* **14** (1924) 459–72; Jeffreys, loc. cit.

[§] Cf. Jacobs, loc. cit., Bullard, loc. cit. Jacobs, *Trans. Amer. Geophys. Union,* **35** (1954) 161–3, remarks that the melting-point-depth curves may be such that solidification may begin at the centre of the Earth, progressing outwards through the inner core, and then transferring to the core-mantle boundary, leaving the core liquid. It may be remarked that calculation shows that the fall in temperature in the core in 10^9 years is quite small.

THE USE OF SOURCES AND SINKS IN CASES OF VARIABLE TEMPERATURE

10.1. Introductory

THE idea† of the *instantaneous point source of heat*, that is, of a finite quantity of heat instantaneously liberated at a given point and time in an infinite solid, has proved most useful in the theory of conduction of heat. One great advantage is that it is based on a very simple physical idea, and this enables the solution of a large number of important problems to be written down immediately from first principles. From the theoretical point of view it has always been recognized that the point source corresponds to the fundamental solution $(1/r)$ of potential theory, and that a complete development of the theory of conduction of heat in bounded regions can be obtained by constructing Green's functions analogous to those of potential theory.‡

We take the solution for the instantaneous point source as fundamental. By integration with respect to the time we obtain the solution for the *continuous point source*, corresponding to release of heat at a given point at a prescribed rate $\phi(t)$ per unit time. If $\phi(t) = Q$, constant, and the supply of heat has gone on for a very long time, the solution becomes in the limit that for the *steady point source* and corresponds to the well-known source solutions of hydrodynamics. Problems on steady sources are discussed in Chapter XVI.

By integrating the solutions for point sources with regard to appropriate space variables, we obtain solutions for instantaneous and continuous line, plane, spherical surface, and cylindrical surface sources, each with its own simple physical interpretation.

Using these solutions, or the corresponding ones of Chapter XIV for finite regions, the solutions of a great many problems can be written down immediately in the form of definite integrals.

† The systematic use of the method is due to Kelvin's 'Compendium of the Fourier mathematics for the conduction of heat in solids and the mathematically allied physical subjects of diffusion of fluids and transmission of electric signals through submarine cables', article 'Heat' in *Encycl. Britt.* (1880) or *Math. and Phys. Papers*, Vol. 2, p. 41.

‡ The fundamental memoirs are: Hobson, *Proc. Lond. Math. Soc.* (1) **19** (1887) 279–94; Sommerfeld, *Math. Ann.* **45** (1894) 263–77; Lord Rayleigh, *Phil. Mag.* (6) **22** (1911) 381–96. See also references in §§ 10.10, 14.1.

10.2. The instantaneous point source

The differential equation of conduction of heat,

$$\frac{\partial^2 v}{\partial x^2}+\frac{\partial^2 v}{\partial y^2}+\frac{\partial^2 v}{\partial z^2}=\frac{1}{\kappa}\frac{\partial v}{\partial t},\tag{1}$$

is satisfied by
$$v=\frac{Q}{8(\pi\kappa t)^{\frac{3}{2}}}e^{-\{(x-x')^2+(y-y')^2+(z-z')^2\}/4\kappa t}.\tag{2}$$

As $t \to 0$ this expression tends to zero at all points except (x',y',z'), where it becomes infinite.

Also the total quantity of heat in the infinite region is

$$\int\limits_{-\infty}^{\infty}\int\limits_{-\infty}^{\infty}\int\limits_{-\infty}^{\infty}\rho c v\, dx dy dz$$

$$=\frac{Q\rho c}{8(\pi\kappa t)^{\frac{3}{2}}}\int\limits_{-\infty}^{\infty}e^{-(x-x')^2/4\kappa t}\,dx\int\limits_{-\infty}^{\infty}e^{-(y-y')^2/4\kappa t}\,dy\int\limits_{-\infty}^{\infty}e^{-(z-z')^2/4\kappa t}\,dz$$

$$=Q\rho c.\tag{3}$$

Thus the solution (2) may be interpreted as the temperature in an infinite solid due to a quantity of heat $Q\rho c$ instantaneously generated at $t=0$ at a point (x',y',z'). The solution (2) is called the temperature due to an instantaneous point source of strength† Q at (x',y',z') at $t=0$.

It may be noted that the temperature at a point distant r from the source has its maximum value at time $t=r^2/6\kappa$. Also, the mean square distance of the heat from the source at time t is $6\kappa t$.

We may now regard the temperature in an infinite solid with initial temperature $f(x,y,z)$ as due to liberation of heat at $t=0$ over the volume of the solid, an amount of heat

$$\rho c f(x',y',z')\,dx'dy'dz'\tag{4}$$

being liberated in the element of volume $dx'dy'dz'$ at (x',y',z'). The temperature, obtained by integration over the volume of the solid, is

$$\frac{1}{8(\pi\kappa t)^{\frac{3}{2}}}\int\limits_{-\infty}^{\infty}\int\limits_{-\infty}^{\infty}\int\limits_{-\infty}^{\infty}f(x',y',z')e^{-\{(x-x')^2+(y-y')^2+(z-z')^2\}/4\kappa t}\,dx'dy'dz',\tag{5}$$

in agreement with 2.2 (8).

---

† The strength is thus the temperature to which the amount of heat liberated would raise unit volume of the substance. This definition has the advantage that an initial distribution of temperature $f(x, y, z)$ may be regarded as due to a distribution of instantaneous sources of strength $f(x',y',z')\,dx'dy'dz'$ in the volume element $dx'dy'dz'$ at (x',y',z') as in (4). But when the number of units of heat liberated by the source is in question, it must always be remembered that this is ρc times the strength of the source.

It is of interest to obtain the solution (2) as the limit of an actual case in which a finite quantity of heat is liberated over a vanishingly small volume. Suppose we take this volume as a sphere of radius a, and consider the case of an infinite medium in which the initial temperature is V in the sphere $0 \leqslant r < a$ and zero in the region $r > a$. Putting $u = vr$ as in § 9.1, the equations for u are

$$\frac{\partial u}{\partial t} = \kappa \frac{\partial^2 u}{\partial r^2}, \quad r > 0,$$

with

$$u = Vr, \quad \text{when } t = 0, \, 0 < r < a,$$

$$u = 0, \quad \text{when } t = 0, \, r > a,$$

$$u = 0, \quad \text{when } r = 0.$$

The solution then follows from 2.4 (1) and is†

$$v = \frac{V}{2r(\pi\kappa t)^{\frac{1}{2}}} \int_0^a r'\{e^{-(r-r')^2/4\kappa t} - e^{-(r+r')^2/4\kappa t}\} \, dr'$$

$$= \frac{V}{2r(\pi\kappa t)^{\frac{1}{2}}} e^{-r^2/4\kappa t} \int_0^a r' e^{-r'^2/4\kappa t}\{e^{rr'/2\kappa t} - e^{-rr'/2\kappa t}\} \, dr'$$

$$= \tfrac{1}{2}V\left\{\operatorname{erf}\frac{r+a}{2(\kappa t)^{\frac{1}{2}}} - \operatorname{erf}\frac{r-a}{2(\kappa t)^{\frac{1}{2}}} - \frac{2(\kappa t)^{\frac{1}{2}}}{r\pi^{\frac{1}{2}}}\left[e^{-(r-a)^2/4\kappa t} - e^{-(r+a)^2/4\kappa t}\right]\right\}. \tag{6}$$

Expanding the integrand in powers of r', and assuming that a is small, we obtain the approximate solution

$$v = \frac{Q}{8(\pi\kappa t)^{\frac{1}{2}}} e^{-r^2/4\kappa t}\left\{1 + \left(\frac{r^2}{\kappa t} - 6\right)\frac{a^2}{40\kappa t}\right\}, \tag{7}$$

where we have written $Q = 4\pi a^3 V/3$.

Now let the radius of the sphere tend to zero, Q remaining constant, and (7) becomes

$$v = \frac{Q}{8(\pi\kappa t)^{\frac{3}{2}}} e^{-r^2/4\kappa t},$$

which is (2) for the case in which the heat is liberated at the origin.

Finally, we note some extensions of the idea of an instantaneous point source to more general systems.

I. *In an anisotropic material* with principal conductivities K_1, K_2, K_3 in the direction of the x, y, and z-axes, (2) is to be replaced by

$$\frac{Q(\rho c)^{\frac{1}{2}}}{8(\pi^3 t^3 K_1 K_2 K_3)^{\frac{1}{2}}} \exp\left\{-\frac{\rho c}{4t}\left[\frac{(x-x')^2}{K_1} + \frac{(y-y')^2}{K_2} + \frac{(z-z')^2}{K_3}\right]\right\}. \tag{8}$$

II. *An instantaneous source of heat in a thin rod*

Suppose the rod is of area ω and perimeter p and it loses heat into a medium at zero through surface conductance H. Then

$$v = \frac{Q'}{2\omega\rho c(\pi\kappa t)^{\frac{1}{2}}} e^{-\nu t - (x-x')^2/4\kappa t}, \tag{9}$$

† The solution (6) has been used by Lovering, *Bull. Geol. Soc. Amer.* 46 (1935) 69–93 for the cooling of a large spherical mass of rock (laccolith). Numerical results for this case are given in Fig. 4 (c).

where $\nu = Hp/\rho c \omega$, satisfies the differential equation 4.2 (2) and represents a quantity of heat Q' liberated at $x = x'$ at time $t = 0$.

III. *An instantaneous source of heat in a thin sheet*

Suppose the sheet lies in the plane $z = 0$ and is of thickness D and that it loses heat over both surfaces into a medium at zero temperature through surface conductance H. The differential equation of the problem is 5.4 (1). Then

$$v = \frac{Q'}{4\pi KDt} e^{-\kappa k^2 t - (x^2 + y^2)/4\kappa t}, \tag{10}$$

where $k^2 = 2H/KD$, satisfies this differential equation and corresponds to a quantity of heat Q' liberated at the origin at time $t = 0$.

10.3. Instantaneous line, plane, and cylindrical and spherical surface sources

Temperatures in these cases can be obtained most simply by integrating the fundamental solution 10.2 (2), or they may be obtained directly by either of the arguments of § 10.2.

I. *Instantaneous line source of strength Q at $t = 0$, parallel to the z-axis and passing through the point (x', y')*

Here we consider a distribution of instantaneous point sources of strength $Q\,dz'$ at z' along the line. The temperature, obtained by integrating 10.2 (2), is

$$v = \frac{Q}{8(\pi\kappa t)^{\frac{3}{2}}} \int_{-\infty}^{\infty} dz' e^{-\{(x-x')^2 + (y-y')^2 + (z-z')^2\}/4\kappa t}$$

$$= \frac{Q}{4\pi\kappa t} e^{-\{(x-x')^2 + (y-y')^2\}/4\kappa t}. \tag{1}$$

In this case the quantity of heat liberated per unit length of the line is $Q\rho c$.

If the polar coordinates of the points (x, y) and (x', y') are (r, θ) and (r', θ') respectively, the distance between these points is given by

$$R^2 = (x-x')^2 + (y-y')^2 = r^2 + r'^2 - 2rr' \cos(\theta - \theta'). \tag{2}$$

In this notation (1) becomes

$$v = \frac{Q}{4\pi\kappa t} e^{-R^2/4\kappa t} = \frac{Q}{2\pi} \int_0^{\infty} \lambda e^{-\kappa\lambda^2 t} J_0(\lambda R)\, d\lambda, \tag{3}$$

where the last form follows from Weber's first integral (*W.B.F.*, § 13.3 (1)).

II. *Instantaneous plane source of strength Q at t = 0, parallel to the plane x = 0, and passing through the point x'*

Here we distribute line sources of strength $Q\,dy'$ along the line $x = x'$. Integrating (1) we have

$$v = \frac{Q}{4\pi\kappa t} \int_{-\infty}^{\infty} dy' e^{-\{(x-x')^2+(y-y')^2\}/4\kappa t} = \frac{Q}{2\sqrt{(\pi\kappa t)}} e^{-(x-x')^2/4\kappa t}. \tag{4}$$

The quantity of heat liberated is $Q\rho c$ per unit area of the plane.

III. *Instantaneous cylindrical surface source at t = 0, of strength Q' and radius r', and with axis along the axis of z*

Here we distribute line sources of strength $Qr'\,d\theta'$ round the circle of radius r'. Then the temperature at the point (r, θ) is

$$\frac{Qr'}{4\pi\kappa t} \int_{0}^{2\pi} e^{-(r^2+r'^2-2rr'\cos\theta')/4\kappa t}\,d\theta' = \frac{Qr'}{4\pi\kappa t} e^{-(r^2+r'^2)/4\kappa t} \int_{0}^{2\pi} e^{(rr'\cos\theta')/2\kappa t}\,d\theta'$$

$$= \frac{Q'}{4\pi\kappa t} e^{-(r^2+r'^2)/4\kappa t} I_0\left(\frac{rr'}{2\kappa t}\right), \tag{5}$$

where $Q' = 2\pi r'Q$. The result follows from *W.B.F.*, § 3.71 (9). The quantity of heat liberated per unit length of the cylinder is $\rho c Q'$.

IV. *Instantaneous spherical surface source at t = 0 of radius r' and strength Q'*

Here, using spherical polar coordinates, we take the point source of strength $Qr'^2\sin\theta'\,d\theta'd\phi'$ at the point (r', θ', ϕ') of the sphere.

Then the temperature at $(r, 0, 0)$ is

$$\frac{2\pi Qr'^2}{8(\pi\kappa t)^{\frac{3}{2}}} \int_{0}^{\pi} e^{-(r^2+r'^2-2rr'\cos\theta')/4\kappa t} \sin\theta'\,d\theta'$$

$$= \frac{Qr'^2}{4\pi^{\frac{1}{2}}(\kappa t)^{\frac{3}{2}}} e^{-(r^2+r'^2)/4\kappa t} \int_{0}^{\pi} e^{rr'\cos\theta'/2\kappa t} \sin\theta'\,d\theta'$$

$$= \frac{Qr'^2}{4\pi^{\frac{1}{2}}(\kappa t)^{\frac{3}{2}}} e^{-(r^2+r'^2)/4\kappa t} \int_{-1}^{1} e^{rr'\mu/2\kappa t}\,d\mu$$

$$= \frac{Q'}{8\pi rr'(\pi\kappa t)^{\frac{1}{2}}} [e^{-(r-r')^2/4\kappa t} - e^{-(r+r')^2/4\kappa t}], \tag{6}$$

where $Q' = 4\pi r'^2 Q$. The quantity of heat liberated on the surface of the sphere is $Q'\rho c$.

V. *Instantaneous ring source at $t = 0$ of strength Q' and radius r' in the plane $z' = 0$*

If instantaneous point sources of strength $Qr'\, d\theta'$ are distributed round the circle $r = r'$ in the plane $z' = 0$, the temperature at time t at the point whose cylindrical coordinates are (r, θ, z) is

$$\frac{Qr'}{8(\pi\kappa t)^{\frac{3}{2}}} \int_0^{2\pi} e^{-[r^2+z^2+r'^2-2rr'\cos(\theta-\theta')]/4\kappa t}\, d\theta' = \frac{Q'}{8(\pi\kappa t)^{\frac{3}{2}}} \exp\left[-\frac{r^2+r'^2+z^2}{4\kappa t}\right] I_0\left(\frac{rr'}{2\kappa t}\right), \quad (7)$$

where $Q' = 2\pi r'Q$, and the total quantity of heat liberated is $\rho c Q'$.

VI. *Instantaneous disc source at $t = 0$ of radius a, in the plane $z = 0$*

If we put $Q' = 2\pi r'q\, dr'$ in (7), and integrate with respect to r' from 0 to a, we get

$$\frac{q}{4(\pi\kappa^3 t^3)^{\frac{1}{2}}} \int_0^a \exp\left[-\frac{r^2+r'^2+z^2}{4\kappa t}\right] I_0\left(\frac{rr'}{2\kappa t}\right) r'\, dr' \tag{8}$$

$$= \frac{qa}{2(\pi\kappa t)^{\frac{1}{2}}} e^{-z^2/4\kappa t} \int_0^\infty e^{-\kappa t\lambda^2} J_0(\lambda r) J_1(\lambda a)\, d\lambda \tag{9}$$

for the temperature when $\pi a^2 q\rho c$ heat units are instantaneously liberated over a disc of radius a. The integral cannot be expressed in terms of tabulated functions except on the axis, $r = 0$, where it becomes

$$\frac{q}{2\sqrt{(\pi\kappa t)}} [1 - e^{-a^2/4\kappa t}] e^{-z^2/4\kappa t}. \tag{10}$$

VII. *The infinite region with initial temperature given by $f(r)$ in cylindrical coordinates*

It follows from (5) that the temperature v at time t is

$$v = \frac{1}{2\kappa t} \int_0^\infty e^{-(r^2+r'^2)/4\kappa t} I_0\left(\frac{rr'}{2\kappa t}\right) f(r')r'\, dr'. \tag{11}$$

If
$$\begin{aligned} f(r) &= V, \text{ constant, } 0 < r < a, \\ f(r) &= 0, \ r > a, \end{aligned}\Bigg\}$$

this becomes
$$v = \frac{V}{2\kappa t} e^{-r^2/4\kappa t} \int_0^a e^{-r'^2/4\kappa t} I_0\left(\frac{rr'}{2\kappa t}\right) r'\, dr'. \tag{12}$$

This integral must be evaluated numerically,† except on the axis $r = 0$, where it has the value
$$V(1 - e^{-a^2/4\kappa t}). \tag{13}$$

Some values of v/V, calculated from (12), are given in Fig. 4 (*b*).

† It is called a P function and is tabulated by Masters, *J. Chem. Phys.* **23** (1955) 1865–74.

10.4. Continuous and periodic sources

I. *The continuous point source*

If heat is liberated at the rate $\phi(t)\rho c$ per unit time from $t = 0$ to $t = t$ at the point (x', y', z'), the temperature at (x, y, z) at time t is, by integrating 10.2 (2),

$$\frac{1}{8(\pi\kappa)^{\frac{3}{2}}} \int_0^t \phi(t') e^{-r^2/4\kappa(t-t')} \frac{dt'}{(t-t')^{\frac{3}{2}}}, \tag{1}$$

where we have written

$$r^2 = (x-x')^2 + (y-y')^2 + (z-z')^2.$$

This distribution of temperature is said to be due to a continuous point source of strength $\phi(t)$ from $t = 0$ onwards.

If $\phi(t)$ is constant and equal to q, we have

$$v = \frac{q}{4(\pi\kappa)^{\frac{3}{2}}} \int_{1/\sqrt{t}}^{\infty} e^{-r^2\tau^2/4\kappa} \, d\tau, \quad \text{on putting } \tau = (t-t')^{-\frac{1}{2}},$$

$$= \frac{q}{4\pi\kappa r} \operatorname{erfc} \frac{r}{\sqrt{(4\kappa t)}}. \tag{2}$$

As $t \to \infty$ this reduces to $v = q/4\pi\kappa r$, a steady temperature distribution in which a constant supply of heat is continually introduced at (x', y', z') and spreads outwards in the infinite solid.

II. *The continuous line source*

We suppose heat to be liberated at the rate $\rho c\phi(t)$ per unit time per unit length of a line parallel to the z-axis and through the point (x', y'). If supply of heat starts at $t = 0$ when the solid is at zero temperature, the temperature at time t is, using 10.3 (1),

$$\frac{1}{4\pi\kappa} \int_0^t \phi(t') e^{-r^2/4\kappa(t-t')} \frac{dt'}{t-t'}, \tag{3}$$

where $r^2 = (x-x')^2 + (y-y')^2$.

If $\phi(t) = q$, constant, this becomes

$$v = \frac{q}{4\pi\kappa} \int_{r^2/4\kappa t}^{\infty} \frac{e^{-u} \, du}{u} \tag{4}$$

$$= -\frac{q}{4\pi\kappa} \operatorname{Ei}\left(-\frac{r^2}{4\kappa t}\right), \tag{5}$$

where
$$-\mathrm{Ei}(-x) = \int_x^\infty \frac{e^{-u}}{u}\, du$$

is the exponential integral.†

For small values of x
$$\mathrm{Ei}(-x) = \gamma + \ln x - x + \tfrac{1}{4}x^2 + O(x^3),$$

where $\gamma = 0.5772...$ is Euler's constant. Thus, for large values of t,

$$v = \frac{q}{4\pi\kappa}\ln\frac{4\kappa t}{r^2} - \frac{\gamma q}{4\pi\kappa}, \tag{6}$$

approximately.

The part of (6) involving r is $(q/2\pi\kappa)\ln(1/r)$, which is the temperature due to steady supply at the rate $q\rho c$ heat units per unit length per unit time. This solution is of great importance. It gives the temperature in an infinite solid which is heated along a line, say by an ideally thin wire carrying electric current, and thus is an approximation to the heating of an infinite solid by a wire carrying current.‡ The effect of the finite diameter of the wire will be discussed in § 13.7.

If a semi-infinite solid has its surface kept at zero temperature and is heated by a line source at distance a from the surface and parallel to it, the solution may be obtained by the method of images as in § 10.10. It is

$$v = -\frac{q}{4\pi\kappa}\left[\mathrm{Ei}\left(-\frac{r^2}{4\kappa t}\right) - \mathrm{Ei}\left(-\frac{r_1^2}{4\kappa t}\right)\right], \tag{7}$$

where r and r_1 are the distances of the point from the line source and its image with respect to the surface.

III. *The continuous cylindrical surface source*

The result follows from 10.3 (5). It is not expressible in terms of tabulated functions.§

IV. *The continuous plane source*

Suppose heat is liberated at the rate $\rho c \phi(t)$ per unit area per unit time in the plane x' starting at time $t = 0$. Then, from 10.3 (4), the temperature at time t is

$$v = \frac{1}{2(\pi\kappa)^{\frac{1}{2}}}\int_0^t e^{-(x-x')^2/4\kappa(t-t')}\frac{\phi(t')\, dt'}{(t-t')^{\frac{1}{2}}}. \tag{8}$$

---

† For numerical values see Jahnke-Emde, *Tables of Functions* (Teubner, edn. 3, 1933) p. 83.

‡ It has frequently been used in the determination of thermal conductivity, cf. Fischer, *Ann. Physik*, (5) **34** (1939) 669; Albrecht, *Met. Z.* **49** (1932) 294.

§ Whitehead, *Electrician*, **99** (1927) 225, uses it as an approximation for the heating outside a buried electric cable.

If $\phi(t) = q$, constant, this becomes

$$v = q\left(\frac{t}{\pi\kappa}\right)^{\frac{1}{2}} e^{-(x-x')^2/4\kappa t} - \frac{q|x-x'|}{2\kappa} \operatorname{erfc} \frac{|x-x'|}{2\sqrt{(\kappa t)}}. \tag{9}$$

V. *The continuous spherical surface source*

If heat is liberated at the rate $\rho c\phi(t)$ per unit time over the surface of a sphere of radius r', starting at $t = 0$, the temperature at the point r at time t is

$$v = \frac{1}{8\pi r r'(\pi\kappa)^{\frac{1}{2}}} \int_0^t \{e^{-(r-r')^2/4\kappa(t-t')} - e^{-(r+r')^2/4\kappa(t-t')}\} \frac{\phi(t')\,dt'}{(t-t')^{\frac{1}{2}}}. \tag{10}$$

If $\phi(t) = q$, constant, this becomes

$$v = \frac{q}{8\pi\kappa r r'} \Big\{ 2(\kappa t/\pi)^{\frac{1}{2}} [e^{-(r-r')^2/4\kappa t} - e^{-(r+r')^2/4\kappa t}] -$$

$$- |r-r'|\operatorname{erfc} \frac{|r-r'|}{2(\kappa t)^{\frac{1}{2}}} + (r+r')\operatorname{erfc} \frac{r+r'}{2(\kappa t)^{\frac{1}{2}}} \Big\}. \tag{11}$$

VI. *The periodic† point source*

If heat is liberated at the rate $\rho c e^{i\omega t}$ from time $t = -\infty$ to $t = t$ so that steady periodic conditions prevail, it follows as in I that

$$v = \frac{1}{8(\pi\kappa)^{\frac{1}{2}}} \int_{-\infty}^t e^{i\omega t' - r^2/4\kappa(t-t')} \frac{dt'}{(t-t')^{\frac{3}{2}}} = \frac{1}{4\pi\kappa r} \exp\{-\mathbf{k}r + i(\omega t - \mathbf{k}r)\}, \tag{12}$$

where
$$\mathbf{k} = (\omega/2\kappa)^{\frac{1}{2}},$$

The periodic plane source may be treated in the same way. The results have already been given in § 2.6.

VII. *The periodic line source*

If heat is liberated at the rate $\rho c e^{i\omega t}$ per unit length from time $t = -\infty$ to $t = t$,

$$v = \frac{1}{4\pi\kappa} \int_{-\infty}^t e^{i\omega t' - r^2/4\kappa(t-t')} \frac{dt'}{t-t'} = \frac{1}{2\pi\kappa} e^{i\omega t} K_0\Big[\left(\frac{i\omega}{\kappa}\right)^{\frac{1}{2}} r\Big], \tag{13}$$

where r is the distance from the line.

10.5. Application to surface heating of a semi-infinite region

Problems in which the region $z > 0$ is heated by constant flux over a portion of its surface are of considerable importance, for example, in high frequency induction heating. They are conveniently solved by integrating the results of § 10.4 for continuous sources. In all cases below, the region $z > 0$ will be taken to be initially at zero temperature and it will be assumed that there is no flow of heat over other portions of the plane.

† Moving periodic sources are discussed by Podolsky, *J. Appl. Phys.* **22** (1951) 581; Ritchie, ibid. **22** (1951) 1389.

I. *Heat supply at the rate q per unit time per unit area for t > 0 over the half-plane* $x < 0$, $-\infty < y < \infty$, $z = 0$.

The temperature v at the point x of the surface at time t is

$$v = \frac{q}{2\pi K} \int_0^t \frac{d\tau}{t-\tau} \int_{-\infty}^0 e^{-(x-x')^2/4\kappa(t-\tau)}\, dx' = \frac{q\kappa^{\frac{1}{2}}}{2K\pi^{\frac{1}{2}}} \int_0^t \frac{d\tau}{(t-\tau)^{\frac{1}{2}}} \operatorname{erfc} \frac{x}{2\kappa^{\frac{1}{2}}(t-\tau)^{\frac{1}{2}}}$$

$$= \frac{2q}{K}\left(\frac{\kappa t}{\pi}\right)^{\frac{1}{2}}\left\{\tfrac{1}{2}\operatorname{erfc}\frac{x}{2(\kappa t)^{\frac{1}{2}}} + \frac{x}{4(\pi\kappa t)^{\frac{1}{2}}}\operatorname{Ei}\left(-\frac{x^2}{4\kappa t}\right)\right\}. \tag{1}$$

By 2.9 (8) it appears that (1) consists of the factor $2q(\kappa t)^{\frac{1}{2}}/K\pi^{\frac{1}{2}}$ for heating over an infinite plane, multiplied by a factor involving x.

The quantity of heat which has crossed a semi-infinite strip $z > 0$ of unit width in the plane $x = 0$ up to time t may be shown to be

$$2q\kappa^{\frac{1}{2}}t^{\frac{3}{2}}/3\pi^{\frac{1}{2}}. \tag{2}$$

This gives the amount of heat which is lost from the region below the heated portion of the surface.

II. *Heat supply at the rate q per unit time per unit area for t > 0 over the infinite strip $-a < x < a$, $-\infty < y < \infty$, $z = 0$*

The temperature v at the point x of the surface† at time t is

$$v = \frac{qaT^{\frac{1}{2}}}{K\pi^{\frac{1}{2}}}\left\{\operatorname{erf}\frac{a+x}{2aT^{\frac{1}{2}}} + \operatorname{erf}\frac{a-x}{2aT^{\frac{1}{2}}} - \frac{a+x}{2a(\pi T)^{\frac{1}{2}}}\operatorname{Ei}\left(-\frac{(a+x)^2}{4a^2 T}\right) - \right.$$

$$\left. - \frac{(a-x)}{2a(\pi T)^{\frac{1}{2}}}\operatorname{Ei}\left(-\frac{(a-x)^2}{4a^2 T}\right)\right\}, \tag{3}$$

where $T = \kappa t/a^2$. Some values of $K\pi^{\frac{1}{2}}v/2qaT^{\frac{1}{2}}$, that is, the ratio of the surface temperature to that in the semi-infinite solid with surface flux q, are shown in Fig. 33.

III. *Heat supply‡ at the rate q per unit time per unit area for t > 0 over the circle* $x^2+y^2 < a^2$, $z = 0$

The temperature at the point whose cylindrical coordinates are (r, θ, z) is

$$\frac{aq}{2K}\int_0^\infty J_0(\lambda r)J_1(\lambda a)\left\{e^{-\lambda z}\operatorname{erfc}\left[\frac{z}{2(\kappa t)^{\frac{1}{2}}} - \lambda(\kappa t)^{\frac{1}{2}}\right] - e^{\lambda z}\operatorname{erfc}\left[\frac{z}{2(\kappa t)^{\frac{1}{2}}} + \lambda(\kappa t)^{\frac{1}{2}}\right]\right\}\frac{d\lambda}{\lambda}. \tag{4}$$

which generalizes the result 8.2 (7). The temperature at the point $(0, 0, z)$ is

$$\frac{2q(\kappa t)^{\frac{1}{2}}}{K}\left\{\operatorname{ierfc}\frac{z}{2(\kappa t)^{\frac{1}{2}}} - \operatorname{ierfc}\frac{(z^2+a^2)^{\frac{1}{2}}}{2(\kappa t)^{\frac{1}{2}}}\right\}. \tag{5}$$

---

† Some numerical results are given by Jaeger, *Aust. J. Sci. Res.* A, **5** (1952) 1–9.
‡ Oosterkamp, *Philips Res. Rep.* **3** (1948) 49; *J. Appl. Phys.* **19** (1948) 1180.

IV. *Steady heat supply at the rate q per unit time per unit area over the rectangle*[†]
$-l < x < l, -a < y < a, z = 0$

The maximum and average temperatures over the rectangle, v_m and v_{av}, are

$$v_m = \frac{2q}{K\pi}\left\{a\sinh^{-1}\frac{l}{a} + l\sinh^{-1}\frac{a}{l}\right\}, \tag{6}$$

$$v_{av} = \frac{2q}{K\pi al}\left\{la^2\sinh^{-1}\frac{l}{a} + al^2\sinh^{-1}\frac{a}{l} + \tfrac{1}{3}[a^3 + l^3 - (l^2+a^2)^{\frac{3}{2}}]\right\}. \tag{7}$$

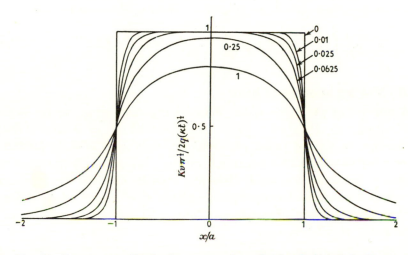

Fig. 33. Temperature distribution across a heated strip of width $2a$ in the surface of a semi-infinite solid. Numbers on the curves are values of $\kappa t/a^2$.

10.6. Application to heat production in an infinite medium

Problems in which heat is generated at a known rate in an infinite medium may be solved by the appropriate integration of the solutions of §§ 10.2, 10.3. The results of § 2.11, for example, may be obtained in this way.

As another example we determine *the temperature at the origin in an infinite solid, initially at zero temperature, if heat is generated for $t > 0$ at the constant rate A, per unit volume per unit time, over the cylindrical region $0 < r < a$, $-b < z < b$.*

The temperature at the origin at time t, due to liberation of the amount $Ar'\,dr'd\theta'dz'dt'$ of heat at the point (r', θ', z') in cylindrical coordinates at time t', is, by 10.2 (2),

$$\frac{Ar'\,dr'd\theta'dz'dt'}{8\rho c\{\pi\kappa(t-t')\}^{\frac{3}{2}}}\,e^{-[r'^2+z'^2]/4\kappa(t-t')}.$$

[†] Results of this type are frequently used in the discussion of cutting-tool temperatures and sliding temperatures. Cf. Loewen and Shaw, *Trans. Amer. Soc. Mech. Engrs.* **76** (1954) 217–31; Trigger and Chao, ibid. **73** (1951) 57–68.

Integrating this with respect to t' from 0 to t, and over the region $0 < r' < a$, $-b < z' < b$, we obtain the solution

$$v = \frac{A}{2\rho c \pi^{\frac{1}{2}} \kappa^{\frac{3}{2}}} \int_0^t \frac{dt'}{(t-t')^{\frac{3}{2}}} \int_0^b e^{-z'^2/4\kappa(t-t')} dz' \int_0^a e^{-r'^2/4\kappa(t-t')} r' \, dr'$$

$$= \frac{A}{\rho c} \int_0^t [1-e^{-a^2/4\kappa u}] \operatorname{erf} \frac{b}{\sqrt{(4\kappa u)}} \, du. \tag{1}$$

The integral is easily evaluated numerically. If $b = \infty$, that is, heat is generated in the infinite cylinder of radius a, (1) gives for the temperature on the axis

$$\frac{A}{\rho c} \int_0^t [1-e^{-a^2/4\kappa u}] \, du = \frac{At}{\rho c} [1-e^{-a^2/4\kappa t}] - \frac{Aa^2}{4K} \operatorname{Ei}\left(-\frac{a^2}{4\kappa t}\right). \tag{2}$$

10.7. Moving sources of heat

A number of problems which may be regarded either as problems in which sources of heat move through a fixed medium, or as cases of heat production at a fixed point past which a uniformly moving medium flows, may be solved by integration of the solutions for instantaneous sources.[†] In this section we shall usually consider only the case in which the medium is infinite: for corresponding problems for a bounded medium the appropriate Green's function of Chapter XIV may be treated in the same way,[‡] cf. VI below.

I. *The moving point source*

Suppose that heat is emitted at the origin for times $t > 0$ at the rate q heat units per unit time, and that an infinite medium moves uniformly past the origin with velocity U parallel to the axis of x. We calculate the temperature at the fixed point (x, y, z) at time t.

In the element of time dt' at t', $q \, dt'$ heat units were emitted at the origin; also the point of the medium which at time t is at (x, y, z), at time t' was at

$$\{x - U(t-t'), y, z\}.$$

[†] Alternatively, the differential equation of conduction of heat in a moving medium 1.7 (2) may be solved with appropriate boundary conditions. Wilson, *Proc. Camb. Phil. Soc.* **12** (1904) 406, gives applications of both methods, one of his examples is given in § 7.11. If $\partial v/\partial t = 0$ in 1.7 (2), a steady state is attained in which (2) and (3) below appear as steady point and line sources. See also Rosenthal, *Trans. Amer. Soc. Mech. Engrs.* **68** (1946) 849–66, who gives many applications; Grosh, Trabant, and Hawkins, *Quart. Appl. Math.* **13** (1955) 161–7.

[‡] For a line source moving in the surface of a cylinder and various technical applications, see Jaeger, *Phil. Mag.* (7) **35** (1944) 169; Muller, *Proc. Roy. Soc.* A, **125** (1929) 507. For a source moving along the axis of a cylinder see Rosenthal and Cameron, *Trans. Amer. Soc. Mech. Engrs.* **69** (1947) 961.

Thus the temperature at t at (x, y, z) due to the heat $q\, dt'$ emitted at t' is, by 10.2 (2),

$$\frac{q\, dt'}{8\rho c[\pi\kappa(t-t')]^{\frac{3}{2}}} \exp\left[-\frac{\{x-U(t-t')\}^2+y^2+z^2}{4\kappa(t-t')}\right].$$

And the temperature at t due to the heat emitted at the origin from time 0 to t is

$$\frac{q}{8\rho c(\pi\kappa)^{\frac{3}{2}}} \int_0^t \frac{e^{-[\{x-U(t-t')\}^2+y^2+z^2]/4\kappa(t-t')}}{(t-t')^{\frac{3}{2}}}\, dt'$$

$$= \frac{q}{2RK\pi^{\frac{1}{2}}} e^{Ux/2\kappa} \int_{R/2\sqrt{(\kappa t)}}^{\infty} e^{-\xi^2-(U^2R^2/16\kappa^2\xi^2)}\, d\xi, \quad (1)$$

where $R^2 = x^2+y^2+z^2$.

This is the solution for supply of heat for a finite time t. If $t \to \infty$, a steady thermal régime is established, and the temperature at (x, y, z) is

$$\frac{q}{4\pi KR} e^{-U(R-x)/2\kappa}. \quad (2)$$

II. *The moving line source (steady conditions)*

If heat is emitted at the rate q' per unit time per unit length along the y-axis, the temperature in the steady state at the point (x, y, z) is found by integration of (2) to be

$$\frac{q'}{4\pi K} \int_{-\infty}^{\infty} \frac{dy'}{[x^2+z^2+y'^2]^{\frac{1}{2}}} e^{-U[(x^2+z^2+y'^2)^{\frac{1}{2}}-x]/2\kappa} = \frac{q'}{2\pi K} e^{Ux/2\kappa} K_0[U(x^2+z^2)^{\frac{1}{2}}/2\kappa], \quad (3)$$

where $K_0(x)$ is the modified Bessel function of the second kind[†] of order zero.

The temperature in an infinite medium which streams with constant velocity U past a heated wire is given by (3). Roberts[‡] has used (2) and (3) as approximations for the distribution of smoke in a medium which flows past a point or line emitting smoke.

III. *The moving plane source (steady conditions)*

If heat is emitted at the rate q'' per unit time per unit area over the plane $x = 0$ and the medium moves across this plane with velocity U in the direction of the x-axis, the steady temperature at the point x is

$$\left.\begin{array}{ll} q''/\rho cU, & \text{if } x > 0 \\ (q''/\rho cU)e^{Ux/\kappa}, & \text{if } x < 0 \end{array}\right\}. \quad (4)$$

[†] Cf. Appendix III. The integral required in (3) becomes *G. and M.*, p. 51 (34), or *W.B.F.*, § 6.22 (7) on putting $y' = (x^2+z^2)^{\frac{1}{2}} \sinh u$.

[‡] Roberts, *Proc. Roy. Soc.* A, **104** (1923) 640.

IV. *A moving point source in a rod which loses heat by radiation*

If q heat units are liberated per unit time at the origin and the rod moves with velocity U along the x-axis, the steady temperature at x is, using 10.2 (9) and the notation of that equation,

$$\frac{q}{\omega\rho c[U^2+4\kappa\nu]^{\frac{1}{2}}}\exp\left\{\frac{Ux}{2\kappa}-\frac{|x|(U^2+4\kappa\nu)^{\frac{1}{2}}}{2\kappa}\right\}. \tag{5}$$

V. *A moving point source in a thin sheet which loses heat by radiation*

The sheet is supposed to lie in the xy-plane and to move with velocity U in the direction of the x-axis. Then, using 10.2 (10) and the notation of that section, the steady temperature at the point x, y due to emission of heat at the rate q' per unit time at the origin is

$$\frac{q'}{2\pi KD}e^{Ux/2\kappa}K_0\left\{\frac{R(U^2+4\kappa^2k^2)^{\frac{1}{2}}}{2\kappa}\right\}, \tag{6}$$

where $R^2 = x^2+y^2$.

VI. *A point source moving on the surface of a slab*

Suppose that the slab is $0 < z < l$ and that it moves with velocity U parallel to the x-axis, there being a point source of heat at the origin which emits q heat units per unit time, and there being no loss of heat from the surface of the slab. This is the problem of I, except for the finite thickness of the slab, and the only change is that the appropriate solution 14.10 (18) for a source in the region $0 < z < l$ must be used in place of 10.2 (2). Thus we get in place of (1)

$$\frac{q}{4\pi\rho cl\kappa}\int_0^t\frac{dt'}{(t-t')}\left\{1+2\sum_{n=1}^{\infty}e^{-\kappa n^2\pi^2(t-t')/l^2}\cos\frac{n\pi z}{l}\right\}\exp\left\{-\frac{[x-U(t-t')]^2+y^2}{4\kappa(t-t')}\right\}.$$

If $t \to \infty$ so that steady conditions prevail, this gives, using Appendix III (30),

$$\frac{q}{2\pi Kl}\left\{K_0\left(\frac{Ur}{2\kappa}\right)+2\sum_{n=1}^{\infty}K_0\left[\frac{Ur}{2\kappa}\left(1+\frac{4\kappa^2n^2\pi^2}{U^2l^2}\right)^{\frac{1}{2}}\right]\cos\frac{n\pi z}{l}\right\}e^{Ux/2\kappa}, \tag{7}$$

where $r^2 = x^2+y^2$.

For a line source along the y-axis emitting q' heat units per unit length per unit time, the corresponding result† is

$$\frac{q'}{\rho clU}\left\{1+\sum_{n=1}^{\infty}\frac{2\cos(n\pi z/l)}{[1+4\kappa^2n^2\pi^2/U^2l^2]^{\frac{1}{2}}}\exp\frac{Ux}{2\kappa}\left[1-\left(1+\frac{4\kappa^2n^2\pi^2}{U^2l^2}\right)^{\frac{1}{2}}\right]\right\}. \tag{8}$$

These results are fundamental for the calculation of temperatures in many practical cases such as frictional heating, grinding, machining, surface heating of a moving object, flame cutting and welding, and so on. In all cases, heat is supplied, not at a point, but over a restricted area: the source solutions (1) to (3) are not applicable too close to this area since they tend to infinity as the point is approached, and therefore integrated source solutions, corresponding to supply of heat over a small area, must be used. In many cases the shape of the area is unknown, and, in fact, its influence on the maximum temperatures attained is not

† For applications and further results cf. Rosenthal, loc. cit.

important. For this reason the shapes for which calculations are most easily made, namely the rectangle and the infinite strip, are usually studied since the results for these cases reduce to single integrals which are easily evaluated numerically.

VII. *The infinite strip source* $-b < x < b$, $-\infty < y < \infty$ *in the plane* $z = 0$. *Heat is supplied at the rate Q per unit time per unit area over the strip, and the surrounding medium moves across it with velocity U in the direction of the x-axis*

Integrating (3) we get for the temperature

$$v = \frac{Q}{2\pi K} \int_{-b}^{b} e^{U(x-x')/2\kappa} K_0\{U[(x-x')^2+z^2]^{\frac{1}{2}}/2\kappa\}\, dx'.$$

Introducing the dimensionless quantities

$$X = \frac{Ux}{2\kappa}, \qquad Z = \frac{Uz}{2\kappa}, \qquad B = \frac{Ub}{2\kappa}, \tag{9}$$

this becomes
$$v = \frac{\kappa Q}{\pi K U} \int_{X-B}^{X+B} e^u K_0 (Z^2+u^2)^{\frac{1}{2}}\, du. \tag{10}$$

The integral (10) is easily evaluated numerically. For the surface, $Z = 0$, results may be written down using the integrals[†]

$$\int_0^a e^{\pm u} K_0(u)\, du = a e^{\pm a}\{K_0(a) \pm K_1(a)\} \mp 1. \tag{11}$$

It may be noted that for large values of B the maximum temperature occurs near $X = B$ and is approximately $Qb/K(\pi B)^{\frac{1}{2}}$ which is the value found in 2.9 (8) for the temperature at the end of time $2b/U$ for heat supply at the rate Q over a plane in the infinite solid.

The temperature in a semi-infinite solid caused by frictional heating[‡] at a non-conducting contact over which it slides with velocity U is just twice the value (10). Some values of the surface temperature are shown in Fig. 34 for various values of X and B.

[†] Luke, *J. Math. Phys.* **29** (1950) 27–30.
[‡] This problem is extensively discussed by Blok, *Instn. Mech. Engrs. General Discussion on Lubrication*, **2** (1937) 222, and Jaeger, *J. Proc. Roy. Soc. N.S.W.* **76** (1943) 203–24. The latter gives numerical information for strip and rectangular sources. For the equivalent problem of surface temperature in grinding see Outwater and Shaw, *Trans. Amer. Soc. Mech. Engrs.* **74** (1952) 73–86.

VIII. *The infinite strip source of VII with its plane inclined at an angle θ to the x-axis†*

The temperature at a point in the strip distant a from its centre is

$$\frac{\kappa Q}{\pi K U} \int_{A-B}^{A+B} e^{u \cos \theta} K_0(|u|) \, du, \tag{12}$$

where $A = Ua/2\kappa$.

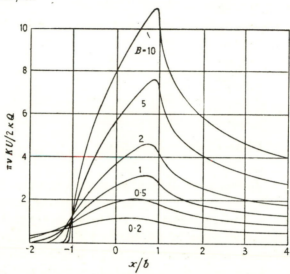

Fig. 34. Surface temperature of a semi-infinite solid caused by friction at a contact of width $2b$ over which it slides with velocity U.

IX. *The rectangular source $-b < x < b$, $-l < y < l$ in the plane $z = 0$. Heat supply at the rate Q per unit time per unit area*

$$v = \frac{Q\kappa}{4KU(2\pi)^{\frac{1}{2}}} \int_0^\infty e^{-Z^2/2u} \left\{ \operatorname{erf} \frac{Y+L}{(2u)^{\frac{1}{2}}} - \operatorname{erf} \frac{Y-L}{(2u)^{\frac{1}{2}}} \right\} \left\{ \operatorname{erf} \frac{X+B+u}{(2u)^{\frac{1}{2}}} - \operatorname{erf} \frac{X-B+u}{(2u)^{\frac{1}{2}}} \right\} \frac{du}{u^{\frac{1}{2}}},$$

where

$$X = Ux/2\kappa, \quad Y = Uy/2\kappa, \quad Z = Uz/2\kappa, \quad L = Ul/2\kappa, \quad B = Ub/2\kappa. \tag{14}$$

10.8. Doublets

We have seen in § 10.2 that

$$v = \frac{Q}{8(\pi\kappa t)^{\frac{3}{2}}} e^{-R^2/4\kappa t}, \tag{1}$$

where

$$R^2 = (x-x')^2 + (y-y')^2 + (z-z')^2, \tag{2}$$

is a solution of the equation of conduction of heat.

† This problem occurs in calculating shear plane temperatures in metal machining, cf. Chao and Trigger, *Trans. Amer. Soc. Mech. Engrs.* **75** (1953) 109–20, who give some numerical values for the integrals (12).

It follows that

$$-\frac{\partial v}{\partial x}, \quad \text{or} \quad \frac{Q(x-x')}{16\pi^{\frac{3}{2}}\kappa^{\frac{5}{2}}t^{\frac{5}{2}}}\, e^{-R^2/4\kappa t} \tag{3}$$

is also a solution.

This can be obtained by combining with a source of strength Q' at $(x'+dx', y', z')$ a sink† of strength $-Q'$ at (x', y', z'), letting $dx' \to 0$ and putting $\lim(Q'\,dx') = Q$.

For the temperature at (x, y, z) due to the source and sink, using the abbreviation (2), is given by

$$v = \frac{Q'}{8(\pi\kappa t)^{\frac{3}{2}}}\left[e^{-\{(x-x'-dx')^2+(y-y')^2+(z-z')^2\}/4\kappa t} - e^{-R^2/4\kappa t}\right]$$

$$= \frac{Q'}{8(\pi\kappa t)^{\frac{3}{2}}}\left[e^{\{2(x-x')dx'-(dx')^2\}/4\kappa t} - 1\right]e^{-R^2/4\kappa t}$$

$$= \frac{Q'(x-x')\,dx'}{16\pi^{\frac{3}{2}}\kappa^{\frac{5}{2}}t^{\frac{5}{2}}}\, e^{-R^2/4\kappa t} + \text{higher powers of } dx'.$$

Proceeding to the limit we obtain the result (3). This distribution of temperature is said to be due to an *instantaneous point doublet of strength Q at (x', y', z'), with its axis parallel to the axis of x.* In the same way differentiations of (1) with respect to y or z give point doublets with axes parallel to these directions; and partial derivatives of higher orders also give solutions of the equation of conduction of heat.

Similarly, starting from the instantaneous line source 10.3 (1), we obtain for the temperature due to an *instantaneous line doublet* at (x', y') whose axis is parallel to the axis of x

$$v = \frac{Q(x-x')}{8\pi\kappa^2 t^2}\exp\left\{-\frac{(x-x')^2+(y-y')^2}{4\kappa t}\right\}. \tag{4}$$

And in linear flow the temperature due to an *instantaneous doublet at the point x'* is from 10.3 (4)

$$v = \frac{Q(x-x')}{4\pi^{\frac{1}{2}}\kappa^{\frac{3}{2}}t^{\frac{3}{2}}}\, e^{-(x-x')^2/4\kappa t}. \tag{5}$$

The extension to the *continuous doublet* of variable or constant strength is obvious. For example, the temperature due to a continuous doublet of strength $\phi(t)$ at the point x', in the case of linear flow, is given by

$$v = \frac{(x-x')}{4\pi^{\frac{1}{2}}\kappa^{\frac{3}{2}}}\int_0^t \frac{\phi(t')\,dt'}{(t-t')^{\frac{3}{2}}}\, e^{-(x-x')^2/4\kappa(t-t')}. \tag{6}$$

---

† When the strength of a *source* as defined in § 10.2 is negative, it is called a *sink*.

Substitute $x-x' = 2\kappa^{\frac{1}{2}}(t-t')^{\frac{1}{2}}\alpha$, and we have

$$v = \frac{1}{\kappa\sqrt{\pi}} \int\limits_{(x-x')/2\sqrt{(\kappa t)}}^{\infty} e^{-\alpha^2}\phi\left(t - \frac{(x-x')^2}{4\kappa\alpha^2}\right) d\alpha, \quad \text{when } x > x',$$

and $\quad v = \frac{1}{\kappa\sqrt{\pi}} \int\limits_{(x-x')/2\sqrt{(\kappa t)}}^{-\infty} e^{-\alpha^2}\phi\left(t - \frac{(x-x')^2}{4\kappa\alpha^2}\right) d\alpha, \quad \text{when } x < x'.$

Thus $\qquad v_{x'+0} = \frac{\phi(t)}{\kappa\sqrt{\pi}} \int\limits_0^{\infty} e^{-\alpha^2} d\alpha = \frac{\phi(t)}{2\kappa}$

and $\qquad\qquad\qquad v_{x'-0} = -\frac{\phi(t)}{2\kappa}.$

It follows that there is a discontinuity in temperature at x' of amount $\phi(t)/\kappa$. Thus, in the case of linear flow in the semi-infinite solid $x > 0$, the plane $x = 0$ can be kept at temperature $\phi(t)$ when $t > 0$, by placing a continuous doublet of strength $2\kappa\phi(t)$ at $x = 0$ (cf. 2.5 (1)).

In two-dimensional problems the boundary $y = 0$ can be kept at temperature $f(x, t)$ by placing a continuous doublet at $(x', 0)$ with its axis parallel to the y-axis and strength $2\kappa f(x', t)\, dx'$, and integrating along the axis of x. A corresponding result holds for the three-dimensional case when the plane $x = 0$ is to be kept at temperature $f(y', z', t)$ (cf. 14.9 (3)).

10.9. The method of wave trains

It was shown in 2.6 (6) that

$$e^{-x(\omega/2\kappa)^{\frac{1}{2}}}\cos\{\omega t - x(\omega/2\kappa)^{\frac{1}{2}} + \epsilon\} \tag{1}$$

is a solution of the equation of linear flow of heat. From its form, it may be regarded as a wave moving to the right. Thus the integral

$$\int\limits_0^{\infty} f(\omega)e^{-x(\omega/2\kappa)^{\frac{1}{2}}}\cos\{\omega t - x(\omega/2\kappa)^{\frac{1}{2}} + \epsilon\}\, d\omega \tag{2}$$

is a general solution of the equation of conduction of heat which may be regarded as a combination of such wave trains. If the sign of x is changed it corresponds to a wave train moving to the left.

It is easy to show that, if $x > 0$,

$$\frac{1}{2\pi\kappa^{\frac{1}{2}}} \int\limits_0^{\infty} e^{-x(\omega/2\kappa)^{\frac{1}{2}}}\cos\{\omega t - x(\omega/2\kappa)^{\frac{1}{2}} - \tfrac{1}{4}\pi\} \frac{d\omega}{\omega^{\frac{1}{2}}} = \frac{1}{2(\pi\kappa t)^{\frac{1}{2}}}e^{-x^2/4\kappa t}, \tag{3}$$

and $\quad \frac{1}{2\pi\kappa} \int\limits_0^{\infty} e^{-x(\omega/2\kappa)^{\frac{1}{2}}}\cos\{\omega t - x(\omega/2\kappa)^{\frac{1}{2}}\}\, d\omega = \frac{x}{4(\pi\kappa^3 t^3)^{\frac{1}{2}}}e^{-x^2/4\kappa t}, \tag{4}$

so that (3) and (4) are, respectively, representations of this type of a unit instantaneous plane source, and a unit plane doublet, at the origin.

Green, and subsequently Robertson, developed a method† based on (2) for the solution of a wide variety of problems in linear flow and have subsequently extended it to cover cylindrical and spherical coordinates. The essential point of the method consists of taking as the elementary solution for a bounded region a combination of waves of type (1) set up so as to satisfy the continuity conditions at the boundaries in the manner familiar in studies of wave motion. Sources or doublets are introduced by using the combinations (3) and (4) of the elementary solutions. In this way, many of the results of Chapter XIV may be obtained.

10.10. The method of images. Linear flow

The method of images, which plays so important a part in the mathematical theory of electricity, is peculiarly adapted to the solution of problems in conduction of heat when the solid is bounded by planes and these are kept at zero temperature.‡ We imagine the solid to be continued in all directions, and then, by the process of taking images in the bounding planes, we obtain a distribution of sources and sinks which gives a temperature function vanishing on the boundaries with the required sources and sinks in the solid. The same method applies when there is no flow of heat over the boundary planes.

It will be seen in Chapter XIV that the Green's function, that is, the temperature due to a single source in the solid, subject to prescribed boundary conditions, is of considerable importance in the solution of the general problem of conduction for that solid. The method of images provides one method of determining the Green's function for certain problems.

The position of the images is the same whether point, or line or plane sources parallel to the reflecting plane, are under consideration. In this section plane sources will be discussed; similar formulae for point sources appear in § 14.10.

I. *Semi-infinite solid $x > 0$. Initial temperature $f(x)$. Zero temperature at the boundary $x = 0$*

Consider the source of strength $f(x')\,dx'$ at the plane x'. We may take the initial temperature as due to a distribution of these sources along the positive axis of x.

---

† Green, *Phil. Mag.* (7) **3** (1927) 784; (7) **5** (1928) 701–20; (7) **9** (1930) 241–60; (7) **12** (1931) 233–55; (7) **18** (1934) 625–40; (7) **38** (1947) 97–115. Robertson, ibid. (7) **15** (1933) 937–57; (7) **18** (1934) 165–76; (7) **18** (1934) 1009–22.

‡ The electrostatic theory was first set out fully in Maxwell's *Electricity and Magnetism* (edn. 1, 1873). In conduction of heat, Lamé, *Leçons sur la théorie de la chaleur* (Paris, 1861) discusses the rectangular parallelepiped, the cylinder whose base is an equilateral triangle, and various tetrahedra. Solutions for the basic problems for regions with rectangular boundaries (including radiation) were given by Hobson, *Proc. Lond. Math. Soc.* (1) **19** (1887) 279.

With the source $f(x')\, dx'$ at x', we associate the sink $-f(x')\, dx'$ at $-x'$, as these two give zero temperature at $x = 0$.

Hence $\qquad v = \dfrac{1}{2\sqrt{(\pi\kappa t)}} \displaystyle\int_0^\infty f(x')\{e^{-(x-x')^2/4\kappa t} - e^{-(x+x')^2/4\kappa t}\}\, dx'.$ $\qquad$ (1)

II. *Finite solid bounded by the planes $x = 0$ and $x = a$. Initial temperature $f(x)$. Bounding planes kept at zero*

Starting with the source $f(x')\, dx'$ at x', we have to take the images of this source in the planes $x = 0$ and $x = a$, a source and a sink alternating so that the boundaries may be kept at zero. In this way we have sources at the points $x' + 2na$ and sinks at the points $-x' + 2na$, where n is zero or any positive or negative integer.

Thus we have finally

$$v = \frac{1}{2\sqrt{(\pi\kappa t)}} \int_0^a f(x')\left(\sum_{n=-\infty}^{\infty} e^{-(x-x'-2na)^2/4\kappa t} - \sum_{n=-\infty}^{\infty} e^{-(x+x'-2na)^2/4\kappa t} \right) dx'. \quad (2)$$

We have already in 3.3 (5) obtained another expression for v in this case, namely,

$$\frac{2}{a} \sum_{n=1}^{\infty} \sin\frac{n\pi x}{a} e^{-\kappa n^2 \pi^2 t/a^2} \int_0^a f(x')\sin\frac{n\pi x'}{a}\, dx',$$

which may be written, when $t > 0$ (cf. 3.3, footnote),

$$\frac{2}{a} \int_0^a f(x') \sum_{n=1}^{\infty} \sin\frac{n\pi x}{a} \sin\frac{n\pi x'}{a} e^{-\kappa n^2 \pi^2 t/a^2}\, dx'. \qquad (3)$$

The fact that these solutions are identical may be demonstrated in several ways: (i) from the properties of theta functions,† (ii) by the use of the Laplace transformation, in which case it appears that solutions of type (3) come from the use of the inversion theorem while solutions of type (2) come from expanding the Laplace transform in a series of negative exponentials, cf. § 14.10 where the present result is obtained in connexion with the point source between parallel planes; (iii) with the aid of the following theorem:

---

† Cf. Poincaré, *Théorie de la propagation de la chaleur*, p. 91; Whittaker and Watson, loc. cit. (edn. 3) p. 475.

If $f(x)$ is an even function of x which can be expanded, as also $f(x \pm 2na)$, in a Fourier series of cosines of multiples of $\pi x/a$, then

$$\sum_{n=-\infty}^{\infty} f(x+2na) = \frac{1}{a} \int_0^{\infty} f(x)\, dx + \frac{2}{a} \sum_{n=1}^{\infty} \cos \frac{n\pi x}{a} \int_0^{\infty} f(x')\cos \frac{n\pi x'}{a}\, dx',$$

provided the integrals are convergent and the series converges.

Since

$$f(x) = \frac{1}{a} \int_0^a f(x')\, dx' + \frac{2}{a} \sum_{n=1}^{\infty} \cos \frac{n\pi x}{a} \int_0^a f(x')\cos \frac{n\pi x'}{a}\, dx',$$

and

$$f(x+2na) = \frac{1}{a} \int_0^a f(x'+2na)\, dx' + \frac{2}{a} \sum_{n=1}^{\infty} \cos \frac{n\pi x}{a} \int_0^a f(x'+2na)\cos \frac{n\pi x'}{a}\, dx'$$

$$= \frac{1}{a} \int_{2na}^{(2n+1)a} f(x')\, dx' + \frac{2}{a} \sum_{n=1}^{\infty} \cos \frac{n\pi x}{a} \int_{2na}^{(2n+1)a} f(x')\cos \frac{n\pi x'}{a}\, dx',$$

and

$$f(x-2na) = \frac{1}{a} \int_{(2n-1)a}^{2na} f(x')\, dx' + \frac{2}{a} \sum_{n=1}^{\infty} \cos \frac{n\pi x}{a} \int_{(2n-1)a}^{2na} f(x')\cos \frac{n\pi x'}{a}\, dx',$$

it follows that

$$\sum_{n=-\infty}^{\infty} f(x+2na) = \frac{1}{a} \int_0^{\infty} f(x')\, dx' + \frac{2}{a} \sum_{n=1}^{\infty} \cos \frac{n\pi x}{a} \int_0^{\infty} f(x')\cos \frac{n\pi x'}{a}\, dx'.$$

Let $f(x) = e^{-x^2/4\kappa t}$, and we have[†]

$$\sum_{n=-\infty}^{\infty} e^{-(x+2na)^2/4\kappa t} = \frac{1}{a} \int_0^{\infty} e^{-x'^2/4\kappa t}\, dx' + \frac{2}{a} \sum_{n=1}^{\infty} \cos \frac{n\pi x}{a} \int_0^{\infty} e^{-x'^2/4\kappa t} \cos \frac{n\pi x'}{a}\, dx'$$

$$= \frac{\sqrt{(\pi\kappa t)}}{a}\left[1 + 2 \sum_{n=1}^{\infty} \cos \frac{n\pi x}{a} e^{-\kappa n^2\pi^2 t/a^2}\right].$$

Therefore[‡]

$$\sum_{n=-\infty}^{\infty} e^{-(x-x'+2na)^2/4\kappa t} = \frac{\sqrt{(\pi\kappa t)}}{a}\left[1 + 2 \sum_{n=1}^{\infty} \cos \frac{n\pi(x-x')}{a} e^{-\kappa n^2\pi^2 t/a^2}\right] \quad (4)$$

and

$$\sum_{n=-\infty}^{\infty} e^{-(x+x'+2na)^2/4\kappa t} = \frac{\sqrt{(\pi\kappa t)}}{a}\left[1 + 2 \sum_{n=1}^{\infty} \cos \frac{n\pi(x+x')}{a} e^{-\kappa n^2\pi^2 t/a^2}\right]. \quad (5)$$

† The integrals required are quoted in § 2.2; also in *F.S.*, p. 213, Ex. 13.
‡ This is Poisson's summation formula.

Using (4) and (5) the solution (2) reduces to

$$v = \frac{2}{a} \int_0^a f(x') \sum_{n=1}^{\infty} e^{-\kappa n^2 \pi^2 t / a^2} \sin \frac{n\pi x}{a} \sin \frac{n\pi x'}{a} \, dx'.$$

III. *The same solid. Initial temperature zero. Boundary $x = 0$ kept at temperature $\phi_1(t)$. Boundary $x = a$ kept at zero*

Starting with the continuous doublet of 10.8 (6) of strength $2\kappa\phi_1(t)$ at $x = 0$,
which would keep $x = 0$ at temperature $\phi_1(t)$ if the solid extended to infinity,
we have to take an equal doublet at $x = 2a$ to keep the plane $x = a$ at zero;
and so on.

Thus we have doublets of strength $2\kappa\phi_1(t)$ at the points $2na$, n being zero, or
any positive or negative integer.

Therefore

$$v = \frac{1}{2\sqrt{(\pi\kappa)}} \int_0^t \frac{\phi_1(t')}{(t-t')^{\frac{3}{2}}} \sum_{n=-\infty}^{\infty} \{(x+2na)e^{-(x+2na)^2/4\kappa(t-t')}\} \, dt'. \tag{6}$$

A corresponding result may be obtained for the case when the boundary $x = 0$
is kept at zero and $x = a$ at $\phi_2(t)$, and by addition of these solutions we are led to
another form for the expression for the temperature in the problem of § 3.5.

IV. *Semi-infinite solid $x > 0$. Initial temperature $f(x)$. No flow of heat at the
boundary $x = 0$*

Here we proceed as in (I) except that, to give no flow of heat at $x = 0$, we associate
a source $f(x') \, dx'$ at $-x'$ with the source $f(x') \, dx'$ at x'. Thus

$$v = \frac{1}{2\sqrt{(\pi\kappa t)}} \int_0^{\infty} f(x')\{e^{-(x-x')^2/4\kappa t} + e^{-(x+x')^2/4\kappa t}\} \, dx'. \tag{7}$$

V. *Finite solid bounded by the planes $x = 0$ and $x = a$. Initial temperature $f(x)$.
No flow of heat at $x = 0$. The plane $x = a$ kept at zero temperature*

Here, proceeding as in (II), there are sources at $\pm 4na \pm x'$, and sinks at

$$\pm(4n+2)a \pm x',$$

and we have finally

$$v = \frac{1}{2\sqrt{(\pi\kappa t)}} \int_0^a f(x') \sum_{n=-\infty}^{\infty} (-1)^n \{e^{-(x-x'-2na)^2/4\kappa t} + e^{-(x+x'-2na)^2/4\kappa t}\} \, dx'. \tag{8}$$

10.11. Applications of the method of images in two or three dimensions

I. *Semi-infinite solid, $x > 0$. Initial temperature $f(x, y)$. Boundary
$x = 0$ kept at zero*

Starting with the line source of strength $f(x', y') \, dx'dy'$ at (x', y'), we
must take an equal sink at $(-x', y')$ to satisfy the condition at the
boundary.

Hence

$$v = \frac{1}{4\pi\kappa t} \int\limits_0^\infty dx' \int\limits_{-\infty}^\infty f(x', y')\{e^{-[(x-x')^2+(y-y')^2]/4\kappa t} - e^{-[(x+x')^2+(y-y')^2]/4\kappa t}\}\, dy'.$$

II. *Semi-infinite solid $x > 0$. Initial temperature $f(x, y, z)$. Boundary $x = 0$ kept at zero*

Starting with the point source of strength $f(x', y', z')\, dx'dy'dz'$ at (x', y', z'), we take an equal sink at $(-x', y', z')$, since these give zero temperature at $x = 0$.

Hence

$$v = \frac{1}{\{2\sqrt{(\pi\kappa t)}\}^3} \int\limits_0^\infty \int\limits_{-\infty}^\infty \int\limits_{-\infty}^\infty f(x', y', z')\{e^{-R^2/4\kappa t} - e^{-R'^2/4\kappa t}\}\, dx'dy'dz',$$

where

$$R^2 = (x-x')^2 + (y-y')^2 + (z-z')^2,$$

and

$$R'^2 = (x+x')^2 + (y-y')^2 + (z-z')^2.$$

III. *The wedge of angle π/m, where m is any positive integer*

The two- and three-dimensional problems given in (I) and (II) are special cases of the wedge of angle π/m, where m is any positive integer. We shall now treat this problem, confining ourselves to the two-dimensional case of a line source at the point (x', y'), the edge of the wedge coinciding with the axis of z. The three-dimensional case of a point source (x', y', z'), and the extension to the general problem of an arbitrary initial temperature offer no difficulty.

Taking cylindrical coordinates, the surface of the wedge is supposed given by the planes $\theta = 0$ and $\theta = \pi/m$; these planes are to be kept at zero temperature.

Within the wedge we have $0 < \theta < \pi/m$.

Let the source be placed at the point P_0 whose coordinates are (a, α).

Let the circle through P_0 with its centre at the origin cut $\theta = 0$ and $\theta = \pi/m$ at A and B (Fig. 35).

Then the angles AOP_0, P_0OB, and AOB are α, β, and γ, where $\beta = \pi/m - \alpha$ and $\gamma = \pi/m$.

Start with the unit source at P_0.

To give zero temperature at OA we put a unit *sink* at P_1, the image of P_0 in OA: i.e. at $(-\alpha)$.

To balance the sink at P_1, in OB, we put a *source* at P_2, the image of P_1 in OB: i.e. at $(\alpha+2\gamma)$.

To balance the source at P_2 in OA, we put a *sink* at P_3, the image of P_2 in OA: i.e. at $-(\alpha+2\gamma)$; and so on.

In this way we have the set of images P_1, P_2,..., where

$$P_0P_2 = P_2P_4 = \ldots = 2\gamma,$$
$$P_1P_3 = P_3P_5 = \ldots = 2\gamma.$$

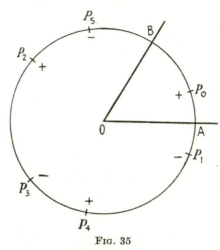

FIG. 35

Also P_{2m-1} lies at $-\{\alpha+2(m-1)\gamma\}$.

Thus $P_0P_{2m-1}+2\beta = 2\alpha+2(m-1)\gamma+2\beta = 2\pi.$

Therefore P_{2m-1} coincides with the image of P in OB, and the set of images is closed, P_{2m-1} being the last.

Also these sources and sinks give, with the source at P_0, zero temperature over the planes $\theta = 0$ and $\theta = \pi/m$.

The temperature at (r, θ) due to this system is

$$v = \sum_{s=0}^{2m-1} (-1)^s v_s, \tag{1}$$

where v_s is the temperature due to a unit source at P_s in the infinite solid.

But we have seen in 10.3 (3) that the temperature at (r, θ) due to a unit source at (r', θ') is

$$\frac{1}{2\pi} \int_0^\infty \lambda e^{-\kappa\lambda^2 t} J_0(\lambda R)\, d\lambda,$$

where $R^2 = r^2+r'^2-2rr' \cos(\theta-\theta').$

Using Neumann's expansion†

$$J_0(\lambda R) = J_0(\lambda r)J_0(\lambda r')+2 \sum_{n=1}^\infty J_n(\lambda r)J_n(\lambda r')\cos n(\theta-\theta'),$$

† Cf. *G. and M.*, p. 73, (54); *W.B.F.*, § 11.2.

this may be written as

$$\frac{1}{2\pi}\int_0^\infty \lambda e^{-\kappa\lambda^2 t}\sum_{n=-\infty}^\infty J_n(\lambda r)J_n(\lambda r')\cos n(\theta-\theta')\,d\lambda,$$

or

$$\frac{1}{2\pi}\sum_{-\infty}^\infty \cos n(\theta-\theta')\int_0^\infty \lambda e^{-\kappa\lambda^2 t}J_n(\lambda r)J_n(\lambda r')\,d\lambda.$$

It follows from (1) that

$$v=\frac{1}{2\pi}\sum_{n=-\infty}^\infty\sum_{s=0}^{m-1}[\cos n(\theta-\alpha-2s\gamma)-\cos n(\theta+\alpha+2s\gamma)]\times$$

$$\times\int_0^\infty \lambda e^{-\kappa\lambda^2 t}J_n(\lambda r)J_n(\lambda a)\,d\lambda.\quad(2)$$

When n is not a multiple of m the series in s has zero for its sum; and when it is a multiple of m, its sum is equal to

$$2m\sin n\theta\sin n\alpha.$$

Thus, from (2), we obtain the solution of our problem of the unit source at the point (a,α) in the wedge $\theta=0$, $\theta=\pi/m$, in the form

$$v=\frac{2}{\gamma}\sum_{p=1}^\infty \sin\frac{p\pi\theta}{\gamma}\sin\frac{p\pi\alpha}{\gamma}\int_0^\infty \lambda e^{-\kappa\lambda^2 t}J_{p\pi/\gamma}(\lambda r)J_{p\pi/\gamma}(\lambda a)\,d\lambda,\quad(3)$$

where, as above, we have written γ for π/m.

For the three-dimensional case, we start with the expression

$$\frac{1}{\{2\sqrt{(\pi\kappa t)}\}^3}e^{-[r^2+r'^2-2rr'\cos(\theta-\theta')+(z-z')^2]/4\kappa t},$$

corresponding to the unit source at (r',θ',z').

Proceeding on the same lines as above, we obtain the solution of our problem in the form

$$v=\frac{e^{-(z-z')^2/4\kappa t}}{\gamma\sqrt{(\pi\kappa t)}}\sum_{p=1}^\infty \sin\frac{p\pi\theta}{\gamma}\sin\frac{p\pi\alpha}{\gamma}\int_0^\infty \lambda e^{-\kappa\lambda^2 t}J_{p\pi/\gamma}(\lambda r)J_{p\pi/\gamma}(\lambda r')\,d\lambda.$$

10.12. Sommerfeld's extension of the method of images

The method of images as used above for the wedge of angle π/m, where m is any positive integer, fails when the angle is $n\pi/m$, where m, n are both positive integers, prime to each other.

For example, when the angle is a right angle, and the given source is at $P_0(r',\theta')$, where $0<\theta'<\frac12\pi$, the images are as follows:

a sink at $P_1(r',-\theta')$:   a source at $P_2(r',\pi+\theta')$;

and   a sink at $P_3(r',-\pi-\theta')$.

In this case no difficulty is encountered (cf. Fig. 36).

But when the angle is $2\pi/3$ and the given source is at $P_0(r', \theta')$, where

$$0 < \theta' < 2\pi/3,$$

FIG. 36

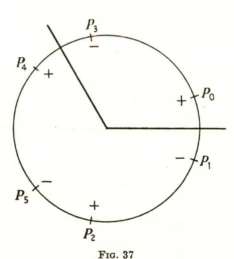

FIG. 37

the successive images are as follows:

a sink at $P_1(r', -\theta')$:   a source at $P_2\left(r', \frac{4\pi}{3} + \theta'\right)$;

a sink at $P_3\left(r', -\frac{4\pi}{3} - \theta'\right)$:   a source at $P_4\left(r', \frac{8\pi}{3} + \theta'\right)$;

and   a sink at $P_5\left(r', -\frac{8\pi}{3} - \theta'\right)$.   (Fig. 37.)

As the expression for the temperature due to a source is periodic of period 2π, and the sink at P_3 gives a singularity corresponding to a sink at $(r', 2\pi/3 - \theta')$, that is to a sink at a point between the planes, the method fails.

However, for the complete representation of the solid we need only the region $0 < \theta < 2\pi/3$, and if we can find a solution of the equation of conduction which has a period 4π and only one singularity in that interval, and this of the proper kind, viz.

$$\frac{e^{-r^2/4\kappa t}}{4\pi\kappa t} \quad \text{or} \quad \frac{e^{-r^2/4\kappa t}}{\{2\sqrt{(\pi\kappa t)}\}^3}, \quad \text{when } r = 0 \text{ and } t = 0,$$

we can use this solution as we did the ordinary expression for the temperature of a source and take the images at the points named above.

This method, as originally introduced by Sommerfeld, really amounts to considering a solution of the equation of conduction on a suitable Riemann's surface (or space). For the angle $n\pi/m$, the Riemann's surface (or space) will be an n-fold one, and the solution will have a period $2n\pi$. The method is of historical interest because, after applying it to the heat problem of a source in the region bounded by the planes $\theta = 0$ and $\theta = 2\pi$, Sommerfeld by its aid gave the first exact solution of the diffraction of waves by a semi-infinite plane (e.g. $\theta = 0$). But simpler methods of treating these questions both for the equation of conduction of heat and for the other partial differential equations of mathematical physics have now been evolved. For this reason it will be sufficient here only to give references to Sommerfeld's and other papers in which the Riemann's surface idea is used.†

We return to the problem of the wedge in § 14.14 and the solution obtained in 10.11 (3) for the angle π/m will be found to be true for the wedge of any angle.‡

† Sommerfeld, *Math. Ann.* **45** (1894) 263, and **47** (1896) 317; *Proc. Lond. Math. Soc.* **28** (1897) 395; Schwarzschild, *Math. Ann.* **55** (1902) 177; Carslaw, *Proc. Lond. Math. Soc.* **30** (1899) 121; *Rep. Brit. Ass.* (1900) 644; Hobson, *Trans. Camb. Phil. Soc.* **18** (1900) 277.

‡ Another solution, using contour integration, is given in *C.H.*, § 90. Cf. also Carslaw, *Proc. Lond. Math. Soc.* (2) **8** (1910) 365; (2) **18** (1920) 291.

CHANGE OF STATE

11.1. Introductory

A most important group of problems is that in which a substance has a transformation point at which it changes from one phase to another with emission or absorption of heat. Such cases arise in many contexts of which the most important are melting and solidification, and, to fix ideas, most of the problems discussed here will be stated in this form. The first published discussion of such problems seems to be that by Stefan† in a study of the thickness of polar ice, and for this reason the problem of freezing is frequently referred to as the 'problem of Stefan'.

The essential new feature of such problems is the existence of a moving surface of separation between the two phases. The way in which this surface moves has to be determined. Heat is liberated or absorbed on it, and the thermal properties of the two phases on different sides of it may be different, so that the problem is one of considerable difficulty. It will be shown below that it is non-linear so that special solutions must be determined and cannot be superposed.

Before considering specific solutions, it is desirable to indicate what exact solutions have been obtained and the methods available. The most important exact solution is that of Neumann for the semi-infinite region $x > 0$ initially at a constant temperature V which is greater than the melting-point, and with the surface $x = 0$ subsequently maintained at zero temperature. There are no closed solutions for other important boundary conditions at $x = 0$ such as constant flux or 'radiation', though there are some for various prescribed values of the surface temperature which are of no great physical importance. An approximation which is frequently made consists of neglecting the heat capacity of the material between $x = 0$ and the surface of separation, that is, of assuming that the flow through this region is of steady type.

Exact solutions are also available for a number of problems on the infinite region in which $x < 0$ is initially solid at constant temperature and $x > 0$ is initially liquid at constant temperature. These solutions

† Stefan, *Ann. Phys. u. Chem.* (Wiedemann) N.F. **42** (1891) 269–86. He gives the results 11.2 (18) and (21) below. In fact the more general result known as Neumann's solution was given by Franz Neumann in his lectures in the 1860's, cf. Riemann–Weber, *Die partiellen Differentialgleichungen der mathematischen Physik* (edn. 5, 1912) Vol. 2, p. 121.

can also be generalized to the case of several transformation temperatures and to the case of a melting range in place of a fixed melting-point.

No exact solutions are available for other regions such as the slab[†] $-a < x < a$ with its surface maintained at zero, or for the region $-a < x < a$ initially liquid and $|x| > a$ initially solid.

For problems on radial flow in cylindrical or spherical coordinates the position is even worse. The only simple exact solution in cylindrical coordinates corresponds to supply or removal of heat by a continuous line source. For the region bounded internally or externally by a circular cylinder with constant surface temperature, only an approximate solution is available.

Apart from the few exact solutions, all problems have to be attacked by numerical methods: digital methods, the differential analyser, and an electrical analogue method called the thermal analyser have all been used. The exact solutions mentioned above are frequently useful as starting solutions in these cases (cf. § 18.5). The systematic use of numerical methods[‡] has the advantage that the variation in the thermal properties with temperature, which is usually considerable over the ranges of temperature involved in problems on melting and solidification, can be taken into account.

Applications of these results are of the greatest practical importance. The problem of ice formation[§] is of the greatest importance both in geophysics and in ice manufacture. A great deal of attention has recently been paid to the solidification of castings.[||] The study of the cooling of large masses of igneous rock is of great importance in geology.[††] Problems on diffusion which lead to similar equations and boundary conditions are also common.[‡‡]

11.2. Melting and solidification in one dimension. Neumann's solution and its generalizations

Throughout this chapter we shall write ρ, c_1, K_1, κ_1 for the thermal constants of material in the solid phase and v_1 for its temperature; the corresponding quantities for the liquid phase will be ρ, c_2, K_2, κ_2, v_2.

[†] Some discussion of this case is given by Lightfoot, *Proc. Lond. Math. Soc.* (2) **31** (1930) 97–116.

[‡] Eyres, Hartree, Ingham, Jackson, Sarjant, and Wagstaff, *Phil. Trans. Roy. Soc.* A, **240** (1948) 1–57; Crank, *Quart. J. Mech. Appl. Math.* **10** (1957) 220.

[§] Plank, *Z. ges. Kälte-Ind. Beih., Reihe* 3, *H.* 10 (1941), ibid. **20** (1913) 109; London and Seban, *Trans. Amer. Soc. Mech. Engrs.* **65** (1943) 771.

[||] For a full account and bibliography see Ruddle, *The Solidification of Castings* (The Institute of Metals, 1950).   [††] Cf. Larsen, *Amer. J. Sci.* **243** A (1945) 399.

[‡‡] *M.D.*, Chap. VII; Danckwerts, *Trans. Faraday Soc.* **46** (1950) 701–12; Hill, *Proc. Roy. Soc.* B, **104** (1928) 39–96.

Except in (VIII), change of volume on solidification will be neglected, so that the density ρ will be the same in both solid and liquid phases.

Suppose that L is the latent heat of fusion of the substance (cal/gm) and that T_1' is its melting-point, then if the surface of separation between the solid and liquid phases is at $X(t)$, one boundary condition to be satisfied at this surface is

$$v_1 = v_2 = T_1', \quad \text{when } x = X(t). \tag{1}$$

A second boundary condition concerns the absorption or liberation of latent heat at this surface. To fix ideas, suppose that the region $x > X(t)$ contains liquid at temperature $v_2(x, t)$, and that $x < X(t)$ contains solid at temperature $v_1(x, t)$. Then, when the surface of separation moves a distance dX, a quantity of heat $L\rho\, dX$ per unit area is liberated and must be removed by conduction. This requires

$$K_1 \frac{\partial v_1}{\partial x} - K_2 \frac{\partial v_2}{\partial x} = L\rho \frac{dX}{dt}. \tag{2}$$

(1) and (2) are the boundary conditions to be satisfied at the surface of separation in this case: it is easy to see that if $x < X$ is liquid at $v_2(x, t)$ and $x > X$ is solid at $v_1(x, t)$, (2) still holds.

Equation (2) may be put into an alternative form by considering the curves of constant temperature $v_1(x, t) = T_1' = v_2(x, t)$ in the xt-plane. On these

$$\frac{\partial v_1}{\partial x}\, dx + \frac{\partial v_1}{\partial t}\, dt = 0 = \frac{\partial v_2}{\partial x}\, dx + \frac{\partial v_2}{\partial t}\, dt,$$

so that (2) may be written

$$K_1 \frac{\partial v_1}{\partial x} - K_2 \frac{\partial v_2}{\partial x} = -L\rho \frac{\partial v_1/\partial t}{\partial v_1/\partial x} = -L\rho \frac{\partial v_2/\partial t}{\partial v_2/\partial x}. \tag{3}$$

In this form, the non-linearity of the problem is apparent. In three dimensions, the boundary condition (3) becomes

$$K_1 |\text{grad } v_1| - K_2 |\text{grad } v_2| = \pm L\rho \frac{\partial v_1/\partial t}{|\text{grad } v_1|} = \pm L\rho \frac{\partial v_2/\partial t}{|\text{grad } v_2|}, \tag{4}$$

where the signs are to be chosen to conform with the problem under consideration.

For linear flow, the temperatures v_1 and v_2 in the solid and liquid regions have to satisfy

$$\frac{\partial^2 v_1}{\partial x^2} - \frac{1}{\kappa_1} \frac{\partial v_1}{\partial t} = 0, \tag{5}$$

$$\frac{\partial^2 v_2}{\partial x^2} - \frac{1}{\kappa_2} \frac{\partial v_2}{\partial t} = 0. \tag{6}$$

In addition to (1), (2), (5), (6), there will be conditions at the fixed boundaries of the region under consideration. We now give the solutions of a number of important problems in linear flow.†

I. *Neumann's solution for the region $x > 0$ initially liquid at constant temperature V with the surface $x = 0$ maintained at zero for $t > 0$*

The equations to be satisfied are (1), (2), (5), (6) with the additional boundary conditions

$$v_2 \to V, \quad \text{as } x \to \infty, \tag{7}$$

$$v_1 = 0, \quad x = 0. \tag{8}$$

It follows from 2.1 (4) that

$$v_1 = A \operatorname{erf} \frac{x}{2(\kappa_1 t)^{\frac{1}{2}}}, \tag{9}$$

where A is a constant, satisfies (5) and (8). Also, if B is a constant,

$$v_2 = V - B \operatorname{erfc} \frac{x}{2(\kappa_2 t)^{\frac{1}{2}}} \tag{10}$$

satisfies (6) and (7). Then (1) requires

$$A \operatorname{erf} \frac{X}{2(\kappa_1 t)^{\frac{1}{2}}} = V - B \operatorname{erfc} \frac{X}{2(\kappa_2 t)^{\frac{1}{2}}} = T_1. \tag{11}$$

Since (11) has to be satisfied for all values of the time, X must be proportional to $t^{\frac{1}{2}}$, that is, say,

$$X = 2\lambda(\kappa_1 t)^{\frac{1}{2}}, \tag{12}$$

where λ is a numerical constant to be determined from the remaining condition (2). Using (9), (10), and (12) in this, gives

$$K_1 A e^{-\lambda^2} - K_2 B(\kappa_1/\kappa_2)^{\frac{1}{2}} e^{-\kappa_1 \lambda^2/\kappa_2} = \lambda L \kappa_1 \rho \pi^{\frac{1}{2}}, \tag{13}$$

or, using (11) and (12),

$$\frac{e^{-\lambda^2}}{\operatorname{erf} \lambda} - \frac{K_2 \kappa_1^{\frac{1}{2}}(V - T_1) e^{-\kappa_1 \lambda^2/\kappa_2}}{K_1 \kappa_2^{\frac{1}{2}} T_1 \operatorname{erfc} \lambda(\kappa_1/\kappa_2)^{\frac{1}{2}}} = \frac{\lambda L \pi^{\frac{1}{2}}}{c_1 T_1}. \tag{14}$$

When λ has been found from (14), v_1 and v_2 can be written down from (9), (10), (11), and (12). They are

$$v_1 = \frac{T_1}{\operatorname{erf} \lambda} \operatorname{erf} \frac{x}{2(\kappa_1 t)^{\frac{1}{2}}}, \tag{15}$$

$$v_2 = V - \frac{(V - T_1)}{\operatorname{erfc} \lambda(\kappa_1/\kappa_2)^{\frac{1}{2}}} \operatorname{erfc} \frac{x}{2(\kappa_2 t)^{\frac{1}{2}}}. \tag{16}$$

The solution above is a particular solution of the differential equations

† For other discussions see Lachmann, *Z. angew. Math. Mech.* **15** (1935) 345; **17** (1937) 379; Huber, ibid. **19** (1939) 1.

and boundary conditions. The initial conditions satisfied by it follow from (12) and (16). They are

$$X = 0, \quad v_2 = V, \quad \text{when } t = 0, \tag{17}$$

that is, the whole of the region $x > 0$ is liquid and at temperature V.

The solution of (14) is easily carried out numerically using tables of the error function. Some values of its roots for the case of water, $K_2 = \kappa_2 = 0 \cdot 00144$, and ice, $K_1 = 0 \cdot 0053$, $\kappa_1 = 0 \cdot 0115$, $L\rho = 73 \cdot 6$, and various values of the initial temperature $(V - T_1')$ of the water and the temperature $(-T_1')$ at $x = 0$ are given in the table.

	$V - T_1 = 0$	1	2	3	4	5
$T_1 = 1$	0·056	0·054	0·053	0·051	0·050	0·049
2	0·079	0·077	0·076	0·074	0·073	0·071
3	0·097	0·095	0·093	0·091	0·090	0·088
4	0·111	0·110	0·108	0·106	0·104	0·103
5	0·124	0·123	0·121	0·119	0·117	0·115

For the important special case in which the liquid is initially at its melting-point so that $V = T_1'$, (14) reduces to

$$\lambda e^{\lambda^2} \operatorname{erf} \lambda = \frac{c_1 T_1}{L\pi^{\frac{1}{2}}}. \tag{18}$$

The roots of (18) may be read off from the graph of $\lambda e^{\lambda^2} \operatorname{erf} \lambda$ in Fig. 38, Curve I. It may be remarked that for materials such as rocks or metals with high melting-points and their surfaces maintained near room temperature, $c_1 T_1 / L\pi^{\frac{1}{2}}$ is of the order of unity. On the other hand, for the case of the freezing of water by conduction of heat into a region a few degrees below zero $c_1 T_1 / L\pi^{\frac{1}{2}}$ is small. In this case, using the first term of Appendix II (4) in (18) gives approximately

$$\lambda^2 = c_1 T_1 / 2L. \tag{19}$$

This result may also be obtained by an interesting physical approximation, namely, assuming that the temperature distribution in the solid is approximately that corresponding to steady flow with $x = X$ at T_1', that is

$$v_1 = xT_1 / X. \tag{20}$$

Substituting in (2) gives

$$K_1 T_1 = L\rho X \frac{dX}{dt}.$$

It follows that

$$X^2 = 2K_1 T_1 t / L\rho, \tag{21}$$

which is equivalent to (19).

II. *The case of supercooled liquid*

Suppose that the melting-point of the solid is T_1, that the region $x > 0$ initially contains liquid at temperature $V < T_1$, and that solidification starts at the plane $x = 0$ and moves to the right, no heat being removed from the solidified material whose temperature will thus have the constant value T_1 throughout.

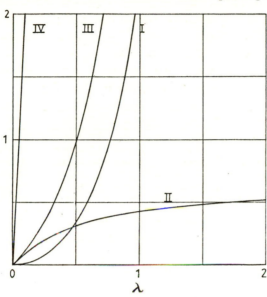

FIG. 38. Curve I, $\lambda e^{\lambda^2}\operatorname{erf}\lambda$. Curve II, $\lambda e^{\lambda^2}\operatorname{erfc}\lambda$. Curve III, $\lambda e^{\lambda^2}(1+\operatorname{erf}\lambda)$. Curve IV, $\lambda e^{\lambda^2}(20+\operatorname{erf}\lambda)$.

If $x = X(t)$ is the surface of separation of the solid and liquid phases, we seek a solution of type

$$X = 2\lambda(\kappa_2 t)^{\frac{1}{2}}, \tag{22}$$

$$v_1 = T_1, \qquad x < X,$$

$$v_2 = V + A \operatorname{erfc} \frac{x}{2(\kappa_2 t)^{\frac{1}{2}}}.$$

The boundary conditions at $x = X$ then give

$$V + A \operatorname{erfc}\lambda = T_1,$$

$$A e^{-\lambda^2} = \lambda L \pi^{\frac{1}{2}}/c_2.$$

It follows that λ is the root of

$$\lambda e^{\lambda^2}\operatorname{erfc}\lambda = (T_1 - V)c_2/L\pi^{\frac{1}{2}}, \tag{23}$$

and can be read off from Curve II of Fig. 38.

III. *Melting in the region $x > 0$*

The problem is now that in which the region $x > 0$ is initially solid at temperature zero, and for $t > 0$ the plane $x = 0$ is maintained at constant temperature $V > T_1$. Then, using as before $K_1, \kappa_1, \rho, c_1, v_1$ for the solid and $K_2, \kappa_2, \rho, c_2, v_2$ for the liquid, the position of the plane of melting is given by

$$X = 2\lambda(\kappa_2 t)^{\frac{1}{2}}, \tag{24}$$

where λ is the root of†

$$\frac{e^{-\lambda^2}}{\operatorname{erf}\lambda} - \frac{K_1\,\kappa_2^{\frac{1}{2}}\,T_1 e^{-\lambda^2\kappa_2/\kappa_1}}{K_2\,\kappa_1^{\frac{1}{2}}\,(V-T_1)\operatorname{erfc}[\lambda(\kappa_2/\kappa_1)^{\frac{1}{2}}]} = \frac{\lambda L\pi^{\frac{1}{2}}}{c_2(V-T_1)}, \tag{25}$$

and the temperatures in the liquid and solid are then

$$v_2 = V - \frac{V-T_1}{\operatorname{erf}\lambda}\operatorname{erf}\frac{x}{2(\kappa_2 t)^{\frac{1}{2}}}, \tag{26}$$

$$v_1 = \frac{T_1}{\operatorname{erfc}\lambda(\kappa_2/\kappa_1)^{\frac{1}{2}}}\operatorname{erfc}\frac{x}{2(\kappa_1 t)^{\frac{1}{2}}}. \tag{27}$$

(25) is the same as (14) with the thermal constants of the solid and liquid interchanged and T_1 and $V-T_1$ also interchanged.

IV. *The region $x > 0$ initially liquid and the region $x < 0$ solid*

Suppose that the region $x < 0$ is initially solid of thermal constants $K_0, \rho_0, c_0, \kappa_0$ at zero temperature and that the region $x > 0$ is initially liquid of thermal constants K_2, ρ, c_2, κ_2 at constant temperature V. The thermal constants K_1, ρ, c_1, κ_1 of the solidified liquid may be different from those of the solid in $x < 0$.

The method is a trivial extension of that used earlier. Writing v_0, v_1, v_2 for the temperatures in the regions $x < 0$, $0 < x < X$, and $x > X$, respectively, where X is the position of the surface of separation between the solid and liquid phases, we assume

$$v_0 = A\left(1 + \operatorname{erf}\frac{x}{2(\kappa_0 t)^{\frac{1}{2}}}\right), \tag{28}$$

$$v_1 = B + C\operatorname{erf}\frac{x}{2(\kappa_1 t)^{\frac{1}{2}}}, \tag{29}$$

$$v_2 = V - D\operatorname{erfc}\frac{x}{2(\kappa_2 t)^{\frac{1}{2}}}. \tag{30}$$

This makes $v_0 \to 0$ as $x \to -\infty$. A, B, C are connected by the continuity relations at $x = 0$ which give

$$A = B, \qquad K_0 A\kappa_1^{\frac{1}{2}} = K_1 C\kappa_0^{\frac{1}{2}}. \tag{31}$$

Proceeding as before, we find

$$X = 2\lambda(\kappa_1 t)^{\frac{1}{2}}, \tag{32}$$

where, now, λ is the root of

$$\frac{K_0\,\kappa_1^{\frac{1}{2}}\,e^{-\lambda^2}}{K_1\,\kappa_0^{\frac{1}{2}}+K_0\,\kappa_1^{\frac{1}{2}}\operatorname{erf}\lambda} - \frac{K_2\,\kappa_1^{\frac{1}{2}}(V-T_1)e^{-\kappa_1\lambda^2/\kappa_2}}{K_1\,\kappa_2^{\frac{1}{2}}\,T_1\operatorname{erfc}[\lambda(\kappa_1/\kappa_2)^{\frac{1}{2}}]} = \frac{\lambda L\pi^{\frac{1}{2}}}{c_1 T_1}. \tag{33}$$

When λ has been found, $v_0, v_1,$ and v_2 are given by

$$v_0 = \frac{K_1\,\kappa_0^{\frac{1}{2}}\,T_1}{K_1\,\kappa_0^{\frac{1}{2}}+K_0\,\kappa_1^{\frac{1}{2}}\operatorname{erf}\lambda}\left(1+\operatorname{erf}\frac{x}{2(\kappa_0 t)^{\frac{1}{2}}}\right), \tag{34}$$

$$v_1 = \frac{T_1}{K_1\,\kappa_0^{\frac{1}{2}}+K_0\,\kappa_1^{\frac{1}{2}}\operatorname{erf}\lambda}\left(K_1\,\kappa_0^{\frac{1}{2}}+K_0\,\kappa_1^{\frac{1}{2}}\operatorname{erf}\frac{x}{2(\kappa_1 t)^{\frac{1}{2}}}\right), \tag{35}$$

$$v_2 = V - \frac{(V-T_1)}{\operatorname{erfc}\lambda(\kappa_1/\kappa_2)^{\frac{1}{2}}}\operatorname{erfc}\frac{x}{2(\kappa_2 t)^{\frac{1}{2}}}. \tag{36}$$

† This solution with $T_1 = 0$, and others with no analogues in conduction of heat, appear in the study of the diffusion of oxygen into muscle in which recovery is necessary at every point before the oxygen can advance, cf. Hill, *Proc. Roy. Soc.* B, **104** (1928) 39–96.

This solution was introduced by Schwarz† as a better approximation than (14)–(16) for the solidification of metal cast in a mould, since the thermal properties of the solid metal and the material of the mould will be very different. It may also be regarded as the fundamental solution for the cooling of an intrusive igneous rock, but in this case, since the thermal properties of rocks do not differ greatly, it is usually permissible to take $K_0 = K_1$, $\kappa_0 = \kappa_1$ and a slight simplification results.

For the case in which the liquid is initially at its melting-point $V = T_1$, (33) reduces to

$$\lambda e^{\lambda^2}\left\{\frac{K_1 \kappa_0^{\frac{1}{2}}}{K_0 \kappa_1^{\frac{1}{2}}}+\operatorname{erf}\lambda\right\} = \frac{c_1 T_1}{L\pi^{\frac{1}{2}}}. \tag{37}$$

As remarked above, for rock materials $(K_1 \kappa_0^{\frac{1}{2}}/K_0 \kappa_1^{\frac{1}{2}})$ is of the order unity, but for cast iron in sand it is as large as 25. Some values of the left-hand side of (37) are given in Fig. 38, Curves III and IV. If

$$V - T_1 > (K_0 \kappa_2^{\frac{1}{2}}/K_2 \kappa_0^{\frac{1}{2}})T_1, \tag{38}$$

the root of (33) becomes negative, corresponding to melting of the solid in $x < 0$ by the hot liquid: if this material is the same as that in $x > 0$ the above solution still holds, if it is different the solution must be rewritten to suit this case.

V. Melting‡ in the region $x > 0$ caused by hot solid in $x < 0$

In this case $x < 0$ is initially solid of thermal constants $K_0, \rho_0, c_0, \kappa_0$ at temperature $V > T_1$. The region $x > 0$ is initially solid of thermal constants K_1, ρ, c_1, κ_1 at temperature zero, and the thermal constants of this material in the liquid state are K_2, ρ, c_2, κ_2.

If $X = 2\lambda(\kappa_2 t)^{\frac{1}{2}}$ is the position of the surface of separation, λ is the root of

$$\frac{K_0 \kappa_2^{\frac{1}{2}}(V-T_1)e^{-\lambda^2}}{K_2 \kappa_0^{\frac{1}{2}}+K_0 \kappa_2^{\frac{1}{2}}\operatorname{erf}\lambda} - \frac{K_1 \kappa_2^{\frac{1}{2}} T_1 e^{-\kappa_2\lambda^2/\kappa_1}}{K_2 \kappa_1^{\frac{1}{2}}\operatorname{erfc}\lambda(\kappa_2/\kappa_1)^{\frac{1}{2}}} = \frac{\lambda L\pi^{\frac{1}{2}}}{c_2}. \tag{39}$$

VI. The case of a melting range

Rocks and alloys do not have fixed melting-points but melting ranges, say from T_2 to T_1, over which the heat of solidification L is liberated. If this heat were liberated uniformly over the range T_2, T_1, its effect could be expressed by adding $L/(T_2-T_1)$ to the specific heat proper, c_2', of the liquid in this range. That is, the effect of heat of solidification in the range T_2, T_1 can be taken into account approximately by using for the specific heat c_2 in this range the value

$$c_2 = c_2'+\frac{L}{T_2-T_1}, \tag{40}$$

and the problem is reduced to one without latent heat but with variable specific heat. Such problems in general may be treated as in (VII), but results for the most important special case, namely that in which the initial temperature of the liquid is exactly T_2 at the top of the melting range,§ may be written down immediately by taking $L = 0$ in (14) and (33), it being understood that c_2 is now the quantity (40).

---

† Arch. Eisenhüttenw. **5** (1931) 139, 177; Z. angew. Math. Mech. **13** (1933) 202.

‡ This is the case of melting of solder by a hot iron. The system solid-steam-water has been discussed in connexion with quenching by Heindlhofer, Phys. Rev. (2) **20** (1922) 221–42.

§ This appears to be the case for most magmatic intrusions. Numerical results for a cooling intrusive sheet are given by Jaeger, Amer. J. Sci. **255** (1957) 306.

Thus, for the region $x > 0$ initially liquid at T_2 and with its surface $x = 0$ subsequently at zero, (14) gives

$$\frac{\exp[(\kappa_1-\kappa_2)\lambda^2/\kappa_2]\mathrm{erfc}[\lambda(\kappa_1/\kappa_2)^{\frac{1}{2}}]}{\mathrm{erf}\,\lambda} = \frac{(T_2-T_1)K_2\,\kappa_1^{\frac{1}{2}}}{T_1K_1\,\kappa_2^{\frac{1}{2}}}. \tag{41}$$

And, for the region $x > 0$ initially liquid at T_2 and the region $x < 0$ initially solid as in (IV), (33) becomes

$$\frac{K_0\,\kappa_1^{\frac{1}{2}}\exp[(\kappa_1-\kappa_2)\lambda^2/\kappa_2]\mathrm{erfc}[\lambda(\kappa_1/\kappa_2)^{\frac{1}{2}}]}{K_1\,\kappa_0^{\frac{1}{2}}+K_0\,\kappa_1^{\frac{1}{2}}\mathrm{erf}\,\lambda} = \frac{(T_2-T_1)K_2\,\kappa_1^{\frac{1}{2}}}{T_1K_1\,\kappa_2^{\frac{1}{2}}}. \tag{42}$$

VII. *Multiphase cases*

The above analysis may easily be extended to a substance which has several transformation temperatures or several ranges over which heat of transformation is liberated. As an illustration we consider the extension of (I) to the case of two transformation temperatures.

Suppose the region $x > 0$ is initially at constant temperature V and that the plane $x = 0$ is maintained at zero temperature for $t > 0$. The material is supposed to have two transition temperatures T_2 and T_1, $V > T_2 > T_1 > 0$, at which heats of transformation L_2 and L_1 are liberated. Change of volume on change of phase will be neglected so that all phases have the same density ρ. Suffixes 1, 2, 3, respectively, will be used to describe the phases in the temperature ranges $(0, T_1)$, (T_1, T_2), and (T_2, V), respectively.

There will be a surface of separation at $x = X_1(t)$ between the phases 1 and 2, and one at $x = X_2(t)$ between the phases 2 and 3, and analysis precisely similar to that of (I) shows that

$$X_1(t) = 2\lambda_1(\kappa_1 t)^{\frac{1}{2}}, \qquad X_2(t) = 2\lambda_2(\kappa_2 t)^{\frac{1}{2}}, \tag{43}$$

where

$$\frac{T_1}{\mathrm{erf}\,\lambda_1}e^{-\lambda_1^2}-\frac{(T_2-T_1)K_2\,\kappa_1^{\frac{1}{2}}}{[\mathrm{erf}\,\lambda_2-\mathrm{erf}\,\lambda_1(\kappa_1/\kappa_2)^{\frac{1}{2}}]K_1\,\kappa_2^{\frac{1}{2}}}e^{-\lambda^2\kappa_1/\kappa_2} = \frac{L_1\lambda_1\pi^{\frac{1}{2}}}{c_1}, \tag{44}$$

$$\frac{T_2-T_1}{\mathrm{erf}\,\lambda_2-\mathrm{erf}\,\lambda_1(\kappa_1/\kappa_2)^{\frac{1}{2}}}e^{-\lambda_2^2}-\frac{(V-T_2)K_3\,\kappa_2^{\frac{1}{2}}}{K_2\,\kappa_3^{\frac{1}{2}}\mathrm{erf}\,\lambda_2(\kappa_2/\kappa_3)^{\frac{1}{2}}}e^{-\lambda_2^2\kappa_2/\kappa_3} = \frac{L_2\lambda_2\pi^{\frac{1}{2}}}{c_2}. \tag{45}$$

When the simultaneous equations (44) and (45) have been solved for λ_1 and λ_2, the temperatures in the three regions are found from

$$v_1 = \frac{T_1\mathrm{erf}[x/2(\kappa_1 t)^{\frac{1}{2}}]}{\mathrm{erf}\,\lambda_1}, \tag{46}$$

$$v_2 = \left[(T_2-T_1)\mathrm{erf}\frac{x}{2(\kappa_2 t)^{\frac{1}{2}}}+T_1\mathrm{erf}\,\lambda_2-T_2\,\mathrm{erf}\,\lambda_1(\kappa_1/\kappa_2)^{\frac{1}{2}}\right][\mathrm{erf}\,\lambda_2-\mathrm{erf}\,\lambda_1(\kappa_1/\kappa_2)^{\frac{1}{2}}]^{-1}, \tag{47}$$

$$v_3 = V-\frac{(V-T_2)\mathrm{erfc}\,x/2(\kappa_3 t)^{\frac{1}{2}}}{\mathrm{erfc}\,\lambda_2(\kappa_2/\kappa_3)^{\frac{1}{2}}}. \tag{48}$$

For an n-phase substance there will be $(n-1)$ phase boundaries given by equations of type (43), and $(n-1)$ equations of type (44), (45) for $\lambda_1,...,\lambda_{n-1}$. Weiner[†] has shown that the solution of these is quite practicable.

VIII. *The effect of change of volume on solidification*

If the densities of the solid and liquid phases are not the same, there will be motion of the liquid. To illustrate the effect of this, the problem of (I) will be discussed, assuming, for definiteness of statement, that the density ρ_1 of the solid is greater than that of the liquid, ρ_2.

† Weiner, *Brit. J. Appl. Phys.* 6 (1955) 361–4.

As in (12) we seek a solution† in which the surface of separation is at $X = 2\lambda(\kappa_1 t)^{\frac{1}{2}}$ and the temperature in the solid is

$$v_1 = A \operatorname{erf} \frac{x}{2(\kappa_1 t)^{\frac{1}{2}}}. \tag{49}$$

Now, when the surface of separation advances a distance dX, the mass of solid $\rho_1 \, dX$ per unit area so formed has been derived from a thickness $\rho_1 \, dX/\rho_2$ of liquid. Thus the liquid moves with velocity u_x along the x-axis given by

$$u_x = -\left(\frac{\rho_1}{\rho_2} - 1\right)\frac{dX}{dt}. \tag{50}$$

The equation of conduction of heat 1.7 (2) in the moving liquid is then

$$\frac{\partial^2 v_2}{\partial x^2} + \frac{(\rho_1 - \rho_2)}{\rho_2 \kappa_2}\frac{dX}{dt}\frac{\partial v_2}{\partial x} - \frac{1}{\kappa_2}\frac{\partial v_2}{\partial t} = 0. \tag{51}$$

It is easy to verify that, with the above value of X, (51) is satisfied by

$$v_2 = V - B \operatorname{erfc}\left\{\frac{x}{2(\kappa_2 t)^{\frac{1}{2}}} + \frac{\lambda(\rho_1 - \rho_2)\kappa_1^{\frac{1}{2}}}{\rho_2 \kappa_2^{\frac{1}{2}}}\right\}, \tag{52}$$

and this makes $v_2 \to V$ as $x \to \infty$. At the boundary $x = X$ we require by (1)

$$A \operatorname{erf} \lambda = V - B \operatorname{erfc}[\lambda \rho_1 \kappa_1^{\frac{1}{2}}/\rho_2 \kappa_2^{\frac{1}{2}}] = T_1. \tag{53}$$

Using these results in‡ (2) gives the equation for λ

$$\frac{e^{-\lambda^2}}{\operatorname{erf}\lambda} - \frac{(V - T_1)K_2 \kappa_1^{\frac{1}{2}} e^{-\lambda^2 \rho_1^2 \kappa_1/\rho_2^2 \kappa_2}}{T_1 K_1 \kappa_2^{\frac{1}{2}} \operatorname{erfc}(\lambda \rho_1 \kappa_1^{\frac{1}{2}}/\rho_2 \kappa_2^{\frac{1}{2}})} = \frac{\lambda L \pi^{\frac{1}{2}}}{c_1 T_1}, \tag{54}$$

which reduces to (14) if $\rho_1 = \rho_2$.

For ice, $\rho_1 = 0\cdot917$, and water, $\rho_2 = 1$, (54) gives $\lambda = 0\cdot116$ for the case $V - T_1 = T_1 = 5°$. This may be compared with the value $\lambda = 0\cdot115$ calculated from (14).

11.3. The region $x > 0$ with other boundary conditions

The method of § 11.2 involves the use of a solution of the equation of linear flow of heat which is a function of $xt^{-\frac{1}{2}}$ only. As remarked in § 2.1, $\operatorname{erf}[x(4\kappa t)^{-\frac{1}{2}}]$ is the only solution of this type, and, since it is not possible to satisfy the 'radiation' or constant flux boundary conditions with it, exact solutions such as those of § 11.2 are not available. Nevertheless, a number of important problems of this type has been discussed.

> I. *Solidification in the region $x > 0$ when the liquid is initially at its melting-point. Constant flux F from the solid at $x = 0$*

In this case, a series solution§ can be found by assuming a power series in t for $X(t)$ and a double power series 2.1 (13) for v_1. Substituting these in 11.2 (1), 11.2 (2), and the boundary condition at $x = 0$, namely

$$K_1 \frac{\partial v}{\partial x} = F, \tag{1}$$

---

† A different treatment and many applications to problems in diffusion are given by Danckwerts, *Trans. Faraday Soc.* 46 (1950) 701–12. See also *M.D.*, Chap. VII.

‡ It may be verified by writing down the heat balance for the region $X, X + dX$ that (2) still holds with ρ replaced by the density of the phase remaining in this region.

§ Evans II, Isaacson, and MacDonald, *Quart. Appl. Math.* 8 (1950) 312–19.

the solution is found to be

$$X(t) = \frac{Ft}{L\rho} - \frac{F^3 t^2}{2\kappa_1 L^3 \rho^3} + \frac{5 F^5 t^3}{6\kappa_1^2 L^5 \rho^5} - \cdots,$$ (2)

$$v_1 = T_1 + \frac{Fx}{K_1} - \frac{F^2(x^2 + 2\kappa_1 t)}{2\kappa_1 K_1 L\rho} + \frac{F^4(x^4 + 12\kappa_1 x^2 t + 12\kappa_1^2 t^2)}{12\kappa_1^3 K_1 L^3 \rho^3} - \cdots.$$ (3)

II. *The problem of* (I), *but with the 'radiation' boundary condition*,

$$\frac{\partial v}{\partial x} = hv, \qquad x = 0.$$ (4)

In this case

$$X(t) = \frac{K_1 h T_1 t}{L\rho} - \frac{K_1^2 h^3 T_1^2 (\kappa_1 L\rho + K_1 T_1) t^2}{2\kappa_1 L^3 \rho^3} + \cdots,$$ (5)

$$v_1 = T_1 + h T_1 x - \frac{K_1 h^2 T_1^2 (x^2 + 2\kappa_1 t)}{2\kappa_1 L\rho} - \frac{K_1 h^3 T_1^2 (x^3 + 6\kappa_1 xt)}{6\kappa_1 L\rho} + \cdots.$$ (6)

Some values of the surface temperature and the position of the surface of separation in these cases, obtained by the thermal analyser, have been given by Kreith and Romie.†

III. *The surface of solidification moving with constant speed*

A simple exact solution for this case is due to Stefan (loc. cit.) Clearly

$$v_1 = T_1 + A(1 - e^{\kappa_1 m^2 t - mx}), \qquad x < \kappa_1 mt,$$ (7)

$$v_2 = T_1, \qquad x > \kappa_1 mt,$$ (8)

where A and m are constants, satisfies 11.2 (1), 11.2 (5), 11.2 (6). It also satisfies 11.2 (2) if

$$A = L/c_1.$$ (9)

It thus represents a solution of the problem of freezing in the region $x > 0$ in which the liquid is initially at its melting-point and the surface of solidification moves with constant speed $\kappa_1 m$. However, since the temperature at $x = 0$,

$$T_1 - (L/c_1)(e^{\kappa_1 m^2 t} - 1),$$

has to be given negative values which increase exponentially with the time, the solution is of no great physical interest.

IV. *Melting with continuous renewal of the melt*

Suppose that a solid is heated by constant flux F at its surface, any melted material being continuously removed, for example by being blown away. This is also the case of sublimation. It has important practical applications and is probably the mechanism of heating of meteorites in the Earth's atmosphere. Here we give only‡ a simple solution for the steady distribution of temperature.

Suppose that the plane at which melting takes place is taken to be $x = 0$, the solid in the region $x > 0$ being regarded as moving towards it with speed U (to be determined). Then, if T_1 is the melting temperature, the temperature v in $x > 0$ is the solution of 1.7 (2) with $u_x = -U$ which tends to T_1 as $x \to 0$, and to zero as $x \to \infty$. That is

$$v = T_1 e^{-Ux/\kappa_1}.$$ (10)

---

† *Proc. Phys. Soc.* B, **68** (1955) 277–91. They also discuss the case (III) and corresponding problems for the cylinder and sphere.

‡ The variable state is discussed by Landau, *Quart. Appl. Math.* **8** (1950) 81–94, and Masters, *J. Appl. Phys.* **27** (1956) 477–85.

Using this value of v it follows from 1.7 (1) that the flux of heat into the solid is zero, so that the rate of supply of heat F from outside must be equal to its rate of removal with the melt, that is

$$F = (L + c_1 T_1)\rho U, \tag{11}$$

and the temperature v is given by (10) with this value of U.

11.4. Integral equation methods. Lightfoot's discussion of solidification

If the thermal properties of the solid and liquid are assumed to be the same, another powerful method becomes available. Suppose that at time t the surface of solidification is at $X(t)$; this surface moves with velocity $\dot{X}(t)$, and heat of solidification is liberated at it at the rate

$$L\rho\dot{X}(t). \tag{1}$$

Thus the liberation of heat of solidification may be represented by a moving source of heat at $X(t)$ of strength given by (1). The temperature at any point can be found by adding terms representing the effects of this moving source and the initial and boundary conditions. The fact that the temperature at $X(t)$ must always be the melting-point T_1 leads to an integral equation† for $X(t)$.

To illustrate the method consider the case of § 11.2 (I), namely that of solidification in the region $x > 0$ with initial temperature $V > T_1$ and with the surface $x = 0$ maintained at zero for $t > 0$. As in § 10.3, the temperature $w(x, t)$ at x at time t due to the moving source (1) is given by

$$w(x, t) = \frac{L}{2c(\pi\kappa)^{\frac{1}{2}}} \int_0^t \frac{\dot{X}(t')\,dt'}{(t-t')^{\frac{1}{2}}} \{e^{-[x-X(t')]^2/4\kappa(t-t')} - e^{-[x+X(t')]^2/4\kappa(t-t')}\}. \tag{2}$$

The temperature $u(x, t)$ at x at time t due to the constant initial temperature V is

$$u(x, t) = V \operatorname{erf} \frac{x}{2(\kappa t)^{\frac{1}{2}}}. \tag{3}$$

The temperature at x and t is the sum of (2) and (3). The condition that the temperature at $x = X(t)$ is the melting-point T_1 is then

$$w\{X(t), t\} + u\{X(t), t\} = T_1, \tag{4}$$

which is an integral equation for $X(t)$.

Lightfoot‡ has solved this problem on the assumption made previously that

$$X(t) = 2\lambda(\kappa t)^{\frac{1}{2}}. \tag{5}$$

With this value of $X(t)$, the integral (2) can be expressed in terms of error functions. Making the change of variables

$$y = x/X(t), \qquad t' = t\left(\frac{1-z^2}{1+z^2}\right)^2, \tag{6}$$

(2) becomes

$$w(x, t) = \frac{2L\lambda}{c\pi^{\frac{1}{2}}}\{I_1 - I_2\}. \tag{7}$$

---

† For further discussion of methods involving integral equations, see Evans II, Isaacson, and MacDonald, *Quart. Appl. Math.* **8** (1950) 312–19.

‡ Lightfoot, *Proc. Lond. Math. Soc.* (2) **31** (1929) 97–116. He derives (8), (9), (10) by series expansion, but they may also be found by other conventional methods.

where

$$I_1 = \int_0^1 \frac{dz}{1+z^2} \exp\left\{-\frac{\lambda^2}{4}\left[\frac{y-1}{z}+(y+1)z\right]^2\right\} = \tfrac{1}{4}\pi e^{\lambda^2}\mathrm{erfc}\,\lambda(1+\mathrm{erf}\,\lambda y), \quad y < 1, \quad (8)$$

$$= \tfrac{1}{4}\pi e^{\lambda^2}\mathrm{erfc}\,\lambda y(1+\mathrm{erf}\,\lambda), \quad y > 1, \quad (9)$$

$$I_2 = \int_0^1 \frac{dz}{1+z^2} \exp\left\{-\frac{\lambda^2}{4}\left[\frac{y+1}{z}+(y-1)z\right]^2\right\} = \tfrac{1}{4}\pi e^{\lambda^2}\mathrm{erfc}\,\lambda y\,\mathrm{erfc}\,\lambda. \quad (10)$$

By (6), $y \gtrless 1$ correspond to the liquid and solid regions, respectively, and $y = 1$ to the plane of solidification. Using (3), and (8) and (10) with $y = 1$, (4) becomes

$$\frac{L\lambda\pi^{\frac{1}{2}}}{c}e^{\lambda^2}\mathrm{erfc}\,\lambda\,\mathrm{erf}\,\lambda + V\,\mathrm{erf}\,\lambda = T_1, \quad (11)$$

which agrees with 11.2 (14) for the case $K_1 = K_2$, $\kappa_1 = \kappa_2$.

Other problems of § 11.2 may be treated in the same way. Lightfoot (loc. cit.) has given an approximate treatment of solidification between parallel planes by considering an infinite series of images in place of the two in (2). Another application of the method has been made by Blevin.[†]

11.5. Solutions in cylindrical and spherical coordinates

The existence of the solutions for linear flow found in § 11.2 depended essentially on the existence of solutions of the equation of conduction of heat which were functions of $xt^{-\frac{1}{2}}$ only. This suggests that solutions in cylindrical or spherical coordinates which are functions of $rt^{-\frac{1}{2}}$ only might lead to useful results.

It is easy to verify that

(i) *in cylindrical coordinates* the function

$$-\mathrm{Ei}\left(-\frac{r^2}{4\kappa t}\right) \quad (1)$$

is a solution of the equation of conduction of heat, and

(ii) *in spherical coordinates* the function

$$\frac{(\kappa t)^{\frac{1}{2}}}{r}e^{-r^2/4\kappa t} - \tfrac{1}{2}\pi^{\frac{1}{2}}\mathrm{erfc}\frac{r}{2(\kappa t)^{\frac{1}{2}}} \quad (2)$$

is a solution of the equation of conduction of heat. Further, Paterson[‡] has shown that these are the only solutions of these types. These solutions have been applied by Frank[§] to the study of radially symmetrical phase growth controlled by diffusion, or, in the present notation, the growth of a spherical or cylindrical solid from a supercooled melt (the linear case has already been discussed in § 11.2 (II)). It will be assumed here that the densities of the solid and liquid phases are the same—if this is not the case there will be motion of the liquid.[||]

† Blevin, *Aust. J. Phys.* **6** (1953) 203–8.
‡ Paterson, *Proc. Glasgow Math. Ass.* **1** (1952–3) 42–47.
§ Frank, *Proc. Roy. Soc.* A, **201** (1950) 586–99.
|| Chambré, *Quart. J. Mech. Appl. Math.* **9** (1956) 224–33.

The following results may be derived as in § 11.2 (II).

I. *The region $r < R$ in cylindrical coordinates consists of solid at its melting-point T_1, and the region $r > R$ of supercooled liquid whose temperature $v_2 \to V < T_1$ as $r \to \infty$*

The position of the surface of separation is given by

$$R = 2\lambda(\kappa_2 t)^{\frac{1}{2}},$$

where λ is the root of

$$\lambda^2 e^{\lambda^2} \mathrm{Ei}(-\lambda^2) + c_2(T_1 - V)/L = 0, \tag{3}$$

and the temperature v_2 in the liquid is given by

$$v_2 = V + \frac{(T_1 - V)\mathrm{Ei}(-r^2/4\kappa_2 t)}{\mathrm{Ei}(-\lambda^2)}. \tag{4}$$

II. *The result for the corresponding problem in spherical coordinates is $R = 2\lambda(\kappa_2 t)^{\frac{1}{2}}$, where λ is a root of*

$$\lambda^2 e^{\lambda^2}\{e^{-\lambda^2} - \lambda\pi^{\frac{1}{2}}\mathrm{erfc}\,\lambda\} = \tfrac{1}{2}c_2(T_1 - V)/L. \tag{5}$$

The temperature in the liquid is

$$v_2 = V + \frac{2\lambda(T_1 - V)}{e^{-\lambda^2} - \lambda\pi^{\frac{1}{2}}\mathrm{erfc}\,\lambda}\left\{\frac{(\kappa_2 t)^{\frac{1}{2}}}{r}e^{-r^2/4\kappa_2 t} - \frac{\pi^{\frac{1}{2}}}{2}\mathrm{erfc}\frac{r}{2(\kappa_2 t)^{\frac{1}{2}}}\right\}. \tag{6}$$

11.6. Problems of melting and freezing with cylindrical symmetry

Despite the importance of such problems in contexts such as the freezing of ice around cylindrical pipes, little general information is available.

We shall consider only the case in which the surface of separation between the solid and liquid phases is at radius $r = R(t)$, the region $r > R$ containing liquid of thermal constants ρ, c_2, K_2, κ_2, and the region $r < R$ solid of thermal constants ρ, c_1, K_1, κ_1. Then, if v_2 and v_1 are the temperatures in the two regions and T_1 is the melting-point, the boundary conditions at $r = R$ are

$$v_1 = v_2 = T_1, \tag{1}$$

$$K_1\frac{\partial v_1}{\partial r} - K_2\frac{\partial v_2}{\partial r} = L\rho\frac{dR}{dt}. \tag{2}$$

I. *The solution for a continuous line source along the axis $r = 0$*

Using the fundamental solution 11.5 (1), it may be shown precisely as in § 11.2 (I) that the differential equations, and boundary conditions (1) and (2), are satisfied by

$$v_1 = T_1 + \frac{Q}{4\pi K_1}\left\{\mathrm{Ei}\left(-\frac{r^2}{4\kappa_1 t}\right) - \mathrm{Ei}(-\lambda^2)\right\}, \qquad 0 < r < R, \tag{3}$$

$$v_2 = V - \frac{(V - T_1)}{\mathrm{Ei}(-\lambda^2\kappa_1/\kappa_2)}\mathrm{Ei}\left(-\frac{r^2}{4\kappa_2 t}\right), \qquad r > R. \tag{4}$$

where
$$R = 2\lambda(\kappa_1 t)^{\frac{1}{2}}, \tag{5}$$
and λ is the root of

$$\frac{Q}{4\pi} e^{-\lambda^2} + \frac{K_2(V-T_1)}{\mathrm{Ei}(-\lambda^2\kappa_1/\kappa_2)} e^{-\lambda^2\kappa_1/\kappa_2} = \lambda^2\kappa_1 L\rho. \tag{6}$$

When $t = 0$, $R = 0$, and $v_2 = V$, so that initially the whole region $r > 0$ is liquid at temperature V. For $t > 0$,

$$\lim_{r \to 0} 2\pi r K_1 \frac{\partial v_1}{\partial r} = Q, \tag{7}$$

so that the solution is that for freezing by a continuous line source along the axis for $t > 0$ which extracts heat at the rate Q per unit time. This is the only simple exact solution available for the cylindrical region.[†]

II. *The region $r > a$ initially liquid at its melting-point T_1, with the surface $r = a$ maintained at zero temperature for $t > 0$*

No exact solution for this case is available. An important approximate solution may be obtained as in 11.2 (20) by assuming that the temperature distribution in the solid is of steady-state type. That is, that

$$v_1 = \frac{T_1 \ln(r/a)}{\ln(R/a)}, \qquad a < r < R. \tag{8}$$

Using this in (2) gives

$$R \ln\left(\frac{R}{a}\right) \frac{dR}{dt} = \frac{K_1 T_1}{L\rho}, \tag{9}$$

and integrating gives

$$2R^2 \ln(R/a) - R^2 + a^2 = 4K_1 T_1 t/L\rho. \tag{10}$$

Equation (10) is, in fact, quite a good approximation for the position of the surface of separation if $L \gg c$ as in the case of the freezing of water. A second approximation has been found by Pekeris and Slichter[‡] using a method of expansion in series. They show that

$$A + B\ln(r/a) + \{(\dot{A} - \dot{B})r^2 + \dot{B}r^2 \ln(r/a)\}/4\kappa + ..., \tag{11}$$

where A and B are functions of t only, satisfies the differential equation of radial flow of heat. This series is then substituted in the boundary conditions, and A and B determined by successive approximation. The case of solidification of the region within the cylinder $r = a$ can be treated similarly. Kreith and Romie (loc. cit.) develop series solutions for the case in which the interface moves with constant speed.

[†] Ingersoll, Adler, Plass, and Ingersoll, *Heat. Pip. Air Condit.* **22** (1950) 113–22; Paterson, *Proc. Glasgow Math. Ass.* **1** (1952–3) 42–47. In the same way, a solution in spherical coordinates may be found from 11.5 (2), but since this corresponds to solidification from a continuous point source whose strength increases linearly with the time, it is not of great physical importance.

[‡] Pekeris and Slichter, *J. Appl. Phys.* **10** (1939) 135–7.

XII

THE LAPLACE TRANSFORMATION: PROBLEMS IN LINEAR FLOW

12.1. Historical

THE methods used in the preceding chapters may be said to be immediate consequences and extensions of Fourier's classical work. An alternative method of dealing with the differential equations of applied mathematics, largely derived from the pioneer work of Heaviside, has recently been developed and is especially well adapted to the solution of problems in conduction of heat. All the solutions previously obtained for conduction of heat in the unsteady state could be found by the new method, but, since in fact the advantage of the method increases with the complexity of the problem, it seems preferable, after a few illustrative applications to problems already discussed, to apply it to more difficult cases which are not so easily dealt with by other means.

Heaviside in the 1890's invented his celebrated operational method for the solution of the systems of ordinary linear differential equations with constant coefficients which appear in electric circuit theory: he gave a rudimentary justification of the method in this case. Subsequently he applied an extension† of his procedure to the partial differential equations of electromagnetism and conduction of heat, and obtained a large number of new solutions—not only solutions of unsolved problems, but new types of solution such as those specially suitable for large or small values of the time. The mathematical value of these solutions was very doubtful, and the necessity of putting the whole theory on a rigorous mathematical basis was evident. The first step in this direction was taken by Bromwich, who in his classical paper‡ formed a contour integral with the Heaviside operational expression as integrand. He then verified that this integral satisfied the differential equation and initial conditions, and subsequently evaluated the integral by the ordinary methods of contour integration. This point of view has been developed by Jeffreys in his Cambridge Tract§ and has been much

---

† A brief account is given in the historical introduction to Carslaw and Jaeger's *Operational Methods in Applied Mathematics* (Oxford, 1948).

‡ 'On normal coordinates in dynamical systems', *Proc. Lond. Math. Soc.* (2) **15** (1916) 401. Wagner, *Arch. Elektrotech.* **4** (1916) 159, developed much the same ideas independently.

§ *Operational Methods in Mathematical Physics* (Cambridge, edn. 2, 1931).

used in the theory of conduction of heat. A related method was developed by Carslaw in which a contour integral is again used,[†] but its integrand is determined independently and not by the Heaviside method.

Another method of justifying Heaviside's work was given by Carson and van der Pol, who showed that the solution could be found from its Heaviside operational representation by solving an integral equation. This integral equation is simply the integral which appears in 12.2 (1) as the definition of the Laplace transform, while Bromwich's contour integral, referred to above, is simply the contour integral which appears in 12.3 (8) in the Inversion Theorem for the Laplace transform.

Thus the Laplace transformation procedure given below unifies the Heaviside, Bromwich, and Carson theory. Its importance was recognized by Doetsch who used it in a series of papers many of which discuss problems in conduction of heat.[‡]

In the following sections we develop the Laplace transformation method briefly *ab initio*, giving statements of the theorems and sketches of the proofs which are adequate for the present requirements; fuller discussion is given in works on the subject.[§] As remarked above, solutions which use the Bromwich–Jeffreys approach will often be found in the literature of conduction of heat; their operational expressions for v are always p times our $\bar{v}$, in the notation used below, and the method of deducing the solution using the theory of contour integration is the same in both cases, so papers written in one notation may easily be read by persons used to the other.

12.2. The Laplace transformation. Fundamental properties

Here we shall always be concerned with the temperature v which is a function of t and whatever space coordinates, say x, y, z, occur in the problem. We write

$$L\{v(x,y,z,t)\} = \bar{v} = \int_0^\infty e^{-pt}v(x,y,z,t)\,dt, \qquad (1)$$

where p is a number whose real part is positive and large enough to make the integral (1) convergent.

[†] *C.H.*, Chaps. X and XI; see also Appendix I below. For an interesting comparison of the two methods see articles by Carslaw and Jeffreys in *Math. Gaz.* **14** (1928) 216, 225.

[‡] 'Probleme aus der Theorie der Wärmeleitung', *Math. Z.* **22** (1925) 285, 293; **25** (1926) 608; **26** (1927) 89; **28** (1928) 567.

[§] Carslaw and Jaeger, *Operational Methods in Applied Mathematics* (Oxford, 1948), referred to subsequently as *C. and J.*; Churchill, *Modern Operational Mathematics in Engineering* (McGraw-Hill, 1944); Doetsch, *Theorie und Anwendung der Laplace-Transformation* (Springer, 1937); Gardner and Barnes, *Transients in Linear Systems*, Vol. I (Wiley, 1942).

The integral (1) is called the Laplace transform of the function v and is a function of p and the space variables x, y, z. Of the two notations given, $L\{v\}$ is convenient when stating theorems, while $\bar{v}$ is a convenient short notation to use in the algebra of the solution.

The apparatus of the method consists of a few elementary theorems and a table of Laplace transforms, that is, of the integrals (1). For example,

if
$$v(t) = 1, \qquad \bar{v}(p) = \int_0^\infty e^{-pt}\, dt = \frac{1}{p},$$

or, if

$$v(t) = \sin \omega t, \qquad \bar{v}(p) = \int_0^\infty e^{-pt} \sin \omega t\, dt = \frac{\omega}{p^2+\omega^2}, \quad \text{etc.}$$

A collection of most of the Laplace transforms useful† in conduction of heat is given in Appendix V.

The theorems we need are given below; in most cases we indicate the proofs briefly without stating the conditions carefully. The exact conditions are in fact not needed, since, as remarked in § 12.3, the analysis at this stage is regarded as formal only and the result as subject to verification.

THEOREM I. $L\{v_1+v_2\} = L\{v_1\}+L\{v_2\}.$

THEOREM II. $L\left\{\dfrac{\partial v}{\partial t}\right\} = pL\{v\}-v_0,$ (2)

where v_0 is the value of $\lim\limits_{t\to+0} v$. In general v_0 will be a function of the space variables x, y, z. The result (2) follows immediately on integrating by parts, since

$$\int_0^\infty e^{-pt}\frac{\partial v}{\partial t}\, dt = [e^{-pt}v]_0^\infty + p\int_0^\infty e^{-pt}v\, dt = -v_0+p\bar{v}.$$

THEOREM III. $L\left\{\dfrac{\partial^n v}{\partial x^n}\right\} = \dfrac{\partial^n \bar{v}}{\partial x^n},$ (3)

with similar results for the other space variables. This is equivalent to

$$\int_0^\infty e^{-pt}\frac{\partial^n v}{\partial x^n}\, dt = \frac{\partial^n}{\partial x^n}\int_0^\infty e^{-pt}v\, dt,$$

---

† Larger tables are to be found in the works referred to above and also in Campbell and Foster, *Fourier Integrals for Practical Applications*, Bell System Technical Monographs, No. B-584, also McLachlan and Humbert, *Formulaire pour le calcul symbolique* (Gauthier–Villars, 1941), and Erdélyi, *Tables of Integral Transforms* (McGraw-Hill, 1954).

and we assume v to be such that the orders of integration and differentiation can be interchanged in this way.

The three theorems above are those of the greatest importance, but we give below a number of useful results which will be referred to occasionally.

THEOREM IV. $$L\left\{\int_0^t v(t')\,dt'\right\} = \frac{1}{p}\,L\{v\},$$ (4)

for, integrating by parts,

$$\int_0^\infty e^{-pt}\,dt\int_0^t v(t')\,dt' = -\left[\frac{1}{p}e^{-pt}\int_0^t v(t')\,dt'\right]_0^\infty + \frac{1}{p}\int_0^\infty e^{-pt}v\,dt = \frac{1}{p}\bar{v}.$$

THEOREM V. *If k is a positive constant, and*

$$L\{v(t)\} = \bar{v}(p), \quad then \quad L\{v(kt)\} = \frac{1}{k}\bar{v}\left(\frac{p}{k}\right),$$ (5)

for $$\int_0^\infty e^{-pt}v(kt)\,dt = \frac{1}{k}\int_0^\infty e^{-(p/k)t'}v(t')\,dt' = \frac{1}{k}\bar{v}\left(\frac{p}{k}\right).$$

THEOREM VI. *If a is any constant and $L\{v\} = \bar{v}(p)$, then*

$$L\{e^{-at}v\} = \bar{v}(p+a),$$ (6)

for $$\int_0^\infty e^{-pt}e^{-at}v\,dt = \int_0^\infty e^{-(p+a)t}v\,dt = \bar{v}(p+a).$$

THEOREM VII. *If $f(t) = H(t-t_0)\phi(t-t_0)$, where $H(t-t_0)$ is Heaviside's unit function defined by*

$$\left.\begin{aligned}H(t-t_0) &= 0, \quad t < t_0,\\ H(t-t_0) &= 1, \quad t > t_0,\end{aligned}\right\}$$ (7)

then $$L\{f(t)\} = e^{-pt_0}L\{\phi(t)\}.$$ (8)

THEOREM VIII. *If $f(t)$ is periodic with period T,*

$$L\{f(t)\} = \frac{1}{1-e^{-pT}}\int_0^T e^{-pt}f(t)\,dt,$$ (9)

for $$L\{f(t)\} = \int_0^\infty e^{-pt}f(t)\,dt = \sum_{n=0}^\infty \int_{nT}^{(n+1)T} f(t)e^{-pt}\,dt$$

$$= \sum_{n=0}^\infty e^{-npT}\int_0^T e^{-pt'}f(t')\,dt' = \frac{1}{1-e^{-pT}}\int_0^T e^{-pt}f(t)\,dt.$$

THEOREM IX. *Lerch's theorem or the uniqueness theorem.*†

If $L\{f_1(t)\} = L\{f_2(t)\}$ *for all* p, *then* $f_1(t) = f_2(t)$ *for all* $t \geqslant 0$, *if the functions are continuous; if the functions have only ordinary discontinuities they can only differ at these points.*

THEOREM X.

$$L\left\{ \int_0^t f_1(\tau) f_2(t-\tau)\, d\tau \right\} = L\left\{ \int_0^t f_2(\tau) f_1(t-\tau)\, d\tau \right\} = L\{f_1(t)\} L\{f_2(t)\}. \quad (10)$$

This is known as the Faltung or Superposition theorem, also as Duhamel's theorem. It is the expression of Duhamel's theorem, § 1.14, in the present notation.

THEOREM XI. *If* $L\{v(t)\} = \bar{v}(p),$

then
$$L\left\{ \frac{1}{\sqrt{(\pi t)}} \int_0^\infty e^{-u^2/4t} v(u)\, du \right\} = \frac{\bar{v}(\sqrt{p})}{\sqrt{p}}. \quad (11)$$

12.3. Solution of the equation of conduction of heat by the Laplace transformation method

Suppose that we have to solve the equation of linear flow

$$\frac{\partial^2 v}{\partial x^2} - \frac{1}{\kappa} \frac{\partial v}{\partial t} = 0, \quad a < x < b, \quad (1)$$

with $v = v_0(x), \quad t = 0, \quad a < x < b,$ (2)

and $v = v_1(t), \quad x = a, \quad t > 0,$ (3)

          $v = v_2(t), \quad x = b, \quad t > 0.$ (4)

We apply the Laplace transformation to (1), that is, multiply by e^{-pt} and integrate with respect to t from 0 to ∞. This gives

$$\int_0^\infty e^{-pt} \frac{\partial^2 v}{\partial x^2}\, dt - \frac{1}{\kappa} \int_0^\infty e^{-pt} \frac{\partial v}{\partial t}\, dt = 0.$$

Using 12.2 (2), (3), this becomes

$$\frac{d^2 \bar{v}}{dx^2} - \frac{p}{\kappa} \bar{v} = -\frac{1}{\kappa} v_0(x), \quad a < x < b. \quad (5)$$

The Laplace transformation thus reduces the partial differential equation (1) to the ordinary differential equation (5). The equation for $\bar{v}$ derived in this way we shall always refer to as the 'subsidiary

† Proofs of Theorems IX and X are to be found in standard works on the Laplace transformation. Theorem XI, and others of the same type, are given in Humbert, *Le Calcul Symbolique* (Actualités Scientifiques et Industrielles, No. 147, 1934).

equation'. The boundary conditions (3) and (4), treated in the same way, give

$$\bar{v} = \bar{v}_1, \quad x = a, \tag{6}$$

$$\bar{v} = \bar{v}_2, \quad x = b. \tag{7}$$

When the subsidiary equation (5) has been solved with the boundary conditions (6) and (7), the Laplace transform $\bar{v}$ of the solution of the problem is known. Before proceeding to the methods of finding v from $\bar{v}$ it may be remarked that more general differential equations, such as that of 4.10 (5), and more general boundary conditions, such as 1.9 (14), lead in precisely the same way to an ordinary differential equation with boundary conditions at a and b, and hence to the value of $\bar{v}$.

If there is more than one space variable, for example, if the general differential equation

$$\nabla^2 v - \frac{1}{\kappa}\frac{\partial v}{\partial t} = 0$$

has to be solved in some region with initial condition $v_0(x, y, z)$, and given boundary conditions, the subsidiary equation, found in the same way, is

$$\nabla^2 \bar{v} - \frac{p}{\kappa}\,\bar{v} = -\frac{1}{\kappa}\,v_0(x, y, z),$$

and is still a partial differential equation, but in three variables instead of four.

Suppose, now, that the subsidiary equation has been solved with the corresponding boundary conditions and thus $\bar{v}$ is known as a function of p (and the space variables). It is next required to find v as a function of the time from $\bar{v}$, and this will be the solution of the original problem.

The simplest method is to look up $\bar{v}(p)$ in the Table of Transforms and to pick out the corresponding function† of t. A large number of problems in linear flow may be solved very simply in this way; examples are given in §§ 12.4, 12.5.

If the transform $\bar{v}$ does not appear in the table, we determine v from $\bar{v}$ by the use of the *Inversion Theorem for the Laplace transformation*. This states that

$$v(t) = \frac{1}{2\pi i} \int\limits_{\gamma - i\infty}^{\gamma + i\infty} e^{\lambda t}\bar{v}(\lambda)\,d\lambda, \tag{8}$$

where γ is to be so large that all the singularities of $\bar{v}(\lambda)$ lie to the left of the line $(\gamma - i\infty, \gamma + i\infty)$. λ is written in place of p in (8) to emphasize the fact that in (8) we are considering the behaviour of $\bar{v}$ regarded as

† It follows from § 12.2, Theorem IX, that the function found in this way is unique.

a function of a complex variable, while in the previous discussion p need not have been complex at all.

There are conditions on $\bar{v}(\lambda)$ or $v(t)$ for the validity of (8), but neither these nor the proof† of the theorem need be considered here, since other assumptions have been made in the course of the work, for example

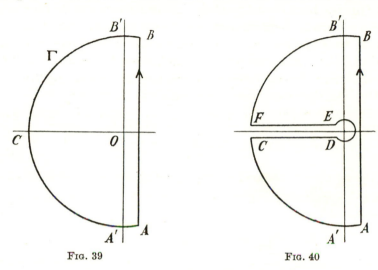

FIG. 39   FIG. 40

that v has a Laplace transform, that the orders of differentiation and integration can be interchanged in 12.2 (3), and so on. Thus from the point of view of strict pure mathematics it is necessary to verify that the solution we obtain does satisfy the differential equation and initial and boundary conditions of the problem. This verification process is most simply performed on the contour integral (8); the method is discussed in Appendix I; all the solutions obtained here can be verified by the process given there.

When the solution has been found as a contour integral (8) it may usually be put in real form by the use of one of two standard methods of procedure.

(i) If $\bar{v}(\lambda)$ is a single-valued function of λ with a row of poles along the negative real axis (and possibly other poles also) we complete the contour by a large circle Γ of radius R, not passing through any pole of the integrand (Fig. 39).

In all the problems we discuss here it can be shown that the integral over the large circle Γ vanishes in the limit as its radius R tends to ∞ (for further discussion see Appendix I). Thus, in the limit, the line

† For proofs see, e.g., *C. and J.*, § 29; Churchill, *Math. Z.* **42** (1937) 567.

integral (8) is equal by Cauchy's theorem to $2\pi i$ times the sum of the residues at the poles of its integrand.† This case usually arises in problems on conduction of heat in finite regions.

(ii) In problems on conduction of heat in semi-infinite regions, $\bar{v}(\lambda)$ usually has a branch point at $\lambda = 0$. In such cases we use the contour of Fig. 40 with a cut along the negative real axis so that $\bar{v}(\lambda)$ is a single-valued function of λ within and on the contour. In the limit as the radius of the large circle tends to infinity the integral round it can be shown to vanish, and the line integral (8) is replaced by a real infinite integral, derived from the integrals along CD and EF, together, possibly, with contributions from the small circle about the origin and any poles of the integrand.

If $\bar{v}(\lambda)$ does not belong to either of the above types, special methods must be devised for dealing with it.

12.4. The semi-infinite region $x > 0$. Solutions obtainable from the table of transforms

Here the subsidiary equation 12.3 (5) is

$$\frac{d^2\bar{v}}{dx^2} - q^2\bar{v} = -\frac{1}{\kappa}\, v_0(x), \quad x > 0, \tag{1}$$

where $v_0(x)$ is the initial temperature, and we write for shortness

$$q^2 = p/\kappa. \tag{2}$$

In this chapter we shall usually consider problems in which $v_0(x)$ is zero so that (1) becomes

$$\frac{d^2\bar{v}}{dx^2} - q^2\bar{v} = 0, \quad x > 0. \tag{3}$$

If $v_0(x)$ is a constant, the modification is trivial; if $v_0(x)$ is a simple function of x, explicit solutions of (1) are easily written down; if it is an arbitrary function, (1) must be solved by Variation of Parameters or some such method and results equivalent to those of Chapter II are obtained finally. However, from the present point of view they are most satisfactorily obtained by the use of the Green's function, Chapter XIV, so their discussion will be deferred to that chapter.

In all cases, v, and thus $\bar{v}$, is to be bounded as $x \to \infty$, so that of the two solutions Be^{+qx} and Ae^{-qx} of (1) or (3) we must use only Ae^{-qx}.

We now consider various boundary conditions at $x = 0$.

---

† This gives a series which would be obtained by Heaviside's expansion theorem.

I. *Prescribed surface temperature. Zero initial temperature*

Suppose we require

$$v = \phi(t), \quad x = 0, \quad t > 0,$$

so that

$$\bar{v} = \bar{\phi}(p), \quad x = 0.$$

The solution of (3) which is finite as $x \to \infty$ and has this value at $x = 0$ is

$$\bar{v} = \bar{\phi}(p)e^{-qx}. \tag{4}$$

(i) If $\phi(t) = V_0$, constant, $\bar{\phi} = V_0/p$, and (4) becomes

$$\bar{v} = \frac{V_0}{p} e^{-qx}.$$

It follows from Appendix V (8) that

$$v = V_0 \operatorname{erfc} \frac{x}{2(\kappa t)^{\frac{1}{2}}}. \tag{5}$$

(ii) If $\phi(t) = V_0 t^{\frac{1}{2}n}$, where n is any positive integer,

$$\bar{\phi} = \frac{V_0 \Gamma(1 + \frac{1}{2}n)}{p^{1 + \frac{1}{2}n}}$$

by Appendix V (2), then from Appendix V (11) we have

$$v = V_0 \Gamma(1 + \tfrac{1}{2}n)(4t)^{\frac{1}{2}n} \, i^n \operatorname{erfc} \frac{x}{2\sqrt{(\kappa t)}}. \tag{6}$$

Cf. 2.5 (5), (7), (8).

(iii) For arbitrary $\phi(t)$ it follows from 12.2 (10) and Appendix V (6) that

$$v = \frac{x}{2\sqrt{(\pi\kappa)}} \int_0^t \phi(\tau) \frac{e^{-x^2/4\kappa(t-\tau)} \, d\tau}{(t-\tau)^{\frac{3}{2}}}, \tag{7}$$

a result already obtained in 2.5 (1).

The result 2.5 (9) for surface temperature $e^{\lambda t}$ follows in the same way from Appendix V (19). Harmonic surface temperature is discussed in § 12.7, and periodic surface temperature in § 15.5.

II. *The radiation boundary condition*

Suppose the initial temperature of the solid is zero, and that it is heated at $x = 0$ by radiation from a medium at V, constant. The boundary condition at $x = 0$ is then

$$\frac{\partial v}{\partial x} = h(v - V), \quad x = 0, \quad t > 0.$$

The boundary condition for the subsidiary equation is

$$\frac{d\bar{v}}{dx} - h\bar{v} = -\frac{hV}{p}, \tag{8}$$

and the solution of (3) which satisfies (8) is

$$\bar{v} = \frac{hVe^{-qx}}{p(q+h)}.$$

Then from Appendix V (14) it follows that

$$v = V\operatorname{erfc}\frac{x}{2\sqrt{(\kappa t)}} - Ve^{hx+h^2\kappa t}\operatorname{erfc}\left(\frac{x}{2\sqrt{(\kappa t)}} + h\sqrt{(\kappa t)}\right).$$

III. *Contact with well-stirred fluid (or perfect conductor)*

The appropriate boundary conditions have been considered in § 1.9 F, but the solution of problems of this type has been deferred to this chapter since the present methods are particularly suitable for them. We write throughout c' for the specific heat of the fluid, u for its temperature, and M for the mass of fluid in contact with unit area of the surface $x = 0$ of the solid.

(i) *The initial temperature of the solid zero, and of the fluid V, constant.† The temperature of the fluid equal to the surface temperature of the solid for $t > 0$.*

Here the boundary conditions at $x = 0$ for $t > 0$ are $u = v$, and

$$Mc'\frac{du}{dt} - K\frac{\partial v}{\partial x} = 0, \quad x = 0, \quad t > 0. \tag{9}$$

Thus the corresponding boundary condition for the subsidiary equation is

$$Mc'p\bar{v} - K\frac{d\bar{v}}{dx} = Mc'V.$$

Then

$$\bar{v} = \frac{Ve^{-qx}}{\kappa q(q+h')},$$

where

$$h' = \frac{K}{Mc'\kappa} = \frac{\rho c}{Mc'}. \tag{10}$$

Therefore, by Appendix V (13),

$$v = Ve^{h'x+h'^2\kappa t}\operatorname{erfc}\left\{\frac{x}{2\sqrt{(\kappa t)}} + h'\sqrt{(\kappa t)}\right\}. \tag{11}$$

(ii) *The same problem, except that the initial temperature of the fluid is zero, and heat is supplied to the fluid at constant rate Q/M per unit mass per unit time.*

In this case the boundary conditions at $x = 0$ are $u = v$, and

$$Mc'\frac{du}{dt} - K\frac{\partial v}{\partial x} = Q, \quad x = 0, \quad t > 0,$$

leading to

$$Mc'p\bar{v} - K\frac{d\bar{v}}{dx} = \frac{Q}{p}, \quad x = 0.$$

Thus

$$\bar{v} = \frac{Qe^{-qx}}{Mc'\kappa pq(q+h')},$$

where h' is given by (10). It follows from Appendix V (15) that

$$v = \frac{2Q}{K}\left(\frac{\kappa t}{\pi}\right)^{\frac{1}{2}}e^{-x^2/4\kappa t} - \frac{Q(1+h'x)}{Kh'}\operatorname{erfc}\frac{x}{2\sqrt{(\kappa t)}} + \frac{Q}{Kh'}e^{h'x+h'^2\kappa t}\operatorname{erfc}\left[\frac{x}{2\sqrt{(\kappa t)}} + h'\sqrt{(\kappa t)}\right]. \tag{12}$$

† Tränkle, *Frequenz*, 8 (1954) 334–8.

(iii) *The problem of* (i) *except that there is heat transfer between the fluid and the surface of the solid at a rate H times their temperature difference.*

Writing u for the temperature of the fluid, the boundary conditions at $x = 0$ now are

$$Mc'\frac{du}{dt} + H(u-v) = 0, \quad x = 0, \quad t > 0, \tag{13}$$

$$-K\frac{\partial v}{\partial x} = H(u-v), \quad x = 0, \quad t > 0. \tag{14}$$

The differential equation for v has to be solved with this boundary condition, and with the initial values V of u, and zero of v. The boundary conditions, derived from (13) and (14), for the subsidiary equation in v are

$$\left.\begin{array}{c} (Mc'p+H)\bar{u} - H\bar{v} = Mc'V \\[2mm] -K\dfrac{d\bar{v}}{dx} + H\bar{v} - H\bar{u} = 0 \end{array}\right\}.$$

It follows in the usual way that

$$\bar{u} = \frac{V(q+h)}{\kappa q(q^2+hq+hh')}, \tag{15}$$

$$\bar{v} = \frac{hVe^{-qx}}{\kappa q(q^2+hq+hh')}, \tag{16}$$

where h' is given by (10), and $h = H/K$. If we write

$$q^2+hq+hh' = (q+\alpha)(q+\beta), \tag{17}$$

(16) becomes

$$\bar{v} = \frac{hVe^{-qx}}{\kappa(\beta-\alpha)q}\left\{\frac{1}{q+\alpha} - \frac{1}{q+\beta}\right\},$$

and, by Appendix V (13),

$$v = \frac{hV}{\beta-\alpha}\left[e^{\alpha x + \kappa t\alpha^2}\operatorname{erfc}\left\{\frac{x}{2\sqrt{(\kappa t)}} + \alpha\sqrt{(\kappa t)}\right\} - e^{\beta x + \kappa t\beta^2}\operatorname{erfc}\left\{\frac{x}{2\sqrt{(\kappa t)}} + \beta\sqrt{(\kappa t)}\right\}\right]. \tag{18}$$

IV. *The semi-infinite solid with heat generated within it*

The present method is especially suitable for such problems, particularly in rather complicated cases. The results given in § 2.11 can all be derived in this way. Here, to illustrate the method, we give some additional ones.

(i) *The region $x > 0$ with zero initial temperature. At $x = 0$ there is radiation into a medium at zero temperature. Heat is produced in the solid for $t > 0$ at the rate $kt^{\frac{1}{2}n}$ per unit time per unit volume, where n may be $-1, 0$, or any positive integer.*

Here the differential equation is

$$\frac{\partial^2 v}{\partial x^2} - \frac{1}{\kappa}\frac{\partial v}{\partial t} = -\frac{k}{K}t^{\frac{1}{2}n}, \quad x > 0, \quad t > 0, \tag{19}$$

with

$$\frac{\partial v}{\partial x} - hv = 0, \quad x = 0, \quad t > 0. \tag{20}$$

Using Appendix V (2) the corresponding subsidiary equation is

$$\frac{d^2\bar{v}}{dx^2} - q^2\bar{v} = -\frac{k\Gamma(1+\frac{1}{2}n)}{Kp^{1+\frac{1}{2}n}}, \quad x > 0, \tag{21}$$

with

$$\frac{d\bar{v}}{dx} - h\bar{v} = 0, \quad x = 0, \tag{22}$$

and, as always, $\bar{v}$ finite as $x \to \infty$. The solution of (21) and (22) is

$$\bar{v} = \frac{\kappa k\Gamma(1+\frac{1}{2}n)}{Kp^{2+\frac{1}{2}n}}\left\{1 - \frac{he^{-qx}}{q+h}\right\}.$$

Thus, using Appendix V (16),

$$v = \frac{k\kappa t^{1+\frac{1}{2}n}}{K(1+\frac{1}{2}n)} + \frac{k\Gamma(1+\frac{1}{2}n)}{K\kappa^{\frac{1}{2}n}(-h)^{n+2}}\left[e^{hx+h^2\kappa t}\operatorname{erfc}\left(\frac{x}{2\sqrt{(\kappa t)}}+h\sqrt{(\kappa t)}\right) - \right.$$
$$\left. - \sum_{r=0}^{n+2}\{-2h\sqrt{(\kappa t)}\}^r\,\mathrm{i}^r\operatorname{erfc}\frac{x}{2\sqrt{(\kappa t)}}\right]. \quad (23)$$

A particular case of this, $n = -1$, corresponding to heat production at a rate proportional to $t^{-\frac{1}{2}}$, gives a rough approximation useful in cases such as the hydrating of cement where the initial rate of heat generation is high.

(ii) *The region* $x > 0$. $x = 0$ *kept at zero for* $t > 0$. *Heat is produced†* at a constant rate k in the region $a < x < b$ for $t > 0$.

Here the subsidiary equations are

$$\frac{d^2\bar{v}}{dx^2} - q^2\bar{v} = 0, \quad \text{in } 0 < x < a \text{ and } x > b, \quad (24)$$

and

$$\frac{d^2\bar{v}}{dx^2} - q^2\bar{v} = -\frac{k}{Kp}, \quad \text{in } a < x < b. \quad (25)$$

These have to be solved with

$$\bar{v} = 0, \quad x = 0, \quad (26)$$

and $\bar{v}$ and $d\bar{v}/dx$ continuous at $x = a$ and $x = b$. (27)

The solutions of (24), (25), and (26) are

$$\bar{v} = Be^{qx} - Be^{-qx}, \quad 0 < x < a,$$

$$\bar{v} = \frac{\kappa k}{Kp^2} + Ce^{qx} + De^{-qx}, \quad a < x < b,$$

$$\bar{v} = Ee^{-qx}, \quad x > b,$$

where the four constants B, C, D, E follow from the four continuity conditions (27). On evaluating these the solution can be written down. For example the temperature gradient at the surface $[\partial v/\partial x]_{x=0}$ has transform

$$2Bq = \frac{k}{Kpq}(e^{-qa} - e^{-qb}),$$

and thus by Appendix V (11) is

$$\frac{2k(\kappa t)^{\frac{1}{2}}}{K}\left\{\operatorname{ierfc}\frac{a}{2\sqrt{(\kappa t)}} - \operatorname{ierfc}\frac{b}{2\sqrt{(\kappa t)}}\right\}. \quad (28)$$

12.5. The finite region $0 < x < l$. Solutions obtainable from the table of transforms. Solutions for small values of the time

I. We consider first *the region* $0 < x < l$ *with zero initial temperature, no flow of heat at* $x = 0$, *and* $x = l$ *maintained at temperature* V, *constant, for* $t > 0$

The subsidiary equation now is

$$\frac{d^2\bar{v}}{dx^2} - q^2\bar{v} = 0, \quad 0 < x < l,$$

† Van Orstrand, *J. Wash. Acad. Sci.* **22** (1932) 529.

with
$$\frac{d\bar{v}}{dx} = 0, \qquad x = 0,$$

and
$$\bar{v} = V/p, \qquad x = l.$$

The solution of these is

$$\bar{v} = \frac{V \cosh qx}{p \cosh ql}. \tag{1}$$

Since (1) does not appear in the Table of Transforms† we derive the solution v by the use of the inversion theorem, 12.3 (8); this will be done in § 12.6 and yields the usual result 3.4 (2). In this section a device is used which gives an alternative form of solution by the use of the Table of Transforms. This form is often more useful than those given earlier, especially for small values of the time.

We express the hyperbolic functions in (1) in terms of negative exponentials, and expand in a series by the binomial theorem. Thus from (1) we get

$$\bar{v} = \frac{V(e^{qx} + e^{-qx})}{pe^{ql}(1 + e^{-2ql})}$$

$$= \frac{V}{p}(e^{-q(l-x)} + e^{-q(l+x)}) \sum_{n=0}^{\infty} (-1)^n e^{-2nql}$$

$$= \frac{V}{p} \sum_{n=0}^{\infty} (-1)^n e^{-q[(2n+1)l-x]} + \frac{V}{p} \sum_{n=0}^{\infty} (-1)^n e^{-[(2n+1)l+x]}. \tag{2}$$

Thus, using Appendix V (8), we get the result already quoted in 3.3 (9),

$$v = V \sum_{n=0}^{\infty} (-1)^n \operatorname{erfc} \frac{(2n+1)l-x}{2\sqrt{(\kappa t)}} + V \sum_{n=0}^{\infty} (-1)^n \operatorname{erfc} \frac{(2n+1)l+x}{2\sqrt{(\kappa t)}}. \tag{3}$$

This series‡ converges quite rapidly except for large values of $\kappa t/l^2$. It is thus complementary to the solution 3.4 (2) which is most rapidly convergent for large values of the time. Within the range of medium $\kappa t/l^2$ in which both (3) and 3.4 (2) are available, the former is probably

---

† It is in fact the Laplace transform of a function which can be expressed in terms of theta-functions. For some examples see Doetsch, *Theorie und Anwendung der Laplace-Transformation* (Springer, 1937) Kap. 20.

‡ The successive terms of the series are the temperatures at depths $l-x$, $l+x$, $3l-x$, $3l+x$,... in the semi-infinite solid. This result is thus analogous to the solutions of problems on vibrations of strings, rods, and transmission lines, in terms of multiply reflected waves.

It may be remarked that these solutions for small values of the time may also be regarded as solutions for small changes in average temperature, cf. Lochs, *Z. angew. Math. Mech.* **34** (1954) 79–80.

a little simpler to use. For example, if $\kappa t/l^2 = 1$ we have from (3) for $x = 0$

$$\frac{v}{2V} = \text{erfc}\,\tfrac{1}{2} - \text{erfc}\,\tfrac{3}{2} + \text{erfc}\,\tfrac{5}{2} - \ldots$$

$$= 0\cdot4795 - 0\cdot0339 + 0\cdot0004 - \ldots = 0\cdot4460.$$

If u is the average temperature of the slab,

$$\bar{u} = \frac{1}{l}\int_0^l \bar{v}\,dx = \frac{V\sinh ql}{lpq\cosh ql} = \frac{V}{lpq}\Big\{1 + 2\sum_{n=1}^{\infty}(-1)^n e^{-2nql}\Big\}.$$

$$u = 2V\Big(\frac{\kappa t}{l^2}\Big)^{\frac{1}{2}}\Big\{\pi^{-\frac{1}{2}} + 2\sum_{n=1}^{\infty}(-1)^n\,\text{ierfc}\,\frac{nl}{\sqrt{(\kappa t)}}\Big\}, \tag{4}$$

using Appendix V (11).

II. *The region $0 < x < l$. Zero initial temperature. $x = 0$ maintained at zero, and $x = l$ at V, constant, for $t > 0$*

Here
$$\bar{v} = \frac{V\sinh qx}{p\sinh ql}. \tag{5}$$

$$v = V\sum_{n=0}^{\infty}\Big\{\text{erfc}\,\frac{(2n+1)l-x}{2\sqrt{(\kappa t)}} - \text{erfc}\,\frac{(2n+1)l+x}{2\sqrt{(\kappa t)}}\Big\}. \tag{6}$$

The total quantity of heat which has crossed the plane $x = 0$ from time $t = 0$ to $t = t$ is

$$\frac{4KVt^{\frac{1}{2}}}{\kappa^{\frac{1}{2}}}\sum_{n=0}^{\infty}\text{ierfc}\,\frac{(2n+1)l}{2\sqrt{(\kappa t)}}. \tag{7}$$

III. *The region $0 < x < l$. Zero initial temperature. No flow of heat at $x = 0$. Constant flux Q into the solid† at $x = l$*

Here
$$\bar{v} = \frac{Q\cosh qx}{Kpq\sinh ql}. $$

$$v = \frac{2Q\sqrt{(\kappa t)}}{K}\sum_{n=0}^{\infty}\Big\{\text{ierfc}\,\frac{(2n+1)l-x}{2\sqrt{(\kappa t)}} + \text{ierfc}\,\frac{(2n+1)l+x}{2\sqrt{(\kappa t)}}\Big\}. \tag{8}$$

IV. *The region $0 < x < l$. The initial temperature V_0, constant. No flow of heat at $x = 0$. At $x = l$ radiation into a medium at zero*

Here the subsidiary equation 12.3 (5) is

$$\frac{d^2\bar{v}}{dx^2} - q^2\bar{v} = -\frac{V_0}{\kappa}, \quad 0 < x < l,$$

with
$$\frac{d\bar{v}}{dx} = 0, \quad x = 0,$$

$$\frac{d\bar{v}}{dx} + h\bar{v} = 0, \quad x = l.$$

† Macey, *Proc. Phys. Soc.* **52** (1940) 625.

The solution of these is

$$\bar{v} = \frac{V_0}{p} - \frac{hV_0 \cosh qx}{p(q \sinh ql + h \cosh ql)} \tag{9}$$

$$= \frac{V_0}{p} - \frac{hV_0 e^{-ql}(e^{qx} + e^{-qx})}{p\{(h+q) + (h-q)e^{-2ql}\}}$$

$$= \frac{V_0}{p} - \frac{hV_0}{p(q+h)}[e^{-q(l-x)} + e^{-q(l+x)}] + \frac{hV_0(h-q)}{p(h+q)^2}[e^{-q(3l-x)} + e^{-q(3l+x)}] + \dots \tag{10}$$

Here the successive exponentials in the series (10) have coefficients which are complicated functions of q, and there is no simple series for v. But the first few terms of v can be written down immediately and give a solution† useful for small values of the time. Thus we have from Appendix V (14)

$$v = V_0 - V_0\left[\operatorname{erfc}\frac{l-x}{2\sqrt{(\kappa t)}} + \operatorname{erfc}\frac{l+x}{2\sqrt{(\kappa t)}}\right] + V_0 e^{h(l-x)+h^2\kappa t}\operatorname{erfc}\left(h\sqrt{(\kappa t)} + \frac{l-x}{2\sqrt{(\kappa t)}}\right) +$$

$$+ V_0 e^{h(l+x)+h^2\kappa t}\operatorname{erfc}\left(h\sqrt{(\kappa t)} + \frac{l+x}{2\sqrt{(\kappa t)}}\right) + \dots. \tag{11}$$

The same procedure is applicable to the problems of §§ 3.11, 3.13.

V. *The region $0 < x < l$. Zero initial temperature. $x = 0$ and $x = l$ maintained at zero for $t > 0$. Heat is produced‡ in the solid for $t > 0$ at the rate* $at^{\frac{1}{2}n}$ *per unit time per unit volume, where* $n = -1, 0, 1, 2, \dots$.

Here the differential equation, 3.14 (1), is

$$\frac{\partial^2 v}{\partial x^2} - \frac{1}{\kappa}\frac{\partial v}{\partial t} = -\frac{a}{K}t^{\frac{1}{2}n}, \quad 0 < x < l, \quad t > 0.$$

The subsidiary equation is

$$\frac{d^2\bar{v}}{dx^2} - q^2\bar{v} = -\frac{a\Gamma(1+\frac{1}{2}n)}{Kp^{1+\frac{1}{2}n}}.$$

This has to be solved with $\bar{v} = 0$ when $x = 0$ and $x = l$. The solution is

$$\bar{v} = \frac{a\kappa\Gamma(1+\frac{1}{2}n)}{Kp^{2+\frac{1}{2}n}}\left\{1 - \frac{\cosh q(\frac{1}{2}l - x)}{\cosh \frac{1}{2}ql}\right\}$$

$$= \frac{a\kappa\Gamma(1+\frac{1}{2}n)}{Kp^{2+\frac{1}{2}n}}\left\{1 - \sum_{m=0}^{\infty}(-1)^m[e^{-q(x+ml)} + e^{-q\{(m+1)l-x\}}]\right\}.$$

$$v = \frac{a\kappa t^{1+\frac{1}{2}n}}{K(1+\frac{1}{2}n)}\left\{1 - \Gamma(2+\frac{1}{2}n)2^{n+2}\sum_{m=0}^{\infty}(-1)^m\left[i^{n+2}\operatorname{erfc}\frac{ml+x}{2\sqrt{(\kappa t)}} + i^{n+2}\operatorname{erfc}\frac{(m+1)l-x}{2\sqrt{(\kappa t)}}\right]\right\}.$$
$$\tag{12}$$

12.6. The finite region $0 < x < l$. Application of the Inversion theorem

I. Considering again the first problem of § 12.5 we proceed to determine v from its transform, 12.5 (1), namely

$$\bar{v} = \frac{V \cosh qx}{p \cosh ql}. \tag{1}$$

† Goldstein, *Z. angew. Math. Mech.* **12** (1932) 234.
‡ Hartree, *Mem. Manchr. Lit. Phil. Soc.* **80** (1935) 85.

The Inversion theorem, 12.3 (8), gives

$$v = \frac{V}{2\pi i} \int_{\gamma-i\infty}^{\gamma+i\infty} \frac{e^{\lambda t} \cosh \mu x \, d\lambda}{\lambda \cosh \mu l}, \tag{2}$$

where γ is to be so large that all the singularities of $\bar{v}(\lambda)$ lie to the left of the line $(\gamma-i\infty, \gamma+i\infty)$, and here, and always in the sequel, we write

$$\mu = \sqrt{\left(\frac{\lambda}{\kappa}\right)}, \tag{3}$$

where the principal value of the square root is taken.

$\bar{v}(\lambda)$ has simple poles at $\lambda = 0$ and at those values of λ which make $\cosh \mu l$ zero. These are given by

$$\mu = \frac{(2n+1)i\pi}{2l}, \quad n = 0, 1, 2,..., \tag{4}$$

that is

$$\lambda = -\frac{\kappa(2n+1)^2\pi^2}{4l^2}, \quad n = 0, 1, 2,.... \tag{5}$$

Since

$$\cosh \mu x = 1 + \frac{\lambda x^2}{2\kappa} + \frac{\lambda^2 x^4}{(4!)\kappa^2} + \cdots$$

is a single-valued function of λ, it follows that $\bar{v}(\lambda)$ is a single-valued function[†] of λ and we use the contour of Fig. 39, where the large circle is not to pass through any of the poles (5) of the integrand. For example, we might take the radius of this circle to be $\kappa m^2\pi^2/l^2$, where m is any large integer.

It follows from Cauchy's theorem that the line integral round the contour is equal to $2\pi i$ times the sum of the residues of the integrand at its poles within the contour.

It is easy to show (cf. Appendix I) that as the radius of the large circle tends to infinity, the integral over the arc ACB of its circumference tends to zero. Thus in the limit the line integral (2) is equal to $2\pi i$ times the sum of the residues of its integrand at the poles $\lambda = 0$ and (5) of its integrand.

The residue at $\lambda = 0$ is unity.

The residue at the pole $\lambda = -\kappa(2n+1)^2\pi^2/4l^2$ is

$$\frac{e^{-\kappa(2n+1)^2\pi^2 t/4l^2} \cosh[i(2n+1)\pi x/2l]}{[\lambda(d/d\lambda)\{\cosh l\sqrt{(\lambda/\kappa)}\}]_{\lambda=-\kappa(2n+1)^2\pi^2/4l^2}}$$

$$= \frac{4(-1)^{n+1}}{\pi(2n+1)} e^{-\kappa(2n+1)^2\pi^2 t/4l^2} \cos \frac{(2n+1)\pi x}{2l}.$$

† All the transforms appearing in this section may be shown to be single-valued functions of λ by using the series for the hyperbolic functions in this way.

Using these results in (2) we have finally

$$v = V - \frac{4V}{\pi} \sum_{n=0}^{\infty} \frac{(-1)^n}{2n+1} e^{-\kappa(2n+1)^2\pi^2t/4l^2} \cos\frac{(2n+1)\pi x}{2l}. \tag{6}$$

If the temperature at $x = l$ is a function of time, $\phi(t)$, we have in place of (1),

$$\bar{v} = \phi \frac{\cosh qx}{\cosh ql}.$$

In this case we find the function $f(t)$ whose Laplace transform is

$$\frac{\cosh qx}{\cosh ql}$$

and use 12.2 (10). Using the contour of Fig. 39 as above we get

$$f(t) = \frac{\kappa\pi}{l^2} \sum_{n=0}^{\infty} (-1)^n(2n+1)e^{-\kappa(2n+1)^2\pi^2t/l^2}\cos\frac{(2n+1)\pi x}{2l}.$$

Thus, from 12.2 (10),

$$v = \frac{\kappa\pi}{l^2} \sum_{n=0}^{\infty} e^{-\kappa(2n+1)^2\pi^2t/l^2}(-1)^n(2n+1)\cos\frac{(2n+1)\pi x}{2l} \int_0^t e^{\kappa(2n+1)^2\pi^2t'/4l^2}\phi(t')\,dt', \tag{7}$$

as in 3.5 (3). For linearly increasing or harmonic surface temperature, v is best found directly from $\bar{v}$.

II. The problem of § 12.5 II, namely *the region $0 < x < l$ with zero initial temperature, and with $x = 0$ maintained at zero and $x = l$ at V for $t > 0$.* Here $\bar{v}$ is given by 12.5 (5), and the Inversion theorem gives

$$v = \frac{V}{2\pi i} \int_{\gamma-i\infty}^{\gamma+i\infty} \frac{e^{\lambda t}\sinh\mu x\,d\lambda}{\lambda\sinh\mu l}. \tag{8}$$

The poles of the integrand of (8) are $\lambda = 0$, with residue x/l, and

$$\mu = n\pi i/l, \qquad n = 1, 2,...,$$

i.e.

$$\lambda = -\kappa n^2\pi^2/l^2, \qquad n = 1, 2, 3,..., \tag{9}$$

with residue

$$\frac{2(-1)^n e^{-\kappa n^2\pi^2t/l^2}\sin(n\pi x/l)}{n\pi}.$$

Thus, using Fig. 39 as before, we find

$$v = \frac{Vx}{l} + \frac{2V}{\pi} \sum_{n=1}^{\infty} \frac{(-1)^n}{n} e^{-\kappa n^2\pi^2t/l^2}\sin\frac{n\pi x}{l}. \tag{10}$$

If Q is the total quantity of heat per unit area which has crossed the plane $x = 0$ from $t = 0$ to $t = t$, so that

$$Q = K \int_0^t \left[\frac{\partial v}{\partial x} \right]_{x=0} dt,$$

we have from 12.5 (5) and 12.2 (4),

$$\bar{Q} = \frac{KVq}{p^2 \sinh ql}. \tag{11}$$

Thus

$$Q = \frac{KV}{2\pi i} \int_{\gamma-i\infty}^{\gamma+i\infty} \frac{e^{\lambda t} \mu \, d\lambda}{\lambda^2 \sinh \mu l}. \tag{12}$$

The integrand of (12) has a double pole at $\lambda = 0$ of residue

$$\frac{1}{l} \left[t - \frac{l^2}{6\kappa} \right]. \tag{13}$$

The other poles are given by (9), and, proceeding as before, we find

$$Q = \frac{KV}{l} \left\{ t - \frac{l^2}{6\kappa} - \frac{2l^2}{\kappa \pi^2} \sum_{n=1}^{\infty} \frac{(-1)^n}{n^2} e^{-\kappa n^2 \pi^2 t/l^2} \right\}. \tag{14}$$

For large values of the time the terms in (14) involving exponentials are negligible, and Q has the value corresponding to steady flow for a time $[t - (l^2/6\kappa)]$. By measuring $l^2/6\kappa$, the time delay in establishing the steady state, κ can be determined experimentally.[†] If this quantity is all that is wanted, only the residue (13) at the pole $\lambda = 0$ is needed. For further discussion of this method see § 15.6.

III. *The region $0 < x < l$. Zero initial temperature. $x = 0$ and $x = l$ kept at zero for $t > 0$. Heat production at the rate $ae^{-\alpha t}$ per unit time per unit volume for $t > 0$*

Here the differential equation is

$$\frac{\partial^2 v}{\partial x^2} - \frac{1}{\kappa} \frac{\partial v}{\partial t} = -\frac{a}{K} e^{-\alpha t}, \quad 0 < x < l, \quad t > 0. \tag{15}$$

Thus the subsidiary equation is

$$\frac{d^2 \bar{v}}{dx^2} - q^2 \bar{v} = -\frac{a}{K(p+\alpha)}. \tag{16}$$

This has to be solved with $\bar{v} = 0$, when $x = 0$ and $x = l$. The solution is

$$\bar{v} = \frac{a\kappa}{Kp(p+\alpha)} \left\{ 1 - \frac{\cosh q(\tfrac{1}{2}l - x)}{\cosh \tfrac{1}{2}ql} \right\}. \tag{17}$$

† Barrer, *Trans. Faraday Soc.* **35** (1939) 628; **36** (1940) 1235; *Phil. Mag.* (7) **28** (1939) 148; (7) **35** (1944) 802.

Thus
$$v = \frac{a\kappa}{K\alpha}(1-e^{-\alpha t}) - \frac{a\kappa}{2\pi i K}\int_{\gamma-i\infty}^{\gamma+i\infty}\frac{e^{\lambda t}\cosh\mu(\tfrac{1}{2}l-x)\,d\lambda}{\lambda(\lambda+\alpha)\cosh\tfrac{1}{2}\mu l}. \tag{18}$$

The poles of the integrand are

$\lambda = 0$, with residue $1/\alpha$,

$\lambda = -\alpha$, with residue $-\dfrac{e^{-\alpha t}\cos(\tfrac{1}{2}l-x)(\alpha/\kappa)^{\frac{1}{2}}}{\alpha\cos\tfrac{1}{2}l(\alpha/\kappa)^{\frac{1}{2}}}$,

$\lambda = -[\kappa(2n+1)^2\pi^2/l^2]$, $n = 0, 1, 2,...$, with residue

$$\frac{4l^2\sin[(2n+1)\pi x/l]e^{-\kappa(2n+1)^2\pi^2 t/l^2}}{\pi(2n+1)[\kappa(2n+1)^2\pi^2-\alpha l^2]}. \tag{19}$$

If α happens to have the value zero or one of the values $\kappa(2n+1)^2\pi^2/l^2$, there is a double pole for $\lambda = -\alpha$, and a separate calculation must be made. If this is not the case we have finally

$$v = \frac{a\kappa}{K\alpha}\left(\frac{\cos(\tfrac{1}{2}l-x)(\alpha/\kappa)^{\frac{1}{2}}}{\cos\tfrac{1}{2}l(\alpha/\kappa)^{\frac{1}{2}}}-1\right)e^{-\alpha t} -$$

$$-\frac{4a\kappa l^2}{\pi K}\sum_{n=0}^{\infty}\frac{e^{-\kappa(2n+1)^2\pi^2 t/l^2}}{(2n+1)[\kappa(2n+1)^2\pi^2-\alpha l^2]}\sin\frac{(2n+1)\pi x}{l}. \tag{20}$$

The case of a constant rate of heat production may be treated in the same way. The result for heat production at a rate which is any function of the time is obtained from 12.2 (10) as in (7).

IV. *The radiation boundary condition*

As an example consider the problem of § 12.5 IV. From the value 12.5 (9) of $\bar{v}$ and the Inversion theorem, the solution is found to be

$$v = V_0 - \frac{hV_0}{2\pi i}\int_{\gamma-i\infty}^{\gamma+i\infty}\frac{e^{\lambda t}\cosh\mu x\,d\lambda}{\lambda(\mu\sinh\mu l+h\cosh\mu l)}. \tag{21}$$

Here the poles of the integrand are $\lambda = 0$, with residue $(1/h)$, and the values

$$\lambda = -\kappa\alpha_n^2, \quad \mu = i\alpha_n, \quad n = 1, 2, 3,..., \tag{22}$$

where $\pm\alpha_n$, $n = 1, 2,...$, are the roots (all real and simple, cf. § 3.10) of

$$\alpha\sin\alpha l = h\cos\alpha l$$

or
$$\alpha\tan\alpha l = h. \tag{23}$$

To find the residues at these poles we need

$$\left[\lambda \frac{d}{d\lambda}(\mu \sinh \mu l + h \cosh \mu l)\right]_{\lambda=-\kappa\alpha_n^2} = \left[\tfrac{1}{2}\mu \frac{d}{d\mu}(\mu \sinh \mu l + h \cosh \mu l)\right]_{\mu=i\alpha_n}$$

$$= -\tfrac{1}{2}\alpha_n\{(1+lh)\sin l\alpha_n + l\alpha_n \cos l\alpha_n\}$$

$$= -\tfrac{1}{2}\{l(h^2+\alpha_n^2)+h\}\cos l\alpha_n.$$

Using this result in (21) we have finally

$$v = 2hV_0 \sum_{n=1}^{\infty} \frac{e^{-\kappa\alpha_n^2 t}\cos \alpha_n x}{[l(h^2+\alpha_n^2)+h]\cos \alpha_n l}. \tag{24}$$

This has been found in 3.11 (12). All the results of §§ 3.11–3.13 may be derived in the same way.

V. *Other boundary conditions: contact with well-stirred fluid*

It was remarked in § 1.9 F that problems in which a solid is in contact with well-stirred fluid at its surface are of some practical importance. Problems of this type for the slab are most simply treated by a routine application of the Laplace transformation method: the classical methods cannot always be used without modification. A number of solutions has already been given without proof in § 3.13.

As an example which illustrates the new features of such problems we consider the following: *The region $0 < x < l$ with no flow of heat at $x = l$, and at $x = 0$ contact with mass M' per unit area of well-stirred fluid of specific heat c'. There is heat transfer between the fluid and the surface of the solid at a rate H times their temperature difference. The initial temperature of the fluid is V_0 and that of the solid zero.*

Writing u for the temperature of the fluid, and v for that in the solid, the boundary conditions 1.9 (14) and 1.9 (16) at $x = 0$ are

$$\left. \begin{array}{ll} M'c' \dfrac{du}{dt} - K \dfrac{\partial v}{\partial x} = 0, & x = 0, \quad t > 0 \\[2mm] K \dfrac{\partial v}{\partial x} + H(u-v) = 0, & x = 0, \quad t > 0 \end{array} \right\}. \tag{25}$$

v satisfies the differential equation

$$\frac{\partial^2 v}{\partial x^2} - \frac{1}{\kappa}\frac{\partial v}{\partial t} = 0, \quad 0 < x < l, \quad t > 0,$$

and the boundary condition at $x = l$ is

$$\frac{\partial v}{\partial x} = 0, \quad x = l, \quad t > 0.$$

The initial conditions are

$$u = V_0, \quad t = 0$$

and
$$v = 0, \quad 0 < x < l, \quad t > 0.$$

Forming the subsidiary equations corresponding to the differential equation and initial and boundary conditions in the usual way, we have

$$
\left.\begin{aligned}
\frac{d^2\bar{v}}{dx^2} - q^2\bar{v} &= 0, \quad 0 < x < l \\
\frac{d\bar{v}}{dx} &= 0, \quad x = l \\
M'c'p\bar{u} - K\frac{d\bar{v}}{dx} &= M'c'V_0, \quad x = 0 \\
K\frac{d\bar{v}}{dx} + H(\bar{u} - \bar{v}) &= 0, \quad x = 0
\end{aligned}\right\}, \tag{26}
$$

where, because of the occurrence of du/dt in the boundary condition (25), a term $M'cV_0$ involving the initial value of u has appeared in the right-hand side of (26).

Solving these in the usual way we get

$$
\bar{v} = \frac{hV_0\cosh q(l-x)}{hp\cosh ql + [p + (hK/M'c')]q\sinh ql}, \tag{27}
$$

$$
\bar{u} = \frac{V_0(h\cosh ql + q\sinh ql)}{hp\cosh ql + [p + (hK/M'c')]q\sinh ql}, \tag{28}
$$

where $h = H/K$. v and u are now found in the usual way from the Inversion theorem to be:

$$
v = \frac{V_0}{1+k} - 2LV_0\sum_{s=1}^{\infty}\frac{\alpha_s^2 - kL}{P_s\cos\alpha_s}e^{-\alpha_s^2 T}\cos\alpha_s(1 - x/l), \tag{29}
$$

$$
u = \frac{V_0}{1+k} + 2kL^2V_0\sum_{s=1}^{\infty}\frac{1}{P_s}e^{-\alpha_s^2 T}, \tag{30}
$$

where $L = lh$, $k = lpc/M'c'$, $T = \kappa t/l^2$, the α_s are the positive roots of

$$
\tan\alpha = \frac{L\alpha}{\alpha^2 - kL},
$$

and $P_s = \alpha_s^4 + (L^2 + L - 2kL)\alpha_s^2 + kL^2(1+k)$.

12.7. The semi-infinite region $x > 0$. Application of the Inversion theorem

As an example of a problem in which the Table of Transforms does not lead to an immediate answer, we consider the case of *the semi-infinite solid $x > 0$ with zero initial temperature, and with its surface $x = 0$ maintained at $V_0\sin(\omega t + \epsilon)$ for $t > 0$.*

Here, the subsidiary equation is

$$
\frac{d^2\bar{v}}{dx^2} - q^2\bar{v} = 0, \quad x > 0. \tag{1}
$$

At $x = 0$ we are to have

$$
v = V_0(\sin\omega t\cos\epsilon + \cos\omega t\sin\epsilon),
$$

so that
$$\bar{v} = \frac{V_0(\omega \cos \epsilon + p \sin \epsilon)}{p^2 + \omega^2}. \tag{2}$$

The solution of (1) and (2) which is finite as $x \to \infty$ is

$$\bar{v} = \frac{V_0(\omega \cos \epsilon + p \sin \epsilon) e^{-qx}}{p^2 + \omega^2}. \tag{3}$$

Since this does not appear in the Table of Transforms we apply the Inversion theorem and obtain

$$v = \frac{V_0}{2\pi i} \int_{\gamma - i\infty}^{\gamma + i\infty} \frac{(\omega \cos \epsilon + \lambda \sin \epsilon)}{\lambda^2 + \omega^2} e^{\lambda t - \mu x} \, d\lambda. \tag{4}$$

The integrand of (4) has a branch point at $\lambda = 0$, and simple poles at $\lambda = \pm i\omega$. We consider

$$\int \frac{(\omega \cos \epsilon + \lambda \sin \epsilon)}{\lambda^2 + \omega^2} e^{\lambda t - \mu x} \, d\lambda \tag{5}$$

round the contour of Fig. 40, within and on which its integrand is a single-valued function of λ. The argument of λ is π on EF and $-\pi$ on CD.

By Cauchy's theorem the integral round this closed contour is $2\pi i$ times the sum of the residues at the poles within it. These poles are at $\lambda = \pm i\omega$.

The residue at the pole $\lambda = i\omega$ is

$$\frac{1}{2i} e^{i\omega t + i\epsilon - x\sqrt{(i\omega/\kappa)}}.$$

Thus the sum of the residues at the poles $\lambda = i\omega$ and $\lambda = -i\omega$ is

$$e^{-x\sqrt{(\omega/2\kappa)}} \sin\{\omega t + \epsilon - x(\omega/2\kappa)^{\frac{1}{2}}\}. \tag{6}$$

Consider now the integral round the contour $ABFEDCA$ in the limit as the radius R of the large circle tends to infinity, and that of the small circle tends to zero. As $R \to \infty$ the integral over the arcs BF and CA tends to zero.† Also the integral over the small circle about the origin tends to zero as the radius of this circle tends to zero. As $R \to \infty$ the integral over AB becomes that of (4). In the integrals along

† For further discussion see Appendix I. In all cases in which we use the Inversion theorem the same result holds and a similar proof applies.

EF and CD we put $\lambda = \rho e^{i\pi}$ and $\lambda = \rho e^{-i\pi}$, respectively, and obtain

$$\int_0^\infty \frac{\omega\cos\epsilon - \rho\sin\epsilon}{\rho^2 + \omega^2}\, e^{-\rho l}\{e^{-ix\sqrt{(\rho/\kappa)}} - e^{ix\sqrt{(\rho/\kappa)}}\}\, d\rho$$

$$= -4i\kappa \int_0^\infty \frac{\omega\cos\epsilon - \kappa u^2\sin\epsilon}{\omega^2 + \kappa^2 u^4}\, e^{-\kappa u^2 l}\sin ux\, u\, du, \quad (7)$$

where we have put $\rho = \kappa u^2$.

Thus, finally, using (4), (6), and (7), we get

$$v = V_0\, e^{-x\sqrt{(\omega/2\kappa)}}\sin\{\omega t + \epsilon - x(\omega/2\kappa)^{\frac{1}{2}}\} +$$

$$+ \frac{2\kappa V_0}{\pi} \int_0^\infty \frac{\omega\cos\epsilon - \kappa u^2\sin\epsilon}{\omega^2 + \kappa^2 u^4}\, e^{-\kappa u^2 l}\sin ux\, u\, du. \quad (8)$$

The first term of (8), derived from the residues (6) at the poles $\pm i\omega$, is the steady state part of the solution, and if only this is needed the integral (7) need not be considered, cf. § 15.5. The results 2.8 (4) and 2.9 (13) may be derived in the same way.

All the results of § 12.4 which have been read off from the Table of Transforms could, of course, have been found by the use of the Inversion theorem as in this section.

12.8. Composite solids

Problems on conduction of heat in composite solids† are usually best solved by the Laplace transformation method. Both the transforms and solutions may be complicated, but no new principles are involved. In § 2.15 some problems on the infinite composite solid were studied; these may also be solved by the present method. Here we consider semi-infinite and finite composite regions.

We take first the semi-infinite region $-l < x < \infty$, of which $-l < x < 0$ is of one medium and $x > 0$ is of another. We write K_1, ρ_1, c_1, κ_1, and v_1 for conductivity, density, specific heat, diffusivity, and

† For composite slabs see Lowan, *Duke Math. J.* **1** (1935) 94; Churchill, ibid. **2** (1936) 405; *Amer. J. Math.* **61** (1939) 651. Composite spheres are discussed by Bromwich, *Proc. Camb. Phil. Soc.* **20** (1921) 411; Carslaw, ibid. **20** (1921) 399; Carslaw and Jaeger, ibid. **35** (1939) 394; Bell, *Proc. Phys. Soc.* **57** (1945) 45. Composite cylinders are discussed by Jaeger, *Phil. Mag.* (7) **32** (1941) 324. Some further references for composite cylinders and spheres are given in §§ 13.8, 13.9. The semi-infinite composite solid with constant surface flux is discussed by Matricon, *J. Phys. Radium,* **12** (1951) 15, and also by Griffith and Horton, *Proc. Phys. Soc.* **58** (1946) 481–7. Sakai, *Sci. Rep. Tohoku Univ.* (1) **11** (1922) 351–78, discusses the slab of n layers with either constant temperature or radiation at its surfaces and gives explicit formulae: see also Jaeger, *Quart. Appl. Math.* **8** (1950) 187–98 and Levy, *Trans. Amer. Soc. Mech. Engrs.* **78** (1956) 1627–35.

temperature in the region $-l < x < 0$, and K_2, ρ_2, c_2, κ_2, v_2 for the corresponding quantities in $x > 0$.

The differential equations to be solved are

$$\frac{\partial^2 v_1}{\partial x^2} - \frac{1}{\kappa_1} \frac{\partial v_1}{\partial t} = 0, \quad -l < x < 0, \quad t > 0,$$

$$\frac{\partial^2 v_2}{\partial x^2} - \frac{1}{\kappa_2} \frac{\partial v_2}{\partial t} = 0, \quad x > 0, \quad t > 0.$$

If we assume that there is no contact resistance at the surface of separation $x = 0$ (cf. § 1.9 G) the boundary conditions there are

$$K_1 \frac{\partial v_1}{\partial x} = K_2 \frac{\partial v_2}{\partial x}, \quad x = 0, \quad t > 0, \tag{1}$$

$$v_1 = v_2, \quad x = 0, \quad t > 0. \tag{2}$$

I. *For the solid described above, with zero initial temperature and* $x = -l$ *kept at constant temperature* V *for* $t > 0$, *the subsidiary equations are*

$$\frac{d^2 \bar{v}_1}{dx^2} - q_1^2 \bar{v}_1 = 0, \quad -l < x < 0, \tag{3}$$

$$\frac{d^2 \bar{v}_2}{dx^2} - q_2^2 \bar{v}_2 = 0, \quad x > 0, \tag{4}$$

where
$$q_1 = (p/\kappa_1)^{\frac{1}{2}}, \quad q_2 = (p/\kappa_2)^{\frac{1}{2}}. \tag{5}$$

These have to be solved with

$$K_1 \frac{d\bar{v}_1}{dx} = K_2 \frac{d\bar{v}_2}{dx}; \quad \bar{v}_1 = \bar{v}_2, \quad \text{at } x = 0, \tag{6}$$

$$\bar{v}_1 = V/p, \quad x = -l, \tag{7}$$

$$\bar{v}_2 \to 0, \quad \text{as } x \to \infty. \tag{8}$$

A solution of (3) which satisfies (7) is

$$\bar{v}_1 = (V/p)\cosh q_1(l+x) + A \sinh q_1(l+x),$$

and a solution of (4) which satisfies (8) is

$$\bar{v}_2 = B e^{-q_2 x}.$$

The unknowns A and B are found from (6), and we get finally

$$\bar{v}_1 = \frac{V(\cosh q_1 x - \sigma \sinh q_1 x)}{p(\cosh q_1 l + \sigma \sinh q_1 l)}, \tag{9}$$

$$\bar{v}_2 = \frac{V}{p(\cosh q_1 l + \sigma \sinh q_1 l)} e^{-q_2 x}, \tag{10}$$

where we use the notation

$$k = \sqrt{\left(\frac{\kappa_1}{\kappa_2}\right)}, \qquad \sigma = \frac{K_2 k}{K_1}, \qquad \alpha = \frac{\sigma-1}{\sigma+1}. \qquad (11)$$

To evaluate v_1 and v_2 we may either use the Inversion Theorem or the expansion procedure of § 12.5. The former method gives

$$v_2 = \frac{V}{2\pi i} \int_{\gamma-i\infty}^{\gamma+i\infty} \frac{e^{\lambda t - k\mu_1 x}\, d\lambda}{\lambda(\cosh \mu_1 l + \sigma \sinh \mu_1 l)}, \qquad (12)$$

where $\mu_1 = \sqrt{(\lambda/\kappa_1)}$. The integrand in (12) has a branch point at $\lambda = 0$, so we use the contour of Fig. 40 as in § 12.7, and the line integral in (12) is found to be equal to the integral over the small circle about the origin together with the integrals over CD and EF.

The small circle gives V.

Putting $\lambda = \kappa_1 u^2 e^{i\pi}$ on EF and $\lambda = \kappa_1 u^2 e^{-i\pi}$ on CD, the contribution from CD and EF becomes

$$\frac{V}{\pi i}\left\{ \int_0^\infty \frac{e^{-\kappa_1 u^2 t - ikxu}\, du}{u[\cos ul + i\sigma \sin ul]} - \text{conjugate} \right\}$$

$$= -\frac{2V}{\pi} \int_0^\infty e^{-\kappa_1 u^2 t}\, \frac{\sigma \cos kux \sin ul + \sin kux \cos ul}{u[\cos^2 ul + \sigma^2 \sin^2 ul]}\, du.$$

Thus

$$v_2 = V - \frac{2V}{\pi} \int_0^\infty e^{-\kappa_1 u^2 t}\, \frac{\sigma \cos kux \sin ul + \sin kux \cos ul}{u[\cos^2 ul + \sigma^2 \sin^2 ul]}\, du. \qquad (13)$$

Similarly

$$v_1 = V - \frac{2\sigma V}{\pi} \int_0^\infty e^{-\kappa_1 u^2 t}\, \frac{\sin u(l+x)\, du}{u[\cos^2 ul + \sigma^2 \sin^2 ul]}. \qquad (14)$$

The method of § 12.5 yields a form of the solution which is often more convenient. Using the notation (11) we have from (10)

$$\bar{v}_2 = \frac{2V}{p(1+\sigma)[1 - \alpha e^{-2q_1 l}]} e^{-kq_1 x - q_1 l}$$

$$= \frac{2V}{(1+\sigma)p} \sum_{n=0}^\infty \alpha^n e^{-q_1[(2n+1)l + kx]}. \qquad (15)$$

It follows from Appendix V (8) that

$$v_2 = \frac{2V}{1+\sigma} \sum_{n=0}^\infty \alpha^n \operatorname{erfc} \frac{(2n+1)l + kx}{2\sqrt{(\kappa_1 t)}}. \qquad (16)$$

Similarly

$$v_1 = V \sum_{n=0}^{\infty} \alpha^n \left\{ \operatorname{erfc} \frac{(2n+1)l+x}{2\sqrt{(\kappa_1 t)}} - \alpha \operatorname{erfc} \frac{(2n+1)l-x}{2\sqrt{(\kappa_1 t)}} \right\}. \tag{17}$$

And in the same way the temperature gradient at the surface is found to be (using Appendix V (7))

$$\left[\frac{\partial v_1}{\partial x} \right]_{x=-l} = -\frac{V}{\sqrt{(\pi \kappa_1 t)}} \left\{ 1 + 2 \sum_{n=1}^{\infty} \alpha^n e^{-n^2 l^2 / \kappa_1 t} \right\}. \tag{18}$$

For very large values of the time, the exponentials in (18) may all be replaced by unity, and we have approximately

$$\left[\frac{\partial v_1}{\partial x} \right]_{x=-l} = -\frac{V}{\sqrt{(\pi \kappa_1 t)}} \left\{ 1 + \frac{2\alpha}{1-\alpha} \right\} = -\frac{V}{\sqrt{(\pi \kappa_1 t)}} \left(\frac{K_2 \rho_2 c_2}{K_1 \rho_1 c_1} \right)^{\frac{1}{2}}. \tag{19}$$

When the surface is kept at zero and the initial temperature of the whole solid is V, it is clear that the temperature gradient at $x = -l$ will be minus the above. This result was used by Perry and Heaviside† in discussion of the age of the Earth. The temperature gradient in Kelvin's classical treatment, § 2.14, was found to be $V/\sqrt{(\pi \kappa_1 t)}$. Now it is known that the Earth contains a core of very different density and physical properties from the outer crust: on this assumption the time required for subsidence to the present temperature gradient is $(K_2 c_2 \rho_2)/(K_1 c_1 \rho_1)$ times that on Kelvin's theory. With the data adopted by Perry and Heaviside this factor is nearly 450 and Kelvin's estimate of 10^8 years would be increased to $4\cdot5 \times 10^{10}$ years.

It is of some interest to consider the case in which l is small, that is, a thin film of another substance on the surface of the semi-infinite solid. Expanding the hyperbolic functions in the denominator of (10) in ascending powers of l we find

$$\bar{v}_2 = \frac{K_1 V e^{-q_2 x}}{p(K_1 + k K_2 l q_1 + \frac{1}{2} K_1 q_1^2 l^2 + \dots)}. \tag{20}$$

Retaining only the first power of l gives

$$\bar{v}_2 = \frac{hV e^{-q_2 l}}{p(h+q_2)},$$

where $h = K_1/K_2 l$. This is the value of $\bar{v}_2$ obtained from the approximate boundary condition of 1.9 (7) which neglected the thermal capacity of the film altogether. As a second approximation, retaining the term in l^2, we have

$$\bar{v}_2 = \frac{hV e^{-q_2 x}}{p(h+q_2+h'p)}, \tag{21}$$

where $h' = K_1 l/2\kappa_1 K_2$. Transforms of this type have appeared in § 12.4 III. (21) is the transform of the solution of the problem of the

† For references see §§ 2.14, 9.14.

semi-infinite solid with zero initial temperature, and boundary condition

$$\frac{\partial v}{\partial x} - h' \frac{\partial v}{\partial t} - hv = -hV, \qquad x = 0, \tag{22}$$

so this may be regarded as an approximate† boundary condition which allows for the heat capacity of the film.

II. *The composite solid above with zero initial temperature and heat produced for $t > 0$ at the constant rate A_0 per unit time per unit volume in $-l < x < 0$ and zero in $x > 0$. $x = -l$ kept at zero temperature*

Here, using the notation (11), the results are

$$v_1 = \frac{A_0(l^2 - x^2)}{2K_1} - \frac{2A_0 \sigma}{\pi K_1} \int_0^\infty e^{-\kappa_1 u^2 t} \frac{(1 - \cos ul)\sin u(l + x)}{u^3[\cos^2 ul + \sigma^2 \sin^2 ul]} \, du \tag{23}$$

$$= \frac{\kappa_1 A_0 t}{K_1} - \frac{4\kappa_1 A_0 t}{K_1} \sum_{n=0}^\infty \alpha^n \left\{ \mathrm{i}^2\mathrm{erfc}\, \frac{(2n+1)l + x}{2\sqrt{(\kappa_1 t)}} - \alpha\, \mathrm{i}^2\mathrm{erfc}\, \frac{(2n+1)l - x}{2\sqrt{(\kappa_1 t)}} + \right.$$
$$\left. + \frac{\sigma}{1+\sigma}\, \mathrm{i}^2\mathrm{erfc}\, \frac{2nl - x}{2\sqrt{(\kappa_1 t)}} - \frac{\sigma}{1+\sigma}\, \mathrm{i}^2\mathrm{erfc}\, \frac{(2n+2)l + x}{2\sqrt{(\kappa_1 t)}} \right\}. \tag{24}$$

$$v_2 = \frac{A_0 l^2}{2K_1} - \frac{2A_0}{\pi K_1} \int_0^\infty e^{-\kappa_1 u^2 t} \frac{(1 - \cos ul)(\sigma \sin ul \cos kux + \cos ul \sin kux)}{u^3(\cos^2 ul + \sigma^2 \sin^2 ul)} \, du \tag{25}$$

$$= \frac{4\kappa_1 t A_0}{K_1(1+\sigma)} \sum_{n=0}^\infty \alpha^n \left\{ \mathrm{i}^2\mathrm{erfc}\, \frac{2nl + kx}{2\sqrt{(\kappa_1 t)}} + \mathrm{i}^2\mathrm{erfc}\, \frac{(2n+2)l + kx}{2\sqrt{(\kappa_1 t)}} - \right.$$
$$\left. - 2\, \mathrm{i}^2\mathrm{erfc}\, \frac{(2n+1)l + kx}{2\sqrt{(\kappa_1 t)}} \right\}. \tag{26}$$

III. *The finite slab with $-l < x < 0$ of one medium, K_1, ρ_1, c_1, κ_1, v_1, and $0 < x < a$ of another, K_2, ρ_2, c_2, κ_2, v_2. Zero initial temperature. $x = -l$ maintained at V, constant, and $x = a$ at zero for $t > 0$*

The subsidiary equations are (3)–(7), but with (8) replaced by

$$\bar{v}_2 = 0, \quad x = a.$$

Solving we find

$$\bar{v}_1 = \frac{V\{\cosh q_1 x \sinh q_2 a - \sigma \sinh q_1 x \cosh q_2 a\}}{p\{\cosh q_1 l \sinh q_2 a + \sigma \sinh q_1 l \cosh q_2 a\}}, \tag{27}$$

$$\bar{v}_2 = \frac{V \sinh q_2(a - x)}{p\{\cosh q_1 l \sinh q_2 a + \sigma \sinh q_1 l \cosh q_2 a\}}. \tag{28}$$

The series expansions of type (15) for these cases are rather complicated, so we consider only the solutions obtained from the Inversion Theorem. The integrands in this case are single-valued functions of λ with simple poles at $\lambda = 0$ and $\lambda = -\kappa_1 \beta_m^2$, where $\pm\beta_m$, $m = 1, 2, 3,...$, are the roots‡ of

$$\cos \beta l \sin k\beta a + \sigma \sin \beta l \cos k\beta a = 0. \tag{29}$$

---

† Fox, *Phil. Mag.* (7) **18** (1934) 209, gives a treatment of the case of a skin of several layers.

‡ It is shown in IV below that these roots are all real and simple.

Applying the Inversion Theorem in the usual way we find

$$v_1 =$$

$$\frac{V(K_1 a - K_2 x)}{K_1 a + K_2 l} - 2V \sum_{m=1}^{\infty} \frac{(\cos \beta_m x \sin ka\beta_m - \sigma \sin \beta_m x \cos ka\beta_m)}{\beta_m \{(l + \sigma ka) \sin \beta_m l \sin ka\beta_m - (\sigma l + ka) \cos \beta_m l \cos ka\beta_m\}} \times$$

$$\times e^{-\kappa_1 \beta_m^2 t}, \quad (30)$$

$$v_2 = \frac{K_1 V(a-x)}{K_1 a + K_2 l} - 2V \sum_{m=1}^{\infty} \frac{\sin k\beta_m(a-x)}{\beta_m \{(l + \sigma ka) \sin \beta_m l \sin ka\beta_m - (\sigma l + ka) \cos \beta_m l \cos ka\beta_m\}} \times$$

$$\times e^{-\kappa_1 \beta_m^2 t}. \quad (31)$$

The series in (30) and (31) may be simplified a little by further use of (29). Here a new point arises: the roots of (29) are the roots of

$$\cot \beta l + \sigma \cot k\beta a = 0, \quad (32)$$

together with the common roots, if any, of

$$\sin \beta l = 0, \quad \text{and} \quad \sin k\beta a = 0. \quad (33)$$

The latter equations have common roots if and only if ka/l is rational. Thus if

$$ka/l = r/s, \quad (34)$$

a rational fraction in its lowest terms, the common positive roots of (33) are $n\pi s/l$, $n = 1, 2, 3,...$, and these roots of (29) give rise to the series

$$-\frac{2V\sigma}{\pi(r+\sigma s)} \sum_{n=1}^{\infty} \frac{\cos ns\pi}{n} \sin \frac{ns\pi x}{l} e^{-\kappa_1 n^2 s^2 \pi^2 t/l^2}, \quad (35)$$

and

$$-\frac{2V}{\pi(r+\sigma s)} \sum_{n=1}^{\infty} \frac{\cos ns\pi}{n} \sin \frac{nr\pi x}{a} e^{-\kappa_1 n^2 s^2 \pi^2 t/l^2}, \quad (36)$$

in v_1 and v_2 respectively. If ka/l is irrational these series do not appear.

In all cases there is a series of terms corresponding to the positive roots† β_n, $n = 1, 2, 3,...$, of (32).

Using (32) the terms of the series (30) and (31) corresponding to these roots may be put in the forms

$$-2V \sum_{n=1}^{\infty} \frac{\sin^2 ka\beta_n \sin \beta_n(l+x)}{\beta_n(l \sin^2 ka\beta_n + \sigma ka \sin^2 l\beta_n)} e^{-\kappa_1 \beta_n^2 t}, \quad (37)$$

$$-2V \sum_{n=1}^{\infty} \frac{\sin l\beta_n \sin ka\beta_n \sin k(a-x)\beta_n}{\beta_n(l \sin^2 ka\beta_n + \sigma ka \sin^2 l\beta_n)} e^{-\kappa_1 \beta_n^2 t}, \quad (38)$$

for v_1 and v_2 respectively.

IV. *The nature of the roots of* (32) *and equations arising in similar problems*

In deriving solutions in the form of infinite series by the use of the Inversion Theorem we usually need to be able to state that the roots of a certain transcendental equation are all real and simple. In the problem of III this equation was (32); for the solid composite sphere it is 13.9 (35); while with more general boundary conditions other types of equation such as 13.9 (25), etc., arise.

† These include the common roots of $\cos \beta l = 0$ and $\cos ka\beta = 0$ which may occur if ka/l is rational. The discussion leading to (37) and (38) is valid in this special case.

A very simple equation has been studied in § 3.9 and extensions of the method†
used there apply in all cases. Here, as an example, we discuss equation (32) above.

Clearly this cannot have a pure imaginary root $\beta = i\eta$ since

$$\coth \eta l + \sigma \coth k\eta l > 0.$$

We proceed to show that it cannot have a complex root of form $\xi \pm i\eta$. Consider
the function U defined as follows:

$$U = U_1 = \sin \beta(l+x), \quad -l < x < 0, \tag{39}$$

$$U = U_2 = \frac{\sin \beta l \sin k\beta(a-x)}{\sin k\beta a}, \quad 0 < x < a, \tag{40}$$

where β is a root of (32).

Then we have

$$\frac{d^2 U_1}{dx^2} + \beta^2 U_1 = 0, \quad -l < x < 0; \qquad \frac{d^2 U_2}{dx^2} + k^2\beta^2 U_2 = 0, \quad 0 < x < a. \tag{41}$$

Also

$$U_1 = 0, \quad \text{when } x = -l; \qquad U_2 = 0, \quad \text{when } x = a; \qquad U_1 = U_2, \quad \text{when } x = 0. \tag{42}$$

And

$$\left[K_1 \frac{dU_1}{dx} - K_2 \frac{dU_2}{dx} \right]_{x=0} = \frac{\beta}{\sin k\beta a} \{ K_1 \cos \beta l \sin k\beta a + K_2 k \sin \beta l \cos k\beta a \} = 0, \tag{43}$$

since β is a root of (32).

Now let β and α be two different roots of (32), and let V_1 and V_2 be the quantities
corresponding to U_1 and U_2 with β replaced by α.

Then, from (41) and the corresponding equations for V_1 and V_2,

$$(\beta^2 - \alpha^2) \int_{-l}^{0} U_1 V_1 \, dx + \int_{-l}^{0} (U_1'' V_1 - V_1'' U_1) \, dx = 0,$$

$$k^2 (\beta^2 - \alpha^2) \int_{0}^{a} U_2 V_2 \, dx + \int_{0}^{a} (U_2'' V_2 - V_2'' U_2) \, dx = 0.$$

Therefore

$$(\beta^2 - \alpha^2)\left\{ K_1 \int_{-l}^{0} U_1 V_1 \, dx + k^2 K_2 \int_{0}^{a} U_2 V_2 \, dx \right\}$$

$$= K_1 \int_{-l}^{0} (U_1 V_1'' - V_1 U_1'') \, dx + K_2 \int_{0}^{a} (U_2 V_2'' - V_2 U_2'') \, dx$$

$$= K_1 [U_1 V_1' - V_1 U_1']_{-l}^{0} + K_2 [U_2 V_2' - V_2 U_2']_{0}^{a} = 0, \tag{44}$$

using (42) and (43). It follows from (44) that α and β cannot be of the form $\xi \pm i\eta$
since U_1, V_1 and U_2, V_2 would be conjugate complex quantities and

$$K_1 \int_{-l}^{0} U_1 V_1 \, dx + k^2 K_2 \int_{0}^{a} U_2 V_2 \, dx$$

would be positive.

We have thus proved that the roots of (32) are all real. That they are symmetrically placed with regard to the origin and not repeated follows in this case

† For the general theory see Ince, *Ordinary Differential Equations* (Longmans, 1927)
Chap. X; Bromwich, *Proc. Camb. Phil. Soc.* **20** (1921) 411.

from consideration of the curves $y = \cot x$ and $y = \cot kax/l$. Other cases may be discussed by an extension of the method above, cf. Ince, loc. cit.

V. *The slab of any number of layers*

In the general case of n layers, there is no difficulty in writing down the Laplace transform of any required quantity. This is most easily done by the matrix method described below. Owing to the complicated results, the discussion of the roots of the denominator, and the numerical evaluation of temperatures, involve extremely heavy algebra. It may be noted that when the Laplace transform has been found, it yields the steady periodic solution immediately, and also, for those cases in which the quantity in question has a linear asymptote of the form $a+bt$, this can be found by a simple numerical routine, cf. § 15.6.

For the determination of the Laplace transform, consider first the slab $0 < x < l$, then if $\bar{v}_x$ and $\bar{f}_x$ are the transforms of the temperature and the flux at the point x, the subsidiary equations give immediately

$$\bar{v}_x = \bar{v}_0 \cosh qx - \bar{f}_0(1/Kq)\sinh qx, \tag{45}$$

$$\bar{f}_x = -\bar{v}_0 Kq \sinh qx + \bar{f}_0 \cosh qx. \tag{46}$$

That is, in the matrix notation defined in § 3.7,

$$\begin{pmatrix} \bar{v}_x \\ \bar{f}_x \end{pmatrix} = \begin{pmatrix} \cosh qx & -(1/Kq)\sinh qx \\ -Kq\sinh qx & \cosh qx \end{pmatrix} \begin{pmatrix} \bar{v}_0 \\ \bar{f}_0 \end{pmatrix}. \tag{47}$$

Now consider a slab of n layers, $(l_1, l_2), (l_2, l_3),..., (l_n, l_{n+1})$; let K_r, κ_r be the conductivity and diffusivity in the rth layer, let $\bar{v}_r$ and $\bar{f}_r$ be the transforms of the temperature and flux at the end $x = l_r$ of this layer, and let $\bar{v}'_r$ and $\bar{f}'_r$ be their values at the end $x = l_{r+1}$ of the layer, then it follows from (47) that

$$\begin{pmatrix} \bar{v}'_r \\ \bar{f}'_r \end{pmatrix} = \begin{pmatrix} A_r & B_r \\ C_r & D_r \end{pmatrix} \begin{pmatrix} \bar{v}_r \\ \bar{f}_r \end{pmatrix}, \tag{48}$$

where $q_r = (p/\kappa_r)^{\frac{1}{2}}$,

$$\begin{rcases} A_r = \cosh(l_{r+1}-l_r)q_r, & B_r = -(1/K_r q_r)\sinh(l_{r+1}-l_r)q_r \\ C_r = -K_r q_r \sinh(l_{r+1}-l_r)q_r, & D_r = \cosh(l_{r+1}-l_r)q_r \end{rcases}, \tag{49}$$

and

$$A_r D_r - B_r C_r = 1.$$

If there is perfect thermal contact between the slabs it follows by repeated application of (49) that

$$\begin{pmatrix} \bar{v}'_n \\ \bar{f}'_n \end{pmatrix} = \begin{pmatrix} A_n & B_n \\ C_n & D_n \end{pmatrix} \cdots \begin{pmatrix} A_1 & B_1 \\ C_1 & D_1 \end{pmatrix} \begin{pmatrix} \bar{v}_1 \\ \bar{f}_1 \end{pmatrix}. \tag{50}$$

If there is linear heat transfer with contact resistance $R_1,..., R_{n+1}$, at $l_1, l_2,..., l_{n+1}$ the medium in $x < l_1$ being at temperature v_0 and that in $x > l_{n+1}$ at temperature v_{n+1},

$$\begin{pmatrix} \bar{v}_{n+1} \\ \bar{f}_{n+1} \end{pmatrix} = \begin{pmatrix} 1 & -R_{n+1} \\ 0 & 1 \end{pmatrix} \begin{pmatrix} A_n & B_n \\ C_n & D_n \end{pmatrix} \begin{pmatrix} 1 & -R_n \\ 0 & 1 \end{pmatrix} \cdots \begin{pmatrix} A_1 & B_1 \\ C_1 & D_1 \end{pmatrix} \begin{pmatrix} 1 & -R_1 \\ 0 & 1 \end{pmatrix} \begin{pmatrix} \bar{v}_0 \\ \bar{f}_0 \end{pmatrix}.$$

In this way, the transforms of the temperature or fluxes at any of $l_1,..., l_{n+1}$ are determined, and values for intermediate points may then be found from (45) and (46). For example, (27) and (28) may be derived in this way.

VI. *The region $0 < x < l$ of one material and $x > l$ of another. No flow of heat at $x = 0$, The initial temperature constant in $0 < x < l$ and zero in $x > l$.*

The solution for this case and numerical values of the temperature at $x = l$ are given by Lovering†.

† *Bull. Geol. Soc. America* **47** (1936) 87.

THE LAPLACE TRANSFORMATION: PROBLEMS ON THE CYLINDER AND SPHERE

13.1. Introductory

In this chapter we consider a number of problems on the sphere and the infinite circular cylinder which are rather more easily handled by the Laplace transformation method than by the classical methods. These include† more complicated boundary conditions, hollow and composite cylinders, solutions useful for small values of the time, solutions for regions bounded internally by cylinders, and corresponding problems for the sphere.

13.2. The circular cylinder $0 \leqslant r < a$ with various boundary conditions

In this section a number of problems already discussed in §§ 7.6–7.9 will be solved briefly as examples of the application of the Laplace transformation to a cylindrical region. The expressions for the transforms will also be needed in § 13.3 where solutions useful for small values of $\kappa t/a^2$ are derived.

I. *Zero initial temperature. Constant surface temperature*

We have to solve

$$\frac{\partial^2 v}{\partial r^2} + \frac{1}{r}\frac{\partial v}{\partial r} - \frac{1}{\kappa}\frac{\partial v}{\partial t} = 0, \quad 0 \leqslant r < a, \quad t > 0, \tag{1}$$

with $v = V$, $r = a$, $t > 0$.

The subsidiary equation is

$$\frac{d^2 \bar{v}}{dr^2} + \frac{1}{r}\frac{d\bar{v}}{dr} - q^2 \bar{v} = 0, \quad 0 \leqslant r < a, \tag{2}$$

where $q^2 = p/\kappa$. This has to be solved with

$$\bar{v} = \frac{V}{p}, \quad r = a, \tag{3}$$

and with $\bar{v}$ finite at $r = 0$. Of the two solutions $I_0(qr)$ and $K_0(qr)$ of (2), the latter tends to infinity as $r \to 0$, and therefore must be excluded.

† Many problems on the cylinder are discussed by Goldstein, *Proc. Lond. Math. Soc.* (2) **34** (1932) 51, and Carslaw and Jaeger, ibid. (2) **46** (1940) 361.

Thus the solution of (2) and (3) is

$$\bar{v} = \frac{V I_0(qr)}{p I_0(qa)}.$$ (4)

Therefore, using the Inversion Theorem,

$$v = \frac{V}{2\pi i} \int_{\gamma-i\infty}^{\gamma+i\infty} e^{\lambda t} \frac{I_0(\mu r)}{\lambda I_0(\mu a)}\, d\lambda,$$ (5)

where, as usual, μ is written for $\sqrt{(\lambda/\kappa)}$.

The integrand of (5) is a single-valued function† of λ, so we use the contour of Fig. 39. The zeros of $I_0(\mu a)$ are at $\lambda = -\kappa \alpha_n^2$, where $\pm \alpha_n$, $n = 1, 2,...$, are the roots of

$$J_0(a\alpha) = 0.$$

Then in the usual way the line integral in (5) is found to be equal to $2\pi i$ times the sum of the residues at the poles of the integrand. The residues are evaluated by using the result (cf. Appendix III (26))

$$\left[\lambda \frac{d}{d\lambda} I_0(\mu a) \right]_{\lambda=-\kappa\alpha_n^2} = [\tfrac{1}{2}\mu a I_1(\mu a)]_{\mu=i\alpha_n} = -\tfrac{1}{2} a\alpha_n J_1(a\alpha_n).$$ (6)

The pole at $\lambda = 0$ has residue 1, since $I_0(z) = 1$ when $z = 0$. Thus, finally,

$$v = V - \frac{2V}{a} \sum_{n=1}^{\infty} e^{-\kappa\alpha_n^2 t} \frac{J_0(r\alpha_n)}{\alpha_n J_1(a\alpha_n)}.$$ (7)

This is the result 7.6 (8) but here we have not made the assumption 7.6 (1), and it can be verified as in Appendix I that (7) satisfies the conditions of the problem. The same remark applies to all the solutions in this chapter.

II. *Zero initial temperature. Surface temperature kt*
We find in the same way

$$v = \frac{k}{2\pi i} \int_{\gamma-i\infty}^{\gamma+i\infty} e^{\lambda t} \frac{I_0(\mu r)}{\lambda^2 I_0(\mu a)}\, d\lambda$$

$$= k\left(t - \frac{a^2 - r^2}{4\kappa} \right) + \frac{2k}{a\kappa} \sum_{n=1}^{\infty} e^{-\kappa\alpha_n^2 t} \frac{J_0(r\alpha_n)}{\alpha_n^3 J_1(a\alpha_n)},$$ (8)

the only change being that there is now a double pole at $\lambda = 0$.

III. *Zero initial temperature. Constant flux F_0 at the surface*
Here the subsidiary equation (2) has to be solved with the boundary condition

$$-K \frac{d\bar{v}}{dr} = \frac{F_0}{p}.$$ (9)

† *W.B.F.*, p. 80, or from the series for $I_0(z)$.

Thus
$$\bar{v} = -\frac{F_0\kappa^{\frac{1}{2}}I_0(qr)}{Kp^{\frac{1}{2}}I_1(qa)}.$$
(10)

The Inversion Theorem gives

$$v = -\frac{F_0 a}{K}\left\{\frac{2\kappa t}{a^2} + \frac{r^2}{2a^2} - \frac{1}{4} - 2\sum_{s=1}^{\infty} e^{-\kappa\alpha_s^2 t/a^2}\frac{J_0(r\alpha_s/a)}{\alpha_s^2 J_0(\alpha_s)}\right\},$$
(11)

where the α_s are the positive roots of

$$J_1(\alpha) = 0.$$
(12)

IV. *Zero initial temperature. Radiation at the surface*

The subsidiary equation is again (2), and, if the cylinder radiates into medium at constant temperature V, the boundary condition is

$$\frac{d\bar{v}}{dr} + h\bar{v} = \frac{hV}{p}, \quad \text{when } r = a.$$
(13)

Therefore the solution is

$$\bar{v} = \frac{hVI_0(qr)}{p\{qI_1(qa) + hI_0(qa)\}}.$$
(14)

Using the Inversion Theorem the solution is found to be

$$\frac{v}{V} = 1 - \sum_{n=1}^{\infty} e^{-\kappa\alpha_n^2 t}\frac{2hJ_0(r\alpha_n)}{a(h^2 + \alpha_n^2)J_0(a\alpha_n)},$$
(15)

where the α_n are the positive roots of

$$\alpha J_1(a\alpha) = hJ_0(a\alpha).$$
(16)

V. *Contact with well-stirred fluid† or perfect conductor*

As an example suppose that *the cylinder* $0 \leqslant r < a$ *is initially at constant temperature V, and at $t = 0$ is placed in contact with mass M per unit length of well-stirred fluid of specific heat c' whose initial temperature is zero. The temperature of the fluid is supposed to be equal to the surface temperature of the solid for $t > 0$, and the fluid loses heat to its surroundings at a rate H times its temperature.*

Here the subsidiary equation is

$$\frac{d^2\bar{v}}{dr^2} + \frac{1}{r}\frac{d\bar{v}}{dr} - q^2\bar{v} = -\frac{V}{\kappa}, \quad 0 \leqslant r < a.$$
(17)

The boundary condition at $r = a$ is, cf. 1.9 (14), (15),

$$-2\pi aK\frac{\partial v}{\partial r} = Mc'\frac{du}{dt} + Hu, \quad r = a,$$
(18)

where u is the temperature of the fluid, and $u = v$ for $t > 0$.

Thus the boundary condition for the subsidiary equation (17) is

$$2\pi aK\frac{d\bar{v}}{dr} + (Mc'p + H)\bar{v} = 0, \quad r = a.$$
(19)

† Lowan, *Phil. Mag.* (7) **17** (1934) 849. Jaeger, *J. Proc. Roy. Soc. N.S.W.* **74** (1940) 342 and **75** (1941) 130, gives solutions of a number of problems of this type for regions bounded internally and externally by a cylinder, and for hollow cylinders. They arise, for example, in the transient heating of single core cables, cf. Goldenberg, *Engineer*, **197** (1954) 779–80.

Therefore $\qquad \bar{v} = \dfrac{V}{p}\left\{1 - \dfrac{(Mc'p+H)I_0(qr)}{2\pi aKqI_1(qa)+(Mc'p+H)I_0(qa)}\right\}$,

and $\qquad v = \displaystyle\sum_{n=1}^{\infty} \dfrac{2V(H'-ka^2\alpha_n^2)J_0(r\alpha_n)}{J_0(a\alpha_n)[(2k+1)a^2\alpha_n^2+(H'-ka^2\alpha_n^2)^2]}\, e^{-\kappa\alpha_n^2 t}$, $\qquad$ (20)

where $\pm\alpha_n$, $n = 1, 2,...$, are the roots of

$$a\alpha J_1(a\alpha) = (H'-ka^2\alpha^2)J_0(a\alpha),\qquad (21)$$

and $\qquad\qquad H' = H/2\pi K, \qquad k = Mc'/2\pi a^2\rho c.$

VI. *Heat production in the cylinder $0 \leqslant r < a$ for $t > 0$. Zero initial temperature*

If the initial temperature of the cylinder is zero and heat is produced for $t > 0$ at the constant rate A_0 per unit volume per unit time, the surface $r = a$ being kept at zero, the subsidiary equation is (cf. 1.6 (7))

$$\frac{d^2\bar{v}}{dr^2}+\frac{1}{r}\frac{d\bar{v}}{dr}-q^2\bar{v} = -\frac{A_0}{Kp}.\qquad (22)$$

The solution of this, with $\bar{v}$ zero at $r = a$ and finite at $r = 0$, is

$$\bar{v} = \frac{\kappa A_0}{Kp^2}-\frac{\kappa A_0 I_0(qr)}{Kp^2 I_0(qa)}.\qquad (23)$$

Therefore, using (8),

$$v = \frac{A_0(a^2-r^2)}{4K}-\frac{2A_0}{aK}\sum_{n=1}^{\infty} e^{-\kappa\alpha_n^2 t}\,\frac{J_0(r\alpha_n)}{\alpha_n^3 J_1(a\alpha_n)},\qquad (24)$$

where the α_n are the positive roots of $J_0(a\alpha) = 0$.

13.3. Solutions useful for small values of the time

I. *Constant surface temperature*

Considering first the problem § 13.2 (I) of constant surface temperature, it was remarked in § 7.6 that the solution 13.2 (7) is not suitable for use with small values of $\kappa t/a^2$, say, for values less than 0·02. The same difficulty was encountered for the slab and sphere, and in these cases alternative solutions can be found, as in § 12.5, by expanding $\bar{v}$ in a series of negative exponentials. In cylindrical problems the procedure is more complicated,† and consists of using the asymptotic expansion of the Bessel functions involved to obtain a form involving exponentials with coefficients which are series in $(1/q)$.

Thus, taking 13.2 (4) and using Appendix III (12) in it, gives, for the case in which (r/a) is not small,

$$\bar{v} = \frac{Va^{\frac{1}{2}}}{pr^{\frac{1}{2}}}e^{-q(a-r)}\frac{[1+(1/8qr)+(9/128q^2r^2)+...]+O[e^{-2qr}]}{[1+(1/8qa)+(9/128q^2a^2)+...]+O[e^{-2qa}]}\qquad (1)$$

$$= \frac{Va^{\frac{1}{2}}}{pr^{\frac{1}{2}}}\left(1+\frac{a-r}{8qar}+\frac{9a^2-7r^2-2ar}{128q^2a^2r^2}+...\right)e^{-q(a-r)},\qquad (2)$$

where in (2) terms in $\exp[-q(3a-r)]$, etc., which correspond to the

† Goldstein, *Proc. Lond. Math. Soc.* (2) **34** (1932) 51–88, gives many examples of this procedure. For a justification, see Carslaw and Jaeger, *Operational Methods in Applied Mathematics*, (Oxford, edn. 2, 1948) § 125.

multiple reflections in § 12.5 (2), have been ignored. Using Appendix V (11), we get from (2),

$$v = \frac{Va^{\frac{1}{2}}}{r^{\frac{1}{2}}}\operatorname{erfc}\frac{a-r}{2(\kappa t)^{\frac{1}{2}}} + \frac{V(a-r)(\kappa t a)^{\frac{1}{2}}}{4ar^{\frac{1}{2}}}\operatorname{ierfc}\frac{a-r}{2(\kappa t)^{\frac{1}{2}}} +$$

$$+ \frac{V(9a^2-7r^2-2ar)\kappa t}{32a^{\frac{3}{2}}r^{\frac{1}{2}}}\operatorname{i^2erfc}\frac{a-r}{2(\kappa t)^{\frac{1}{2}}} + ... \quad (3)$$

Because of the neglect of later terms, the series (3) is not applicable over so wide a range of values of $(\kappa t/a^2)$ as those of § 12.5, but it is quite satisfactory in the range $\kappa t/a^2 < 0\cdot02$ provided that r/a is not too small.

For the temperature at the centre, $r = 0$, we have $I_0(qr) = 1$ in 13.2 (4), and, using the asymptotic expansion of $I_0(qa)$ as before, we find

$$\bar{v} = \frac{V(2\pi a)^{\frac{1}{2}}}{p^{\frac{1}{2}}\kappa^{\frac{1}{2}}\{1+(1/8qa)+(9/128q^2a^2)+...\}}e^{-qa}$$

$$= \frac{V(2\pi a)^{\frac{1}{2}}}{\kappa^{\frac{1}{2}}p^{\frac{1}{2}}}e^{-qa} - \frac{V(2\pi)^{\frac{1}{2}}\kappa^{\frac{1}{2}}}{8p^{\frac{1}{2}}a^{\frac{1}{2}}}e^{-qa} - ....$$

Using Appendix V (20), the first term in the expansion of v is found to be[†]

$$\frac{Va}{\sqrt{(\pi\kappa t)}}e^{-a^2/8\kappa t}K_{\frac{1}{2}}\left(\frac{a^2}{8\kappa t}\right). \quad (4)$$

II. *Constant surface flux*

Treating 13.2 (10) in the same way, gives

$$v = -\frac{F_0}{K}\left\{2\left(\frac{\kappa at}{r}\right)^{\frac{1}{2}}\operatorname{ierfc}\frac{a-r}{2(\kappa t)^{\frac{1}{2}}} + \frac{\kappa t(a+3r)}{2a^{\frac{1}{2}}r^{\frac{1}{2}}}\operatorname{i^2erfc}\frac{a-r}{2(\kappa t)^{\frac{1}{2}}} + ...\right\}, \quad (5)$$

provided, as before, that r/a is not small.

III. *Radiation at the surface*

Assuming, as before, that r/a is not small, and using the asymptotic expansions of I_0 and I_1 in 13.2 (14), gives

$$\bar{v} = \frac{hVa^{\frac{1}{2}}}{pqr^{\frac{1}{2}}}e^{-q(a-r)}\frac{1+(1/8qr)+(9/128q^2r^2)+...}{1+[ah-(3/8)]/aq+[ah-(15/16)]/8a^2q^2+...}$$

$$= \frac{hVa^{\frac{1}{2}}}{pqr^{\frac{1}{2}}}e^{-q(a-r)}\left\{1+\frac{1}{q}\left[\frac{1}{8r}+\frac{3}{8a}-h\right]+...\right\}.$$

Therefore

$$v = \frac{2hV(a\kappa t)^{\frac{1}{2}}}{r^{\frac{1}{2}}}\operatorname{ierfc}\frac{a-r}{2\sqrt{(\kappa t)}} + \frac{4hVa^{\frac{1}{2}}\kappa t}{r^{\frac{1}{2}}}\left[\frac{1}{8r}+\frac{3}{8a}-h\right]\operatorname{i^2erfc}\frac{a-r}{2\sqrt{(\kappa t)}} + ... \quad (6)$$

If r/a is small the method leading to (4) must be used.

† Carsten and McKerrow, *Phil. Mag.* (7) **35** (1944) 812, discuss in detail this case and its extension to small values of r. They give tables of the Bessel functions $K_{\frac{1}{4}}$, $K_{\frac{3}{4}}$, $K'_{\frac{1}{4}}$, $K'_{\frac{3}{4}}$, which are needed in the solution.

13.4. The hollow cylinder $a < r < b$

We consider this region with zero initial temperature† and boundary conditions

$$k_1 \frac{\partial v}{\partial r} - k_2 v = k_3, \quad r = a \ \Bigg\} ,$$
$$k_1' \frac{\partial v}{\partial r} + k_2' v = k_3', \ \cdot r = b \ \Bigg\}$$

$$(1)$$

where k_1, k_2, k_1', k_2' are constants which may be positive or zero (provided k_1 and k_2, or k_1' and k_2', do not both vanish) and k_3 and k_3' are constants. By choice of these constants the general result includes all combinations of constant temperature, constant flux, zero flux, or radiation at either surface.

The subsidiary equation is

$$\frac{d^2 \bar{v}}{dr^2} + \frac{1}{r} \frac{d\bar{v}}{dr} - q^2 \bar{v} = 0, \quad a < r < b,$$

and the solution will be of the form

$$\bar{v} = A I_0(qr) + B K_0(qr),$$

$$(2)$$

where A and B are chosen so that $\bar{v}$ satisfies the transforms of (1), namely

$$k_1 \frac{d\bar{v}}{dr} - k_2 \bar{v} = \frac{k_3}{p}, \quad r = a \ \Bigg\} .$$
$$k_1' \frac{d\bar{v}}{dr} + k_2' \bar{v} = \frac{k_3'}{p}, \quad r = b \ \Bigg\}$$

Substituting (2) in these and solving for A and B we get finally

$$p\Delta(p)\bar{v} = k_3\{[k_1' q K_1(qb) - k_2' K_0(qb)]I_0(qr) + [k_1' q I_1(qb) + k_2' I_0(qb)]K_0(qr)\} - $$
$$- k_3'\{[k_1 q K_1(qa) + k_2 K_0(qa)]I_0(qr) + [k_1 q I_1(qa) - k_2 I_0(qa)]K_0(qr)\}, \quad (3)$$

where

$$\Delta(p) = [k_1 q I_1(qa) - k_2 I_0(qa)][k_1' q K_1(qb) - k_2' K_0(qb)] - $$
$$- [k_1 q K_1(qa) + k_2 K_0(qa)][k_1' q I_1(qb) + k_2' I_0(qb)]. \quad (4)$$

v is now determined by the Inversion Theorem. The integrand is a single-valued function of λ with a simple‡ pole at $\lambda = 0$, and simple poles at $\lambda = -\kappa \alpha_n^2$, where $\pm \alpha_n$ are the roots (all real and simple)§ of

$$[k_1 \alpha J_1(a\alpha) + k_2 J_0(a\alpha)][k_1' \alpha Y_1(b\alpha) - k_2' Y_0(b\alpha)] - $$
$$- [k_1 \alpha Y_1(a\alpha) + k_2 Y_0(a\alpha)][k_1' \alpha J_1(b\alpha) - k_2' J_0(b\alpha)] = 0. \quad (5)$$

---

† Arbitrary initial temperature is discussed in § 14.8.

‡ If $k_2 = k_2' = 0$ there is a double pole at $\lambda = 0$. We exclude this case which is easily treated along the same lines.

§ Carslaw and Jaeger, loc. cit. Note that we use repeatedly the results $I_0(ix) = J_0(x)$, etc., quoted in Appendix III, (25) and (26).

We use the contour of Fig. 39 in the usual way. The residue at the pole $\lambda = 0$ is

$$\frac{-ak_3\{k_1' - bk_2' \ln(r/b)\} + bk_3'\{k_1 + ak_2 \ln(r/a)\}}{ak_2 k_1' + bk_1 k_2' + abk_2 k_2' \ln(b/a)}. \tag{6}$$

To find the residue at the pole $\lambda = -\kappa\alpha_n^2$ we need

$$\left[\lambda \frac{d\Delta}{d\lambda}\right]_{\lambda = -\kappa\alpha_n^2} = \left[\tfrac{1}{2}\mu \frac{d\Delta}{d\mu}\right]_{\mu = i\alpha_n}$$

$$= \tfrac{1}{2}i\alpha_n \begin{bmatrix} a[k_1 \mu I_0(\mu a) - k_2 I_1(\mu a)][k_1' \mu K_1(\mu b) - k_2' K_0(\mu b)] - \\ - b[k_1 \mu I_1(\mu a) - k_2 I_0(\mu a)][k_1' \mu K_0(\mu b) - k_2' K_1(\mu b)] + \\ + a[k_1 \mu K_0(\mu a) + k_2 K_1(\mu a)][k_1' \mu I_1(\mu b) + k_2' I_0(\mu b)] - \\ - b[k_1 \mu K_1(\mu a) + k_2 K_0(\mu a)][k_1' \mu I_0(\mu b) + k_2' I_1(\mu b)] \end{bmatrix}_{\mu = i\alpha_n},$$

where we have used (4) and the recurrence formulae, Appendix III (13) and (15). To simplify this we notice that, when $\mu = i\alpha_n$,

$$\frac{k_1 \mu I_1(\mu a) - k_2 I_0(\mu a)}{k_1' \mu I_1(\mu b) + k_2' I_0(\mu b)} = \frac{k_1 \mu K_1(\mu a) + k_2 K_0(\mu a)}{k_1' \mu K_1(\mu b) - k_2' K_0(\mu b)}$$

$$= \frac{k_1 \alpha_n J_1(a\alpha_n) + k_2 J_0(a\alpha_n)}{k_1' \alpha_n J_1(b\alpha_n) - k_2' J_0(b\alpha_n)} = \frac{k_1 \alpha_n Y_1(a\alpha_n) + k_2 Y_0(a\alpha_n)}{k_1' \alpha_n Y_1(b\alpha_n) - k_2' Y_0(b\alpha_n)} = \rho, \quad \text{say.} \tag{7}$$

Using this result and the Wronskian relation, Appendix III (22), we find

$$\left[\lambda \frac{d\Delta}{d\lambda}\right]_{\lambda = -\kappa\alpha_n^2} = \tfrac{1}{2}\rho(k_1'^2\alpha_n^2 + k_2'^2) - \frac{1}{2\rho}(k_1^2\alpha_n^2 + k_2^2)$$

$$= \frac{F(\alpha_n)}{2[k_1 \alpha_n J_1(a\alpha_n) + k_2 J_0(a\alpha_n)][k_1' \alpha_n J_1(b\alpha_n) - k_2' J_0(b\alpha_n)]}, \tag{8}$$

where

$$F(\alpha_n) = (k_1'^2\alpha_n^2 + k_2'^2)[k_1 \alpha_n J_1(a\alpha_n) + k_2 J_0(a\alpha_n)]^2 - \\ - (k_1^2\alpha_n^2 + k_2^2)[k_1' \alpha_n J_1(b\alpha_n) - k_2' J_0(b\alpha_n)]^2. \tag{9}$$

Thus, we get finally

$$v = \frac{-ak_3[k_1' - bk_2' \ln(r/b)] + bk_3'[k_1 + ak_2 \ln(r/a)]}{ak_2 k_1' + bk_1 k_2' + abk_2 k_2' \ln(b/a)} -$$

$$- \pi \sum_{n=1}^{\infty} e^{-\kappa\alpha_n^2 t} \frac{1}{F(\alpha_n)} \{k_1' \alpha_n J_1(b\alpha_n) - k_2' J_0(b\alpha_n)\} C_0(r, \alpha_n) \times$$

$$\times [k_3\{k_1' \alpha_n J_1(b\alpha_n) - k_2' J_0(b\alpha_n)\} - k_3'\{k_1 \alpha_n J_1(a\alpha_n) + k_2 J_0(a\alpha_n)\}], \tag{10}$$

where the α_n are the roots of (5), $F(\alpha_n)$ is defined in (9), and

$$C_0(r, \alpha_n) = J_0(r\alpha_n)[k_1 \alpha_n Y_1(a\alpha_n) + k_2 Y_0(a\alpha_n)] - \\ - Y_0(r\alpha_n)[k_1 \alpha_n J_1(a\alpha_n) + k_2 J_0(a\alpha_n)]. \tag{11}$$

Numerous special cases[†] follow from this general result or may be derived directly by the same method. As an example suppose we have constant flux, F_0, at $r = a$, and zero temperature at $r = b$. Then in (1) we have $k_2 = k_1' = k_3' = 0$; $k_1 = 1$; $k_3 = -F_0/K$; $k_2' = 1$, and the solution (10) becomes[‡]

$$v = \frac{aF_0}{K}\ln\frac{b}{r} + \frac{\pi F_0}{K}\sum_{n=1}^{\infty} e^{-\kappa\alpha_n^2 t}\frac{J_0^2(b\alpha_n)[J_0(r\alpha_n)Y_1(a\alpha_n) - Y_0(r\alpha_n)J_1(a\alpha_n)]}{\alpha_n[J_1^2(a\alpha_n) - J_0^2(b\alpha_n)]},$$

(12)

where the α_n are the positive roots of

$$J_1(a\alpha)Y_0(b\alpha) - Y_1(a\alpha)J_0(b\alpha) = 0. \tag{13}$$

More complicated boundary conditions, such as 1.9 (14), corresponding to contact with well-stirred fluid or perfect conductor at one or both faces[§] may be discussed in precisely the same way. Periodic surface temperatures are discussed by Awbery.[||] Heat production in an insulated wire carrying electric current[††] is properly a problem on a finite cylinder, but various approximate solutions based on heat flow in a hollow cylinder may be used.

Some values of the roots of (5) for the case $k_1 = k_1' = 0$ are given in Appendix IV. Roots of (5) for the case $k_2 = k_1' = 0$, that is, roots of (13), are given by Bogert.[‡‡] The case $k_2 = 0$ has also been discussed numerically.[§§]

13.5. The region bounded internally by the circular cylinder $r = a$

Despite its obvious importance in connexion with flow of heat from buried pipes and cables, cooling of mines, etc., this region has been studied only comparatively recently. Nicholson[||||] first gave the solution (6) below, but his reasoning is open to criticism; Titchmarsh (loc. cit.,

---

[†] Walters, *Phil. Mag.* (7) **38** (1947) 70–78; Comenetz, *Quart. Appl. Math.* **5** (1947) 503–10.

[‡] Fischer, *Ann. Physik*, (5) **34** (1939) 669, applies this solution to determination of thermal conductivity.

[§] The result for this case is given by Jaeger, *Proc. Roy. Soc. N.S.W.* **74** (1940) 342. Some experimental arrangements involve problems of this type, e.g. Stephens, *Phil. Mag.* (7) **15** (1933) 857.

[||] *Phil. Mag.* (7) **28** (1939) 447.

[††] Buchholz, *Z. angew. Math. Mech.* **9** (1929) 280.

[‡‡] *J. Math. Phys.* **30** (1951) 102.

[§§] The roots of (5) in this case are tabulated by Lipow and Zwick, *J. Math. Phys.* **34** (1955) 308–15. Some numerical values for the temperature are given by Geckler, *Jet Propulsion*, **25** (1955) 31–35.

[||||] *Proc. Roy. Soc.* A, **100** (1921) 226.

§ 10.10) uses the Fourier integral; Smith† uses the contour integral method developed in *C.H.* A number of solutions using the operational and Laplace transformation methods has been given by Goldstein (loc. cit.) and Carslaw and Jaeger (loc. cit.). Some numerical results are given by Jaeger.‡

I. *Initial temperature zero. The surface $r = a$ at constant temperature V*

Here the subsidiary equation is

$$\frac{d^2\bar{v}}{dr^2} + \frac{1}{r}\frac{d\bar{v}}{dr} - q^2\bar{v} = 0, \quad r > a, \tag{1}$$

to be solved with $\bar{v}$ finite as $r \to \infty$, and

$$\bar{v} = V/p, \quad r = a. \tag{2}$$

The solution is
$$\bar{v} = \frac{VK_0(qr)}{pK_0(qa)}. \tag{3}$$

Then, using the Inversion Theorem, we get

$$v = \frac{V}{2\pi i}\int_{\gamma-i\infty}^{\gamma+i\infty} e^{\lambda t}\frac{K_0(\mu r)\,d\lambda}{K_0(\mu a)\lambda}, \tag{4}$$

where $\mu = \sqrt{(\lambda/\kappa)}$. The integrand of (4) has a branch point at $\lambda = 0$, so we use the contour of Fig. 40; it is known (*W.B.F.*, § 15.7) that there are no zeros of $K_0(\mu a)$ in this contour, so that the line integral in (4) may be replaced by the sum of the integrals over *CD*, *EF* and the small circle about the origin.

The integral round the small circle gives $2\pi i$ in the limit as its radius tends to zero.

On *EF* we put $\lambda = \kappa u^2 e^{i\pi}$, and obtain for the integral

$$2\int_0^\infty e^{-\kappa u^2 t}\frac{K_0(rue^{\frac{1}{2}i\pi})\,du}{K_0(aue^{\frac{1}{2}i\pi})\,u} = 2\int_0^\infty e^{-\kappa u^2 t}\frac{J_0(ur)-iY_0(ur)\,du}{J_0(ua)-iY_0(ua)\,u}, \tag{5}$$

since $\quad K_0(ze^{\frac{1}{2}i\pi}) = -\tfrac{1}{2}\pi i H_0^{(2)}(z) = -\tfrac{1}{2}\pi i[J_0(z)-iY_0(z)].$

The integral over *CD* gives minus the conjugate of (5). Combining these results we have finally

$$v = V + \frac{2V}{\pi}\int_0^\infty e^{-\kappa u^2 t}\frac{J_0(ur)Y_0(ua)-Y_0(ur)J_0(ua)}{J_0^2(au)+Y_0^2(au)}\frac{du}{u}. \tag{6}$$

---

† *J. Appl. Phys.* **8** (1937) 441.
‡ *Proc. Roy. Soc. Edin.* A, **61** (1942) 223; also Jaeger and Clarke, ibid., p. 229.

For small values of the time we proceed as in § 13.3, using the asymptotic expansions of the Bessel functions in (3), and obtain

$$\bar{v} = \frac{V}{p}\left(\frac{a}{r}\right)^{\frac{1}{2}} e^{-q(r-a)}\left\{1 + \frac{(r-a)}{8arq} + \frac{(9a^2 - 2ar - 7r^2)}{128a^2r^2q^2}\cdots\right\}.$$

Thus from Appendix V (11),

$$v = \frac{Va^{\frac{1}{2}}}{r^{\frac{1}{2}}}\text{erfc}\,\frac{r-a}{2\sqrt{(\kappa t)}} + \frac{V(r-a)(\kappa t)^{\frac{1}{2}}}{4a^{\frac{1}{2}}r^{\frac{3}{2}}}\,\text{ierfc}\,\frac{r-a}{2\sqrt{(\kappa t)}} +$$

$$+ \frac{V(9a^2 - 2ar - 7r^2)\kappa t}{32a^{\frac{3}{2}}r^{\frac{5}{2}}}\,\text{i}^2\text{erfc}\,\frac{r-a}{2\sqrt{(\kappa t)}} + \dots. \quad (7)$$

Values of the temperature as a function of r for various values of the time are shown† in Fig. 41.

The quantity of greatest practical importance is the flux at the surface

$$f = -K\left[\frac{\partial v}{\partial r}\right]_{r=a} = \frac{4VK}{a\pi^2}\int\limits_0^\infty e^{-\kappa u^2 t}\,\frac{du}{u[J_0^2(au) + Y_0^2(au)]}. \quad (8)$$

Writing $T = \kappa t/a^2$, the flux f at the surface is also given by

$$f = \frac{KV}{a}\{(\pi T)^{-\frac{1}{2}} + \tfrac{1}{2} - \tfrac{1}{4}(T/\pi)^{\frac{1}{2}} + \tfrac{1}{8}T\dots\}, \quad (9)$$

for small values of T, and by

$$f = \frac{2VK}{a}\left\{\frac{1}{\ln(4T) - 2\gamma} - \frac{\gamma}{[\ln(4T) - 2\gamma]^2} - \dots\right\}, \quad (10)$$

for large values of T. In (10), $\gamma = 0\cdot57722\dots$ is Euler's constant.

Numerical values of the integral (8) have been tabulated by Jaeger and Clarke (loc. cit.) and are shown in Fig. 42.

II. *Initial temperature V, constant. Radiation at $r = a$ into a medium at zero temperature*

The subsidiary equation is

$$\frac{d^2\bar{v}}{dr^2} + \frac{1}{r}\frac{d\bar{v}}{dr} - q^2\bar{v} = -\frac{V}{\kappa}, \quad (11)$$

---

† The most complete set of values is given by Jaeger, *J. Math. Phys.* **34** (1956) 316–21, who gives results to 3D for $(r/a) = 1, (0\cdot1), 2, (1), 10, (10), 100$, and $\kappa t/a^2 = 0\cdot001$, $(0\cdot001), 0\cdot01, (0\cdot01), 0\cdot1, (0\cdot1), 1, (1), 10, (10), 100, (100), 1000$. A more accurate table to 5D for $(r/a) = 1, (1), 10$ and $\kappa t/a^2 = 0\cdot1, (0\cdot1), 1, (1), 10$ has been given in *Problem Report* No. 76 (1954) of the Computation Laboratory of Harvard University. Perry and Berggren, *Univ. California Publ. in Engng.* **5** (1944) 59, give an extensive discussion of flow in cylindrical regions including some numerical values obtained by Schmidt's method. Goldenberg, *Proc. Phys. Soc.* B, **69** (1956) 256–60, gives results for $r/a \doteq 2, 10$, 100 and values of the integrated heat flux.

and the boundary condition is

$$\frac{d\bar{v}}{dr} = h\bar{v}, \quad \text{at } r = a. \tag{12}$$

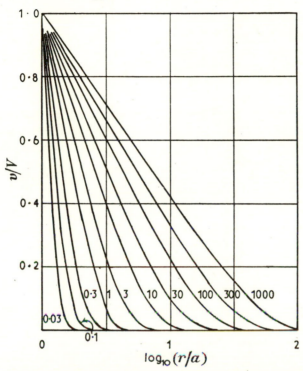

Fig. 41. Temperatures in the region bounded internally by the cylinder $r = a$, with zero initial temperature and constant surface temperature V. The numbers on the curves are the values of $\kappa t/a^2$.

Thus
$$\bar{v} = \frac{V}{p}\left\{1 + \frac{hK_0(qr)}{qK_0'(qa) - hK_0(qa)}\right\}, \tag{13}$$

and
$$v = V + \frac{hV}{2\pi i}\int_{\gamma-i\infty}^{\gamma+i\infty} e^{\lambda t}\frac{K_0(\mu r)\, d\lambda}{\lambda[\mu K_0'(\mu a) - hK_0(\mu a)]}. \tag{14}$$

The integrand has a branch point at $\lambda = 0$, so we use the contour of Fig. 40. There are no poles within† or on this contour. Then proceeding as in I above we obtain

$$v = -\frac{2hV}{\pi}\int_0^\infty e^{-\kappa u^2 t}\frac{J_0(ur)[uY_1(ua) + hY_0(ua)] - Y_0(ur)[uJ_1(ua) + hJ_0(ua)]}{[uJ_1(ua) + hJ_0(ua)]^2 + [uY_1(ua) + hY_0(ua)]^2}\frac{du}{u}. \tag{15}$$

† Carslaw and Jaeger, loc. cit., or Erdélyi and Kermack, *Proc. Camb. Phil. Soc.* **41** (1945) 74.

Values of the surface temperature are shown in Fig. 43 for the values 0·2, 0·5, 1, 2·5, 5, 10 of ah.

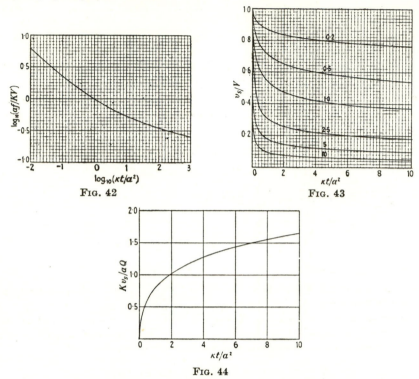

FIG. 42

FIG. 43

FIG. 44

FIG. 42. The flux f at the surface of the region bounded internally by a circular cylinder of radius a, with zero initial temperature and constant surface temperature V.

FIG. 43. Surface temperature v_s of the region bounded internally by a circular cylinder of radius a, with constant initial temperature V and radiation at its surface into a medium at zero. The numbers on the curves are values of ah.

FIG. 44. Surface temperature v_s of the region bounded internally by a circular cylinder of radius a, with zero initial temperature and constant flux Q at the surface.

III. *Zero initial temperature. Constant flux Q heat units, per unit time per unit area at $r = a$*

The solution is†

$$\bar{v} = \frac{QK_0(qr)}{KpqK_1(qa)},\tag{16}$$

$$v = -\frac{2Q}{\pi K}\int_0^\infty (1-e^{-\kappa u^2 t})\frac{J_0(ur)Y_1(ua)-Y_0(ur)J_1(ua)}{u^2[J_1^2(ua)+Y_1^2(ua)]}\,du.\tag{17}$$

† In this case the contour of Fig. 40 can be used, but the result must be left as a loop integral. The form above is obtained by applying the usual procedure to $p\bar{v}$ and then using 12.2 (4).

The result useful for small values of $\kappa t/a^2$ is

$$v = \frac{2Q}{K}\left(\frac{\kappa at}{r}\right)^{\frac{1}{2}}\left\{\operatorname{ierfc}\frac{r-a}{2(\kappa t)^{\frac{1}{2}}} - \frac{(3r+a)(\kappa t)^{\frac{1}{2}}}{4ar}\,\mathrm{i}^2\operatorname{erfc}\frac{r-a}{2(\kappa t)^{\frac{1}{2}}} + \ldots\right\}. \quad (18)$$

For large values of the time, it follows by the method of § 13.6 that

$$v = \frac{Qa}{2K}\ln\frac{4\kappa t}{Cr^2} + O\left(\frac{a^2}{\kappa t}\right), \quad (19)$$

where $\gamma = \ln C$, and $\gamma = 0\cdot57722\ldots$ is Euler's constant. (19) is also the expression 10.4 (6) for a line source emitting $2\pi aQ$ units of heat per unit time per unit length.

Some values of the surface temperature† are given in Fig. 44.

IV. *Zero initial temperature. Surface temperature* $\sin(\omega t+\epsilon)$

$$v = \frac{N_0(\omega'r)}{N_0(\omega'a)}\sin\{\omega t+\epsilon+\phi_0(\omega'r)-\phi_0(\omega'a)\} -$$

$$-\frac{2\kappa}{\pi}\int_0^\infty e^{-\kappa u^2 t}\frac{[J_0(ur)Y_0(ua)-J_0(ua)Y_0(ur)](\omega\cos\epsilon-\kappa u^2\sin\epsilon)u\,du}{[J_0^2(ua)+Y_0^2(ua)](\omega^2+\kappa^2 u^4)}, \quad (20)$$

where $\omega' = \sqrt{(\omega/\kappa)}$, and $K_0(ze^{i\pi/4}) = \ker z + i\ker z = N_0(z)e^{i\phi_0(z)}$.

V. *The regions* $z > 0$, $r > a$, *and* $0 < z < l$, $r > a$. *Unit initial temperature. Product solutions‡*

Clearly solutions for these regions can be written down as in § 8.4 using the method of § 1.15. Thus for the region $z > 0$, $r > a$, with unit initial temperature and zero surface temperature, the solution is

$$v = -\frac{2}{\pi}\operatorname{erf}\frac{z}{2\sqrt{(\kappa t)}}\int_0^\infty e^{-\kappa u^2 t}\frac{J_0(ur)Y_0(ua)-Y_0(ur)J_0(ua)}{J_0^2(au)+Y_0^2(au)}\frac{du}{u}. \quad (21)$$

Numerical values of the temperature may be found from Fig. 41.

13.6. Solutions suitable for large values of the time

In problems involving finite regions, solutions have usually been obtained in the form of series which converge more rapidly, the greater the value of t. For problems on infinite regions, on the other hand, solutions have usually taken the form

$$v = \int_0^\infty e^{-u^2 T}f(u)\,du, \quad (1)$$

where T is a parameter such as $\kappa t/a^2$, and these become more instead of less difficult to evaluate for large values of T. It is usually important to have available approximate results for large values of T: these may be obtained by manipulation of the

---

† Further values as well as some results for r/a, 2, 5, and 10 are given by Ingersoll, Adler, Plass, and Ingersoll, *Heat. Pip. Air Condit.* **22** (1950) 113–22. They also discuss in detail the comparison of results of this type with those for the continuous line source in an infinite medium.

‡ Other problems for the region $r > a$ involving non-radial flow are discussed by Blackwell, *Canad. J. Phys.* **31** (1953) 472–9.

integral (1), but, for many important practical problems the following method†
may be used as a routine procedure. If $\bar{v}(p)$ is the Laplace transform of v, v is found
from the Inversion Theorem

$$v(t) = \frac{1}{2\pi i} \int_{\gamma-i\infty}^{\gamma+i\infty} e^{\lambda t} \bar{v}(\lambda) \, d\lambda. \tag{2}$$

In the class of problem under consideration, $\bar{v}(\lambda)$ has a branch point at the origin,
and the first step in the evaluation of v is to transform the contour $(\gamma-i\infty, \gamma+i\infty)$
of (2) into the contour $CDEF$ of Fig. 40, which begins at $-\infty$ in the lower half-
plane, passes once round the branch point at the origin in the positive direction,
and ends at $-\infty$ in the upper half plane and for this reason may be denoted by
$(-\infty, 0+)$. Thus we obtain from (2)

$$v(t) = \frac{1}{2\pi i} \int_{-\infty}^{(0+)} e^{\lambda t} \bar{v}(\lambda) \, d\lambda. \tag{3}$$

In all the problems with which we have been concerned, this transformation
is legitimate, that is, it can be shown that the integrals over the portions $AA'C$
and $FB'B$ of the large circle of Fig. 40 tend to zero as the radius of this circle tends
to infinity. By writing down explicit expressions for $\bar{v}(\lambda)$ on the path $CDEF$, the
contour integral (3) is reduced to the real infinite integral (1). To determine solutions suitable for large values of the time, we start with the form (3), expand $\bar{v}(\lambda)$
in ascending powers of λ, and integrate this series term by term, assuming that this
process can be justified. Typical integrals which will be needed are

$$\frac{1}{2\pi i} \int_{-\infty}^{(0+)} \lambda^{-\nu-1} e^{\lambda t} \, d\lambda = \frac{t^\nu}{\Gamma(1+\nu)}; \qquad \frac{1}{2\pi i} \int_{-\infty}^{(0+)} \lambda^n e^{\lambda t} \, d\lambda = 0, \quad n = 0, 1, 2, ..., \tag{4}$$

$$\frac{1}{2\pi i} \int_{-\infty}^{(0+)} \frac{1}{\lambda} e^{\lambda t} \ln(k\lambda) \, d\lambda = -\ln(Ct/k), \tag{5}$$

$$\frac{1}{2\pi i} \int_{-\infty}^{(0+)} e^{\lambda t} [\ln(k\lambda)]^2 \, d\lambda = \frac{2}{t} \ln(Ct/k), \tag{6}$$

$$\frac{1}{2\pi i} \int_{-\infty}^{(0+)} e^{\lambda t} [\ln(k\lambda)]^3 \, d\lambda = \frac{\pi^2}{2t} - \frac{3}{t} [\ln(Ct/k)]^2, \tag{7}$$

where $\ln C = \gamma = 0.5772...$ is Euler's constant, k is supposed to be real and
positive, and ν is unrestricted. Of the above, (4) follows from the definition of the
gamma function, (5) from the Laplace transform of $\ln t$, and (6) and (7) by the usual
method using the path $CDEF$. Other necessary results follow by differentiation

† This is essentially one of the fundamental procedures of Heaviside's operational
calculus, cf. McLachlan, *Complex variable and Operational Calculus* (Cambridge, 1939);
the contours of (2) and (3) are his Br_1 and Br_2. Goldstein has applied the method to a
wide variety of problems in cylindrical coordinates, cf. *Proc. Lond. Math. Soc.* (2) **34**
(1932) 51.

with regard to t as a parameter, thus (5) gives

$$\frac{1}{2\pi i} \int_{-\infty}^{(0+)} e^{\lambda t} \ln(k\lambda) \, d\lambda = -1/t, \tag{8}$$

$$\frac{1}{2\pi i} \int_{-\infty}^{(0+)} e^{\lambda t} \lambda^n \ln(k\lambda) \, d\lambda = \frac{(-1)^{n+1}n!}{t^{n+1}}, \tag{9}$$

and so on.

As an example consider *the region $r > a$ with constant flux Q*.

Here from 13.5 (16), writing $\mu = (\lambda/\kappa)^{\frac{1}{2}}$, and using Appendix III (9), (10),

$$v = \frac{Q}{2\pi i K} \int_{-\infty}^{(0+)} \frac{e^{\lambda t} K_0(\mu r) \, d\lambda}{\lambda \mu K_1(\mu a)} \tag{10}$$

$$= -\frac{Qa}{2\pi i K} \int_{-\infty}^{(0+)} \frac{e^{\lambda t} \{\ln(\frac{1}{2}C r\mu) + \frac{1}{4} r^2 \mu^2 [\ln(\frac{1}{2}C\mu r) - 1] + ...\} \, d\lambda}{\lambda \{1 + \frac{1}{2}\mu^2 a^2 [\ln(\frac{1}{2}C\mu a) - \frac{1}{2}] + ...\}}$$

$$= -\frac{Qa}{2\pi i K} \int_{-\infty}^{(0+)} \frac{e^{\lambda t}}{\lambda} \{\ln(\frac{1}{2}C r\mu) + \frac{1}{4} r^2 \mu^2 [\ln(\frac{1}{2}C r\mu) - 1] -$$

$$- \frac{1}{2}\mu^2 a^2 \ln(\frac{1}{2}C r\mu)[\ln(\frac{1}{2}C r\mu) + \ln(a/r) - \frac{1}{2}] + ...\} \, d\lambda$$

$$= -\frac{Qa}{2\pi i K} \int_{-\infty}^{(0+)} \frac{e^{\lambda t}}{\lambda} \left\{ \frac{1}{2}\ln\left(\frac{C^2 r^2 \lambda}{4\kappa}\right) + \frac{r^2 \lambda}{8\kappa}\ln\left(\frac{C^2 r^2 \lambda}{4\kappa}\right) - \frac{r^2 \lambda}{4\kappa} - \right.$$

$$\left. - \frac{\lambda a^2}{8\kappa}\left[\ln\left(\frac{C^2 r^2 \lambda}{4\kappa}\right)\right]^2 - \frac{\lambda a^2}{4\kappa}\left\{\ln\left(\frac{a}{r}\right) - \frac{1}{2}\right\}\ln\left(\frac{C^2 r^2 \lambda}{4\kappa}\right) + ...\right\} \, d\lambda,$$

where $\gamma = \ln C$, and $\gamma = 0.5772...$ is Euler's constant.

Using (4), (5), (6), (8), this gives

$$v = \frac{Qa}{2K}\left\{\ln\frac{4\kappa t}{Cr^2} + \frac{a^2}{2\kappa t}\ln\frac{4\kappa t}{Cr^2} + \frac{1}{4\kappa t}\left[a^2 + r^2 - 2a^2\ln\frac{a}{r}\right] + ...\right\}. \tag{11}$$

The result 13.7 (18), and similar results for all the problems of § 13.7, may be derived in the same way. Additional theory needed to treat 13.5 (3) and 13.5 (13) is given by Ritchie and Sakakura.†

13.7. The region $r > a$ bounded internally by a circular cylinder of perfect conductor

In many problems of practical importance, for example, in the heating of buried electric cables and in 'probe' methods of measuring thermal conductivity, a cylinder of a metallic conductor is surrounded by an infinite medium such as soil or rock of much lower thermal conductivity. To a good approximation, the metal may be regarded as being a perfect conductor of heat—the effect of finite conductivity may be estimated from the more complicated results of § 13.8. In most problems interest is centred on the temperature of the material within the cylinder: this usually is either initially at unit temperature (the region outside it being

† *J. Appl. Phys.* **27** (1956) 1453.

at zero) or has heat supplied to it at a constant rate. Similar problems arise if the region $r < a$ contains well-stirred fluid.

In many of the practical problems there is a contact resistance between the cylinder and the surrounding medium: this will be discussed in III and IV below, but the simpler case of no contact resistance will be discussed first.

I. *The region $r > a$ initially at zero temperature. At $r = a$ it is in contact with a perfect conductor of heat capacity S per unit length of the cylinder, initially at temperature V_0. No contact resistance at $r = a$*†

Writing v for the temperature in $r > a$ we get as in § 13.2 V

$$\bar{v} = \frac{aV_0 K_0(qr)}{\kappa q[aqK_0(aq) + \alpha K_1(aq)]}, \tag{1}$$

where
$$\alpha = 2\pi a^2 \rho c / S \tag{2}$$

is a parameter which is twice the ratio of the heat capacity of an equivalent volume of the medium to that of the perfect conductor. It follows that

$$v = \frac{2V_0}{\pi} \int_0^\infty e^{-\kappa t u^2/a^2} \{J_0(ru/a)[uY_0(u) - \alpha Y_1(u)] -$$

$$-Y_0(ur/a)[uJ_0(u) - \alpha J_1(u)]\} \frac{du}{\Delta(u)}, \tag{3}$$

where
$$\Delta(u) = [uJ_0(u) - \alpha J_1(u)]^2 + [uY_0(u) - \alpha Y_1(u)]^2. \tag{4}$$

The temperature V in $r < a$, which is the value of (3) for $r = a$, is

$$V = \frac{4\alpha V_0}{\pi^2} \int_0^\infty e^{-\kappa t u^2/a^2} \frac{du}{u\Delta(u)}. \tag{5}$$

Values of V as a function of $\kappa t/a^2$ for various values of α are shown in Fig. 45.

II. *The problem of* I, *except that the cylinder is initially at zero temperature and heat is supplied to it for $t > 0$ at the rate Q per unit length per unit time*

$$\bar{v} = \frac{aQK_0(qr)}{\kappa Spq[aqK_0(aq) + \alpha K_1(aq)]}, \tag{6}$$

$$v = \frac{2Qa^2}{\pi \kappa S} \int_0^\infty (1 - e^{-\kappa u^2 t/a^2}) \{J_0(ur/a)[uY_0(u) - \alpha Y_1(u)] -$$

$$-Y_0(ur/a)[uJ_0(u) - \alpha J_1(u)]\} \frac{du}{u^2 \Delta(u)}, \tag{7}$$

† Bullard, *Proc. Roy. Soc.* A **222** (1954) 408–29, and Jaeger, *Aust. J. Phys.* **9** (1956) 167–79, give short tables of the integral (3).

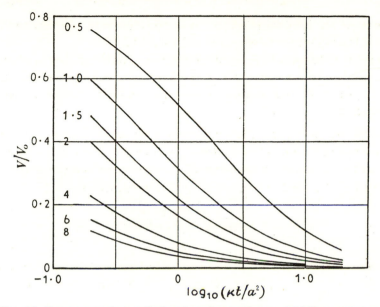

FIG. 45. Temperature in a cylinder of perfect conductor, initially at temperature V_0, in an infinite medium initially at zero. The numbers on the curves are values of the parameter α.

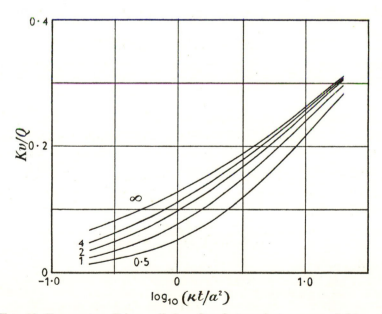

FIG. 46. Temperature V in a cylinder of perfect conductor surrounded by an infinite medium if heat is supplied to the cylinder at a constant rate. The numbers on the curves are values of α.

where α and $\Delta(u)$ are defined in (2) and (4), respectively. In determining (7) from (6), the method leading to 13.5 (17) has to be used. The temperature V in the cylinder, which is the value of (7) when $r = a$, is

$$V = \frac{2Q\alpha^2}{\pi^3 K} \int_0^\infty \frac{(1 - e^{-\kappa t u^2/a^2})\, du}{u^3 \Delta(u)}. \tag{8}$$

V is plotted in Fig. 46 as a function of $\kappa t/a^2$ for various values of α.

III. *Heating of a buried electric cable*

An electric cable consists essentially of three parts, the current-carrying *core* which is separated by the *insulation* from a protecting metal *sheath*. Thus the simplest idealization of a practical cable† will regard the core and sheath as perfect conductors of thermal capacities S_1 and S_2, respectively, per unit length of the cable, and will regard the insulation as of negligible heat capacity and thermal resistance R per unit length of cable. The outside radius of the sheath is supposed to be a, and the cable is supposed to be buried in soil of thermal properties K, κ, ρ, c, contact resistance between the sheath and the soil being neglected.

Two problems are of practical importance, (i) heating by steady current in the core, and (ii) short-circuit heating in which the temperature of the core is suddenly raised by a finite amount and it is required to study the way in which this heat is dissipated.

Short-circuit heating. If the temperature of the core is initially V_0 and that of the sheath and the surrounding material is initially zero, the temperature V in the core at time t is

$$V = V_0\, F(h, \alpha_1, \alpha_2, \tau), \tag{9}$$

where

$$h = 2\pi RK, \qquad \alpha_1 = 2\pi a^2 \rho c/S_1, \qquad \alpha_2 = 2\pi a^2 \rho c/S_2, \qquad \tau = \kappa t/a^2, \tag{10}$$

and

$$F(h, \alpha_1, \alpha_2, \tau) = \frac{4\alpha_1 \alpha_2^2}{\pi^2} \int_0^\infty \frac{\exp(-u^2\tau)\, du}{u\Delta_1(u)}, \tag{11}$$

and

$$\Delta_1(u) = [u(\alpha_1 + \alpha_2 - hu^2)J_0(u) - \alpha_2(\alpha_1 - hu^2)J_1(u)]^2 +$$
$$+ [u(\alpha_1 + \alpha_2 - hu^2)Y_0(u) - \alpha_2(\alpha_1 - hu^2)Y_1(u)]^2. \tag{12}$$

Constant-current heating. If the whole system is initially at zero temperature and heat is supplied at the rate Q per unit length per unit time to the core, the temperature V in the core at time t is

$$V = (Q/K)G(h, \alpha_1, \alpha_2, \tau), \tag{13}$$

where

$$G(h, \alpha_1, \alpha_2, \tau) = \frac{2\alpha_1^2 \alpha_2^2}{\pi^3} \int_0^\infty \frac{[1 - \exp(-u^2\tau)]\, du}{u^3 \Delta_1(u)}, \tag{14}$$

where the symbols are defined in (10) and (12) above.

† A very full discussion from the engineering point of view is given by Whitehead and Hutchings, *J. Instn. Elect. Engrs.* **83** (1938) 517. Whitehead, *Proc. Phys. Soc.* **56** (1944) 358–66, gives some approximations for the case of short-circuit heating (effectively to I above). The question is discussed from the present point of view by Jaeger (loc. cit.), and Jaeger and Newstead, *Instn. Elect. Engrs. Monograph* No. 253 S (1957). See also Schubert, *Z. angew. Phys.* **2** (1950) 174–9.

The integrals (11) and (14) have not been tabulated, but Jaeger (loc. cit.) gives graphs of $F(1, \alpha, \infty, \tau)$, $F(2, \alpha, \infty, \tau)$, $G(1, \alpha, \infty, \tau)$, and $G(2, \alpha, \infty, \tau)$ which correspond to an unsheathed cable.

If the cable, instead of being surrounded by an infinite medium, is buried at depth a below the surface of the ground, an 'image' must be added at an equal distance above the surface. This image can be represented adequately by a continuous line source as in 10.4 (7).

IV. 'Probe' methods of determining thermal conductivity

If a cylinder of radius a of perfect conductor is surrounded by an infinite medium and is heated at the rate Q per unit length per unit time for $t > 0$, all initial temperatures being zero, the temperature V of the perfect conductor at time t is by (13)

$$V = (Q/K)G(h, \alpha_1, \infty, \tau),   \tag{15}$$

if there is contact resistance R per unit length between the cylinder and the surrounding medium. For small values of τ, it follows as in § 13.3 that

$$V = \frac{Q\alpha_1}{2\pi K}\left\{\tau - \frac{\alpha_1 \tau^2}{2h} + O(\tau^{\frac{5}{2}})\right\}, \quad \text{if } h \neq 0,   \tag{16}$$

or

$$V = \frac{Q\alpha_1}{2\pi K}\left\{\tau - \frac{4\alpha_1}{3\pi^{\frac{1}{2}}}\tau^{\frac{3}{2}} + O(\tau^2)\right\}, \quad \text{if } h = 0.   \tag{17}$$

For large values of τ, as in § 13.6,

$$V = \frac{Q}{4\pi K}\left\{2h + \ln\frac{4\tau}{C} - \frac{(4h - \alpha_1)}{2\alpha_1 \tau} + \frac{\alpha_1 - 2}{2\alpha_1 \tau}\ln\frac{4\tau}{C} + ...\right\}   \tag{18}$$

where $C = 1 \cdot 7811 = \exp\gamma$, and $\gamma = 0 \cdot 5772...$ is Euler's constant.

It follows from (18) that, in all cases, a curve of V against $\ln t$ has a linear asymptote of slope Q/K, and thus if Q is known K is determined immediately. If $\kappa = 0 \cdot 01$, $a = 0 \cdot 1, \tau = \kappa t/a^2$ has the value 600 at the end of 10 minutes so that the asymptote is rapidly attained: on the other hand, if $a = 1$, the value is only 6 so that much longer times are necessary.

Probe methods[†] are very attractive for measurements in granular materials, soils, and rocks—in the latter case, since the probe must be inserted in a hole drilled in the rock, radii of at least 1 cm are to be expected, and a substantial thermal resistance exists in the air space between the probe and the rock material. Blackwell, loc. cit., gives a full discussion of this case and indicates the possibility of determining all three of R, κ, and K from experimental results by using (16) and (18).

13.8. Composite cylindrical regions

Problems on conduction of heat in composite circular or hollow circular cylinders are easily treated by the Laplace transformation, but

---

† de Vries, *Soil Sci.* **73** (1952) 83–89; van der Held, Hardebol, and Kalshoven, *Physica*, **19** (1953) 208–16; Buettner, *Trans. Amer. Geophys. Union*, **36** (1955) 831–7; Blackwell, *J. Appl. Phys.* **25** (1954) 137–44. The effect of finite length of the heating probe is discussed by Blackwell, *Canad. J. Phys.* **31** (1953) 472–9, ibid. **34** (1956) 412–17, who concludes that the length should be at least 25 to 30 times the diameter.

the results are very complicated.† Here we study two cases of the infinite composite region.

I. *The region $0 \leqslant r < a$ in cylindrical coordinates is of one substance K_1, κ_1, etc., and the region $r > a$ of another K_2, κ_2, etc. Initial temperature V, constant, in the region $0 \leqslant r < a$, and zero in the region $r > a$*

Writing v_1 and v_2 for the temperatures in the two regions, the subsidiary equations are

$$\frac{d^2\bar{v}_1}{dr^2} + \frac{1}{r}\frac{d\bar{v}_1}{dr} - q_1^2\bar{v}_1 = -\frac{V}{\kappa_1}, \quad 0 \leqslant r < a, \tag{1}$$

$$\frac{d^2\bar{v}_2}{dr^2} + \frac{1}{r}\frac{d\bar{v}_2}{dr} - q_2^2\bar{v}_2 = 0, \qquad r > a, \tag{2}$$

where $q_1 = \surd(p/\kappa_1)$ and $q_2 = \surd(p/\kappa_2)$.

If we assume that there is no contact resistance (cf. § 1.9 G) at the boundary $r = a$, the boundary conditions there are

$$\bar{v}_1 = \bar{v}_2, \quad \text{and} \quad K_1\frac{d\bar{v}_1}{dr} = K_2\frac{d\bar{v}_2}{dr}, \quad \text{when } r = a. \tag{3}$$

These have to be solved subject to the conditions that $\bar{v}_1$ is finite at $r = 0$, and that $\bar{v}_2$ is bounded as $r \to \infty$. The solutions are

$$\bar{v}_1 = \frac{V}{p} - \frac{VK_2\kappa_1^{\frac{1}{2}}}{p\Delta}K_1(q_2 a)I_0(q_1 r), \tag{4}$$

$$\bar{v}_2 = \frac{VK_1\kappa_2^{\frac{1}{2}}I_1(q_1 a)K_0(q_2 r)}{p\Delta}, \tag{5}$$

where
$$\Delta = K_2\kappa_1^{\frac{1}{2}}I_0(q_1 a)K_1(q_2 a) + K_1\kappa_2^{\frac{1}{2}}I_1(q_1 a)K_0(q_2 a). \tag{6}$$

Then, using the Inversion Theorem with the contour of Fig. 40, we get finally

$$v_1 = \frac{4VK_1K_2\kappa_2}{\pi^2 a}\int_0^\infty e^{-\kappa_1 u^2 t}\frac{J_0(ur)J_1(ua)\,du}{u^2[\phi^2(u)+\psi^2(u)]}, \tag{7}$$

$$v_2 = \frac{2VK_1\kappa_2^{\frac{1}{2}}}{\pi}\int_0^\infty e^{-\kappa_1 u^2 t}\frac{J_1(ua)[J_0(\kappa ur)\phi(u)-Y_0(\kappa ur)\psi(u)]\,du}{u[\phi^2(u)+\psi^2(u)]}, \tag{8}$$

where $\kappa = \surd(\kappa_1/\kappa_2)$, and

$$\left.\begin{aligned}\psi(u) &= K_1\kappa_2^{\frac{1}{2}}J_1(au)J_0(\kappa au) - K_2\kappa_1^{\frac{1}{2}}J_0(au)J_1(\kappa au)\\ \phi(u) &= K_1\kappa_2^{\frac{1}{2}}J_1(au)Y_0(\kappa au) - K_2\kappa_1^{\frac{1}{2}}J_0(au)Y_1(\kappa au)\end{aligned}\right\}. \tag{9}$$

† Various problems are discussed by Jaeger, *Phil. Mag.* (7) **32** (1941) 324; Penner and Sherman, *J. Chem. Phys.* **15** (1947) 569–74; Schaaf, *Quart. Appl. Math.* **3** (1945) 356–60; Tranter, *Phil. Mag.* (7) **38** (1947) 131–4; Thiruvenkatachar and Ramakrishna, *Quart. Appl. Math.* **10** (1952) 255–62; Levy, *Trans. Amer. Soc. Mech. Engrs.* **78** (1956) 1627–35. The Green's function for the infinite composite region is given by Jaeger, *Phil. Mag.* (7) **35** (1944) 169.

II. *The same regions as in* I. *Zero initial temperature in both regions.*
Heat production at the constant rate A_0 per unit time per unit volume in
the region $0 \leqslant r < a$ for $t > 0$

In this case (1) is replaced by

$$\frac{d^2\bar{v}_1}{dr^2} + \frac{1}{r}\frac{d\bar{v}_1}{dr} - q_1^2\,\bar{v}_1 = -\frac{A_0}{K_1 p}, \quad 0 \leqslant r < a, \tag{10}$$

the other equations being the same as before. (4) and (5) are replaced by

$$\bar{v}_1 = \frac{\kappa_1 A_0}{K_1 p^2} - \frac{A_0 K_2 \kappa_1^{\frac{3}{2}}}{K_1 p^2 \Delta} K_1(q_2\,a) I_0(q_1 r), \tag{11}$$

$$\bar{v}_2 = \frac{A_0 \kappa_1 \kappa_2^{\frac{1}{2}} I_1(q_1 a) K_0(q_2 r)}{p^2 \Delta}. \tag{12}$$

And†

$$v_1 = \frac{4A_0 K_2 \kappa_2}{\pi^2 a} \int_0^\infty \frac{(1 - e^{-\kappa_1 u^2 t}) J_0(ur) J_1(ua)\,du}{u^4[\phi^2(u) + \psi^2(u)]}, \tag{13}$$

$$v_2 = \frac{2A_0 \kappa_2^{\frac{1}{2}}}{\pi} \int_0^\infty \frac{(1 - e^{-\kappa_1 u^2 t}) J_1(ua)[J_0(\kappa ur)\phi(u) - Y_0(\kappa ur)\psi(u)]\,du}{u^3[\phi^2(u) + \psi^2(u)]}, \tag{14}$$

where $\phi(u)$ and $\psi(u)$ are defined in (9).

These results are the exact solutions of the problems treated approximately in § 13.7 I and II, so they enable the appropriateness of the approximations of § 13.7 to be checked in special cases.

There is no difficulty in writing down the Laplace transforms of the temperatures in a composite cylindrical region using the matrix method developed for slabs in § 12.8. The analysis is the same as that of § 7.3 for steady periodic temperatures in composite cylinders with $ki^{\frac{1}{2}}$ replaced by q. Results suitable for large or small values of the time may be obtained by the method of §§ 13.6, 13.3.

13.9. The sphere. Radial flow

Since problems on radial flow in the sphere are reduced by the substitution $u = vr$ to problems on linear flow in the rod, and the latter have been very fully treated, there is little point in reproducing the corresponding work here. However, if results for the sphere are needed and the corresponding solutions for the rod are not available, it is more satisfactory to apply the Laplace transformation directly to the problem for the sphere: some examples are given below.

† The results are derived from (7) and (8) by 12.2 (4): The Inversion Theorem if applied to (11) and (12) only gives a loop integral $(-\infty, 0+)$, cf. § 13.5 III, footnote.

I. *The sphere* $0 \leqslant r < a$. *Initial temperature* V, *constant. Zero surface temperature*

The subsidiary equation is

$$\frac{d^2\bar{v}}{dr^2} + \frac{2}{r}\frac{d\bar{v}}{dr} - q^2\bar{v} = -\frac{V}{\kappa}, \quad 0 \leqslant r < a, \tag{1}$$

or

$$\frac{d^2(r\bar{v})}{dr^2} - q^2(r\bar{v}) = -\frac{rV}{\kappa}, \quad 0 \leqslant r < a, \tag{2}$$

to be solved with $\bar{v} = 0, \quad$ when $r = a,$ \hfill (3)

and $\bar{v}$ finite at $r = 0$.

The general solution of (2) is

$$r\bar{v} = A \sinh qr + B \cosh qr + \frac{rV}{p}, \tag{4}$$

and we must take $B = 0$ in order to have $\bar{v}$ finite as $r \to 0$. Thus

$$\bar{v} = \frac{V}{p} - \frac{aV \sinh qr}{rp \sinh qa}. \tag{5}$$

Applying the Inversion Theorem gives the results of § 9.3. Alternatively we may proceed as in § 12.5 to find solutions suitable for small values of the time. (5) gives

$$\bar{v} = \frac{V}{p} - \frac{aV}{rp} \sum_{n=0}^{\infty} \{e^{-q[(2n+1)a-r]} - e^{-q[(2n+1)a+r]}\},$$

and therefore

$$v = V - \frac{aV}{r} \sum_{n=0}^{\infty} \left\{ \text{erfc}\,\frac{(2n+1)a-r}{2\sqrt{(\kappa t)}} - \text{erfc}\,\frac{(2n+1)a+r}{2\sqrt{(\kappa t)}} \right\}. \tag{6}$$

If $r = 0$ this becomes

$$v = V - \frac{aV}{\sqrt{(\pi\kappa t)}} \sum_{n=0}^{\infty} \exp\left[-\frac{(2n+1)^2 a^2}{4\kappa t} \right].$$

II. *Heat is produced for* $t > 0$ *at the constant rate* A_0 *per unit time per unit volume in the region* $0 \leqslant r < a$ *of infinite medium whose initial temperature is zero*[†]

Here, as in (2), the subsidiary equations are

$$\frac{d^2(r\bar{v})}{dr^2} - q^2(r\bar{v}) = -\frac{rA_0}{Kp}, \quad 0 \leqslant r < a, \tag{7}$$

$$\frac{d^2(r\bar{v})}{dr^2} - q^2(r\bar{v}) = 0, \quad r > a. \tag{8}$$

---

† Goldenberg, *Brit. J. Appl. Phys.* **2** (1951) 233. Goldenberg and Tranter, ibid. **3** (1952) 296–8, discuss the case in which the media in $r > a$ and $r < a$ have different conductivities, and give some numerical results.

The solution of (7) which is finite as $r \to 0$ is

$$r\bar{v} = \frac{r\kappa A_0}{Kp^2} + A \sinh qr. \tag{9}$$

Also the solution of (8) which is finite as $r \to \infty$ is

$$r\bar{v} = Be^{-qr}. \tag{10}$$

A and B are to be found from the condition that $\bar{v}$ and $d\bar{v}/dr$ are to be continuous at $r = a$. These give

$$A = -\frac{a\kappa A_0}{Kp^2}\left(1 + \frac{1}{qa}\right)e^{-qa}, \qquad B = \frac{a\kappa A_0}{Kp^2}\left(\cosh qa - \frac{1}{qa}\sinh qa\right). \tag{11}$$

From (9) and (11) we find for the temperature at $r = 0$,

$$\bar{v}_{r=0} = \frac{\kappa A_0}{Kp^2} - \frac{\kappa A_0}{Kp^2}(1 + qa)e^{-qa},$$

so that

$$v_{r=0} = \frac{a^2 A_0}{2K}\left\{1 + \left(\frac{2\kappa t}{a^2} - 1\right)\mathrm{erf}\frac{a}{2(\kappa t)^{\frac{1}{2}}} - 2\left(\frac{\kappa t}{\pi a^2}\right)^{\frac{1}{2}}e^{-a^2/4\kappa t}\right\}. \tag{12}$$

Also, for $0 < r < a$,

$$v = \frac{\kappa A_0 t}{K}\left\{1 - \frac{2a}{r}\,\mathrm{i}^2\mathrm{erfc}\frac{a-r}{2(\kappa t)^{\frac{1}{2}}} + \frac{2a}{r}\,\mathrm{i}^2\mathrm{erfc}\frac{a+r}{2(\kappa t)^{\frac{1}{2}}} - \right.$$
$$\left. - \frac{4(\kappa t)^{\frac{1}{2}}}{r}\,\mathrm{i}^3\mathrm{erfc}\frac{a-r}{2(\kappa t)^{\frac{1}{2}}} + \frac{4(\kappa t)^{\frac{1}{2}}}{r}\,\mathrm{i}^3\mathrm{erfc}\frac{a+r}{2(\kappa t)^{\frac{1}{2}}}\right\}, \tag{13}$$

and for $r > a$,

$$v = \frac{2\kappa A_0 ta}{rK}\left\{\mathrm{i}^2\mathrm{erfc}\frac{r-a}{2(\kappa t)^{\frac{1}{2}}} + \mathrm{i}^2\mathrm{erfc}\frac{r+a}{2(\kappa t)^{\frac{1}{2}}} - \frac{2(\kappa t)^{\frac{1}{2}}}{a}\,\mathrm{i}^3\mathrm{erfc}\frac{r-a}{2(\kappa t)^{\frac{1}{2}}} + \frac{2(\kappa t)^{\frac{1}{2}}}{a}\,\mathrm{i}^3\mathrm{erfc}\frac{r+a}{2(\kappa t)^{\frac{1}{2}}}\right\}. \tag{14}$$

Results for the case in which heat is generated at the rate $A_0 t^n$, where n may be $-\frac{1}{2}, 0, \frac{1}{2}, 1,\ldots$, may be obtained in the same way.

III. *The region $0 < r < a$ contains mass M_1 of perfect conductor of specific heat c_1, it is surrounded by an infinite region of thermal conductivity K and diffusivity κ, there being contact resistance $1/H$ per unit area over the surface $r = a$. Zero initial temperature. Heat is supplied within $r = a$ at the rate Q per unit time*

Writing v_1 for the temperature of the perfect conductor, and v for the temperature in $r > a$, we find

$$\bar{v}_1 = \frac{Qa^2[1 + ah + aq]}{M_1 c_1 \kappa p[a^3 q^3 + a^2 q^2(1 + ah) + ka^2 hq + kah]}, \tag{15}$$

$$\bar{v} = \frac{hQa^4}{r\kappa M_1 c_1 p[a^3 q^3 + a^2 q^2(1 + ah) + ka^2 hq + kah]}e^{-q(r-a)}, \tag{16}$$

where

$$k = 4\pi a^3 \rho c/M_1 c_1, \quad \text{and} \quad h = H/K. \tag{17}$$

If k and ah are known numerically, the denominators of (15) and (16) can be factorized and $\bar{v}_1$ and $\bar{v}$ expressed in partial fractions with denominators of type $p(q+b)$; v_1 and v can then be found from Appendix V (12) and (14).

Alternatively, the Inversion Theorem can be used, giving

$$v_1 = \frac{Q}{4\pi a K}\left\{\frac{1 + ah}{ah} - \frac{2a^2 k^2 h^2}{\pi}\int_0^\infty \frac{e^{-\kappa u^2 t/a^2}\,du}{[u^2(1 + ah) - kah]^2 + [u^3 - kahu]^2}\right\}. \tag{18}$$

For small values of the time, it follows as in § 12.5 that

$$v_1 = \frac{Q}{M_1 c_1}\left\{t - \frac{khkt^2}{2a} + \cdots\right\},$$ (19)

while for large values of the time, the method of § 13.6 gives

$$\frac{Q}{4\pi Ka}\left\{\frac{1+ah}{ah} - \frac{a}{(\pi\kappa t)^{\frac{1}{2}}} - \frac{a^2[2+ah(2-k)]}{2hk\pi^{\frac{1}{2}}(\kappa t)^{\frac{1}{2}}} + \cdots\right\}.$$ (20)

This problem corresponds to that of § 13.7 with a sphere replacing a cylinder. It has been used in the same way for the determination of thermal conductivity.

IV. *The problem of* III† *except that* $Q = 0$ *and the initial temperature of the sphere is* V_0

The temperature v_1 of the sphere at time t is

$$v_1 = \frac{2ka^2h^2V_0}{\pi}\int\limits_0^\infty \frac{e^{-\kappa u^2 t/a^2}u^2\,du}{[u^2(1+ah)-kah]^2+[u^3-kahu]^2}.$$ (21)

For small values of the time, it is

$$v_1 = V_0\left\{1 - \frac{khkt}{a} + \cdots\right\},$$ (22)

and for large values of the time,

$$v_1 = \frac{V_0 a^3}{2k\pi^{\frac{1}{2}}(\kappa t)^{\frac{3}{2}}} + \frac{3V_0 a^4[2+ah(2-k)]}{4hk^2\pi^{\frac{1}{2}}(\kappa t)^{\frac{3}{2}}} + \cdots.$$ (23)

V. *The region* $0 < r < a$ *contains mass* M_1 *of perfect conductor of specific heat* c_1. *The region* $a < r < b$ *consists of solid of conductivity* K *and diffusivity* κ. *There is no contact resistance at the boundary* $r = a$. *The initial temperature of the whole is* V_0, *constant, and for* $t > 0$ *the surface* $r = b$ *is kept at zero*

If v is the temperature in $a \leqslant r \leqslant b$,

$$v = \frac{4kbV_0}{r}\sum_{n=1}^\infty \frac{\sin\alpha_n(b-r)}{2k(b-a)\alpha_n+4a\alpha_n\sin^2\alpha_n(b-a)-k\sin 2(b-a)\alpha_n}e^{-\kappa\alpha_n^2 t},$$ (24)

where α_n, $n = 1, 2,...$, are the positive roots of

$$ka\alpha\cos\alpha(b-a) = (a^2\alpha^2-k)\sin\alpha(b-a),$$ (25)

and $$k = 4\pi a^3\rho c/M_1 c_1.$$ (26)

VI. *The problem of* V, *except that the initial temperature is zero and heat is supplied in the sphere* $r = a$ *at the rate* Q *per unit time*

$$v = \frac{Q}{4\pi K}\left(\frac{1}{r} - \frac{1}{b}\right) - $$

$$- \frac{Qk}{r\pi Ka}\sum_{n=1}^\infty \frac{\sin\alpha_n(b-a)\sin\alpha_n(b-r)}{\alpha_n[2k(b-a)\alpha_n+4a\alpha_n\sin^2\alpha_n(b-a)-k\sin 2\alpha_n(b-a)]}e^{-\kappa\alpha_n^2 t},$$ (27)

where the α_n are the positive roots of (25).

† Paterson, *Proc. Glasgow Math. Ass.* **1** (1953) **164–9**, considers the case in which heat is generated in the region $r > a$.

VII. *Composite spherical solids*†

We consider the problem of the solid sphere of radius b, of which an inner core, $0 \leqslant r < a$, has conductivity, diffusivity, and temperature K_1, κ_1, v_1, and in an outer shell, $a < r < b$, the corresponding quantities are K_2, κ_2, v_2. It is assumed that there is no contact resistance at $r = a$.

Suppose that initially the solid is at constant temperature V, and that for $t > 0$ the outer surface, $r = b$, is kept at zero temperature.

Writing, as usual, $u_1 = rv_1$, $u_2 = rv_2$, we have to solve the equations

$$\frac{\partial u_1}{\partial t} = \kappa_1 \frac{\partial^2 u_1}{\partial r^2}, \quad 0 \leqslant r < a, \quad t > 0; \quad \frac{\partial u_2}{\partial t} = \kappa_2 \frac{\partial^2 u_2}{\partial r^2}, \quad a < r < b, \quad t > 0, \quad (28)$$

with

$$u_1 = u_2, \quad r = a, \quad t > 0, \tag{29}$$

$$K_1\left(\frac{1}{r}\frac{\partial u_1}{\partial r} - \frac{u_1}{r^2}\right) = K_2\left(\frac{1}{r}\frac{\partial u_2}{\partial r} - \frac{u_2}{r^2}\right), \quad r = a, \quad t > 0, \tag{30}$$

$$u_2 = 0, \quad r = b, \quad t > 0,$$

$$u_1 = rV, \quad u_2 = rV, \quad t = 0,$$

and v_1 finite at $r = 0$.

It should be remarked that, because of the form of the boundary condition (30), results for composite spheres do not follow immediately from corresponding results for composite slabs.

The subsidiary equations are, writing $q_1^2 = p/\kappa_1$, $q_2^2 = p/\kappa_2$,

$$\frac{d^2\bar{u}_1}{dr^2} - q_1^2\bar{u}_1 = -\frac{rV}{\kappa_1}, \quad 0 \leqslant r < a; \quad \frac{d^2\bar{u}_2}{dr^2} - q_2^2\bar{u}_2 = -\frac{rV}{\kappa_2}, \quad a < r < b,$$

with

$$\bar{u}_1 = \bar{u}_2, \quad r = a; \quad \bar{u}_2 = 0, \quad r = b,$$

$$K_1\left(a\frac{d\bar{u}_1}{dr} - \bar{u}_1\right) = K_2\left(a\frac{d\bar{u}_2}{dr} - \bar{u}_2\right), \quad r = a,$$

and $\bar{v}_1$ finite at $r = 0$.

Solving these equations we get

$$\bar{v}_1 = \frac{V}{p} - \frac{abVK_2 q_2 \sinh rq_1}{rp\{K_2\psi_2(b)\sinh q_1 a + K_1\psi_1(a)\sinh q_2(b-a)\}}, \tag{31}$$

$$\bar{v}_2 = \frac{V}{p} - \frac{bV\{K_2\psi_2(r)\sinh q_1 a + K_1\psi_1(a)\sinh q_2(r-a)\}}{rp\{K_2\psi_2(b)\sinh q_1 a + K_1\psi_1(a)\sinh q_2(b-a)\}}, \tag{32}$$

where

$$\psi_1(r) = aq_1\cosh q_1 r - \sinh q_1 r, \tag{33}$$

$$\psi_2(r) = aq_2\cosh q_2(r-a) + \sinh q_2(r-a). \tag{34}$$

The discussion now follows closely that for the composite slab, § 12.8. The Inversion Theorem is applied to the second terms of (31) and (32), giving integrals whose integrands have simple poles at $\lambda = 0$ (giving contributions of $-V$), and simple poles at $\lambda = -\kappa_1\alpha_m^2$, where $\pm\alpha_m$, $m = 1, 2,...$, are the roots of

$$K_2\{ka\alpha\cos k(b-a)\alpha + \sin k(b-a)\alpha\}\sin a\alpha +$$

$$+ K_1\{a\alpha\cos a\alpha - \sin a\alpha\}\sin k(b-a)\alpha = 0, \tag{35}$$

and

$$k = \sqrt{(\kappa_1/\kappa_2)}, \quad \sigma = K_1/kK_2. \tag{36}$$

The roots of (35) are the roots of

$$K_2\{ka\alpha\cot k(b-a)\alpha + 1\} + K_1\{a\alpha\cot a\alpha - 1\} = 0. \tag{37}$$

† For references see § 12.8.

together with the common roots of

$$\sin a\alpha = 0, \qquad \sin k(b-a)\alpha = 0. \tag{38}$$

If $k(b-a)/a$ is irrational, the equations (38) have no common roots and we get from (31) and (32)

$$v_1 = \frac{2bV}{r} \sum_{n=1}^{\infty} \frac{1}{\phi(\alpha_n)} \sin r\alpha_n \sin a\alpha_n \sin k(b-a)\alpha_n e^{-\kappa_1 \alpha_n^2 t}, \tag{39}$$

$$v_2 = \frac{2bV}{r} \sum_{n=1}^{\infty} \frac{1}{\phi(\alpha_n)} \sin^2 a\alpha_n \sin k(b-r)\alpha_n e^{-\kappa_1 \alpha_n^2 t}, \tag{40}$$

where $\pm\alpha_n$, $n = 1, 2,...$, are the roots of (37), and

$$\phi(\alpha_n) = \sigma a\alpha_n \sin^2 k\alpha_n(b-a) + k(b-a)\alpha_n \sin^2 a\alpha_n + \frac{1-k\sigma}{ka\alpha_n}\sin^2 a\alpha_n \sin^2 k(b-a)\alpha_n. \tag{41}$$

If $k(b-a)/a$ is rational, suppose in its lowest terms it is equal to l/m. Then the equations (38) have the common positive roots

$$\alpha = \frac{mn\pi}{a}, \quad n = 1, 2, 3,... . \tag{42}$$

These roots of (35) give rise to additional terms

$$-\frac{2bV}{r\pi(l\sigma+m)} \sum_{n=1}^{\infty} \frac{(-1)^{(l+m)n}}{n} \sin\frac{mn\pi r}{a} e^{-\kappa_1 m^2 n^2 \pi^2 t/a^2}, \tag{43}$$

and

$$\frac{2bV\sigma}{r\pi(l\sigma+m)} \sum_{n=1}^{\infty} \frac{1}{n}\sin\frac{n\pi l(b-r)}{b-a} e^{-\kappa_1 m^2 n^2 \pi^2 t/a^2}, \tag{44}$$

in the expressions for v_1 and v_2 respectively.

XIV

THE USE OF GREEN'S FUNCTIONS IN THE SOLUTION OF THE EQUATION OF CONDUCTION

14.1. Introductory

THE use of Green's functions in the theory of potential is well known. The function is most conveniently defined for the closed surface S as the potential which vanishes over the surface, and is infinite as $1/r$, when r is zero, at the point $P(x', y', z')$ inside the surface. If this solution of the equation $\nabla^2 u = 0$ is denoted by $G(P)$, the solution with no infinity inside S and an arbitrary value V over the surface is given by

$$u = \frac{1}{4\pi} \int\int \frac{\partial}{\partial n} G(P) V \, dS,$$

$\partial/\partial n$ denoting differentiation along the outward-drawn normal.†

We proceed to show how a similar function may be employed with advantage in the mathematical theory of the conduction of heat. In this case we shall take *the Green's function as the temperature at (x, y, z) at the time t due to an instantaneous point source of strength unity generated at the point $P(x', y', z')$ at the time τ, the solid being initially at zero temperature, and the surface being kept at zero temperature.*

This solution may be written

$$u = F(x, y, z, x', y', z', t-\tau) \quad (t > \tau)$$

and u satisfies the equation

$$\frac{\partial u}{\partial t} = \kappa \nabla^2 u \quad (t > \tau).$$

However, since t only enters in the form $(t-\tau)$, we have also

$$\frac{\partial u}{\partial \tau} + \kappa \nabla^2 u = 0 \quad (\tau < t).$$

Further, $\lim_{t \to \tau}(u) = 0$ at all points inside S, except at the point (x', y', z'), where the solution takes the form

$$\frac{1}{8[\pi\kappa(t-\tau)]^{\frac{3}{2}}} e^{-[(x-x')^2+(y-y')^2+(z-z')^2]/4\kappa(t-\tau)}.$$

Finally, at the surface S, $u = 0$ $(\tau < t)$.

† Cf. Clerk Maxwell, *Electricity and Magnetism*, Vol. I, § 97 (b); Webster, *Electricity and Magnetism*, p. 290; G. and M., Chap. IX; Sommerfeld, *Partial Differential Equations in Physics* (New York, 1949) Chap. III.

　　　　　　　　　　　　A a

Let v be the temperature at the time t in this solid due to the surface temperature $\phi(x, y, z, t)$ and the initial temperature $f(x, y, z)$.

Then v satisfies the equations

$$\frac{\partial v}{\partial t} = \kappa \nabla^2 v \quad (t > 0),$$

$$v = f(x, y, z) \quad \text{initially, inside } S,$$

and $$v = \phi(x, y, z, t) \quad \text{at } S, \text{ when } t > 0.$$

Also, since the time τ of our former equations lies within the interval for t, we have

$$\frac{\partial v}{\partial \tau} = \kappa \nabla^2 v \quad (\tau < t),$$

$$v = \phi(x, y, z, \tau) \quad \text{at the surface.}$$

Therefore $$\frac{\partial}{\partial \tau}(uv) = u \frac{\partial v}{\partial \tau} + v \frac{\partial u}{\partial \tau} = \kappa [u \nabla^2 v - v \nabla^2 u],$$

and

$$\int_0^{t-\epsilon} \left[\iiint \frac{\partial}{\partial \tau}(uv) \, dx\,dy\,dz \right] d\tau = \kappa \int_0^{t-\epsilon} \left[\iiint (u \nabla^2 v - v \nabla^2 u) \, dx\,dy\,dz \right] d\tau,$$

the triple integration being taken through the solid, and ϵ being any positive number less than t, as small as we please.

Interchanging the order of integration on the left-hand side of this equation and applying Green's theorem to the right-hand side, we have

$$\iiint (uv)_{\tau=t-\epsilon} \, dx\,dy\,dz - \iiint (uv)_{\tau=0} \, dx\,dy\,dz$$

$$= \kappa \int_0^{t-\epsilon} \left[\iint \left(u \frac{\partial v}{\partial n} - v \frac{\partial u}{\partial n} \right) dS \right] d\tau = \kappa \int_0^{t-\epsilon} \left[\iint v \left(\frac{\partial u}{\partial n} \right)_i dS \right] d\tau,$$

where $\partial/\partial n_i$ denotes differentiation along the inward-drawn normal, and we have used the condition that u vanishes at the surface.

Now take the limit as ϵ tends to zero. The left-hand side gives

$$[v_P]_t \left[\iiint u_{\tau=t-0} \, dx\,dy\,dz \right] - \iiint (u)_{\tau=0}(v)_{\tau=0} \, dx\,dy\,dz,$$

the first integral being taken through an element of volume including the point $P(x', y', z')$, where the function u becomes infinite at $t = \tau$, the second integral being taken through the solid; and $[v_P]_t$ stands for the value of v at the point $P(x', y', z')$ at the time t.

But since u is the temperature at the time t due to a unit source at (x', y', z') at the time τ,

$$\iiint (u)_{\tau=t-0}\, dx\,dy\,dz = 1,$$

and we have

$$[v_P]_t = \iiint (u)_{\tau=0}(v)_{\tau=0}\, dx\,dy\,dz + \kappa \int_0^t \left[\iint v\frac{\partial u}{\partial n_i}\, dS\right] d\tau$$

$$= \iiint (u)_{\tau=0} f(x,y,z)\, dx\,dy\,dz + \kappa \int_0^t \left[\iint \phi(x,y,z,\tau)\frac{\partial u}{\partial n_i}\, dS\right] d\tau$$

$$(1)$$

as the temperature at (x', y', z') at the time t due to the initial distribution $f(x,y,z)$ and the surface temperature $\phi(x,y,z,t)$.†

In the case of radiation at the surface, the Green's function u *is taken as the temperature at (x, y, z) at time t due to an instantaneous point source of strength unity generated at (x', y', z') at time τ, radiation taking place at the surface into a medium at zero temperature.*

The temperature at $P(x', y', z')$ at the time t due to an initial distribution $f(x,y,z)$ and radiation at the surface into a medium at temperature $\phi(x,y,z,t)$ follows from a discussion similar to that given above. We find in the end

$$[v_P]_t = \iiint (u)_{\tau=0} f(x,y,z)\, dx\,dy\,dz + h\kappa \int_0^t \left[\iint u\phi(x,y,z,\tau)\, dS\right] d\tau$$

$$= \iiint (u)_{\tau=0} f(x,y,z)\, dx\,dy\,dz + \kappa \int_0^t \left[\iint \left(\frac{\partial u}{\partial n}\right)_i \phi(x,y,z,\tau)\, dS\right] d\tau,$$

$$(2)$$

since at the surface
$$\frac{\partial u}{\partial n_i} = hu,$$

and our result takes the same form as in (1).

The solution of the general problems in conduction of heat is thus

† This discussion is due to Minnigerode, and was published in his Göttingen Dissertation, *Über die Wärmeleitung in Krystallen* (Göttingen, 1862). Cf. also Betti, *Ann. Univ. tosc. Pisa*, **10** (1868) 143; *Ann. Mat. pura appl., Milano*, **1** (1868) 373; *Mem. Soc. ital. Sci.*, Firenze (3) **1** (1868) 373; *Collectanea Mathematica inedita in Memoriam Domenici Chelini* (Milano, 1881) 238; Sommerfeld, *Math. Ann.* **45** (1894) 274; Weber–Riemann, loc. cit., Bd. II, § 52; Carslaw, *Phil. Mag.* (6) **4** (1902) 162; *Proc. Edin. Math. Soc.* **21** (1903) 40; *Proc. Lond. Math. Soc.* (2) **8** (1910) 365; (2) **13** (1914) 236. See also for application to the equation $(\nabla^2 + \kappa^2)u = 0$, Pockels, *Über die Partielle-Differentialgleichung* $(\nabla^2 + \kappa^2)u = 0$, Tl. IV, § 4 (Leipzig, 1891); Schwarzschild, *Math. Ann.* **55** (1902) 177.

reduced to the determination of the Green's function for the solid in which the temperature is required.

In the case of linear or two-dimensional flow of heat, results similar to (1) and (2) can be obtained at once. Instead of an infinity of order

$$\frac{1}{\{2\sqrt{(\pi\kappa t)}\}^3}e^{-R^2/4\kappa t},$$

we have $\dfrac{1}{2\sqrt{(\pi\kappa t)}}e^{-R^2/4\kappa t}$ and $\dfrac{1}{4\pi\kappa t}e^{-R^2/4\kappa t}$,

respectively. With this change the equations which correspond to (1) and (2) will be

$$[v_P]_t = \int (u)_{\tau=0}f(x)\,dx + \kappa\int_0^t \phi(\tau)\frac{\partial u}{\partial n_i}\,d\tau, \tag{3}$$

and $$[v_P]_t = \iint (u)_{\tau=0}f(x,y)\,dxdy + \kappa\int_0^t\left[\int\phi(x,y,\tau)\frac{\partial u}{\partial n_i}\,ds\right]d\tau, \tag{4}$$

where the integration with regard to s is along the bounding arc.

These results have very simple and important physical interpretations. Thus (1) states that the temperature at time t in the solid with initial temperature $f(x,y,z)$ and zero surface temperature is that obtained from a distribution of instantaneous sources at $t = 0$ over its volume, an amount of heat $\rho c f(x,y,z)\,dxdydz$ being liberated in the element of volume $dxdydz$ at (x,y,z). This may be regarded as obvious from the physical point of view. In the same way, if heat is generated in the solid, the temperature is found from a distribution of continuous sources over its volume. Again from (1), the temperature at time t in the solid with zero initial temperature and prescribed surface temperature is that obtained from a distribution of continuous doublets over the surface with their axes normal to the surface (cf. § 10.8).

In this chapter we shall determine the Green's functions for a number of important regions and boundary conditions. In a few cases they can be written down directly, but generally we shall use the Laplace transformation; as remarked in Appendix I it can be verified that Green's functions found by this method do satisfy the required conditions.

When the Green's function is known, the solution of the problem of conduction of heat for the given region and boundary conditions, and for initial temperature an arbitrary function of position, can be written down immediately using the results of this section. Some of these solutions have already been discussed by other methods in which it has been necessary to assume the possibility of the expansion of the arbitrary function in the form required in the solution. This assumption will not now be necessary.†

The solutions obtained in this way are valid for $t > 0$, and as $t \to 0$ tend to the given initial value. By putting $t = 0$ in them we obtain formal expansions of arbitrary functions. For example, putting $t = 0$ in 14.8 (6) gives

$$f(r) = \frac{2}{a^2} \sum_{n=1}^{\infty} \frac{J_0(r\alpha_n)}{J_1^2(a\alpha_n)} \int_0^a r'f(r')J_0(r'\alpha_n)\, dr', \quad 0 < r < a, \qquad (5)$$

where the α_n are the positive roots of $J_0(\alpha a) = 0$. This is the Fourier–Bessel expansion which was assumed in § 7.6. Each Green's function yields an expansion theorem (or an integral theorem) in this way; many of these have not yet been studied.‡

14.2. Linear flow. The semi-infinite solid $x > 0$

I. *Initial temperature $f(x)$. Boundary kept at $\phi(t)$*

In this case the Green's function, or the temperature at x at the time t due to the unit instantaneous plane source at x' at the time τ, is (cf. 10.10 I)

$$\frac{1}{2\sqrt{\{\pi\kappa(t-\tau)\}}} \{e^{-(x-x')^2/4\kappa(t-\tau)} - e^{-(x+x')^2/4\kappa(t-\tau)}\}. \qquad (1)$$

It follows from 14.1 (3), with a slight change in the notation, that the temperature at x at the time t is given by

$$v = \frac{1}{2\sqrt{(\pi\kappa t)}} \int_0^{\infty} f(x')\{e^{-(x-x')^2/4\kappa t} - e^{-(x+x')^2/4\kappa t}\}\, dx' +$$

$$+ \frac{x}{2\sqrt{(\pi\kappa)}} \int_0^t \phi(\tau) \frac{e^{-x^2/4\kappa(t-\tau)}}{(t-\tau)^{\frac{3}{2}}}\, d\tau, \qquad (2)$$

which agrees with the results of § 2.5.

---

† The pure mathematician would study first the form of the expansion and the conditions under which it is possible. The applied mathematician will often assume both the form of the expansion and its validity. But it is not always clear what the form is, and it is easy to assume a wrong one. Cf. *C.H.*, § 57 (II), where a constant term is omitted. See above 7.8 (3) and § 14.8 (II). One of the advantages of the Laplace transformation method is that errors of this kind are avoided.

‡ To make the result (5) rigorous it would be necessary to justify the interchange of orders of integration and summation made in deducing 14.8 (6) from 14.8 (5), and also to justify putting $t = 0$ in 14.8 (6).

II. *Unit instantaneous plane source at x' at t = 0. Radiation at x = 0 into a medium at zero*

We start with the solution for a unit instantaneous source at $t = 0$ at x' in infinite medium. This is

$$u = \frac{1}{2\sqrt{(\pi\kappa t)}}e^{-(x-x')^2/4\kappa t}.$$

(3)

The solution v of our problem will be of the form

$$v = u+w,$$

where w satisfies the equation of conduction of heat in $x > 0$ and vanishes when $t = 0$, and is such that

$$\frac{\partial v}{\partial x} = hv, \quad \text{when } x = 0, t > 0.$$

(4)

We use the Laplace transformation. From Appendix V (7) it follows that

$$\bar{u} = \frac{1}{2\kappa q}e^{-q|x-x'|},$$

(5)

where, as usual, $q = \sqrt{(p/\kappa)}$. The subsidiary equation for w is

$$\frac{d^2\bar{w}}{dx^2} - q^2\bar{w} = 0.$$

The solution of this which is bounded as $x \to \infty$ is Ae^{-qx}, thus

$$\bar{v} = \frac{1}{2\kappa q}e^{-q|x-x'|}+Ae^{-qx},$$

where A is a constant to be determined from the condition

$$\frac{d\bar{v}}{dx} = h\bar{v}, \quad \text{at } x = 0.$$

Thus we have

$$\bar{v} = \frac{1}{2\kappa q}e^{-q|x-x'|} + \frac{(q-h)}{2\kappa q(q+h)}e^{-q(x+x')}$$

$$= \frac{1}{2\kappa q}e^{-q|x-x'|} + \frac{1}{2\kappa q}e^{-q(x+x')} - \frac{h}{\kappa q(q+h)}e^{-q(x+x')}.$$

Therefore, from Appendix V (7) and (13),

$$v = \frac{1}{2\sqrt{(\pi\kappa t)}}\{e^{-(x-x')^2/4\kappa t}+e^{-(x+x')^2/4\kappa t}\}-he^{\kappa th^2+h(x+x')}\,\mathrm{erfc}\Big\{\frac{x+x'}{2\sqrt{(\kappa t)}}+h\sqrt{(\kappa t)}\Big\}.$$

(6)

The last term in (6) may be transformed into

$$-\frac{h}{\sqrt{(\pi\kappa t)}}\int_0^\infty e^{-h\xi}e^{-(x+x'+\xi)^2/4\kappa t}\,d\xi,$$

and represents the temperature due to a line of sinks extending from $-x'$ to $-\infty$.†

III. *The region $x > 0$. Initial temperature $f(x)$. Radiation at the surface into medium at $\phi(t)$*

It follows from (6) and 14.1 (3) that

$$v = \int_0^\infty \left\{\frac{1}{2\sqrt{(\pi\kappa t)}}\left[e^{-(x-x')^2/4\kappa t}+e^{-(x+x')^2/4\kappa t}\right]-\right.$$

$$\left.-he^{\kappa t h^2+h(x+x')}\,\mathrm{erfc}\left[\frac{x+x'}{2\sqrt{(\kappa t)}}+h\sqrt{(\kappa t)}\right]\right\}f(x')\,dx'+$$

$$+\kappa h\int_0^t\left\{\frac{e^{-x^2/4\kappa(t-\tau)}}{\sqrt{\{\pi\kappa(t-\tau)\}}}-he^{\kappa h^2(t-\tau)+hx}\,\mathrm{erfc}\left[\frac{x}{2\sqrt{\{\kappa(t-\tau)\}}}+h\sqrt{\{\kappa(t-\tau)\}}\right]\right\}\phi(\tau)\,d\tau.$$

$$(7)$$

14.3. Linear flow in the region $0 < x < a$

I. *Unit instantaneous plane source at x' at $t = 0$. The boundaries $x = 0$ and $x = a$ at zero*

As before we put $v = u+w$, where

$$u = \frac{1}{2\sqrt{(\pi\kappa t)}}e^{-(x-x')^2/4\kappa t}$$

is the solution for unit instantaneous source at x' at $t = 0$ in infinite solid, and w is a solution of the equation of conduction of heat which is zero at $t = 0$ and is chosen so that v vanishes on both boundaries. The subsidiary equation for w is

$$\frac{d^2\bar{w}}{dx^2}-q^2\bar{w} = 0.$$

Thus, using the value 14.2 (5) of $\bar{u}$,

$$\bar{v} = \frac{1}{2\kappa q}e^{-q|x-x'|}+A\sinh qx+B\cosh qx.$$

† The solution was first given by Bryan, *Proc. Camb. Phil. Soc.* **7** (1891) 246. See also Bryan, *Proc. Lond. Math. Soc.* (1) **22** (1891) 424; Carslaw, *Phil. Mag.* (6) **4** (1902) 162; Hobson, *Proc. Lond. Math. Soc.* (1) **19** (1887) 279.

Here A and B are to be chosen so that $\bar{v} = 0$ at $x = 0$ and $x = a$, that is

$$\frac{1}{2\kappa q}e^{-qx'} + B = 0, \qquad \frac{1}{2\kappa q}e^{-q(a-x')} + A\sinh qa + B\cosh qa = 0.$$

Solving and substituting we find

$$\bar{v} = \frac{\cosh q(a+x-x') - \cosh q(a-x-x')}{2\kappa q\sinh qa}, \quad \text{when } x < x', \qquad (1)$$

and when $x' < x < a$ the result is the same with x and x' interchanged. Evaluating v in the usual way by the Inversion Theorem we find†

$$v = \frac{2}{a}\sum_{n=1}^{\infty} e^{-\kappa n^2\pi^2 t/a^2}\sin\frac{n\pi x}{a}\sin\frac{n\pi x'}{a}. \qquad (2)$$

The temperature at time t, when the initial temperature is $f(x)$, and the surfaces $x = 0$ and $x = a$ are kept at $\phi_1(t)$ and $\phi_2(t)$, follows from 14.1 (3) in the form 3.5 (2).

II. *Unit instantaneous plane source at $t = 0$ at x'. Boundary conditions*

$$k_1\frac{\partial v}{\partial x} - h_1 v = 0, \quad x = 0, \qquad k_2\frac{\partial v}{\partial x} + h_2 v = 0, \quad x = a, \qquad (3)$$

where k_1, h_1, k_2, h_2 are constants which are positive or zero (k_1 and h_1 or k_2 and h_2 are not both to vanish). By choice of these constants all cases of zero temperature, radiation, or no flow of heat at either surface may be included. The result, proved as in I, is

$$v = \sum_{n=1}^{\infty} Z_n(x)Z_n(x')e^{-\kappa\alpha_n^2 t}, \qquad (4)$$

where

$$Z_n(x) = \frac{[2(k_2^2\alpha_n^2+h_2^2)]^{\frac{1}{2}}(k_1\alpha_n\cos\alpha_n x + h_1\sin\alpha_n x)}{\{(k_1^2\alpha_n^2+h_1^2)[a(k_2^2\alpha_n^2+h_2^2)+k_2 h_2] + k_1 h_1(k_2^2\alpha_n^2+h_2^2)\}^{\frac{1}{2}}}, \qquad (5)$$

and $\pm\alpha_n$, $n = 1, 2, \ldots$, are the roots of

$$\tan\alpha a = \frac{\alpha(k_1 h_2 + k_2 h_1)}{k_1 k_2\alpha^2 - h_1 h_2}. \qquad (6)$$

If $h_1 = h_2 = 0$, a term $(1/a)$ is to be added to (5). This result is set out explicitly in (7). Many of the results of §§ 3.10, 3.12 follow from (4) and 14.1 (3).

---

† This result can be obtained by the method of images as in § 10.10.

III. *Unit instantaneous plane source at $t = 0$ at x'. No flow of heat over the surfaces $x = 0$ and $x = a$*

$$v = \frac{1}{a}\left\{1 + 2\sum_{n=1}^{\infty} e^{-\kappa n^2 \pi^2 t/a^2} \cos\frac{n\pi x}{a}\cos\frac{n\pi x'}{a}\right\}. \tag{7}$$

14.4. Two-dimensional problems. Rectangular boundaries

The Green's function in such cases is the temperature at time t due to a unit instantaneous line source through (x', y') at time τ. That is, we need a solution v of the equation of conduction of heat which satisfies the prescribed boundary conditions and as $t \to \tau$ behaves like

$$u = \frac{1}{4\pi\kappa(t-\tau)}\, e^{-[(x-x')^2 + (y-y')^2]/4\kappa(t-\tau)}$$

$$= \left[\frac{1}{2\sqrt{\{\pi\kappa(t-\tau)\}}}\, e^{-(x-x')^2/4\kappa(t-\tau)}\right] \times \left[\frac{1}{2\sqrt{\{\pi\kappa(t-\tau)\}}}\, e^{-(y-y')^2/4\kappa(t-\tau)}\right]. \tag{1}$$

Since u can be expressed as the product (1) it follows by the reasoning of § 1.15 that, for any of the boundary conditions 14.3 (3), the solution v for the two-dimensional region can be expressed as the product of the corresponding one-variable solutions.

I. *The rectangular corner $x > 0$, $y > 0$ with zero surface temperature*

The temperature due to an instantaneous unit line source at (x', y') time τ is, by 14.2 (1),

$$\frac{1}{4\pi\kappa(t-\tau)}\{e^{-(x-x')^2/4\kappa(t-\tau)} - e^{-(x+x')^2/4\kappa(t-\tau)}\} \times \{e^{-(y-y')^2/4\kappa(t-\tau)} - e^{-(y+y')^2/4\kappa(t-\tau)}\}. \tag{2}$$

The solution for this region with prescribed initial and surface temperatures can then be written down by 14.1 (4).

Similarly, if there is radiation at the surface into medium at zero, the Green's function is the product of two expressions of type 14.2 (6).

II. *The rectangle $0 < x < a$, $0 < y < b$ with zero surface temperature*

From 14.3 (2) the Green's function for zero surface temperature is found to be

$$\frac{4}{ab}\sum_{m=1}^{\infty}\sum_{n=1}^{\infty}\exp\left\{-\kappa\pi^2(t-\tau)\left[\frac{m^2}{a^2}+\frac{n^2}{b^2}\right]\right\}\sin\frac{m\pi x}{a}\sin\frac{m\pi x'}{a}\sin\frac{n\pi y}{b}\sin\frac{n\pi y'}{b}. \tag{3}$$

III. *The rectangle $0 < x < a$, $0 < y < b$, with no flow of heat over the surface*

By 14.3 (7) the Green's function is

$$\frac{1}{ab}\left\{1+2\sum_{n=1}^{\infty} e^{-\kappa n^2\pi^2 t/a^2}\cos\frac{n\pi x}{a}\cos\frac{n\pi x'}{a}\right\}\times$$

$$\times\left\{1+2\sum_{n=1}^{\infty} e^{-\kappa n^2\pi^2 t/b^2}\cos\frac{n\pi y}{b}\cos\frac{n\pi y'}{b}\right\}. \quad (4)$$

14.5. The rectangular parallelepiped $0 < x < a$, $0 < y < b$, $0 < z < c$

As in § 14.4, for any of the boundary conditions 14.3 (3), the Green's function can be written down as the product of three one-dimensional solutions. Results for the region $x > 0$, $y > 0$, $z > 0$ and various semi-infinite regions may also be obtained in the same way. Here we consider only the finite region.

The Green's function for zero surface temperature is by 14.3 (3)

$$\frac{8}{abc}\sum_{l=1}^{\infty}\sum_{m=1}^{\infty}\sum_{n=1}^{\infty}\sin\frac{l\pi x}{a}\sin\frac{l\pi x'}{a}\sin\frac{m\pi y}{b}\sin\frac{m\pi y'}{b}\sin\frac{n\pi z}{c}\sin\frac{n\pi z'}{c}\times$$

$$\times\exp\left\{-\kappa\pi^2(t-\tau)\left[\frac{l^2}{a^2}+\frac{m^2}{b^2}+\frac{n^2}{c^2}\right]\right\}. \quad (1)$$

Thus the solution for initial temperature $f(x,y,z)$ and zero surface temperature is

$$\frac{8}{abc}\sum_{l=1}^{\infty}\sum_{m=1}^{\infty}\sum_{n=1}^{\infty}\sin\frac{l\pi x}{a}\sin\frac{m\pi y}{b}\sin\frac{n\pi z}{c}\exp\left\{-\kappa\pi^2(t-\tau)\left[\frac{l^2}{a^2}+\frac{m^2}{b^2}+\frac{n^2}{c^2}\right]\right\}\times$$

$$\times\int_0^a\int_0^b\int_0^c f(x',y',z')\sin\frac{l\pi x'}{a}\sin\frac{m\pi y'}{b}\sin\frac{n\pi z'}{c}\,dx'dy'dz'. \quad (2)$$

If the face $x = 0$ is maintained at temperature $\phi(y,z,t)$ and the other faces at zero, the initial temperature being zero, the solution is

$$\frac{8\kappa\pi}{a^2bc}\sum_{l=1}^{\infty}\sum_{m=1}^{\infty}\sum_{n=1}^{\infty} l\sin\frac{l\pi x}{a}\sin\frac{m\pi y}{b}\sin\frac{n\pi z}{c}\int_0^t d\tau\int_0^b\int_0^c\sin\frac{m\pi y'}{b}\sin\frac{n\pi z'}{c}\phi(x',y',\tau)\times$$

$$\times\exp\left\{-\kappa\pi^2(t-\tau)\left[\frac{l^2}{a^2}+\frac{m^2}{b^2}+\frac{n^2}{c^2}\right]\right\}\,dx'dy'. \quad (3)$$

If heat is produced for $t > 0$ in the solid at the rate $A(x, y, z, t)$ per unit time per unit volume, and the surface is kept at zero, the temperature is

$$\frac{8\kappa}{abcK} \sum_{l=1}^{\infty} \sum_{m=1}^{\infty} \sum_{n=1}^{\infty} \sin\frac{l\pi x}{a} \sin\frac{m\pi y}{b} \sin\frac{n\pi z}{c} \times$$

$$\times \int_0^t \int_0^a \int_0^b \int_0^c \sin\frac{l\pi x'}{a} \sin\frac{m\pi y'}{b} \sin\frac{n\pi z'}{c} A(x', y', z', \tau) \exp\left\{-\kappa\pi^2(t-\tau)\left[\frac{l^2}{a^2}+\frac{m^2}{b^2}+\frac{n^2}{c^2}\right]\right\} \times$$

$$\times dx' dy' dz' d\tau. \quad (4)$$

If $A(x, y, z, t) = A_0$, constant, this becomes

$$\frac{64\kappa A_0}{K\pi^3} \sum_{l=0}^{\infty} \sum_{m=0}^{\infty} \sum_{n=0}^{\infty} \frac{\sin\{(2l+1)\pi x/a\}\sin\{(2m+1)\pi y/b\}\sin\{(2n+1)\pi z/c\}(1-e^{-\alpha_{l,m,n}t})}{(2l+1)(2m+1)(2n+1)\alpha_{l,m,n}},$$

$$(5)$$

where

$$\alpha_{l,m,n} = \kappa\pi^2\left[\frac{(2l+1)^2}{a^2}+\frac{(2m+1)^2}{b^2}+\frac{(2n+1)^2}{c^2}\right].$$

14.6. Linear flow. Composite regions

Green's functions for composite regions may be found by the method previously used.

I. *The region $x > 0$ of one substance K_1, κ_1, ρ_1, c_1, v_1, and $x < 0$ of another K_2, κ_2, ρ_2, c_2, v_2. No contact resistance at the boundary. Unit instantaneous plane source at $t = 0$ at x' in the region $x > 0$†*

The temperature due to a unit instantaneous source at $t = 0$ at x' in an infinite medium of thermal constants K_1, κ_1 is

$$u_1 = \frac{1}{2\sqrt{(\pi\kappa_1 t)}} e^{-(x-x')^2/4\kappa_1 t}, \quad (1)$$

so that

$$\bar{u}_1 = \frac{1}{2\kappa_1 q_1} e^{-q_1|x-x'|}, \quad (2)$$

where we write $q_1 = \sqrt{(p/\kappa_1)}$ and subsequently $q_2 = \sqrt{(p/\kappa_2)}$.

In the region $x > 0$ we assume

$$v_1 = u_1 + w_1,$$

where w_1 is a solution of the equation of linear flow which vanishes when $t = 0$, and as $x \to \infty$, that is

$$\bar{v}_1 = \frac{1}{2\kappa_1 q_1} e^{-q_1|x-x'|} + Ae^{-q_1 x}, \quad x > 0. \quad (3)$$

For $x < 0$ we need a solution of the equation of linear flow which vanishes as $x \to -\infty$. This is

$$\bar{v}_2 = Be^{q_2 x}, \quad x < 0. \quad (4)$$

† Sommerfeld, *Math. Ann.* **45** (1894) 263.

Here A and B are to be chosen so that

$$\bar{v}_1 = \bar{v}_2; \qquad K_1 \frac{d\bar{v}_1}{dx} = K_2 \frac{d\bar{v}_2}{dx}, \quad \text{when } x = 0. \tag{5}$$

Thus we find

$$\bar{v}_1 = \frac{1}{2\kappa_1 q_1} e^{-q_1|x-x'|} + \frac{(K_1 \kappa_2^{\frac{1}{2}} - K_2 \kappa_1^{\frac{1}{2}})}{2\kappa_1 q_1 (K_2 \kappa_1^{\frac{1}{2}} + K_1 \kappa_2^{\frac{1}{2}})} e^{-q_1(x+x')}, \tag{6}$$

$$\bar{v}_2 = \frac{K_1 \kappa_1^{-\frac{1}{2}}}{(K_1 \kappa_2^{\frac{1}{2}} + K_2 \kappa_1^{\frac{1}{2}})q_2} e^{q_2 x - q_1 x'}. \tag{7}$$

Therefore

$$v_1 = \frac{1}{2\sqrt{(\pi\kappa_1 t)}} e^{-(x-x')^2/4\kappa_1 t} + \frac{K_1 \kappa_2^{\frac{1}{2}} - K_2 \kappa_1^{\frac{1}{2}}}{2(K_1 \kappa_2^{\frac{1}{2}} + K_2 \kappa_1^{\frac{1}{2}})\sqrt{(\pi\kappa_1 t)}} e^{-(x+x')^2/4\kappa_1 t}, \tag{8}$$

$$v_2 = \frac{K_1 \kappa_2 \kappa_1^{-\frac{1}{2}}}{(K_1 \kappa_2^{\frac{1}{2}} + K_2 \kappa_1^{\frac{1}{2}})\sqrt{(\pi\kappa_2 t)}} e^{-\{x - x'\sqrt{(\kappa_2/\kappa_1)}\}^2/4\kappa_2 t}. \tag{9}$$

That is, the solution in the region $x > 0$ is that due to the original source together with a source of strength

$$(K_1 \kappa_2^{\frac{1}{2}} - K_2 \kappa_1^{\frac{1}{2}})/(K_1 \kappa_2^{\frac{1}{2}} + K_2 \kappa_1^{\frac{1}{2}})$$

at the 'image' position, $-x'$, in infinite medium of thermal constants K_1, κ_1.

The solution in $x < 0$ is that for a source of strength

$$2K_1 \kappa_2 \kappa_1^{-\frac{1}{2}}/(K_1 \kappa_2^{\frac{1}{2}} + K_2 \kappa_1^{\frac{1}{2}})$$

at the point $x'\sqrt{(\kappa_2/\kappa_1)}$ in infinite solid of thermal constants K_2, κ_2.

II. *The problem of I except that there is a contact resistance†at $x = 0$, so that the boundary condition there is*

$$K_1 \frac{\partial v_1}{\partial x} = K_2 \frac{\partial v_2}{\partial x} = H(v_1 - v_2), \quad x = 0. \tag{10}$$

The solution is

$$v_1 = \frac{1}{2\sqrt{(\pi\kappa_1 t)}} \{e^{-(x-x')^2/4\kappa_1 t} + e^{-(x+x')^2/4\kappa_1 t}\} - $$
$$- \frac{H}{K_1} e^{h_1(x+x') + h_1^2 \kappa_1 t} \operatorname{erfc}\left\{\frac{(x+x')}{2\sqrt{(\kappa_1 t)}} + h_1 \sqrt{(\kappa_1 t)}\right\}, \tag{11}$$

$$v_2 = \frac{kH}{K_2} e^{h_2^2 \kappa_2 t + h_2(kx' - x)} \operatorname{erfc}\left\{\frac{kx' - x}{2\sqrt{(\kappa_2 t)}} + h_2 \sqrt{(\kappa_2 t)}\right\}, \tag{12}$$

where $\quad k = \sqrt{(\kappa_2/\kappa_1)}, \quad h_2 = \dfrac{H(K_1 \kappa_2^{\frac{1}{2}} + K_2 \kappa_1^{\frac{1}{2}})}{K_1 K_2 \kappa_2^{\frac{1}{2}}}, \quad h_1 = kh_2.$

† Cf. Mersman, *Trans. Amer. Math. Soc.* **53** (1943) 14. Schaaf, *Quart. Appl. Math.* **5** (1947) 107–11, considers a source in the plane $x = 0$; in this case there are several possibilities, since heat may be supplied in either medium or between them.

III. *The region* $-l < x < 0$ *of one medium,* $K_1, \rho_1, c_1, \kappa_1, v_1,$ *and* $0 < x < a$ *of another,* $K_2, \rho_2, c_2, \kappa_2, v_2.$ *No contact resistance at the boundary. Unit instantaneous source at $t = 0$ at x' in the region $0 < x < a$. The surfaces $x = a$ and $x = -l$ kept at zero temperature for $t > 0$*

$$v_1 = 2k\sigma \sum_{n=1}^{\infty} \frac{\sin k(a-x')\beta_n \sin(x+l)\beta_n \sin l\beta_n \sin ka\beta_n}{l \sin^2 ka\beta_n + \sigma ka \sin^2 l\beta_n} e^{-\kappa_1 \beta_n^2 t} +$$

$$+ \frac{2k\sigma s}{l(r+\sigma s)} \sum_{n=1}^{\infty} \sin\frac{nr\pi x'}{a} \cdot \sin\frac{ns\pi x}{l} e^{-\kappa_1 n^2 s^2 \pi^2 t/l^2}, \quad (13)$$

$$v_2 = 2k\sigma \sum_{n=1}^{\infty} \frac{\sin k(a-x')\beta_n \sin^2 l\beta_n \sin k(a-x)\beta_n}{l \sin^2 ka\beta_n + \sigma ka \sin^2 l\beta_n} e^{-\kappa_1 \beta_n^2 t} +$$

$$+ \frac{2ks}{l(r+\sigma s)} \sum_{n=1}^{\infty} \sin\frac{nr\pi x'}{a} \sin\frac{nr\pi x}{a} e^{-\kappa_1 n^2 s^2 \pi^2 t/l^2}, \quad (14)$$

where $k = \sqrt{(\kappa_1/\kappa_2)}, \sigma = kK_2/K_1, \pm\beta_n, n = 1, 2, \ldots,$ are the roots of

$$\cot \beta l + \sigma \cot k\beta a = 0,$$

and the second series in (13) and (14) occur only if $ka/l = r/s$, a rational fraction in its lowest terms, cf. § 12.8.

IV. *The region $-l < x < 0$ of one medium, $K_1, \rho_1, c_1, \kappa_1, v_1,$ and $x > 0$ of another, $K_2, \rho_2, c_2, \kappa_2, v_2.$ No contact resistance at the boundary. Unit instantaneous source at $t = 0$ at x' in the region $x > 0$. The surface $x = -l$ kept at zero for $t > 0$*

As in § 12.8 the solutions may be expressed either as infinite integrals, or as infinite series corresponding to an infinite series of images.

Thus, for example,

$$v_1 = \frac{2\sigma k}{\pi} \int_0^{\infty} \frac{\{\sin kux' \cos ul + \sigma \cos kux' \sin ul\}\sin u(l+x)}{\cos^2 ul + \sigma^2 \sin^2 ul} e^{-\kappa_1 u^2 t} \, du \quad (15)$$

$$= \frac{\sigma}{(1+\sigma)\sqrt{(\pi\kappa_2 t)}} \sum_{n=0}^{\infty} \alpha^n \left\{ \exp\left[-\frac{(kx'-x+2nl)^2}{4\kappa_1 t}\right] - \exp\left[-\frac{\{kx'+x+2(n+1)l\}^2}{4\kappa_1 t}\right] \right\},$$

$$\quad (16)$$

where $\alpha = (\sigma-1)/(\sigma+1)$.

14.7. The sphere. Radial flow

The fundamental solutions for radial flow in spherical regions† are those for the unit instantaneous spherical surface source of radius r' at

---

† The Green's function for the sphere defined in § 14.1 is the solution for a unit instantaneous point source in the sphere (§ 14.16). The solution for the instantaneous spherical surface source given here may be derived by integrating 14.16 (8) for sources distributed uniformly over a sphere, but problems in radial flow are so important that it is worth while to derive it directly, particularly as the procedure corresponds to that of § 14.2 for the instantaneous plane source. A similar remark applies to the solutions of § 14.8.

time $t = 0$. The solution for this source in an infinite medium is, by 10.3 (6),

$$u = \frac{1}{8\pi r r'(\pi\kappa t)^{\frac{1}{2}}}[e^{-(r-r')^2/4\kappa t}-e^{-(r+r')^2/4\kappa t}], \tag{1}$$

so that

$$\bar{u} = \frac{1}{8\pi r r'\kappa q}\{e^{-q|r-r'|}-e^{-q(r+r')}\}. \tag{2}$$

As before we seek a solution

$$v = u+w,$$

where w is a solution of the equation of conduction of heat,

$$\frac{\partial^2(rw)}{\partial r^2}-\frac{1}{\kappa}\frac{\partial(rw)}{\partial t} = 0, \tag{3}$$

which vanishes for $t = 0$ and is such that v satisfies the boundary conditions.

I. *The sphere $0 \leqslant r < a$ with zero surface temperature. Unit instantaneous spherical surface source at $r = r'$ at $t = 0$*

Here $\bar{w}$ is to satisfy the subsidiary equation

$$\frac{d^2(r\bar{w})}{dr^2}-q^2(r\bar{w}) = 0, \quad 0 \leqslant r < a, \tag{4}$$

and to be finite at $r = 0$. Thus we must have

$$\bar{w} = \frac{A \sinh qr}{r}, \tag{5}$$

where A is determined from the condition that $\bar{v} = 0$ at $r = a$, that is

$$\frac{1}{8\pi a r'\kappa q}\{e^{-q(a-r')}-e^{-q(a+r')}\}+\frac{A \sinh qa}{a} = 0.$$

Using this value of A we find

$$\bar{v} = \frac{\sinh qr \sinh q(a-r')}{4\pi r r'\kappa q \sinh qa}, \quad 0 < r < r', \tag{6}$$

and, if $a > r > r'$, r and r' have to be interchanged in (6).

Using the Inversion Theorem we find in the usual way

$$v = \frac{1}{2\pi a r r'}\sum_{n=1}^{\infty} e^{-\kappa n^2 \pi^2 t/a^2}\sin\frac{n\pi r}{a}\sin\frac{n\pi r'}{a}. \tag{7}$$

Taking a source of strength $4\pi r'^2 f(r')\, dr'$ on the sphere $r = r'$, and integrating with respect to r' from 0 to a, gives

$$\frac{2}{ar}\sum_{n=1}^{\infty} e^{-\kappa n^2 \pi^2 t/a^2}\sin\frac{n\pi r}{a}\int_0^a r'f(r')\sin\frac{n\pi r'}{a}\, dr' \tag{8}$$

for the temperature in the sphere with initial temperature $f(r)$ and zero surface temperature. The result for surface temperature $\phi(t)$ follows also as in § 14.1.

Alternatively, expanding (6) in a series of negative exponentials, we get

$$v = \frac{1}{8rr'\pi(\pi\kappa t)^{\frac{1}{2}}} \sum_{n=-\infty}^{\infty} \{e^{-(2na-r+r')^2/4\kappa t} - e^{-(2na+r+r')^2/4\kappa t}\}, \tag{9}$$

leading to the solution 9.3 (21) for the temperature in a sphere whose initial temperature is $f(r)$.

II. *The sphere $0 \leqslant r < a$ with radiation† at its surface into medium at zero. Unit instantaneous spherical surface source at $t = 0$ at $r = r'$*

$$v = \frac{1}{2\pi arr'} \sum_{n=1}^{\infty} \frac{(ah-1)^2 + a^2\alpha_n^2}{a^2\alpha_n^2 + ah(ah-1)} \sin r\alpha_n \sin r'\alpha_n \, e^{-\kappa\alpha_n^2 t}, \tag{10}$$

where the α_n are the positive roots of

$$a\alpha \cot a\alpha + (ah-1) = 0. \tag{11}$$

III. *The spherical shell $a < r < b$. Unit instantaneous spherical surface source at $r = r'$ at $t = 0$. Boundary conditions*

$$\left.\begin{aligned} k_1 \frac{\partial v}{\partial r} - h_1 v = 0, & \quad r = a \\ k_2 \frac{\partial v}{\partial r} + h_2 v = 0, & \quad r = b \end{aligned}\right\}, \tag{12}$$

where k_1, h_1, k_2, h_2 are positive or zero

$$v = \frac{1}{2\pi rr'} \sum_{n=1}^{\infty} e^{-\kappa\alpha_n^2 t} R_n(r) R_n(r'), \tag{13}$$

where

$$R_n(r)$$
$$= \frac{(H^2 + b^2 k_2^2 \alpha_n^2)^{\frac{1}{2}}[G\sin(r-a)\alpha_n + ak_1\alpha_n\cos(r-a)\alpha_n]}{\{(b-a)(a^2k_1^2\alpha_n^2 + G^2)(b^2k_2^2\alpha_n^2 + H^2) + (Hak_1 + Gbk_2)(GH + abk_1 k_2\alpha_n^2)\}^{\frac{1}{2}}}, \tag{14}$$

$$G = ah_1 + k_1, \qquad H = bh_2 - k_2,$$

the α_n are the positive roots of

$$(GH - abk_1 k_2 \alpha^2)\sin(b-a)\alpha + \alpha(ak_1 H + bk_2 G)\cos(b-a)\alpha = 0, \tag{15}$$

† If $h = 0$, that is, no flow of heat at the surface, a term $3/(4\pi a^3)$ is to be added to the right-hand side of (10).

and if $h_1 = h_2 = 0$, a term

$$\frac{3}{4\pi(b^3-a^3)}$$

is to be added to the right-hand side of (13).

IV. *The region bounded internally by the sphere* $r = a$. *Unit instantaneous spherical surface source at* $r = r'$ *at* $t = 0$. *Boundary condition at* $r = a$

$$k\frac{\partial v}{\partial r} - hv = 0,$$

where $k \geqslant 0$, $h \geqslant 0$

$$v = \frac{1}{8\pi r r'(\pi\kappa t)^{\frac{1}{2}}}\left[e^{-(r-r')^2/4\kappa t} + e^{-(r+r'-2a)^2/4\kappa t} - \right.$$

$$- \frac{ah+k}{ak}(4\pi\kappa t)^{\frac{1}{2}}\exp\left\{\kappa t\left(\frac{ah+k}{ak}\right)^2 + (r+r'-2a)\frac{ah+k}{ak}\right\} \times$$

$$\left. \times \operatorname{erfc}\left\{\frac{r+r'-2a}{2\sqrt{(\kappa t)}} + \frac{ah+k}{ak}\sqrt{(\kappa t)}\right\}\right]. \quad (16)$$

14.8. The cylinder. Radial flow

In this case we start from the solution, 10.3 (5), for a unit instantaneous cylindrical surface source at $t = 0$ and $r = r'$ in an infinite medium,

$$u = \frac{1}{4\pi\kappa t}e^{-(r^2+r'^2)/4\kappa t}I_0\left(\frac{rr'}{2\kappa t}\right), \quad (1)$$

so that,† using Appendix V (22) with $\nu = 0$,

$$\left.\begin{array}{l} \bar{u} = \dfrac{1}{2\pi\kappa}I_0(qr')K_0(qr), \quad \text{when } r > r' \\[3mm] \bar{u} = \dfrac{1}{2\pi\kappa}I_0(qr)K_0(qr'), \quad \text{when } r < r' \end{array}\right\}. \quad (2)$$

As usual we seek a solution $v = u+w$, where $\bar{w}$ satisfies

$$\frac{d^2\bar{w}}{dr^2} + \frac{1}{r}\frac{d\bar{w}}{dr} - q^2\bar{w} = 0, \quad (3)$$

and is to be such that $\bar{v}$ satisfies the boundary conditions.

I. *The region* $0 \leqslant r < a$ *with zero surface temperature. Unit instantaneous cylindrical surface source at* $r = r'$ *at* $t = 0$

Here the solution of (3) which is finite at the origin is $AI_0(qr)$, where A is to be determined from the fact that $\bar{v} = 0$ on $r = a$, that is

$$AI_0(qa) + \frac{1}{2\pi\kappa}I_0(qr')K_0(qa) = 0.$$

† Equivalent results are given in *W.B.F.*, § 13.7 (2); *G. and M.*, p. 74 (57).

Using this value of A we find

$$\bar{v} = \frac{I_0(qr')}{2\pi\kappa I_0(qa)}\{I_0(qa)K_0(qr) - I_0(qr)K_0(qa)\}, \quad r > r', \tag{4}$$

and its value for $0 < r < r'$ is obtained by interchanging r and r' in (4). Using the Inversion Theorem in the usual way gives

$$v = \frac{1}{\pi a^2} \sum_{n=1}^{\infty} \frac{J_0(r\alpha_n)J_0(r'\alpha_n)}{J_1^2(a\alpha_n)} e^{-\kappa\alpha_n^2 t}, \tag{5}$$

where $\pm\alpha_n$, $n = 1, 2, \ldots$, are the roots of

$$J_0(a\alpha) = 0.$$

By taking a source of strength $2\pi r'f(r')\,dr'$ at r', integrating from 0 to a with respect to r', and assuming that $f(r)$ is such that the orders of integration and summation may be interchanged, we obtain the solution

$$\frac{2}{a^2} \sum_{n=1}^{\infty} e^{-\kappa\alpha_n^2 t} \frac{J_0(r\alpha_n)}{J_1^2(a\alpha_n)} \int_0^a r'f(r')J_0(r'\alpha_n)\,dr' \tag{6}$$

for the cylinder with zero surface temperature and initial temperature $f(r)$. Problems on arbitrary initial and surface temperatures in the other regions discussed below may be treated in the same way using § 14.1.

II. *The cylinder $0 \leqslant r < a$. Unit instantaneous cylindrical surface source at $t = 0$ and $r = r'$. Boundary condition at $r = a$*

$$k\frac{\partial v}{\partial r} + hv = 0,$$

where $k \geqslant 0$, $h \geqslant 0$

This includes the result of I and also the cases of radiation and no flow of heat at the surface.

$$v = \frac{1}{\pi a^2} \sum_{n=1}^{\infty} e^{-\kappa\alpha_n^2 t} \frac{J_0(r\alpha_n)J_0(r'\alpha_n)}{J_0^2(a\alpha_n) + J_1^2(a\alpha_n)}, \tag{7}$$

where the α_n are the positive roots of

$$k\alpha J_1(a\alpha) - hJ_0(a\alpha) = 0,$$

and, if $h = 0$, a term $1/\pi a^2$ is to be added to the right-hand side of (7).

III. *The problem of II except that the boundary condition† at $r = a$ is*

$$k_1\frac{\partial v}{\partial t} + k_2\frac{\partial v}{\partial r} + k_3 v = 0, \tag{8}$$

---

† For this type of boundary condition see § 1.9 F. Similar generalizations of IV and V below are easily made.

where k_1, k_2, k_3 are positive constants.

$$v = \frac{k_2}{\pi a^2} \sum_{n=1}^{\infty} e^{-\kappa \alpha_n^2 t} \frac{J_0(r\alpha_n)J_0(r'\alpha_n)}{k_2 J_1^2(a\alpha_n) + (k_2 + 2\kappa k_1/a)J_0^2(a\alpha_n)}, \tag{9}$$

where the α_n are the positive roots of

$$(k_3 - k_1\kappa\alpha^2)J_0(a\alpha) - k_2\alpha J_1(a\alpha) = 0. \tag{10}$$

IV. *The hollow cylinder $a < r < b$. Unit instantaneous cylindrical surface source at $t = 0$ and $r = r'$. Boundary conditions*

$$\left. \begin{aligned} k_1\frac{\partial v}{\partial r} - k_2 v &= 0, \quad r = a \\ k_1'\frac{\partial v}{\partial r} + k_2' v &= 0, \quad r = b \end{aligned} \right\}.$$

$$v = \frac{\pi}{4} \sum_{n=1}^{\infty} \frac{1}{F(\alpha_n)} \alpha_n^2 [k_1'\alpha_n J_1(b\alpha_n) - k_2' J_0(b\alpha_n)]^2 C_0(r, \alpha_n)C_0(r', \alpha_n)e^{-\kappa\alpha_n^2 t}, \tag{11}$$

where the α_n are the positive roots of 13.4 (5), and $F(\alpha_n)$ and $C_0(r, \alpha_n)$ are defined in 13.4 (9), (11). Here k_1, k_2, k_1', k_2' are to be $\geqslant 0$; if $k_2 = k_2' = 0$, a term $1/\pi(b^2 - a^2)$ is to be added to the right-hand side of (11).

V. *The region bounded internally by the cylinder $r = a$. Unit instantaneous cylindrical surface source at $r = r'$ at $t = 0$. The boundary condition at $r = a$*

$$k\frac{\partial v}{\partial r} - hv = 0, \quad k \geqslant 0, \quad h \geqslant 0.$$

$$v = \frac{1}{2\pi} \int_0^{\infty} e^{-\kappa u^2 t} C(u, r)C(u, r')u \, du, \tag{12}$$

where

$$C(u, r) = \frac{J_0(ur)[kuY_1(au) + hY_0(au)] - Y_0(ur)[kuJ_1(au) + hJ_0(au)]}{\{[kuJ_1(au) + hJ_0(au)]^2 + [kuY_1(au) + hY_0(au)]^2\}^{\frac{1}{2}}}. \tag{13}$$

14.9. The semi-infinite solid $x > 0$. Three-dimensional problems

Here the Green's function, that is, the temperature at time t at (x, y, z) due to a unit instantaneous point source at x', y', z' at time τ, the boundary $x = 0$ being kept at zero, is, as in § 10.11 II,

$$u = \frac{1}{8[\pi\kappa(t-\tau)]^{\frac{3}{2}}} \left\{ \exp\left[-\frac{(x-x') + (y-y')^2 + (z-z')^2}{4\kappa(t-\tau)} \right] - \right.$$
$$\left. - \exp\left[-\frac{(x+x')^2 + (y-y')^2 + (z-z')^2}{4\kappa(t-\tau)} \right] \right\}. \tag{1}$$

Also, in the notation of § 14.1,

$$\left(\frac{\partial u}{\partial n} \right)_i = \left(\frac{\partial u}{\partial x} \right)_{x=0} = \frac{x'}{8\pi^{\frac{3}{2}}\kappa^{\frac{5}{2}}(t-\tau)^{\frac{5}{2}}} \exp\left[-\frac{x'^2 + (y-y')^2 + (z-z')^2}{4\kappa(t-\tau)} \right]. \tag{2}$$

Thus if the initial temperature of the solid is $f(x, y, z)$ and the surface

$x = 0$ is maintained at temperature $F(y, z, t)$ for $t > 0$, the solution is, by 14.1 (1),

$$v = \frac{1}{8(\pi\kappa t)^{\frac{3}{2}}} \int_0^\infty \int_{-\infty}^\infty \int_{-\infty}^\infty f(x', y', z')[e^{-(x-x')^2/4\kappa t} - e^{-(x+x')^2/4\kappa t}] \times$$

$$\times e^{-[(y-y')^2+(z-z')^2]/4\kappa t} f(x', y', z') \, dx'dy'dz' +$$

$$+ \frac{x}{8(\pi\kappa)^{\frac{3}{2}}} \int_0^t \int_{-\infty}^\infty \int_{-\infty}^\infty \frac{F(y', z', \tau)}{(t-\tau)^{\frac{5}{2}}} \exp\left\{-\frac{x^2+(y-y')^2+(z-z')^2}{4\kappa(t-\tau)}\right\} d\tau dy'dz'. \tag{3}$$

For the two-dimensional problem in which all quantities are independent of z the solution is obtained either by evaluating the integral with respect to z' in (3), or by using the corresponding results for the instantaneous line source in place of (1) and (2). (3) proves the statement at the end of § 10.8 about continuous doublets.

If there is *radiation at the boundary* $x = 0$ *into a medium at zero*, the Green's function is

$$u = \frac{1}{8[\pi\kappa(t-\tau)]^{\frac{3}{2}}} \{e^{-(x-x')^2/4\kappa(t-\tau)} + e^{-(x+x')^2/4\kappa(t-\tau)}\} e^{-[(y-y')^2+(z-z')^2]/4\kappa(t-\tau)} -$$

$$- \frac{h}{4\pi\kappa(t-\tau)} \operatorname{erfc}\left\{\frac{x+x'}{2\sqrt{[\kappa(t-\tau)]}} + h\sqrt{[\kappa(t-\tau)]}\right\} \times$$

$$\times \exp\left\{h(x+x') + \kappa h^2(t-\tau) - \frac{(y-y')^2+(z-z')^2}{4\kappa(t-\tau)}\right\}. \tag{4}$$

To verify this statement we have only to notice that (4) satisfies the differential equation; also it is infinite in the required way at x', y', z'; it is zero everywhere else as $t \to 0$; and it satisfies the boundary condition at $x = 0$, since 14.2 (6) has these properties. Alternatively the result can be found by the methods of the next section. The solution for the region $x > 0$ with initial temperature $f(x, y, z)$ and radiation into a medium at temperature $\phi(y, z, t)$ can be written down by 14.1 (2).

14.10. The region bounded by two parallel planes

We first give some representations for the Laplace transform of the temperature due to an instantaneous point source which are fundamental for the treatment of problems on cylindrical regions.

The temperature due to a unit source at (x', y', z') in an infinite medium is

$$u = \frac{1}{8(\pi\kappa t)^{\frac{3}{2}}} e^{-[R^2+(z-z')^2]/4\kappa t}, \tag{1}$$

where $\quad R^2 = (x-x')^2 + (y-y')^2 = r'^2 + r^2 - 2rr'\cos(\theta-\theta'), \tag{2}$

and (r, θ), (r', θ') are the polar coordinates of the points (x, y), (x', y').

Using Appendix V (6) we have

$$\bar{u} = \frac{e^{-q\sqrt{[R^2+(z-z')^2]}}}{4\pi\kappa\sqrt{[R^2+(z-z')^2]}} \tag{3}$$

$$= \frac{1}{2\pi^2\kappa} \int_0^\infty \cos\xi(z-z')K_0(\eta R)\,d\xi \tag{4}$$

$$= \frac{1}{4\pi\kappa} \int_0^\infty e^{-\eta|z-z'|}\frac{J_0(\xi R)}{\eta}\xi\,d\xi, \tag{5}$$

where

$$\eta = \sqrt{(\xi^2+q^2)}. \tag{6}$$

The two representations† (4) and (5) of $\bar{u}$ are useful in various problems on cylindrical regions: here we need the latter.

I. *The planes at zero temperature*

We suppose the planes to be $z = 0$ and $z = l$, and determine the temperature at (x, y, z) due to a unit instantaneous point source at (x', y', z') at time $t = 0$.

As usual we seek a solution $v = u+w$, where w is to satisfy the differential equation of conduction of heat, and to vanish for $t = 0$, and to be such that v satisfies the boundary conditions. The subsidiary equation for w is

$$\frac{1}{r^2}\frac{\partial^2\bar{w}}{\partial\theta^2} + \frac{\partial^2\bar{w}}{\partial r^2} + \frac{1}{r}\frac{\partial\bar{w}}{\partial r} + \frac{\partial^2\bar{w}}{\partial z^2} - q^2\bar{w} = 0, \tag{7}$$

and this is satisfied by

$$\bar{w} = \frac{1}{4\pi\kappa} \int_0^\infty \frac{\xi}{\eta} J_0(\xi R)\{A\sinh\eta z + B\sinh\eta(l-z)\}\,d\xi, \tag{8}$$

where A and B are arbitrary functions of ξ. These are to be chosen so that $\bar{v}$, which is the sum of (5) and (8), vanishes on $z = 0$ and $z = l$. Thus

$$B\sinh\eta l = -e^{-\eta z'}, \quad \text{and} \quad A\sinh\eta l = -e^{-\eta(l-z')}.$$

Using these values, and Appendix III (25), we get

$$\bar{v} = \frac{1}{2\pi\kappa} \int_0^\infty \frac{\xi J_0(\xi R)\sinh\eta(l-z')\sinh\eta z}{\eta\sinh\eta l}\,d\xi, \tag{9}$$

for $0 < z < z'$, and if $z' < z < l$ the result is the same with z and z'

† They are the cases $\mu = -\frac{1}{2}$, $\nu = 0$, and $\mu = 0$, $\nu = \frac{1}{2}$, respectively, of *W.B.F.*, § 13.47 (2).

interchanged. To determine v from $\bar{v}$ we may proceed in either of two ways. In the first of these we rewrite the integral (9) in the form

$$\frac{1}{2\pi^2 i\kappa} \int_{-\infty i}^{\infty i} \frac{\xi K_0(\xi R)\sinh(l-z')(q^2-\xi^2)^{\frac{1}{2}}\sinh z(q^2-\xi^2)^{\frac{1}{2}} \, d\xi}{(q^2-\xi^2)^{\frac{1}{2}}\sinh l(q^2-\xi^2)^{\frac{1}{2}}}, \qquad (10)$$

and evaluate it by completing the path by portion of a large circle to the right. The integrand has poles at $\xi = q_m$, where $\dot{q}_m = (q^2+m^2\pi^2/l^2)^{\frac{1}{2}}$, $m = 1, 2,...$. Evaluating the residues at these, we get finally

$$\bar{v} = \frac{1}{\pi\kappa l} \sum_{m=1}^{\infty} \sin\frac{m\pi z}{l} \sin\frac{m\pi z'}{l} K_0(q_m R). \qquad (11)$$

Using Appendix V (23) and 12.2 (6), it follows that

$$v = \frac{e^{-R^2/4\kappa t}}{2\pi\kappa t l} \sum_{m=1}^{\infty} e^{-\kappa m^2\pi^2 t/l^2} \sin\frac{m\pi z}{l} \sin\frac{m\pi z'}{l}. \qquad (12)$$

Alternatively, we may write down the function whose Laplace transform is

$$\frac{\sinh(l-z')(q^2+\xi^2)^{\frac{1}{2}}\sinh z(q^2+\xi^2)^{\frac{1}{2}}}{(q^2+\xi^2)^{\frac{1}{2}}\sinh l(q^2+\xi^2)^{\frac{1}{2}}} \qquad (13)$$

in either of the forms

$$\frac{2\kappa}{l} \sum_{n=1}^{\infty} \sin\frac{n\pi z}{l} \sin\frac{n\pi z'}{l} e^{-\kappa\xi^2 t - \kappa n^2\pi^2 t/l^2}, \qquad (14)$$

or

$$\frac{1}{2}\left(\frac{\kappa}{\pi t}\right)^{\frac{1}{2}} e^{-\kappa\xi^2 t} \sum_{n=-\infty}^{\infty} \{e^{-(2nl+z'-z)^2/4\kappa t} - e^{-(2nl-z'-z)^2/4\kappa t}\}. \qquad (15)$$

Assuming that the order of integration and inverting the Laplace transform in (9) may be interchanged, using (14), and Appendix III (29) to evaluate the integral in ξ, we again get (12). On the other hand, using the form (15) in the same way gives the alternative expression

$$v = \frac{1}{8(\pi\kappa t)^{\frac{3}{2}}} e^{-R^2/4\kappa t} \sum_{n=-\infty}^{\infty} \{e^{-(2nl+z'-z)^2/4\kappa t} - e^{-(2nl-z'-z)^2/4\kappa t}\}, \qquad (16)$$

which might have been written down directly as the infinite series of images in the planes $z = 0$ and $z = l$. The equivalence of (14) and (15) gives the alternative proof of this result referred to in § 10.10.

II. *The case of radiation at $z = 0$ and $z = l$ into a medium at zero temperature* may be treated in the same way. The solution for a unit instantaneous source at (x', y', z') at $t = 0$ is

$$v = \frac{e^{-R^2/4\kappa t}}{2\pi\kappa t} \sum_{n=1}^{\infty} \frac{(\alpha_n \cos\alpha_n z + h\sin\alpha_n z)(\alpha_n \cos\alpha_n z' + h\sin\alpha_n z')}{l(\alpha_n^2+h^2)+2h} e^{-\kappa\alpha_n^2 t}, \qquad (17)$$

where the α_n are the positive roots of

$$\tan \alpha l = \frac{2\alpha h}{\alpha^2 - h^2}.$$

III. *If there is no flow of heat across the planes $z = 0$ and $z = l$*, the solution for an instantaneous point source at (x', y', z') at $t = 0$ is

$$v = \frac{1}{4\pi l\kappa t}e^{-R^2/4\kappa t}\left\{1 + 2\sum_{n=1}^{\infty}e^{-\kappa n^2\pi^2 t/l^2}\cos\frac{n\pi z}{l}\cos\frac{n\pi z'}{l}\right\}, \tag{18}$$

$$= \frac{1}{8(\pi\kappa t)^{\frac{3}{2}}}e^{-R^2/4\kappa t}\sum_{n=-\infty}^{\infty}\{e^{-(2nl+z'-z)^2/4\kappa t} + e^{-(2nl-z'-z)^2}\}, \tag{19}$$

where the latter could have been written down from the infinite series of images. It should be noted that (18) cannot be deduced by putting $h = 0$ in (17).

IV. *Rectangular arrays of sources in an infinite medium*

Taking $z' = \frac{1}{2}l$ in (18) gives the temperature at (x, y, z) due to a line of instantaneous point sources at $z' = \frac{1}{2}l \pm nl$. In the same way, the temperature due to a doubly infinite array at the points $x' = \frac{1}{2}a \pm na$, $z' = \frac{1}{2}l \pm nl$, is†

$$\frac{1}{2al(\pi\kappa t)^{\frac{1}{2}}}e^{-(y-y')^2/4\kappa t}\left\{1 + 2\sum_{n=1}^{\infty}(-1)^n e^{-4\kappa n^2\pi^2 t/l^2}\cos\frac{2n\pi z}{l}\right\} \times$$

$$\times \left\{1 + 2\sum_{n=1}^{\infty}(-1)^n e^{-4\kappa n^2\pi^2 t/a^2}\cos\frac{2n\pi x}{a}\right\}. \tag{20}$$

The corresponding result for a three-dimensional array may be written down in the same way.

For a rectangular array of line sources parallel to the y-axis, integrating (20) gives

$$\frac{1}{al}\left\{1 + 2\sum_{n=1}^{\infty}(-1)^n e^{-4\kappa n^2\pi^2 t/l^2}\cos\frac{2n\pi z}{l}\right\}\left\{1 + 2\sum_{n=1}^{\infty}(-1)^n e^{-4\kappa n^2\pi^2 t/a^2}\cos\frac{2n\pi x}{a}\right\}. \tag{21}$$

14.11. The semi-infinite solid $z > 0$ with a thin surface skin of much better conductor‡ in the plane $z = 0$. Unit instantaneous source at $(0, 0, z')$

We have to solve

$$\frac{\partial^2 v}{\partial r^2} + \frac{1}{r}\frac{\partial v}{\partial r} + \frac{\partial^2 v}{\partial z^2} - \frac{1}{\kappa}\frac{\partial v}{\partial t} = 0, \quad r > 0, \quad z > 0, \tag{1}$$

with the boundary condition 1.9 (21),

$$\frac{\partial^2 v}{\partial r^2} + \frac{1}{r}\frac{\partial v}{\partial r} + h\frac{\partial v}{\partial z} - \frac{1}{\kappa_1}\frac{\partial v}{\partial t} = 0, \quad r > 0, \quad z = 0, \tag{2}$$

---

† The simplest way of deducing (20) is to write down the series of images in the x-direction and use the general result which follows from the equality of (18) and (19).

‡ This is the Green's function for the semi-infinite solid with surface diffusion. For references to other problems of this type, see § 1.9 H.

where $h = K/dK_1$. Subtracting (1) and (2), the boundary condition becomes

$$\frac{\partial^2 v}{\partial z^2} - h\frac{\partial v}{\partial z} + \left(\frac{1}{\kappa_1} - \frac{1}{\kappa}\right)\frac{\partial v}{\partial t} = 0, \quad r > 0, \quad z = 0. \tag{3}$$

As in § 14.10 we seek a solution $v = u + w$, where $\bar{u}$ is given by 14.10 (5), and

$$\bar{w} = \frac{1}{4\pi\kappa}\int_0^\infty \frac{A\xi J_0(\xi r)}{\eta}e^{-\eta z}\,d\xi, \tag{4}$$

where A is to be determined from (3). Proceeding as in § 14.10, we get finally, for $z' > z > 0$,

$$\bar{v} = \frac{1}{2\pi\kappa}\int_0^\infty \frac{\xi J_0(\xi r)[(p + \kappa_1\xi^2)\sinh\eta z + \kappa_1 h\eta\cosh\eta z]}{\eta(p + \kappa_1\xi^2 + \kappa_1 h\eta)}e^{-\eta z'}\,d\xi. \tag{5}$$

For simplicity, we shall only evaluate v for the surface $z = 0$. In this case, (5) becomes

$$\bar{v} = \frac{h\kappa_1}{2\pi\kappa}\int_0^\infty \frac{\xi J_0(\xi r)e^{-\eta z'}\,d\xi}{\kappa_1\xi^2 + \kappa_1 h\eta + p}$$

$$= \frac{h\kappa_1}{2\pi\kappa}\int_0^\infty \xi J_0(\xi r)\,d\xi\int_1^\infty e^{-\eta z' - (\zeta-1)[\kappa_1\xi^2 + \kappa_1 h\eta + p]}\,d\zeta. \tag{6}$$

We now apply the Inversion Theorem to (6), remembering that $\eta = [(p/\kappa) + \xi^2]^{\frac{1}{2}}$ and assuming, here and subsequently, that orders of integration can be interchanged freely. Using § 12.2 Theorems VI and VII and Appendix V (6), we get

$$v = \frac{h\kappa_1}{2\pi\kappa}\int_0^\infty \xi J_0(\xi r)\,d\xi\int_1^\infty \exp\left\{-\xi^2[\kappa t + (\kappa_1 - \kappa)(\zeta - 1)] - \frac{[z' + (\zeta - 1)h\kappa_1]^2}{4\kappa(t + 1 - \zeta)}\right\} \times$$

$$\times \frac{[z' + h\kappa_1(\zeta - 1)]H(t - \zeta + 1)}{2(\pi\kappa)^{\frac{1}{2}}(t + 1 - \zeta)^{\frac{3}{2}}}\,d\zeta.$$

Using Appendix III (29) to evaluate the integral in ξ, this gives finally

$$v = \frac{hk}{8(k-1)^{\frac{1}{2}}(\pi\kappa t)^{\frac{3}{2}}}\int_1^k \exp\left\{-\frac{[z'(k-1) + hk\kappa t(x-1)]^2}{4\kappa t(k-1)(k-x)} - \frac{r^2}{4\kappa tx}\right\} \times$$

$$\times \frac{[z'(k-1) + hk\kappa t(x-1)]\,dx}{x(k-x)^{\frac{3}{2}}}, \tag{7}$$

where $k = \kappa_1/\kappa$.

14.12. The infinite composite solid. Unit instantaneous point source† at $(0, 0, z')$

Suppose that the region $z > 0$ is of one material $K_1, \rho_1, c_1, \kappa_1$, and $z < 0$ of another, $K_2, \rho_2, c_2, \kappa_2$, and that there is no contact resistance at the boundary $z = 0$. We write v_1 for the temperature in the region $z > 0$, and v_2 for that in $z < 0$. Then, if there is a unit source at the point $(0, 0, z')$ in the region $z > 0$, it follows by the

---

† This problem is of considerable importance since it appears in many practical contexts, e.g. frictional heating and the ageing of neutrons. The treatment given here is that of Bellman, Marshak, and Wing, *Phil. Mag.* (7) **40** (1949) 297–308.

method of § 14.10 that the transforms of the temperatures in the two regions are

$$\bar{v}_1 = \frac{1}{4\pi\kappa_1} \int_0^\infty \left\{ \frac{e^{-\eta_1|z-z'|}}{\eta_1} - \frac{e^{-\eta_1(z+z')}}{\eta_1} + \frac{2K_1\,e^{-\eta_1(z+z')}}{K_1\eta_1+K_2\eta_2} \right\} \xi J_0(\xi r)\,d\xi, \tag{1}$$

$$\bar{v}_2 = \frac{K_1}{2\pi\kappa_1} \int_0^\infty \frac{e^{\eta_2 z - \eta_1 z'}}{K_1\eta_1+K_2\eta_2} \xi J_0(\xi r)\,d\xi, \tag{2}$$

where $$\eta_1 = [\xi^2 + (p/\kappa_1)]^{\frac{1}{2}}, \qquad \eta_2 = [\xi^2 + (p/\kappa_2)]^{\frac{1}{2}}. \tag{3}$$

To determine v_2 we remove the quantity $K_1\eta_1+K_2\eta_2$ from the denominator of (2) by using the result

$$\frac{1}{K_1\eta_1+K_2\eta_2} = \int_0^\infty e^{-\zeta(K_1\eta_1+K_2\eta_2)}\,d\zeta.$$

Assuming that orders of integration can be interchanged freely, v_2 can now be written down by 12.2 (10) and Appendix V (6). This gives

$$v_2 = \frac{K_1}{2\pi\kappa_1} \int_0^\infty \xi J_0(\xi r)\,d\xi \int_0^\infty d\zeta \int_0^t \frac{(\zeta K_1 + z')(\zeta K_2 - z)\,d\tau}{4\pi(\kappa_1\kappa_2)^{\frac{1}{2}}\tau^{\frac{3}{2}}(t-\tau)^{\frac{3}{2}}} \times$$

$$\times \exp\left\{ -\kappa_1\xi^2\tau - \kappa_2\xi^2(t-\tau) - \frac{(\zeta K_1 + z')^2}{4\kappa_1\tau} - \frac{(\zeta K_2 - z)^2}{4\kappa_2(t-\tau)} \right\}.$$

The integral in ξ can be written down from Appendix III (29), and the integral in ζ is elementary but complicated.

In terms of the dimensionless quantities

$$Z = -\frac{z}{(\kappa_1 t)^{\frac{1}{2}}}, \qquad Z' = \frac{z'}{(\kappa_1 t)^{\frac{1}{2}}}, \qquad R = \frac{r}{(\kappa_1 t)^{\frac{1}{2}}}, \qquad \sigma = \frac{K_2\kappa_1^{\frac{1}{2}}}{K_1\kappa_2^{\frac{1}{2}}}, \qquad k = \frac{\kappa_1^{\frac{1}{2}}}{\kappa_2^{\frac{1}{2}}}, \tag{4}$$

the final result is

$$v_2 = \frac{k^2}{8\pi^2(\kappa_1 t)^{\frac{3}{2}}} \int_0^1 \frac{\exp[-k^2R^2/4(k^2u+1-u)]}{(k^2u+1-u)u^{\frac{1}{2}}(1-u)^{\frac{1}{2}}} f(Z, Z', k, \sigma, u)\,du, \tag{5}$$

where

$$f(Z, Z', k, \sigma, u) = \frac{kZ(1-u)+\sigma^3 uZ'}{(1-u+\sigma^2 u)^2} \exp\left\{ -\frac{Z'^2(1-u)+k^2Z^2u}{4u(1-u)} \right\} +$$

$$+ \frac{\sigma\pi^{\frac{1}{2}}u^{\frac{1}{2}}(1-u)^{\frac{1}{2}}}{(1-u+\sigma^2 u)^{\frac{3}{2}}}\left\{ 1 - \frac{(kZ-\sigma Z')^2}{2(1-u+\sigma^2 u)} \right\} \times$$

$$\times \mathrm{erfc}\left\{ \frac{(1-u)Z'+\sigma kZu}{2u^{\frac{1}{2}}(1-u)^{\frac{1}{2}}(1-u+\sigma^2 u)^{\frac{1}{2}}} \right\} \exp\left\{ -\frac{(\sigma Z' - kZ)^2}{4(1-u+\sigma^2 u)} \right\}. \tag{6}$$

An expression for v_1 of the same type can be found in the same way.

14.13. The cylinder $r = a$

I. *We consider first the temperature at the point* (r, θ, z) *in cylindrical coordinates due to a unit instantaneous point source at* $t = 0$ *at the point* $(r', \theta', 0)$, *the surface* $r = a$ *being kept at zero*

In this case we start with the form 14.10 (4) of $\bar{u}$, and use the Addition Theorem† for the Bessel function $K_0(\eta R)$ which occurs in it, viz.

$$K_0(\eta R) = \sum_{n=-\infty}^{\infty} \cos n(\theta - \theta') I_n(\eta r) K_n(\eta r'), \quad r < r' \Bigg\}$$
$$K_0(\eta R) = \sum_{n=-\infty}^{\infty} \cos n(\theta - \theta') I_n(\eta r') K_n(\eta r), \quad r > r' \Bigg\} \tag{1}$$

As before $v = u + w$, where $\bar{w}$ is to satisfy 14.10 (7), and to be bounded at $r = 0$, and such that $\bar{u} + \bar{w} = 0$ when $r = a$. Thus

$$\bar{w} = -\frac{1}{2\pi^2\kappa} \sum_{n=-\infty}^{\infty} \cos n(\theta - \theta') \int_0^\infty \cos \xi z \frac{I_n(\eta r) I_n(\eta r') K_n(\eta a)}{I_n(\eta a)} \, d\xi, \tag{2}$$

where η is defined in 14.10 (6). It follows that if $r > r'$

$$\bar{v} = \frac{1}{2\pi^2\kappa} \sum_{n=-\infty}^{\infty} \cos n(\theta - \theta') \int_0^\infty \cos \xi z \frac{I_n(\eta r') F_n(\eta a, \eta r) \, d\xi}{I_n(\eta a)}, \tag{3}$$

where
$$F_n(x, y) = I_n(x) K_n(y) - K_n(x) I_n(y).$$

If $r < r'$ we interchange r and r' in (3).

The integrals in (3) are evaluated by considering

$$\int e^{i\xi z} \frac{I_n(\eta r') F_n(\eta a, \eta r) \, d\xi}{I_n(\eta a)}$$

round a contour consisting of the real axis and a large semicircle in the upper half-plane (for $z > 0$) not passing through any pole of the integrand. These poles are at $\xi = i\sqrt{(q^2 + \alpha^2)}$, where α is a positive root of

$$J_n(a\alpha) = 0. \tag{4}$$

Writing $\sum_\alpha$ to denote a summation over these roots, and evaluating the residues at these poles we get finally

$$\bar{v} = \frac{1}{2\pi\kappa a^2} \sum_{n=-\infty}^{\infty} \cos n(\theta - \theta') \sum_\alpha \frac{e^{-|z|\sqrt{(q^2+\alpha^2)}}}{\sqrt{(q^2+\alpha^2)}} \frac{J_n(\alpha r) J_n(\alpha r')}{[J_n'(\alpha a)]^2}. \tag{5}$$

Therefore, by Appendix V (7),

$$v = \frac{e^{-z^2/4\kappa t}}{2\pi a^2 \sqrt{(\pi\kappa t)}} \sum_{n=-\infty}^{\infty} \cos n(\theta - \theta') \sum_\alpha e^{-\kappa\alpha^2 t} \frac{J_n(\alpha r) J_n(\alpha r')}{[J_n'(\alpha a)]^2}. \tag{6}$$

The result for a line source parallel to the axis of z follows by integrating (6) with respect to z, or it may be obtained directly in the same way. The result 14.8 (5) for the cylindrical surface source follows by integration with respect to θ'.

† *W.B.F.*, § 11.41; *G. and M.*, p. 74 (59).

II. *The problem of* I *except that at* $r = a$ *there is radiation into a medium at zero*

$$v = \frac{e^{-z^2/4\kappa t}}{2\pi a^2 \sqrt{(\pi \kappa t)}} \sum_{n=-\infty}^{\infty} \cos n(\theta - \theta') \sum_{\alpha} e^{-\kappa \alpha^2 t} \frac{\alpha^2 J_n(\alpha r) J_n(\alpha r')}{[\alpha^2 + h^2 - (n^2/a^2)] J_n^2(\alpha a)}, \quad (7)$$

where the summation in α is over the positive roots of $\alpha J_n'(\alpha a) + h J_n(\alpha a) = 0$, and if $h = 0$, an additional term

$$\frac{e^{-z^2/4\kappa t}}{2a^2\pi(\pi \kappa t)^{\frac{1}{2}}}$$

has to be added to (7).

III. *Unit instantaneous point source at* $(r', \theta', 0)$ *at time* $t = 0$ *in the region bounded internally by the cylinder* $r = a$. *The surface* $r = a$ *kept at zero temperature*

$$v = \frac{e^{-z^2/4\kappa t}}{4\pi\sqrt{(\pi\kappa t)}} \sum_{n=-\infty}^{\infty} \cos n(\theta - \theta') \int_0^\infty \alpha e^{-\kappa\alpha^2 t} \frac{U_n(\alpha r) U_n(\alpha r')}{J_n^2(\alpha a) + Y_n^2(\alpha a)} \, d\alpha, \quad (8)$$

where
$$U_n(\alpha r) = J_n(\alpha r) Y_n(\alpha a) - J_n(\alpha a) Y_n(\alpha r). \quad (9)$$

IV. *The problem of* III *except that there is radiation at* $r = a$ *into medium at zero*

$$v = \frac{e^{-z^2/4\kappa t}}{4\pi\sqrt{(\pi\kappa t)}} \sum_{n=-\infty}^{\infty} \cos n(\theta - \theta') \int_0^\infty C_n(ur) C_n(ur') u e^{-\kappa u^2 t} \, du,$$

where
$$C_n(z) = \frac{J_n(z)[uY_n'(au) - hY_n(au)] - Y_n(z)[uJ_n'(au) - hJ_n(au)]}{\{[uJ_n'(au) - hJ_n(au)]^2 + [uY_n'(au) - hY_n(au)]^2\}^{\frac{1}{2}}}.$$

14.14. The wedge $0 < \theta < \theta_0$

I. *Zero surface temperature*

We consider the case of a unit instantaneous point source at $(r', \theta', 0)$ at $t = 0$. We use the form 14.10 (4) of $\bar{u}$ together with the result ($G.$ *and* $M.$,† p. 101) that if r, r', and λ are real and positive, $0 < \theta - \theta' < 2\pi$, and $r > r'$,

$$K_0(\lambda R) = P \int_{-\infty i}^{\infty i} \frac{\cos\nu(\pi - \theta + \theta')}{\sin\nu\pi} K_\nu(\lambda r) I_\nu(\lambda r') i \, d\nu, \quad (1)$$

where P implies that the principal value at the origin is taken. Thus when $\theta_0 > \theta > \theta'$, $r > r'$,

$$\bar{u} = \frac{i}{2\pi^2\kappa} \int_0^\infty \cos\xi z \left[P \int_{-\infty i}^{\infty i} \frac{\cos\nu(\pi - \theta + \theta')}{\sin\nu\pi} I_\nu(\eta r') K_\nu(\eta r) \, d\nu \right] d\xi. \quad (2)$$

Then we find

$$\bar{w} = -\frac{i}{2\pi^2\kappa} \int_0^\infty \cos\xi z \left[P \int_{-\infty i}^{\infty i} \frac{\cos\nu(\pi - \theta_0 + \theta')\sin\nu\theta + \cos\nu(\pi - \theta')\sin\nu(\theta_0 - \theta)}{\sin\nu\pi \sin\nu\theta_0} \times \right.$$
$$\left. \times I_\nu(\eta r') K_\nu(\eta r) \, d\nu \right] d\xi,$$

---

† It is easily proved by completing the contour by a large semicircle in the right-hand half-plane and using Cauchy's theorem.

since this satisfies the equation for $\bar{w}$, and makes $\bar{v} = \bar{u} + \bar{w}$ zero on $\theta = 0$ and $\theta = \theta_0$. Also

$$\bar{v} = -\frac{i}{\pi^2 \kappa} \int_0^\infty \cos \xi z \left[\int_{-\infty i}^{\infty i} \frac{\sin \nu \theta' \sin \nu(\theta_0 - \theta)}{\sin \nu \theta_0} I_\nu(\eta r') K_\nu(\eta r) \, d\nu \right] d\xi, \quad (3)$$

since the path can now be completed at the origin. If $0 < \theta < \theta'$, θ and θ' are to be interchanged in (3); and if $0 < r < r'$, r and r' are to be interchanged.

The inner integral is evaluated by completing the path $(-\infty i, \infty i)$ by a large semicircle in the right-hand half-plane and evaluating the residues at the poles $n\pi/\theta_0$ of the integrand. We write

$$s = n\pi/\theta_0 \quad (4)$$

and use $\sum\limits_s$ for a summation over the values $n = 1$ to ∞. Then, for $r > r'$ and $0 < \theta < \theta_0$, we find

$$\bar{v} = \frac{2}{\pi\kappa\theta_0} \sum_s \sin s\theta \sin s\theta' \int_0^\infty \cos \xi z I_s(\eta r') K_s(\eta r) \, d\xi. \quad (5)$$

Now from Appendix V (22) and 12.2 (6) it follows that, if $r > r'$, $I_s(\eta r') K_s(\eta r)$ is the Laplace transform of

$$\frac{1}{2t} \exp\left\{ -\kappa \xi^2 t - \frac{r^2 + r'^2}{4\kappa t} \right\} I_s\left(\frac{rr'}{2\kappa t}\right). \quad (6)$$

It follows from (5) that, for all values of r and θ,

$$v = \frac{1}{\pi\kappa\theta_0 t} \sum_s \sin s\theta \sin s\theta' I_s\left(\frac{rr'}{2\kappa t}\right) e^{-(r^2 + r'^2)/4\kappa t} \int_0^\infty \cos \xi z e^{-\kappa \xi^2 t} \, d\xi$$

$$= \frac{1}{2\theta_0 \pi^{\frac{1}{2}}(\kappa t)^{\frac{3}{2}}} e^{-z^2/4\kappa t} \sum_s e^{-(r^2 + r'^2)/4\kappa t} I_s\left(\frac{rr'}{2\kappa t}\right) \sin s\theta \sin s\theta'. \quad (7)$$

II. *No flow of heat over the surface*

It follows in the same way that the solution for a unit instantaneous point source at $(r', \theta', 0)$ at $t = 0$ is

$$v = \frac{1}{4\theta_0 \pi^{\frac{1}{2}}(\kappa t)^{\frac{3}{2}}} e^{-(z^2 + r^2 + r'^2)/4\kappa t} \left\{ I_0\left(\frac{rr'}{2\kappa t}\right) + 2 \sum_s \cos s\theta \cos s\theta' I_s\left(\frac{rr'}{2\kappa t}\right) \right\}. \quad (8)$$

III. *The wedge of angle 2π*

In this special case,† in which the bounding surface is the half-plane $y = 0$, $x > 0$, there is a simple closed expression for the Green's function, namely

$$\frac{1}{16(\pi\kappa t)^{\frac{3}{2}}}\left\{e^{-(z^2+R^2)/4\kappa t}\left[1+\mathrm{erf}\left\{\left(\frac{rr'}{\kappa t}\right)^{\frac{1}{2}}\cos\tfrac{1}{2}(\theta-\theta')\right\}\right]\mp\right.$$

$$\left.\mp e^{-(z^2+R'^2)/4\kappa t}\left[1+\mathrm{erf}\left\{\left(\frac{rr'}{\kappa t}\right)^{\frac{1}{2}}\cos\tfrac{1}{2}(\theta+\theta')\right\}\right]\right\},\quad (9)$$

where

$$R^2 = r^2+r'^2-2rr'\cos(\theta-\theta'),\qquad R'^2 = r^2+r'^2-2rr'\cos(\theta+\theta'),\qquad (10)$$

the negative sign being taken for the case in which the half-plane is at zero temperature, and the positive sign for the case in which there is no flow of heat across it.

14.15. Cylindrical regions

The Green's functions for any of the regions bounded by the surfaces of the cylindrical coordinate system may be derived by the methods of the last few sections.‡

For example, for the temperature at (r, θ, z) due to a unit instantaneous point source at $t = 0$ at (r', θ', z') in the region $0 < z < l$, $0 \leqslant r < a$, the surface being kept at zero, we have

$$\bar{v} = \frac{1}{\pi\kappa l}\sum_{m=1}^{\infty}\sin\frac{m\pi z}{l}\sin\frac{m\pi z'}{l}\sum_{n=-\infty}^{\infty}\frac{I_n(rq_m)F_n(aq_m, r'q_m)}{I_n(aq_m)}\cos n(\theta-\theta'),\quad (1)$$

where $q_m = \sqrt{(q^2+m^2\pi^2/l^2)}$, $r < r'$, and $F_n(x, y) = I_n(x)K_n(y)-K_n(x)I_n(y)$.

This gives, for $r \gtrless r'$,

$$v = \frac{2}{\pi a^2 l}\sum_{m=1}^{\infty}e^{-\kappa m^2\pi^2 t/l^2}\sin\frac{m\pi z}{l}\sin\frac{m\pi z'}{l}\sum_{n=-\infty}^{\infty}\cos n(\theta-\theta')\sum_{\alpha}e^{-\kappa\alpha^2 t}\frac{J_n(\alpha r)J_n(\alpha r')}{[J_n'(\alpha a)]^2},\quad (2)$$

where the α are the positive roots of $J_n(\alpha a) = 0$.

Again, if the surface of the region $a < r < b$, $0 < z < l$, $0 < \theta < \theta_0$ is kept at zero, and there is a unit instantaneous point source at $t = 0$ at (r', θ', z'), the temperature at (r, θ, z) at time t is

$$\frac{2\pi^2}{l\theta_0}\sum_{m=1}^{\infty}e^{-\kappa m^2\pi^2 t/l^2}\sin\frac{m\pi z}{l}\sin\frac{m\pi z'}{l}\sum_{s}\sin s\theta\sin s\theta'\sum_{\alpha}\frac{\alpha^2 e^{-\kappa\alpha^2 t}J_s^2(\alpha b)U_s(\alpha r)U_s(\alpha r')}{J_s^2(\alpha a)-J_s^2(\alpha b)},\quad (3)$$

where s and $U_s(\alpha r)$ are defined in 14.14 (4) and 14.13 (9), respectively, and the α are the positive roots of $U_s(\alpha b) = 0$.

14.16. The sphere $r = a$

We require the temperature at the point whose spherical polar coordinates are (r, θ, ϕ) due to a unit instantaneous point source, which we first take to be at $(r', 0, 0)$.

† Carslaw, *Proc. Lond. Math. Soc.* (1) **30** (1899), 121–63. This result follows very readily by the method of images in a Riemann's space referred to in § 10.12.

‡ Carslaw and Jaeger, *J. Lond. Math. Soc.* **15** (1940) 273.

The temperature u at (r, θ, ϕ) due to such a source in an infinite medium is

$$u = \frac{1}{8(\pi \kappa t)^{\frac{3}{2}}} e^{-R^2/4\kappa t}, \tag{1}$$

where

$$R^2 = r^2 + r'^2 - 2rr' \cos \theta. \tag{2}$$

Then, from Appendix V (6),

$$\bar{u} = \frac{1}{4\pi \kappa R} e^{-qR}. \tag{3}$$

By using one of the addition theorems for Bessel functions ($W.B.F.$, § 11.41 (11)) this may be expressed in a form suitable for use with spherical polar coordinates, viz.

$$\left.\begin{aligned}
\bar{u} &= \frac{1}{4\pi \kappa \sqrt{(rr')}} \sum_{n=0}^{\infty} (2n+1) K_{n+\frac{1}{2}}(qr') I_{n+\frac{1}{2}}(qr) P_n(\mu), \quad \text{when } r < r' \\
\bar{u} &= \frac{1}{4\pi \kappa \sqrt{(rr')}} \sum_{n=0}^{\infty} (2n+1) K_{n+\frac{1}{2}}(qr) I_{n+\frac{1}{2}}(qr') P_n(\mu), \quad \text{when } r > r'
\end{aligned}\right\}, \tag{4}$$

where $\mu = \cos \theta$.

I. *The source in the sphere $0 \leqslant r < a$. Surface at zero temperature*

As usual we let $v = u + w$, where w is to satisfy the equation of conduction of heat, and to vanish at $t = 0$, and is to be such that $v = u + w$ vanishes at $r = a$.

$\bar{w}$ is to satisfy the subsidiary equation

$$\frac{\partial^2 \bar{w}}{\partial r^2} + \frac{2}{r} \frac{\partial \bar{w}}{\partial r} + \frac{1}{r^2 \sin \theta} \frac{\partial}{\partial \theta}\left(\sin \theta \frac{\partial \bar{w}}{\partial \theta}\right) - q^2 \bar{w} = 0. \tag{5}$$

The solution of this which is finite at the origin is

$$\bar{w} = \sum_{n=1}^{\infty} A_n r^{-\frac{1}{2}} I_{n+\frac{1}{2}}(qr) P_n(\mu), \tag{6}$$

where the coefficients A_n are to be determined from the fact that $\bar{u} + \bar{w}$ vanishes when $r = a$. This gives, using (4),

$$A_n = -\frac{(2n+1) K_{n+\frac{1}{2}}(qa) I_{n+\frac{1}{2}}(qr')}{4\pi \kappa \sqrt{(r')} I_{n+\frac{1}{2}}(qa)}.$$

Thus, when $0 < r < r'$,

$$\bar{v} = \frac{1}{4\pi \kappa \sqrt{(rr')}} \sum_{n=0}^{\infty} \frac{K_{n+\frac{1}{2}}(qr') I_{n+\frac{1}{2}}(qa) - I_{n+\frac{1}{2}}(qr') K_{n+\frac{1}{2}}(qa)}{I_{n+\frac{1}{2}}(qa)} \times$$
$$\times (2n+1) I_{n+\frac{1}{2}}(qr) P_n(\mu), \tag{7}$$

and when $r' < r < a$ we interchange r and r' in (7).

v is found by applying the Inversion Theorem in the usual way to the terms of (7). This gives

$$v = \frac{1}{2\pi a^2 \sqrt{(rr')}} \sum_{n=0}^{\infty} \sum_{\alpha} e^{-\kappa \alpha^2 t} \frac{J_{n+\frac{1}{2}}(\alpha r) J_{n+\frac{1}{2}}(\alpha r')}{[J'_{n+\frac{1}{2}}(\alpha a)]^2} (2n+1) P_n(\mu), \qquad (8)$$

where the α are the positive roots of

$$J_{n+\frac{1}{2}}(a\alpha) = 0. \qquad (9)$$

To find the temperature at P, (r, θ, ϕ), due to an instantaneous unit source at P', (r', θ', ϕ'), we use (8) with μ replaced by $\cos\gamma$, where γ is the angle POP'. Also we know that

$$P_n(\cos\gamma) = P_n(\mu)P_n(\mu') + 2 \sum_{m=1}^{n} \frac{(n-m)!}{(n+m)!} P_n^m(\mu) P_n^m(\mu') \cos m(\phi-\phi'), \qquad (10)$$

where $\mu = \cos\theta$ and $\mu' = \cos\theta'$. Using these results the temperature in the sphere $0 \leqslant r < a$ with arbitrary initial and surface temperatures can be written down.

II. *The problem of* I *except that at* $r = a$ *there is radiation into medium at zero temperature*

$$v = \frac{1}{2\pi\sqrt{(rr')}} \sum_{n=0}^{\infty} (2n+1) P_n(\mu) \sum_{\alpha} \frac{\alpha^2 e^{-\kappa \alpha^2 t} J_{n+\frac{1}{2}}(r\alpha) J_{n+\frac{1}{2}}(r'\alpha)}{[(ah-\frac{1}{2})^2 + a^2\alpha^2 - (n+\frac{1}{2})^2] J_{n+\frac{1}{2}}^2(a\alpha)}, \qquad (11)$$

where the α are the positive roots of

$$(ah-\tfrac{1}{2}) J_{n+\frac{1}{2}}(a\alpha) + a\alpha J'_{n+\frac{1}{2}}(a\alpha) = 0. \qquad (12)$$

For the case of no flow of heat, a term $(3/4\pi a^3)$ is to be added to the value of (11) for $h = 0$.

III. *The region bounded internally by* $r = a$. *Unit instantaneous point source at* $t = 0$ *at* $(r', 0, 0)$. *The surface* $r = a$ *kept at zero temperature*

$$v = \frac{1}{4\pi\sqrt{(rr')}} \sum_{n=0}^{\infty} (2n+1) P_n(\mu) \int_0^{\infty} \frac{C_{n+\frac{1}{2}}(ur) C_{n+\frac{1}{2}}(ur')}{J_{n+\frac{1}{2}}^2(au) + Y_{n+\frac{1}{2}}^2(ua)} e^{-\kappa u^2 t} u \, du, \qquad (13)$$

where $C_{n+\frac{1}{2}}(z) = J_{n+\frac{1}{2}}(z) Y_{n+\frac{1}{2}}(ua) - Y_{n+\frac{1}{2}}(z) J_{n+\frac{1}{2}}(ua)$.

IV. *The problem of* III *except that at* $r = a$ *there is radiation into medium at zero temperature*

$$v = \frac{1}{4\pi\sqrt{(rr')}} \sum_{n=0}^{\infty} (2n+1) P_n(\mu) \int_0^{\infty} e^{-\kappa u^2 t} F_{n+\frac{1}{2}}(ur) F_{n+\frac{1}{2}}(ur') u \, du, \qquad (14)$$

where

$$F_\nu(ur) = \frac{(2ah+1)[J_\nu(ur) Y_\nu(ua) - Y_\nu(ur) J_\nu(ua)] - 2ua[J_\nu(ur) Y'_\nu(ua) - Y_\nu(ur) J'_\nu(ua)]}{\{[(2ah+1) J_\nu(ua) - 2ua J'_\nu(ua)]^2 + [(2ah+1) Y_\nu(ua) - 2ua Y'_\nu(ua)]^2\}^{\frac{1}{2}}}.$$

$$(15)$$

14.17. The cone $0 \leqslant \theta < \theta_0$

Suppose we have a unit instantaneous point source at $t = 0$ at $(r', 0, 0)$. As before we start with u and $\bar{u}$ given by 14.16 (1), (3), and we require an expression for $\bar{u}$ suitable for use with a conical boundary: this is provided by the following integral for e^{-qR}/R or $(2q/\pi R)^{\frac{1}{2}} K_{\frac{1}{2}}(qR)$, analogous to that of 14.14 (1) for $K_0(qR)$ which was developed for the corresponding problem on the wedge.

The result required† is that, if $r > r'$ and $\mathbf{R}(q) > 0$,

$$\int_C \frac{(2\nu+1)I_{\nu+\frac{1}{2}}(qr')K_{\nu+\frac{1}{2}}(qr)P_\nu(-\mu)\, d\nu}{\sin \nu\pi} = 2i \sum_{n=0}^{\infty} (2n+1)K_{n+\frac{1}{2}}(qr)I_{n+\frac{1}{2}}(qr')P_n(\mu),$$

(1)

where C is a contour beginning at infinity in the first quadrant, passing through the point $(-\frac{1}{2})$, and ending at infinity in the fourth quadrant.

To find the temperature at (r, θ, ϕ) *due to the source when the surface of the cone* $\theta = \theta_0$ *is kept at zero,* we start with the value given by (1) and 14.16 (4), namely

$$\bar{u} = \frac{1}{8\pi i\kappa\sqrt{(rr')}} \int_C \frac{(2\nu+1)I_{\nu+\frac{1}{2}}(qr')K_{\nu+\frac{1}{2}}(qr)P_\nu(-\mu)\, d\nu}{\sin \nu\pi}$$

(2)

for $r > r'$, and for $r < r'$ we have to interchange r and r' in (2). $\bar{w}$ has to satisfy 14.16 (5), so we take

$$\bar{w} = \frac{1}{8\pi i\kappa\sqrt{(rr')}} \int_C \frac{(2\nu+1)I_{\nu+\frac{1}{2}}(qr')K_{\nu+\frac{1}{2}}(qr)P_\nu(\mu)}{\sin \nu\pi} f(\nu)\, d\nu,$$

where $f(\nu)$ is to be chosen so that $\bar{u}+\bar{w} = 0$ when $\mu = \mu_0 = \cos\theta_0$.

This requires $\qquad f(\nu) = -P_\nu(-\mu_0)/P_\nu(\mu_0),$

and thus we get, for $r > r'$,

$$\bar{v} = \frac{1}{8\pi i\kappa\sqrt{(rr')}} \int_C (2\nu+1)I_{\nu+\frac{1}{2}}(qr')K_{\nu+\frac{1}{2}}(qr)\frac{[P_\nu(-\mu)P_\nu(\mu_0)-P_\nu(-\mu_0)P_\nu(\mu)]}{\sin \nu\pi P_\nu(\mu_0)}\, d\nu,$$

(3)

and if $r < r'$ we interchange r and r' in (3).

The integral in (3) is now evaluated by completing the contour by portion of a large circle in the right-hand half-plane and using Cauchy's

---

† Carslaw, *Math. Ann.* **75** (1914) **133**. The result is obtained by completing the contour in the right-hand half-plane by portion of a large circle and using Cauchy's theorem. Other results needed below, such as (5), are also proved in this paper.

theorem. The integrand has poles at the zeros† of $P_\nu(\mu_0)$ regarded as a function of ν; it does not have poles at the integral values of ν, since for these $P_n(\mu) = (-1)^n P_n(-\mu)$. Evaluating the residues at the poles we get finally

$$\bar{v} = -\frac{1}{4\kappa\sqrt{(rr')}} \sum_\nu \frac{(2\nu+1)I_{\nu+\frac{1}{2}}(qr')K_{\nu+\frac{1}{2}}(qr)}{\sin\nu\pi} \frac{P_\nu(-\mu_0)P_\nu(\mu)}{(d/d\nu)[P_\nu(\mu_0)]},$$

the summation in ν being over the roots greater than $-\frac{1}{2}$ of

$$P_\nu(\mu_0) = 0. \tag{4}$$

But, for these values of ν,

$$(1-\mu_0^2)P_\nu(-\mu_0)\left[\frac{d}{d\mu}P_\nu(\mu)\right]_{\mu=\mu_0} = \frac{2}{\pi}\sin\nu\pi. \tag{5}$$

Thus

$$\bar{v} = -\frac{1}{2\pi\kappa\sqrt{(rr')}} \sum_\nu (2\nu+1)I_{\nu+\frac{1}{2}}(qr')K_{\nu+\frac{1}{2}}(qr)P_\nu(\mu)\times$$

$$\times \left[(1-\mu_0^2)\left\{\frac{d}{d\mu_0}P_\nu(\mu_0)\right\}\left\{\frac{d}{d\nu}P_\nu(\mu_0)\right\}\right]^{-1}. \tag{6}$$

And, finally, using Appendix V (22),

$$v = -\frac{1}{4\pi\kappa t\sqrt{(rr')}} \sum_\nu e^{-(r^2+r'^2)/4\kappa t}(2\nu+1)I_{\nu+\frac{1}{2}}\left(\frac{rr'}{2\kappa t}\right)P_\nu(\mu)\times$$

$$\times \left[(1-\mu_0^2)\left\{\frac{d}{d\mu_0}P_\nu(\mu_0)\right\}\left\{\frac{d}{d\nu}P_\nu(\mu_0)\right\}\right]^{-1}, \tag{7}$$

for $r \gtrless r'$, the summation being over the roots of (4).

If the source is at the point (r', θ', ϕ') *instead of on the axis*, $P_n(\mu)$ in $\bar{u}$ has to be replaced by $P_n(\cos\gamma)$, where γ is the angle between the radii vectores from the origin to the points (r,θ,ϕ) and (r',θ',ϕ').

Writing $\mu = \cos\theta$ and $\mu' = \cos\theta'$, the Addition Theorem for Spherical Harmonics gives for $\theta > \theta'$

$$P_\nu(-\cos\gamma) = \sum_{m=0}^{\infty} \epsilon_m \frac{\Gamma(\nu+m+1)\Gamma(m-\nu)}{\Gamma(\nu+1)\Gamma(-\nu)} P_\nu^{-m}(-\mu)P_\nu^{-m}(\mu')\cos m(\phi-\phi'), \tag{8}$$

where

$$\begin{aligned}\epsilon_m &= 1, \quad m = 0 \\ &= 2, \quad m = 1, 2, 3,...\end{aligned}\Bigg\}. \tag{9}$$

If $\theta < \theta'$ we have to interchange θ and θ' in (8).

---

† It is known that the zeros of $P_\nu(\mu)$ and $P_\nu^{-m}(\mu)$ are all real and distinct, cf. Macdonald, *Proc. Lond. Math. Soc.* (1) **31** (1900) 265. Also, since $P_\nu^{-m}(\mu) = P_{-\nu-1}^{-m}(\mu)$, they are symmetrically disposed about the point $\nu = -\frac{1}{2}$ through which the contour C passes.

Introducing (8) in (2) and proceeding as before, except that in place of (5) its generalization (Carslaw, loc. cit.),

$$-(1-\mu_0^2)P_\nu^{-m}(-\mu_0)\frac{d}{d\mu_0}P_\nu^{-m}(\mu_0) = \frac{2}{\Gamma(m+\nu+1)\Gamma(m-\nu)}, \quad (10)$$

when ν is a root of (12), is needed, we get finally

$$v = -\frac{1}{4\pi\kappa t\sqrt{(rr')}}\sum_{m=0}^{\infty}\sum_{\nu} \epsilon_m e^{-(r^2+r'^2)/4\kappa t}I_{\nu+\frac{1}{2}}\left(\frac{rr'}{2\kappa t}\right)(2\nu+1)P_\nu^{-m}(\mu)\cos m(\phi-\phi')\times$$

$$\times P_\nu^{-m}(\mu')\left[(1-\mu_0)^2\frac{d}{d\mu_0}P_\nu^{-m}(\mu_0)\frac{d}{d\nu}P_\nu^{-m}(\mu_0)\right]^{-1}, \quad (11)$$

where the summation in ν is over the roots greater than $-\frac{1}{2}$ of

$$P_\nu^{-m}(\mu_0) = 0. \quad (12)$$

Using (11) the temperature at any point of a cone with arbitrary initial and surface temperatures can be written down. Problems on solids bounded by other surfaces of the spherical polar coordinate system may be treated in the same way.

14.18. Continuous sources

Solutions for continuous point or line sources in the regions considered in this chapter can be written down by integration of the appropriate Green's function. They can, however, be very simply obtained directly. As an example, consider the *continuous line source, emitting Q heat units per unit length per unit time for* $t > 0$, *parallel to the z-axis through the point* $(r', 0)$ *in the cylinder* $r = a$. *The cylinder is at zero temperature at* $t = 0$, *and there is no flow of heat across its boundary.*

We require a solution of the subsidiary equation

$$\frac{\partial^2\bar{v}}{\partial r^2}+\frac{1}{r}\frac{\partial\bar{v}}{\partial r}+\frac{1}{r^2}\frac{\partial^2\bar{v}}{\partial\theta^2}-q^2\bar{v} = 0, \quad 0\leqslant r < a, \quad (1)$$

with

$$\frac{\partial\bar{v}}{\partial r} = 0, \quad r = a, \quad (2)$$

and

$$\lim_{R\to 0} R\frac{\partial\bar{v}}{\partial R} = -\frac{Q}{2\pi Kp}, \quad (3)$$

where

$$R = \sqrt{(r^2+r'^2-2rr'\cos\theta)}. \quad (4)$$

A solution of (1) and (3) is $\dfrac{Q}{2\pi Kp}K_0(qR).$

By the addition theorem, 14.13 (1), this is equal to

$$\frac{Q}{2\pi Kp}\sum_{n=0}^{\infty}\epsilon_n I_n(qr)K_n(qr')\cos n\theta, \quad \text{when } r < r', \quad (5)$$

where $\epsilon_n = 1$, if $n = 0$, and $\epsilon_n = 2$, if $n = 1, 2,...$. If $r > r'$ we interchange r and r' in (5). Thus for the solution of (1)–(4) we assume

$$\bar{v} = \frac{Q}{2\pi Kp}\sum_{n=0}^{\infty}\epsilon_n\{I_n(qr)K_n(qr')+a_n I_n(qr)\}\cos n\theta, \quad 0\leqslant r < r'$$

$$\left.\vphantom{\sum}\right\}, \quad (6)$$

$$\bar{v} = \frac{Q}{2\pi Kp}\sum_{n=0}^{\infty}\epsilon_n\{I_n(qr')K_n(qr)+a_n I_n(qr)\}\cos n\theta, \quad a > r > r'$$

C c

where the a_n are to be chosen so that (6) satisfies (2). This requires

$$a_n = -\frac{I_n(qr')K_n'(qa)}{I_n'(qa)}.$$

Therefore

$$\bar{v} = \frac{Q}{2\pi Kp} \sum_{n=0}^{\infty} \epsilon_n \cos n\theta \, \frac{I_n(qr)[K_n(qr')I_n'(qa)-I_n(qr')K_n'(qa)]}{I_n'(qa)}, \tag{7}$$

when $0 \leqslant r < r'$, and when $a > r > r'$ we interchange r and r' in (7).

From the Inversion Theorem we get

$$v = \frac{Q}{4\pi^2 iK} \sum_{n=0}^{\infty} \epsilon_n \cos n\theta \int_{\gamma-i\infty}^{\gamma+i\infty} \frac{I_n(\mu r)[K_n(\mu r')I_n'(\mu a)-I_n(\mu r')K_n'(\mu a)]e^{\lambda t} \, d\lambda}{\lambda I_n'(\mu a)}. \tag{8}$$

If $n = 0$, there is a double pole at the origin of residue

$$\frac{2\kappa t}{a^2} + \frac{r^2+r'^2}{2a^2} + \ln\frac{a}{r'} - \frac{3}{4}. \tag{9}$$

If $n > 0$, there are simple poles at the origin of residue

$$\frac{r^n(r'^{2n}+a^{2n})}{2na^{2n}r'^n}. \tag{10}$$

Thus the poles at $\lambda = 0$ give

$$\frac{Q}{2\pi K}\left\{\frac{2\kappa t}{a^2} + \frac{r^2+r'^2}{2a^2} + \ln\frac{a}{r'} - \frac{3}{4} + \sum_{n=1}^{\infty} \frac{r^n(r'^{2n}+a^{2n})}{na^{2n}r'^n}\cos n\theta\right\}$$

$$= \frac{Q}{2\pi K}\left\{\frac{2\kappa t}{a^2} + \frac{r^2+r'^2}{2a^2} - \frac{3}{4} - \tfrac{1}{2}\ln\left(1-\frac{2rr'}{a^2}\cos\theta+\frac{r^2r'^2}{a^4}\right)-\right.$$

$$\left.-\tfrac{1}{2}\ln\left(\frac{r'^2}{a^2}-\frac{2rr'}{a^2}\cos\theta+\frac{r^2}{a^2}\right)\right\}. \tag{11}$$

The other zeros of the denominators in (8) give

$$-\frac{Q}{\pi K}\sum_{n=0}^{\infty}\epsilon_n\cos n\theta\sum_{m=1}^{\infty}e^{-\kappa\alpha_{n,m}^2 t}\frac{J_n(r\alpha_{n,m})J_n(r'\alpha_{n,m})}{(a^2\alpha_{n,m}^2-n^2)J_n^2(a\alpha_{n,m})}, \tag{12}$$

where the $\alpha_{n,m}$ are the positive roots of

$$J_n'(a\alpha) = 0.$$

The complete solution is the sum of (11) and (12).

FURTHER APPLICATIONS OF THE LAPLACE TRANSFORMATION

15.1. Introductory

IN this chapter we discuss a number of miscellaneous problems which do not fall into any of the previous categories, and whose only common property is that the Laplace transformation method is well suited to their study, and, usually, leads to rather more complicated transforms than those discussed earlier. We indicate briefly the application of the method to problems on conduction of heat in moving solids, the theory of heat exchangers, cases in which heat is generated in the solid, the calculation of steady periodic temperatures, flow of heat in non-homogeneous materials, and a number of other problems. The direct application of the Laplace transformation to problems with several space variables, as an alternative to the methods previously discussed, is also considered.

15.2. Conduction of heat in a moving solid

Comparatively few transient problems of this type have been solved. These are mostly for the semi-infinite solid which moves with velocity u_x along the x-axis and has various surface conditions at $x = 0$. Positive values of u_x correspond to an accreting medium (such as a snowfield which is being supplemented by continuous falls).† Negative values of u_x correspond to removal of material at $x = 0$ by erosion,‡ melting or sublimation, mining operations,§ or similar processes. A number of problems on the flow of liquid through tubes has been considered by Owen‖ and Somers†† on the assumption that the velocity of the liquid is constant over the cross-section of the tube. Problems on moving sources of heat can also be solved by the present method, but direct integration as in § 10.7 is more powerful and easier to follow. We proceed to solve a number of typical problems.

† Benfield, *Mon. Not. R. Astr. Soc. Geophys. Suppl.* **6** (1951) 139–47; *J. Glaciol.* **2** (1952) 250–4.

‡ Benfield, *J. Appl. Phys.* **20** (1949) 66–70; *Quart. Appl. Math.* **7** (1949) 436–9; ibid. **6** (1948) 439–43. The effect of heat generation in the moving medium is considered also.

§ Wiles, *S. Afr. J. Sci.* **39** (1943) 95–97.

‖ Owen, *Proc. Lond. Math. Soc.* (2) **23** (1925) 238–49.

†† Somers, *Proc. Phys. Soc.* **25** (1912) 74–76.

I. *The region $x > 0$ moves with velocity U. Its initial temperature is $V_0 + ax$ and the surface $x = 0$ is maintained at temperature $V_1 + bt$ for $t > 0$. There is heat generation in the solid at the rate A_0 per unit time per unit volume*

This case, for various values of the constants V_0, V_1, a, b, A_0, covers many of the applications referred to above. The results are valid for both signs of U. The differential equation to be solved is, by 1.7 (2),

$$\frac{\partial^2 v}{\partial x^2} - \frac{U}{\kappa}\frac{\partial v}{\partial x} - \frac{1}{\kappa}\frac{\partial v}{\partial t} = -\frac{A_0}{K}, \tag{1}$$

with

$$v = V_0 + ax, \quad x > 0, \, t = 0, \tag{2}$$

$$v = V_1 + bt, \quad x = 0, \, t > 0. \tag{3}$$

The subsidiary equation is

$$\frac{d^2\bar{v}}{dx^2} - \frac{U}{\kappa}\frac{d\bar{v}}{dx} - \frac{p}{\kappa}\bar{v} = -\frac{A_0}{Kp} - \frac{V_0 + ax}{\kappa}, \tag{4}$$

to be solved with $\bar{v} = \dfrac{V_1}{p} + \dfrac{b}{p^2},$ when $x = 0$. $\tag{5}$

The solution is

$$\bar{v} = \left[\frac{V_1 - V_0}{p} + \frac{b + aU - (\kappa A_0/K)}{p^2}\right]\exp\left\{\frac{Ux}{2\kappa} - x\left(\frac{U^2}{4\kappa^2} + \frac{p}{\kappa}\right)^{\frac{1}{2}}\right\} +$$

$$+ \left(\frac{\kappa A_0}{K} - aU\right)\frac{1}{p^2} + \frac{V_0 + ax}{p}, \quad (6)$$

and, by Appendix V (19) and (29) and § 12.2, Theorem VI,

$$v = V_0 + ax + (\kappa A_0 t/K) - aUt +$$

$$+ \tfrac{1}{2}(V_1 - V_0)\left\{\operatorname{erfc}\frac{x - Ut}{2(\kappa t)^{\frac{1}{2}}} + e^{Ux/\kappa}\operatorname{erfc}\frac{x + Ut}{2(\kappa t)^{\frac{1}{2}}}\right\} +$$

$$+ \frac{1}{2U}\left(b + aU - \frac{\kappa A_0}{K}\right)\left\{(x + Ut)e^{Ux/\kappa}\operatorname{erfc}\frac{x + Ut}{2(\kappa t)^{\frac{1}{2}}} + (Ut - x)\operatorname{erfc}\frac{x - Ut}{2(\kappa t)^{\frac{1}{2}}}\right\}. \quad (7)$$

II. *The region $x > 0$ moves with velocity U. Its initial temperature is V_0. The boundary condition at $x = 0$ is*

$$\frac{\partial v}{\partial x} - hv = 0. \tag{8}$$

In this case we have for the solution of the differential equation (1) with $A_0 = 0$ and boundary condition (8),

$$\bar{v} = \frac{V_0}{p} - \frac{hV_0}{p\{[(p/\kappa) + (U^2/4\kappa^2)]^{\frac{1}{2}} + [h - (U/2\kappa)]\}}\exp\left\{\frac{Ux}{2\kappa} - x\left(\frac{U^2}{4\kappa^2} + \frac{p}{\kappa}\right)^{\frac{1}{2}}\right\}. \tag{9}$$

It follows from Appendix V (31) that

$$v = V_0 - \tfrac{1}{2}V_0\left\{\operatorname{erfc}\frac{x-Ut}{2(\kappa t)^{\frac{1}{2}}} + \frac{\kappa h}{\kappa h - U}e^{Ux/\kappa}\operatorname{erfc}\frac{x+Ut}{2(\kappa t)^{\frac{1}{2}}}\right\} +$$
$$+ \frac{V_0(2\kappa h - U)}{2(\kappa h - U)}e^{hx - ht(U-\kappa h)}\operatorname{erfc}\frac{x+(2\kappa h - U)t}{2(\kappa t)^{\frac{1}{2}}}. \quad (10)$$

This result is valid for both signs of U and $2\kappa h - U$. It appears in connexion with the diffusion of impurities in molten metal.†

III. *Steady periodic temperature in the region $x > 0$ moving with velocity U and with temperature $V_0\cos(\omega t + \theta)$ at $x = 0$*

Assuming a solution of form

$$v = \mathbf{R}\{V(x)e^{i(\omega t + \theta)}\}, \quad (11)$$

$V(x)$ must satisfy

$$\frac{d^2V}{dx^2} - \frac{U}{\kappa}\frac{dV}{dx} - \frac{i\omega}{\kappa}V = 0. \quad (12)$$

The solution of this which remains finite as $x \to \infty$ is

$$\exp\left\{\frac{Ux}{2\kappa} - x\left(\frac{U^2}{4\kappa^2} + \frac{i\omega}{\kappa}\right)^{\frac{1}{2}}\right\}.$$

Writing

$$\left(\frac{U^2}{4\kappa^2} + \frac{i\omega}{\kappa}\right) = ae^{i\phi}, \quad (13)$$

the required solution is

$$v = V_0\exp\left\{\frac{Ux}{2\kappa} - xa^{\frac{1}{2}}\cos\tfrac{1}{2}\phi\right\}\cos(\omega t - xa^{\frac{1}{2}}\sin\tfrac{1}{2}\phi + \theta). \quad (14)$$

IV. *Combined radial flow and diffusion in two dimensions*

Suppose that in steady radial flow of incompressible fluid, mass m crosses any circle per second. Then the radial velocity u_r of the fluid is

$$u_r = \frac{m}{2\pi\rho r}. \quad (15)$$

Under these conditions, the equation of conduction of heat becomes

$$\frac{\partial^2 v}{\partial r^2} + \frac{1}{r}\frac{\partial v}{\partial r} - \frac{u_r}{\kappa}\frac{\partial v}{\partial r} - \frac{1}{\kappa}\frac{\partial v}{\partial t} = 0, \quad (16)$$

with u_r given by (15). The subsidiary equation for (16) with zero initial temperature is

$$\frac{d^2\bar{v}}{dr^2} + \frac{1}{r}(1 - 2\nu)\frac{d\bar{v}}{dr} - q^2\bar{v} = 0, \quad (17)$$

where $\nu = m/4\pi\kappa\rho$. The solution of (17) which is finite as $r \to \infty$ is

$$r^\nu K_\nu(qr). \quad (18)$$

As one application of the solution (18) we notice that, by Appendix V (33), the function whose Laplace transform is

$$\frac{Q(qr)^\nu}{2^{\nu+1}\pi K\Gamma(\nu+1)}K_\nu(qr) \quad (19)$$

is

$$\frac{Q}{4\pi Kt\Gamma(\nu+1)}\left(\frac{r^2}{4\kappa t}\right)^\nu e^{-r^2/4\kappa t}. \quad (20)$$

The solution (20) has the property that the total quantity of heat in the region $r > 0$ is Q, constant. Thus it is the analogue of the line source of § 10.3 for the present case in which the medium moves radially.

† Hulme, *Proc. Phys. Soc.* B **68** (1955) 393–400. See also Smith, Tiller, and Rutter, *Canad. J. Phys.* **33** (1955) 723–45.

As a second example, it follows as in § 13.5 that the solution of (16) in the region $r > a$ with zero initial temperature and constant temperature V_0 over $r = a$ is

$$V_0 + \frac{2V_0}{\pi}\left(\frac{r}{a}\right)^\nu \int_0^\infty \frac{e^{-\kappa u^2 t}[J_\nu(ur)Y_\nu(ua) - Y_\nu(ur)J_\nu(ua)]\,du}{u[J_\nu^2(ua) + Y_\nu^2(ua)]}. \tag{21}$$

V. *An infinite solid, initially at constant temperature V_0, is moving with velocity* $-U$ *along the x-axis. For times $t > 0$ the cylinder $x^2 + y^2 = a^2$ is maintained at zero temperature*†

The differential equation is

$$\frac{\partial^2 v}{\partial x^2} + \frac{\partial^2 v}{\partial y^2} + \frac{U}{\kappa}\frac{\partial v}{\partial x} - \frac{1}{\kappa}\frac{\partial v}{\partial t} = 0. \tag{22}$$

Putting
$$v = ue^{-Ux/2\kappa} + V_0, \tag{23}$$

it becomes
$$\frac{\partial^2 u}{\partial x^2} + \frac{\partial^2 u}{\partial y^2} - \frac{U^2}{4\kappa^2}u - \frac{1}{\kappa}\frac{\partial u}{\partial t} = 0. \tag{24}$$

Transforming to polar coordinates in the xy-plane, the subsidiary equation is

$$\frac{\partial^2 \bar{u}}{\partial r^2} + \frac{1}{r}\frac{\partial \bar{u}}{\partial r} + \frac{1}{r^2}\frac{\partial^2 \bar{u}}{\partial \theta^2} - q^2 \bar{u} = 0, \tag{25}$$

where
$$q^2 = \frac{p}{\kappa} + \frac{U^2}{4\kappa^2}. \tag{26}$$

The boundary condition is

$$\bar{u} = -\frac{V_0}{p}\exp[Ur\cos\theta/2\kappa], \quad \text{when } r = a;$$

$$= -\frac{V_0}{p}\sum_{n=0}^\infty \epsilon_n I_n\left(\frac{Ua}{2\kappa}\right)\cos n\theta, \tag{27}$$

where
$$\epsilon_0 = 1, \qquad \epsilon_n = 2 \quad \text{if } n \geqslant 1. \tag{28}$$

(25) is satisfied by
$$\bar{u} = \sum_{n=0}^\infty a_n \cos n\theta\, K_n(qr), \tag{29}$$

where the constants a_n are found by substituting (29) in (27). This gives finally

$$\bar{u} = -\frac{V_0}{p}\sum_{n=0}^\infty \frac{\epsilon_n I_n(Ua/2\kappa)K_n(qr)}{K_n(qa)}\cos n\theta. \tag{30}$$

It follows that

$$u = -V_0\sum_{n=0}^\infty \frac{\epsilon_n I_n(Ua/2\kappa)K_n(Ur/2\kappa)}{K_n(Ua/2\kappa)}\cos n\theta - \frac{2V_0}{\pi}\sum_{n=0}^\infty e^{-U^2 t/4\kappa}\epsilon_n \cos n\theta\, I_n\left(\frac{Ua}{2\kappa}\right) \times$$

$$\times \int_0^\infty \frac{e^{-\kappa u^2 t}[J_n(ur)Y_n(ua) - Y_n(ur)J_n(ua)]u\,du}{[u^2 + (U^2/4\kappa^2)][J_n^2(ua) + Y_n^2(ua)]}. \tag{31}$$

† An example due to Concer is that of a cylindrical hole maintained at zero temperature: material being removed from one side of the hole and replaced at the other.

VI. *A thin rod moves†* in $0 < x < l$ *with velocity U. The initial temperature of the rod is zero. For $t > 0$ the end $x = l$ is maintained at constant temperature V,'and the end $x = 0$ at zero. There is radiation at the surface of the rod into medium at zero*

In this case the differential equation 4.9 (1) becomes

$$\frac{\partial^2 v}{\partial x^2} - \frac{U}{\kappa}\frac{\partial v}{\partial x} - \frac{\nu}{\kappa}v - \frac{1}{\kappa}\frac{\partial v}{\partial t} = 0, \quad 0 < x < l, \tag{32}$$

to be solved with $v = V$ at $x = l$, $v = 0$ at $x = 0$, and zero initial temperature. The subsidiary equation is

$$\frac{d^2\bar{v}}{dx^2} - \frac{U}{\kappa}\frac{d\bar{v}}{dx} - \frac{p+\nu}{\kappa}\bar{v} = 0, \quad 0 < x < l,$$

with $\bar{v} = V/p$ at $x = l$, and $\bar{v} = 0$ at $x = 0$. Thus

$$\bar{v} = \frac{V\sinh\{x[U^2+4\kappa(p+\nu)]^{\frac{1}{2}}/2\kappa\}}{p\sinh\{l[U^2+4\kappa(p+\nu)]^{\frac{1}{2}}/2\kappa\}}e^{U(x-l)/2\kappa}. \tag{33}$$

Using the Inversion Theorem we find

$$v = \frac{V}{2\pi i}e^{U(x-l)/2\kappa}\int_{\gamma-i\infty}^{\gamma+i\infty} e^{\lambda t}\frac{\sinh\{x[U^2+4\kappa(\lambda+\nu)]^{\frac{1}{2}}/2\kappa\}}{\sinh\{l[U^2+4\kappa(\lambda+\nu)]^{\frac{1}{2}}/2\kappa\}}\frac{d\lambda}{\lambda}. \tag{34}$$

The integrand of (34) has simple poles at $\lambda = 0$ and for

$$l[U^2+4\kappa(\lambda+\nu)]^{\frac{1}{2}}/2\kappa = n\pi i, \quad n = 1, 2,...,$$

that is,

$$\lambda = -\nu - \frac{U^2}{4\kappa} - \frac{\kappa n^2\pi^2}{l^2}, \quad n = 1, 2,.... \tag{35}$$

Evaluating the residues at these poles we get finally

$$v = V\frac{\sinh x(U^2+4\kappa\nu)^{\frac{1}{2}}/2\kappa}{\sinh l(U^2+4\kappa\nu)^{\frac{1}{2}}/2\kappa}e^{U(x-l)/2\kappa} +$$

$$+ \frac{2V\pi}{l^2}e^{U(x-l)/2\kappa}\sum_{n=1}^{\infty}\frac{(-1)^n n\sin(n\pi x/l)}{[(\nu/\kappa)+(U^2/4\kappa^2)+(n^2\pi^2/l^2)]}e^{-[\nu+(U^2/4\kappa)+(\kappa n^2\pi^2/l^2)]t}. \tag{36}$$

15.3. Heat regenerators and heat exchangers

The transfer of heat from a moving hot fluid to a cold fluid is of the greatest practical importance. Broadly speaking, there are two general types of system by which this is effected, heat exchangers and heat regenerators.

In a heat exchanger the hot fluid and the cold fluid flow on either side of a thin partition whose function is merely to separate the two fluids; heat is transferred from one to the other through this partition, which, in the ideal case, will be so thin as to have negligible heat capacity and will simply act as a contact resistance to heat transfer (§ 1.9 D). The fluids

† A number of problems of this type is studied by Owen, *Proc. Lond. Math. Soc.* (2) **23** (1925) 238.

may either flow in the same direction (parallel flow) or opposite directions (counterflow); a steady state is rapidly attained for which solutions are given in all works on heat transfer.[†]

In heat regenerators the hot and cold fluids are passed alternately over a solid wall: the wall absorbs heat when the hot fluid flows over it, and subsequently gives up this heat to cold fluid, the process being repeated cyclically. Here the storage of heat in the wall is fundamental so that the flow of heat in the wall must be studied carefully, and the problem is one of considerable difficulty. In practical applications it is the final steady periodic state which is of interest.

Here we give some applications of the Laplace transformation method to non-steady states in idealized systems of the above types.

We suppose the surface of the wall to be the plane $z = 0$, and the fluid to be flowing uniformly in the region $z < 0$ with velocity U in the direction of the axis of x. The fluid is supposed to be well mixed, so that its temperature is constant in any plane perpendicular to the direction of flow, but there is supposed to be no conduction of heat in the fluid in the direction of flow. Let M' be the mass of fluid in contact with unit area of the wall, c' the specific heat of the fluid, u its temperature at the point x and time t, v_s the surface temperature of the wall at x and t, and H the surface conductance of the wall.

As in § 4.9 the equation satisfied by the temperature in the fluid is found to be

$$\frac{\partial u}{\partial t} + U \frac{\partial u}{\partial x} + b'(u - v_s) = 0, \tag{1}$$

where

$$b' = H/M'c'. \tag{2}$$

Equation (1) and

$$K \frac{\partial v}{\partial z} = H(v - u), \quad z = 0, \tag{3}$$

will be the boundary conditions for the equation of conduction of heat,

$$\frac{\partial^2 v}{\partial x^2} + \frac{\partial^2 v}{\partial z^2} - \frac{1}{\kappa} \frac{\partial v}{\partial t} = 0,$$

in the solid. Clearly these general equations[‡] present considerable difficulty, and in practice various simplifying assumptions have been made. Usually the solid has been assumed to have zero conductivity parallel to the flow, and either finite[§] or infinite conductivity in the perpendicular direction.

[†] e.g. Schack, Goldschmidt, and Partridge, *Industrial Heat Transfer* (Wiley, 1933).

[‡] Schmeidler, *Z. angew. Math. Mech.* **8** (1928) 385, discusses the steady periodic condition in regenerators for this case.

[§] Ackermann, ibid. **11** (1931) 192.

I. *The case† of infinite conductivity of the solid perpendicular to the direction of flow and zero conductivity in that direction*

If M is the mass of solid per unit area of the wall and c its specific heat, the temperature over the mass having the surface value $v_s(x, t)$, this temperature satisfies the differential equation

$$\frac{\partial v_s}{\partial t} + b(v_s - u) = 0, \tag{4}$$

where
$$b = H/Mc. \tag{5}$$

The differential equations of the problem are (1) and (4). We consider‡ the region $x > 0$ with both solid and fluid initially at zero, and with the fluid kept at unit temperature at $x = 0$ for $t > 0$.

The subsidiary equations derived from (1) and (4) are

$$U\frac{d\bar{u}}{dx} + (p+b')\bar{u} - b'\bar{v}_s = 0, \quad x > 0, \tag{6}$$

$$(p+b)\bar{v}_s - b\bar{u} = 0, \quad x > 0. \tag{7}$$

These have to be solved with $\bar{u} = 1/p$ at $x = 0$. Thus we find

$$\bar{u} = \frac{1}{p}\exp\left\{-\frac{x(p+b')}{U} + \frac{xbb'}{U(p+b)}\right\}, \tag{8}$$

$$\bar{v}_s = \frac{b}{p(p+b)}\exp\left\{-\frac{x(p+b')}{U} + \frac{xbb'}{U(p+b)}\right\}. \tag{9}$$

To find v_s and u we notice that from Appendix V (24)

$$\frac{1}{p}e^{ax/p} = L\{I_0[2(axt)^{\frac{1}{2}}]\}. \tag{10}$$

Thus, by 12.2 (6),

$$\frac{1}{p+b}e^{ax/(p+b)} = L\{e^{-bt}I_0[2(axt)^{\frac{1}{2}}]\}. \tag{11}$$

---

† This case is approximately realized in practice when fluid flows through granulated medium whose particles are so small that their temperature may be supposed to be constant over their volume.

‡ This is the problem of the initial heating of the ideal regenerator. It has been discussed by Anzelius, *Z. angew. Math. Mech.* **6** (1926) 291–4; Schumann, *J. Franklin Inst.* **208** (1929) 405–16; Nusselt, *Z. Ver. dtsch. Ing.* **71** (1927) 85; Kronig and van Gijn, *Physica*, **12** (1946) 118–28. A general review and references are given by Thiele, *Ind. Eng. Chem.* **38** (1946) 646–50. Nusselt, *Z. Ver. dtsch. Ing.* **72** (1928) 1052, and Hausen, *Z. angew. Math. Mech.* **11** (1931) 105, derive the solution of the steady periodic case from this in the form of an integral equation and discuss its solution. Hausen, ibid. **9** (1929) 173, gives another treatment of the steady periodic case; approximations to his solutions are made by Schultz, *Appl. Sci. Res.* A, **3** (1953) 165–73. The application of numerical analysis is discussed by Allen, *Quart. J. Mech. Appl. Math.* **5** (1952) 455–61.

Then, by § 12.2, Theorem IV,

$$\frac{1}{b}\left[\frac{1}{p}-\frac{1}{p+b}\right]e^{ax/(p+b)} = \frac{1}{p(p+b)}e^{ax/(p+b)} = L\left\{\int_0^t e^{-b\tau}I_0[2(ax\tau)^{\frac{1}{2}}]\,d\tau\right\},$$

$$(12)$$

and from (11) and (12) it follows that

$$\frac{1}{p}e^{ax/(p+b)} = L\left\{e^{-bt}I_0[2(axt)^{\frac{1}{2}}]+b\int_0^t e^{-b\tau}I_0[2(ax\tau)^{\frac{1}{2}}]\,d\tau\right\} \qquad (13)$$

$$= L\left\{1+(ax)^{\frac{1}{2}}\int_0^t e^{-b\tau}I_1[2(ax\tau)^{\frac{1}{2}}]\tau^{-\frac{1}{2}}\,d\tau\right\} \qquad (14)$$

$$= L\left\{e^{ax/b}-\frac{a}{b}e^{(ax/b)-bt}\int_0^x e^{-ax'/b}I_0[2(ax't)^{\frac{1}{2}}]\,dx'\right\}, \qquad (15)$$

where (14) follows from (13) by integrating by parts, and the important alternative form (15) is easily obtained by multiplying both sides of (11) by $\exp(-ax/b)$ and integrating with respect to x from 0 to x.

Using (12) and (14), together with § 12.2, Theorem VII, it follows from (8) and (9) that u and v_s are zero if $t < x/U$, and if $t > x/U$

$$u = e^{-b'x/U}\left\{1+(ax)^{\frac{1}{2}}\int_0^{t-x/U} e^{-b\tau}I_1[2(ax\tau)^{\frac{1}{2}}]\tau^{-\frac{1}{2}}\,d\tau\right\}, \qquad (16)$$

$$v_s = be^{-b'x/U}\int_0^{t-x/U} e^{-b\tau}I_0[2(ax\tau)^{\frac{1}{2}}]\,d\tau, \qquad (17)$$

where $a = bb'/U$. Results for other values of the input temperature may be written down by Duhamel's theorem.

Equations similar to those of this problem arise in the theory of ion exchange columns and have recently attracted considerable attention.† Because of its frequent occurrence, the function

$$J(x,y) = 1-e^{-y}\int_0^x e^{-\tau}I_0[2(\tau y)^{\frac{1}{2}}]\,d\tau \qquad (18)$$

which appears in (12), (13), (15) is regarded as fundamental, and the equality of the two latter expressions gives the important result

$$J(x,y)+J(y,x) = 1+e^{-x-y}I_0[2(xy)^{\frac{1}{2}}]. \qquad (19)$$

† Goldstein, *Proc. Roy. Soc.* A, **219** (1953) 151–85; Thomas, *J. Amer. Chem. Soc.* **66** (1944) 1664–6. It may be noted that Thomas uses in place of (18) the function

$$\phi(x,y) = [1-J(x,y)]\exp(x+y),$$

and some other authors have followed him. For further references, see the bibliography in Goldstein, loc. cit.

Goldstein, loc. cit., gives a full discussion of the properties of this function.

II. *The case of heat generation in the solid*†

Suppose that heat is generated in the solid for $t > 0$ at the constant rate Q per unit time per unit mass of solid, the initial temperature of both solid and fluid being zero, and the fluid being kept at zero temperature at $x = 0$ for $t > 0$.

In this case (4) is replaced by

$$\frac{\partial v_s}{\partial t} + b(v_s - u) = \frac{Q}{c}. \tag{20}$$

The solution of (1) and (20) with $u = v_s = 0$ when $t = 0$, $x > 0$, and $u = 0$ when $x = 0$, $t > 0$, is

$$\bar{u} = \frac{Qb'}{c(b+b')^2} \left\{ \frac{b+b'}{p^2} - \frac{1}{p} + \frac{1}{p+b+b'} \right\} \left\{ 1 - \exp\left[-\frac{b'x}{U} - \frac{px}{U} + \frac{ax}{p+b} \right] \right\}, \tag{21}$$

where

$$a = bb'/U. \tag{22}$$

To evaluate $\bar{u}$ we need the results

$$\frac{1}{p^2} e^{ax/(p+b)} = \left(\frac{1}{p} + \frac{b}{p^2} \right) \frac{1}{p+b} e^{ax/(p+b)} = L\left\{ \int_0^t [1+bt-b\tau] e^{-b\tau} I_0[2(ax\tau)^{\frac{1}{2}}] \, d\tau \right\}, \tag{23}$$

$$\frac{1}{p+b'} e^{ax/p} = \left[\frac{1}{p} - \frac{b'}{p(p+b')} \right] e^{ax/p} = L\left\{ I_0[2(axt)^{\frac{1}{2}}] - b'e^{-b't} \int_0^t e^{b'\tau} I_0[2(ax\tau)^{\frac{1}{2}}] \, d\tau \right\}, \tag{24}$$

which follow from (11) and § 12.2, Theorem X. Using them and Theorems VI, VII of § 12.2, we find from (21) that

$$u = Qb'\{(b+b')t - 1 + \exp[-(b+b')t]\}/c(b+b')^2, \quad \text{if } t < x/U, \tag{25}$$

and if $t > x/U$,

$$u = Qb'\{(b+b')t - 1 + \exp[-(b+b')t]\}/c(b+b')^2 -$$

$$- \frac{Qb'}{c(b+b')^2} e^{-b'x/U} \int_0^{t-(x/U)} \{[b(b+b')[t-\tau-(x/U)] + b']e^{-b\tau} -$$

$$- b'e^{-(b+b')[t-(x/U)]+b'\tau}\} I_0[2(ax\tau)^{\frac{1}{2}}] \, d\tau. \tag{26}$$

If the rate of heat generation is zero‡ up to time x/U, and Q, constant, if $t > x/U$ (that is, after the fluid which left the origin at time $t = 0$ has reached the point x), (21) is replaced by

$$\bar{u} = \frac{Q}{cp^2} e^{-px/U} \left\{ 1 - \exp\left[-\frac{b'x}{U} + \frac{bb'x}{U(p+b)} \right] \right\}, \tag{27}$$

---

† Heat generation in spheres immersed in moving fluid is discussed by Monro and Amundson, *Ind. Eng. Chem.* **42** (1950) 1481–8.

‡ This case is discussed by Brinkley, *J. Appl. Phys.* **18** (1947) 582–5. It occurs when heat is generated by a chemical reaction. The case in which heat production is a linear function of v_s, $Q = Q_0(1 + \beta v_s)$, is also readily treated in the same way.

and by (23) and § 12.2, Theorem VII, u is zero if $t < x/U$, and if $t > x/U$

$$u = \frac{Q}{c}\left\{t - \frac{x}{U} - e^{-b'x/U}\int_0^{t-(x/U)}\{b[t-(x/U)-\tau]+1\}e^{-b\tau}I_0[2(bb'x\tau/U)^{\frac{1}{2}}]\,d\tau\right\}. \quad (28)$$

The result can alternatively be expressed in terms of the function $J(x, y)$ defined in (19).

III. *The case of finite conductivity K of the solid perpendicular to the direction of flow, and zero conductivity in that direction*

As an example of this case which leads to a simple solution, we consider the case of semi-infinite solid $z > 0$, and suppose the surface temperature of the solid equal to the temperature of the fluid at any point (i.e. the case of very large H in (1)).

In the solid the temperature v has to satisfy

$$\frac{\partial^2 v}{\partial z^2} - \frac{1}{\kappa}\frac{\partial v}{\partial t} = 0, \quad z > 0, x > 0, t > 0, \quad (29)$$

and at the surface
$$v = u, \quad x > 0, t > 0, z = 0, \quad (30)$$

$$K\frac{\partial v}{\partial z} = M'c'\left(\frac{\partial u}{\partial t} + U\frac{\partial u}{\partial x}\right), \quad x > 0, t > 0, z = 0, \quad (31)$$

where u is the temperature at x in the fluid, and (31) follows from (1) and (3). As before we consider the case of zero initial temperature in the region $z > 0, x > 0$, with $u = 1$ at $x = 0$ for $t > 0$. The subsidiary equations are

$$\frac{\partial^2 \bar{v}}{\partial z^2} - q^2\bar{v} = 0, \quad z > 0, x > 0, \quad (32)$$

$$\bar{v} = \bar{u}, \quad z = 0, x > 0, \quad (33)$$

$$U\frac{d\bar{u}}{dx} + p\bar{u} - k\frac{\partial \bar{v}}{\partial z} = 0, \quad x > 0, z = 0, \quad (34)$$

where
$$k = K/M'c'. \quad (35)$$

These have to be solved with $\bar{u} = 1/p$ when $x = 0$, $\bar{v}$ finite as $z \to \infty$, and $\bar{u}$ and $\bar{v}$ finite as $x \to \infty$. The solutions are

$$\bar{u} = \frac{1}{p}\exp\left\{-\frac{px}{U} - \frac{kxp^{\frac{1}{2}}}{U\kappa^{\frac{1}{2}}}\right\}, \quad (36)$$

$$\bar{v} = \frac{1}{p}\exp\left\{-\frac{px}{U} - \left(\frac{kx}{U}+z\right)\frac{p^{\frac{1}{2}}}{\kappa^{\frac{1}{2}}}\right\}. \quad (37)$$

Thus, from Appendix V (8) and § 12.2, Theorem VII, u and v are zero for $t < x/U$, and for $t > x/U$

$$u = \mathrm{erfc}\left[\frac{kx}{2U\{\kappa(t-x/U)\}^{\frac{1}{2}}}\right], \quad (38)$$

$$v = \mathrm{erfc}\left[\frac{kx+zU}{2U\{\kappa(t-x/U)\}^{\frac{1}{2}}}\right]. \quad (39)$$

There is no difficulty in solving the subsidiary equations for the more general boundary conditions (1) and (3), and for cases in which the solid is in the form of a slab or a cylinder, but rather more difficult Laplace transforms are obtained.

IV. *The problem of* III *but with H finite*

Writing $h = H/K$, $k = K/M'c'$, we find

$$\bar{u} = \frac{1}{p}\exp\left\{-\frac{px}{U} - \frac{kxhq}{U(q+h)}\right\}, \tag{40}$$

$$u = 1 - \frac{2}{\pi}\int_0^\infty \frac{d\xi}{\xi}\exp\left\{-\kappa\xi^2\left(t - \frac{x}{U}\right) - \frac{kxh\xi^2}{U(h^2+\xi^2)}\right\}\sin\frac{kh^2x\xi}{U(h^2+\xi^2)}, \tag{41}$$

which reduces to (38) by a known integral if $h \to \infty$. The integrand of (41) is finite when $u = 0$, and the integral is not difficult to evaluate. An expression for small values of the time may be obtained by expanding the exponential in (40) in powers of $(q+h)^{-1}$ and using results of type Appendix V (14) and (18).

V. *The problem of* III *except that heat is generated in the fluid for* $t > 0$ *at the constant rate* Q *per unit time per unit mass. The temperature of the fluid zero at* $x = 0$ *for* $t > 0$. *The initial temperatures of the solid and fluid zero*

The temperature u in the fluid is given by

$$c'\kappa k'^2 u/Q = 2k'(\kappa t/\pi)^{\frac{1}{2}} - 1 + e^{\kappa t k'^2}\,\mathrm{erfc}\,[k'(\kappa t)^{\frac{1}{2}}] -$$

$$- H\left(t - \frac{x}{U}\right)\left\{\left[\frac{4k'^2\kappa(Ut-x)}{U\pi}\right]^{\frac{1}{2}}e^{-k^2x^2/4\kappa U(Ut-x)} - \frac{kk'x+U}{U}\,\mathrm{erfc}\,\frac{kx}{2[\kappa U(Ut-x)]^{\frac{1}{2}}} +$$

$$+ e^{\kappa k'^2 t}\,\mathrm{erfc}\left[\frac{kx}{2[\kappa U(Ut-x)]^{\frac{1}{2}}} + k'[\kappa(Ut-x)/U]^{\frac{1}{2}}\right]\right\}, \tag{42}$$

where k is defined in (35) and $k' = k/\kappa = \rho c/M'c'$.

VI. *The unsteady state in heat exchangers*

As an example of the conditions in a heat exchanger before the steady state is attained we consider the case of counterflow in the region $x > 0$. Suppose that on one side of a thin partition (of negligible heat capacity) in the plane $z = 0$, fluid flows with velocity U in the direction of the axis of x; let M' be the mass per unit area and c' the specific heat of this fluid, and let u be its temperature at the point x at time t. On the other side of the partition let there be mass M_1 per unit area of fluid of specific heat c_1, let u_1 be its temperature at x at time t and let $-U_1$ be its velocity in the direction of x. Let H be the overall coefficient of heat transfer† of the wall so that the rate of flow of heat across the wall at x is

$$H(u-u_1). \tag{43}$$

The differential equation for u is (1), namely

$$\frac{\partial u}{\partial t} + U\frac{\partial u}{\partial x} + b'(u-u_1) = 0, \quad x > 0, \tag{44}$$

and that for u_1 is the corresponding one with $-U_1$ in place of U, viz.

$$\frac{\partial u_1}{\partial t} - U_1\frac{\partial u_1}{\partial x} - b_1(u-u_1) = 0, \quad x > 0, \tag{45}$$

where $\qquad\qquad b' = H/M'c', \qquad b_1 = H/M_1 c_1. \tag{46}$

† This will be K/d for a single layer, where K is its conductivity and d its thickness, cf. § 1.9.

We suppose the initial temperature of both fluids to be zero, and that $u = 1$ at $x = 0$ for $t > 0$. The subsidiary equations are

$$U\frac{d\bar{u}}{dx} + (p+b')\bar{u} - b'\bar{u}_1 = 0, \quad x > 0, \tag{47}$$

$$U_1\frac{d\bar{u}_1}{dx} - (p+b_1)\bar{u}_1 + b_1\bar{u} = 0, \quad x > 0. \tag{48}$$

These have to be solved with $\bar{u} = 1/p$ at $x = 0$, and $\bar{u}_1$ zero as $x \to \infty$. Solving for $\bar{u}$ we get

$$\bar{u} = \frac{1}{p}\exp\left\{-\frac{(p+b')x}{U} + kx[(p+\rho) - \sqrt{\{(p+\rho)^2 - \sigma^2\}}]\right\}, \tag{49}$$

where $\qquad \rho = \dfrac{Ub_1 + U_1 b'}{U + U_1}, \qquad \sigma^2 = \dfrac{4UU_1 b'b_1}{(U+U_1)^2}, \qquad k = \dfrac{U+U_1}{2UU_1}, \tag{50}$

u is found from Appendix V (25) using analysis of the same type as that in (11)–(16) above. The result is

$$u = 0, \quad \text{if } t < x/U,$$

$$u = e^{-b'x/U} + \sigma kxe^{-b'x/U}\int_0^{t-x/U} e^{-\rho\tau}I_1\{\sigma\sqrt{[\tau(\tau+2kx)]}\}\frac{d\tau}{[\tau(\tau+2kx)]^{\frac{1}{2}}}, \tag{51}$$

if $t > x/U$. u_1 is determined in the same way; it is zero if $t < x/U$, and if $t > x/U$

$$u_1 = \frac{\sigma kU}{b'}e^{-b'x/U}\int_0^{t-x/U} e^{-\rho\tau}\left\{\frac{I_1\{\sigma[\tau(\tau+2kx)]^{\frac{1}{2}}\}}{[\tau(\tau+2kx)]^{\frac{1}{2}}} + \frac{\sigma kxI_2\{\sigma[\tau(\tau+2kx)]^{\frac{1}{2}}\}}{\tau+2kx}\right\}d\tau.$$

Counterflow in a finite region $0 < x < l$ is treated in the same way; u will have a prescribed value at $x = 0$, and u_1 at $x = l$. In parallel flow in $x > 0$, both u and u_1 are specified at $x = 0$.

15.4. Heat flow through systems in parallel

It has frequently been remarked previously that in composite systems involving a good conductor as well as a bad conductor, a simple and adequate approximation is obtained by regarding the good conductor as a perfect conductor of finite heat capacity. The same result applies if n relatively bad conductors of temperatures $v_1,..., v_n$. and areas $\omega_1,..., \omega_n$ are all connected to a single perfect conductor of mass M, temperature v, and specific heat c'.

The boundary conditions at the surface of the perfect conductor are then

$$v_1 = v_2 = ... = v_n = v, \tag{1}$$

$$Mc'\frac{dv}{dt} + \omega_1 f_1 + ... + \omega_n f_n = Q, \tag{2}$$

where $f_1,..., f_n$ are the fluxes into the various conductors and Q is the rate of heat production in the perfect conductor.

(1) and (2), together with the boundary conditions at the other ends of the conductors, supply sufficient equations to determine $v_1,..., v_n$. Some examples are given below.

I. *The regions $0 < x < a$ and $x > a$ are of the same material K, ρ, c, κ. There is no flow of heat at $x = 0$. At $x = a$ both regions are in contact with mass M per unit area of perfect conductor of specific heat c'. The initial temperature of the perfect conductor is V_0 and that of the surrounding regions is zero.*

The temperature of the perfect conductor is

$$\frac{4kV_0}{\pi} \int_0^\infty \frac{\cos^2 u \, du}{1 + 2ku \sin 2u + 4k^2 u^2 \cos^2 u} e^{-\kappa u^2 t/a^2} \, du, \tag{3}$$

where
$$k = Mc'/2\rho ca. \tag{4}$$

II. *The region $0 \leqslant r < a$ contains material K_1, ρ_1, c_1, κ_1, v_1; the region $a < r < b$ contains perfect conductor ρ, c, v; the region $r > b$ contains material K_2, ρ_2, c_2, κ_2, v_2. The region $a < r < b$ is initially at temperature V_0 and the others at zero*†

The temperature v_1 in $0 \leqslant r \leqslant a$ is given by

$$v_1 = \frac{4V_0 k'' K_2}{\pi^2 K_1} \int_0^\infty \frac{e^{-\kappa_1 u^2 t/a^2} J_0(u) J_0(ur/a) \, du}{u \Delta(u)}, \tag{5}$$

where
$$k = \frac{b \kappa_1^{\frac{1}{2}}}{a \kappa_2^{\frac{1}{2}}}, \qquad k' = \frac{b K_2 \kappa_1^{\frac{1}{2}}}{a K_1 \kappa_2^{\frac{1}{2}}}, \qquad k'' = \frac{(b^2 - a^2)\rho c}{2a^2 \rho_1 c_1}, \tag{6}$$

and

$$\Delta(u) = [J_1(u) J_0(ku) - k' J_0(u) J_1(ku) + k'' u J_0(u) J_0(ku)]^2 +$$
$$+ [J_1(u) Y_0(ku) - k' J_0(u) Y_1(ku) + k'' u J_0(u) Y_0(ku)]^2. \tag{7}$$

If the initial temperature of the whole system is zero and heat is supplied to the region $a < r < b$ at the rate Q per unit time,

$$v_1 = \frac{2K_2 Q}{K_1^2 \pi^3} \int_0^\infty \frac{(1 - e^{-\kappa_1 u^2 t/a^2}) J_0(u) J_0(ur/a) \, du}{u^3 \Delta(u)}. \tag{8}$$

15.5. Steady periodic temperature

In many practical problems a periodic temperature or flux is applied to a body, and it is desired to find the steady periodic variation in temperature which exists after the transient involving the initial conditions has died away. This may be done by resolving the prescribed temperature into its Fourier components and treating these separately as in § 3.6, but in practice the resulting Fourier series is usually slowly convergent near the most interesting values of the time and devices such as that of §§ 3.6, 4.8 have long been used to get more useful forms of the solution.

One convenient method of obtaining such solutions is by the use of the Laplace transformation: the method given below is almost exactly that

† This system has been introduced by Grunberg, *J. Phys. U.S.S.R.* **4** (1941) 463, and Grunberg and Sontz, ibid. **4** (1941) 97, as an approximation to an oil-filled cable. They give some approximations for small values of the time.

of the steady-state operational calculus of Waidelich† but assumes no previous knowledge of this.

For definiteness, we shall consider only the most important case of an applied rectangular wave form, but other cases may be treated in the same way. In the calculations we assume the periodic temperature or flux to be applied for $t > 0$, the solid being initially at zero temperature. Thus the applied temperature or flux may be represented by the function

$$
\left.
\begin{aligned}
\phi(t) &= 0, \quad t < 0 \\
\phi(t) &= 1, \quad nT < t < nT+T_1, \quad n = 0, 1,\dots \\
\phi(t) &= 0, \quad nT+T_1 < t < (n+1)T, \quad n = 0, 1,\dots
\end{aligned}
\right\}.
\tag{1}
$$

For example, flux $F_0\,\phi(t)$ represents a flux F_0 which is 'on' for time T_1 and 'off' for time $T-T_1$, and so on, with period T. It follows from § 12.2, Theorem VIII, that

$$
\bar{\phi} = \frac{1-e^{-pT_1}}{p(1-e^{-pT})}.
\tag{2}
$$

To illustrate the method, we consider the problem already discussed in § 3.6, namely, *the region $0 < x < l$ with $v = 0$ when $x = 0$, and $v = \phi(t)$ when $x = l$.*

As in § 12.6, we get, using (2),

$$
\bar{v} = \frac{(1-e^{-pT_1})\sinh qx}{p(1-e^{-pT})\sinh ql}
\tag{3}
$$

$$
= (1-e^{-pT_1}+e^{-pT}-e^{-p(T+T_1)}+\dots)\frac{\sinh qx}{p\sinh ql}.
\tag{4}
$$

We first need the values of v during the first period after $t = 0$: it follows from (4), 12.6 (10), and § 12.2, Theorem VII, that

$$
v = \frac{x}{l}+\frac{2}{\pi}\sum_{n=1}^{\infty}\frac{(-1)^n}{n}e^{-\kappa n^2\pi^2 t/l^2}\sin\frac{n\pi x}{l}, \quad 0 < t < T_1,
\tag{5}
$$

$$
v = \frac{2}{\pi}\sum_{n=1}^{\infty}\frac{(-1)^n}{n}\{e^{-\kappa n^2\pi^2 t/l^2}-e^{-\kappa n^2\pi^2(t-T_1)/l^2}\}\sin\frac{n\pi x}{l}, \quad T_1 < t < T.
\tag{6}
$$

These results, of course, might have been obtained in Chapter XII and have as yet no obvious relevance to the steady periodic solution. To find this, we apply the Inversion Theorem to (3) and get

$$
v = \frac{1}{2\pi i}\int_{\gamma-i\infty}^{\gamma+i\infty}\frac{e^{\lambda t}(1-e^{-\lambda T_1})\sinh\mu x\,d\lambda}{\lambda(1-e^{-\lambda T})\sinh\mu l},
\tag{7}
$$

† Waidelich, *J. Appl. Phys.* **13** (1942) 706–12, *Proc. Inst. Radio Engrs.* **34** (1946) 78 P. Carslaw and Jaeger, *Operational Methods in Applied Mathematics* (Oxford, edn. 2, 1948) §§ 128, 129.

where, as usual, $\mu = (\lambda/\kappa)^{\frac{1}{2}}$. The integrand of (7) has a simple pole at $\lambda = 0$, which gives a term (xT_1/lT), and simple poles at

$$\lambda = 2m\pi i/T, \qquad m = \pm 1, \pm 2,... \tag{8}$$

which give a series of oscillations of periods $T, \frac{1}{2}T,...$. These comprise the Fourier series for the steady periodic part of the temperature; we do not evaluate this but write v_s for its sum which we wish to find in a more convenient form. Finally, the integrand has a set of poles at

$$\lambda = -\kappa n^2\pi^2/l^2, \qquad n = 1, 2,... \tag{9}$$

and, evaluating the residues at these poles and combining the results obtained, we get

$$v = v_s + \frac{2}{\pi}\sum_{n=1}^{\infty} \frac{(-1)^n[e^{-\kappa n^2\pi^2 t/l^2} - e^{-\kappa n^2\pi^2(t-T_1)/l^2}]}{n(1-e^{\kappa n^2\pi^2 T/l^2})}\sin\frac{n\pi x}{l}, \tag{10}$$

consisting of the steady periodic part and a transient part.

Now (10) is a general solution valid for all values of the time, so it must agree with (5) if $0 < t < T_1$, and with (6) if $T_1 < t < T$. Thus, equating (10) and (5) we get for $0 < t < T_1$

$$v_s = \frac{x}{l} + \frac{2}{\pi}\sum_{n=1}^{\infty} \frac{(-1)^n(e^{\kappa n^2\pi^2(T_1-t)/l^2} - e^{\kappa n^2\pi^2(T-t)/l^2})}{n(1-e^{\kappa n^2\pi^2 T/l^2})}\sin\frac{n\pi x}{l}. \tag{11}$$

Since the left-hand side of (11) is periodic with period T, it follows that the right-hand side is its value at time $rT+t$ where r is any integer. Similarly, using (6), the steady temperature at time $rT+t$ where $T_1 < t < T$ is

$$v_s = \frac{2}{\pi}\sum_{n=1}^{\infty} \frac{(-1)^n[e^{\kappa n^2\pi^2(T+T_1-t)/l^2} - e^{\kappa n^2\pi^2(T-t)/l^2}]}{n(1-e^{\kappa n^2\pi^2 T/l^2})}\sin\frac{n\pi x}{l}. \tag{12}$$

These results agree with 3.6 (17), (19) obtained by the use of Duhamel's theorem. Some further results of the same type will now be given.

I. *The semi-infinite solid $x > 0$ heating by pulsed† surface flux $F_0\phi(t)$*

In this case the surface temperature is ultimately

$$\frac{2F_0 T_1}{KT}\left(\frac{\kappa t}{\pi}\right)^{\frac{1}{2}} + v_P, \tag{13}$$

consisting of a periodic part v_P superposed on the rising temperature due to the

† This problem appears as an approximation to many practical ones, for example, the heating of machine guns and the heating of a rotating cylinder by friction over portion of its surface. For the heating of a rotating anode X-ray tube see Oosterkamp, *Philips Res. Rep.* **3** (1948) 161. Pulsed surface flux over a circular area in the surface of a semi-infinite solid is discussed by Jaeger, *Quart. Appl. Math.* **11** (1953) 132–7.

average surface flux $F_0 T_1/T$. The value of v_P at time bT after the beginning of a heating period is

$$v_P = \frac{2F_0 T_1}{Ka}\left(\frac{\kappa}{\pi T}\right)^{\frac{1}{2}}\{(1-a)b^{\frac{1}{2}} - \pi^{-\frac{1}{2}}I(a,b)\}, \quad 0 < b < a, \tag{14}$$

$$v_P = \frac{2F_0 T_1}{Ka}\left(\frac{\kappa}{\pi T}\right)^{\frac{1}{2}}\{(1-a)b^{\frac{1}{2}} - (b-a)^{\frac{1}{2}} - \pi^{-\frac{1}{2}}I(a,b)\}, \quad a < b < 1, \tag{15}$$

where $a = T_1/T$, and

$$I(a,b) = \int_0^\infty \frac{e^{-b\xi^2}[(1-a)e^{-\xi^2} - e^{-(1-a)\xi^2} + a]\,d\xi}{\xi^2(1-e^{-\xi^2})}. \tag{16}$$

The integrals (16) are easy to evaluate. Some values are given by Jaeger, loc. cit.

II. *The pulsed point source in an infinite medium*

Suppose that heat is supplied at the origin at the rate $Q\phi(t)$. Then, when steady conditions have been attained, the temperature at distance r from the origin at time bT after the beginning of a heating period is $(Qa/4\pi Kr) + v_P$, where the periodic part v_P is

$$v_P = \frac{Q}{4\pi Kr}\left\{\text{erfc}\,\frac{C}{2b^{\frac{1}{2}}} - a + \frac{2}{\pi}\int_0^\infty \frac{e^{-b\xi^2}[e^{-(1-a)\xi^2} - e^{-\xi^2}]\sin C\xi}{\xi(1-e^{-\xi^2})}\,d\xi\right\}, \quad 0 < b < a, \tag{17}$$

where

$$a = T_1/T, \qquad C = r(\kappa T)^{-\frac{1}{2}}. \tag{18}$$

III. *The pulsed line source in an infinite medium*

Suppose that heat is supplied along a line at the rate $Q\phi(t)$ per unit length. Then, for large values of the time, the temperature at distance r from the line is

$$-\frac{Qa}{4\pi K}\text{Ei}\left(-\frac{r^2}{4\kappa t}\right) + v_P, \tag{19}$$

where the periodic part v_P is

$$v_P = -\frac{Q(1-a)}{4\pi K}\text{Ei}\left(-\frac{C^2}{4b}\right) - \frac{Q}{2\pi K}\int_0^\infty \frac{e^{-b\xi^2}J_0(C\xi)[(1-a)e^{-\xi^2} - e^{-(1-a)\xi^2} + a]\,d\xi}{\xi(1-e^{-\xi^2})}, \tag{20}$$

for $0 < b < a$, where a, b and C are defined in II.

15.6. Linear asymptotes and time-lag

When the Laplace transform $\bar{y}$ of any quantity has the form

$$\bar{y} = \frac{f(p)}{p^2\Delta(p)}, \tag{1}$$

where $f(p)/\Delta(p)$ is regular at $p = 0$, y is given by

$$y = \frac{f(0)}{\Delta(0)}(t-L) + \text{transient terms}, \tag{2}$$

where

$$L = \frac{f(0)\Delta'(0) - \Delta(0)f'(0)}{f(0)\Delta(0)}. \tag{3}$$

That is, for large values of the time, y is asymptotically a straight line of slope $f(0)/\Delta(0)$ and intercept on the t-axis or 'time-lag' L. Many quantities in conduction of heat which behave in this way suggest themselves, for example, (i) the total flow of heat through a wall whose surfaces are kept at constant temperatures,

(ii) the temperature in a body with linearly increasing surface temperature, and
(iii) the temperature in a closed, thermally insulated system to which heat is
supplied at a constant rate.

These linear asymptotes are of considerable importance, both because they give
an approximation to the temperature which is adequate for many purposes, and
because the measurement of both slope and intercept provides a method for
simultaneous determination† of K and κ. The time-lag also provides a convenient
measure of possible errors by measuring apparatus exposed to varying temperatures.

Many examples for simple cases have been given earlier, but the method is also
available for systems which are so complicated that explicit evaluation of the
complete solution is impracticable. For example, a simple routine has been given
for calculating time-lags for composite walls of any number of layers.‡

We give below some results for the time-lags in temperature for various closed
systems heated at a constant rate. In all cases K, ρ, c, κ are the thermal constants
of the solid.

I. *The slab $0 < x < a$. No flow of heat at $x = 0$. At $x = a$ contact with mass M
per unit area of well-stirred fluid or perfect conductor of specific heat c' to which
heat is supplied at a constant rate for $t > 0$. Contact resistance $1/hK$ at the surface
$x = a$. $\mu = Mc'/a\rho c$*

The time-lags L_x in the solid and L_f in the fluid are

$$L_x = \frac{a^2}{\kappa(1+\mu)}\left(\frac{\mu}{2}+\frac{1}{6}+\frac{\mu}{ah}\right)-\frac{x^2}{2\kappa}, \qquad L_f = -\frac{a^2}{\kappa(1+\mu)}\left(\frac{1}{3}+\frac{1}{ah}\right). \tag{4}$$

II. *The cylinder $0 \leqslant r < a$ in contact with mass M per unit length of well-stirred
fluid. Heat supply and contact resistance as in I. $\mu = Mc'/\pi a^2 \rho c$*

$$L_r = \frac{a^2}{\kappa(1+\mu)}\left(\frac{\mu}{4}+\frac{1}{8}+\frac{\mu}{2ah}\right)-\frac{r^2}{4\kappa}, \qquad L_f = -\frac{a^2}{\kappa(1+\mu)}\left(\frac{1}{8}+\frac{1}{2ah}\right). \tag{5}$$

III. *The sphere $0 \leqslant r < a$ in contact with mass M of well-stirred fluid. Heat
supply and contact resistance as in I. $\mu = 3Mc'/4\pi a^3 \rho c$*

$$L_r = \frac{a^2}{\kappa(1+\mu)}\left(\frac{\mu}{6}+\frac{1}{10}+\frac{\mu}{3ah}\right)-\frac{r^2}{6\kappa}, \qquad L_f = -\frac{a^2}{\kappa(1+\mu)}\left(\frac{1}{15}+\frac{1}{3ah}\right). \tag{6}$$

It may be noted that, while these time-lags depend markedly on h, particularly
if it is small, the quantity $L_c+\mu L_f$, where L_c is the time-lag at the centre ($x = 0$ or
$r = 0$), is independent of h, being $a^2/6\kappa$, $a^2/8\kappa$, and $a^2/10\kappa$ for the slab, cylinder,
and sphere, respectively. This result may be shown numerically to be nearly true
also for the finite cylinder discussed below.

IV. *The finite cylinder of radius a and length l in contact over the whole of its
surface with mass M of well-stirred fluid of specific heat c' to which heat is supplied*

† Barrer, *Trans. Faraday Soc.* **35** (1939) 628, gives formulae for the quantity of heat
flowing through a slab with its surfaces at constant temperature. He gives corresponding results for the sphere in *Phil. Mag.* (7) **35** (1944) 802, and for the cylinder in *Trans.
Faraday Soc.* **36** (1940) 1235. For the latter case see also Jaeger, ibid. **42** (1946) 615.
‡ Jaeger, *Quart. Appl. Math.* **8** (1950) 187–98. Levy, *Trans. Amer. Soc. Mech. Engrs.*
78 (1956) 1627–35, discusses the cylindrical case.

at a constant rate for $t > 0$. No contact resistance. $\mu = Mc'/\pi a^2 l\rho c$, $\lambda = 1/(1+\mu)$

The time-lags L_f in the fluid and L_c at the centre of the cylinder are given† in the Table.

l/a	$\kappa L_c/a^2$	$-\kappa L_f/a^2$	$\kappa(L_c+\mu L_f)/a^2$
1	$0.102-0.040\lambda$	0.040λ	0.062
1.5	$0.162-0.061\lambda$	0.061λ	0.101
2	$0.201-0.075\lambda$	0.075λ	0.126
2.5	$0.223-0.084\lambda$	0.084λ	0.139
3	$0.235-0.091\lambda$	0.091λ	0.144
4	$0.246-0.099\lambda$	0.099λ	0.147
6	$0.248-0.108\lambda$	0.108λ	0.140
∞	$0.25-0.125\lambda$	0.125λ	0.125

15.7. Heat generation

A number of problems involving heat generation at a rate which is either constant or a simple function of position or time have already been solved. All these could have been solved by direct application of the Laplace transformation method, and we indicate here its application to some more complicated problems, in particular, to some involving heat production at a rate which is a linear function of the temperature and heat production at a rate determined by a solution of the diffusion equation.

I. *The slab $0 < x < l$ with no flow of heat at $x = 0$ and zero temperature at $x = l$. Zero initial temperature. Heat production at the rate $K(A+Bv)$ for $t > 0$*

The equation to be solved is

$$\frac{\partial^2 v}{\partial x^2} - \frac{1}{\kappa}\frac{\partial v}{\partial t} + Bv = -A. \tag{1}$$

The subsidiary equation is

$$\frac{d^2\bar{v}}{dx^2} - q^2\bar{v} = -\frac{A}{p}, \tag{2}$$

where

$$q^2 = \frac{p}{\kappa} - B. \tag{3}$$

This has to be solved with

$$\bar{v} = 0, \quad x = l; \qquad d\bar{v}/dx = 0, \quad x = 0. \tag{4}$$

The solution is

$$\bar{v} = \frac{A}{pq^2}\left\{1 - \frac{\cosh qx}{\cosh ql}\right\}, \tag{5}$$

$$v = \frac{A}{B}\left\{\frac{\cos xB^{\frac{1}{2}}}{\cos lB^{\frac{1}{2}}} - 1\right\} +$$

$$+ \frac{16Al^2}{\pi}\sum_{n=0}^{\infty}\frac{(-1)^n\exp\{[-(2n+1)^2\pi^2+4Bl^2]\kappa t/4l^2\}\cos(2n+1)\pi x/2l}{[4Bl^2-(2n+1)^2\pi^2](2n+1)}. \tag{6}$$

† The solution is given in Carslaw and Jaeger, *Operational Methods in Applied Mathematics* (edn. 2, 1948) § 127. The case of contact resistance at the surface, referred to above, may be treated in the same way.

This solution holds for either sign of B. If B is negative (for example, in heating by electric current in a material such as carbon with a negative temperature coefficient of resistance, or in the case of removal of heat from a rod by radiation or convection) the steady state solution involves hyperbolic functions. If B is positive (as in most cases of heating by electric current), the exponential terms in (6) all tend to zero as $t \to \infty$ if

$$B < \pi^2/4l^2, \tag{7}$$

and, if this inequality holds, a steady temperature distribution given by†

$$\frac{A}{B}\left\{\frac{\cos x B^{\frac{1}{2}}}{\cos l B^{\frac{1}{2}}} - 1\right\} \tag{8}$$

exists. If $B > \pi^2/4l^2$, heat is produced at too great a rate for its removal to be possible, and no steady distribution exists.

II. *The problem of* I *but with boundary condition*

$$\frac{\partial v}{\partial x} + hv = 0, \qquad x = l. \tag{9}$$

$$v = \frac{hA\cos x B^{\frac{1}{2}}}{B\{h\cos l B^{\frac{1}{2}} - B^{\frac{1}{2}}\sin l B^{\frac{1}{2}}\}} - \frac{A}{B} + 2Ahl^3 \sum_{n=1}^{\infty} \frac{\cos(x\alpha_n/l)\exp[(Bl^2 - \alpha_n^2)\kappa t/l^2]}{(Bl^2 - \alpha_n^2)[\alpha_n^2 + lh(lh+1)]\cos\alpha_n}, \tag{10}$$

where α_n, $n = 1, 2,...$, are the positive roots of

$$\alpha\tan\alpha = lh. \tag{11}$$

In this case, the condition for the existence of a steady state is

$$B < \alpha_1^2/l^2. \tag{12}$$

III. *The cylinder* $0 \leqslant r < a$ *with zero initial temperature and* $r = a$ *maintained at zero for* $t > 0$. *Heat production at the rate* $K(A+Bv)$ *for* $t > 0$

$$v = \frac{A}{B}\frac{J_0(rB^{\frac{1}{2}})}{J_0(aB^{\frac{1}{2}})} - \frac{A}{B} + 2a^2A \sum_{n=1}^{\infty} \frac{e^{(-\alpha_n^2 + a^2B)\kappa t/a^2}J_0(r\alpha_n/a)}{\alpha_n(a^2B - \alpha_n^2)J_1(\alpha_n)}, \tag{13}$$

where α_n, $n = 1, 2,...$, are the positive roots of

$$J_0(\alpha) = 0. \tag{14}$$

If B is positive, a steady state solution exists only if $B < \alpha_1^2/a^2$.

IV. *Steady temperature in a slab with heat generation at a rate which is an exponential function of the temperature*

The solution of this problem in steady temperature‡ is of interest in connexion with the upper limit to possible rates of heat production which appeared in (7) and (12). By a simple change of variables the equation to be solved may be put in the form

$$\frac{d^2v}{dx^2} + \beta e^v = 0, \quad 0 < x < 1, \tag{15}$$

---

† Numerical values for the steady temperature for the slab, cylinder, and sphere are given by Jakob, *Trans. Amer. Soc. Mech. Engrs.* **65** (1943) 593 (for the case $B > 0$) and ibid. **70** (1948) 25–30 (for the case $B < 0$).

‡ The variable state is discussed by Copple, Hartree, Porter, and Tyson, *J. Instn. Elect. Engrs.* **85** (1939) 56–66, using the differential analyser.

with
$$\frac{dv}{dx} = 0, \qquad x = 0, \tag{16}$$

and
$$v = 0, \qquad x = 1. \tag{17}$$

Suppose v_0 is the value of v when $x = 0$: we first solve (15) and (16) with $v = v_0$, when $x = 0$. The first integral of (15), subject to these conditions, is

$$\frac{dv}{dx} = -\{2\beta(e^{v_0} - e^v)\}^{\frac{1}{2}}. \tag{18}$$

Integrating again gives

$$v = v_0 - 2\ln\cosh[x(\tfrac{1}{2}\beta e^{v_0})^{\frac{1}{2}}]. \tag{19}$$

If this is to satisfy (17), v_0 must be given by

$$e^{\frac{1}{2}v_0} = \cosh(\tfrac{1}{2}\beta e^{v_0})^{\frac{1}{2}},$$

or, putting
$$z = (\tfrac{1}{2}\beta e^{v_0})^{\frac{1}{2}},$$

$$\cosh z = (\tfrac{1}{2}\beta)^{-\frac{1}{2}}z. \tag{20}$$

If $0 < \beta < 0.88...$, this equation has two roots, corresponding to two possible values of v_0 and thus to two possible steady state solutions. If $\beta > 0.88...$ it has no real roots and there is no solution.

V. *Heat production by an irreversible first order reaction*

In this case, assuming the reaction velocity to be independent of the temperature, the rate of heat production may be taken to be KAC, where A is a constant and C is the concentration of the diffusing substance. This latter is determined by the differential equation

$$\frac{\partial^2 C}{\partial x^2} - \frac{1}{D}\frac{\partial C}{\partial t} - \frac{k}{D}C = 0, \tag{21}$$

where D is the diffusion coefficient and k is a constant (cf. *M.D.*, Chap. VIII). (21) is, in fact, of the form being considered in this section and may be solved by the methods of this section or 1.14 (25). When C has been found, the temperature is given by

$$\frac{\partial^2 v}{\partial x^2} - \frac{1}{\kappa}\frac{\partial v}{\partial t} + AC = 0. \tag{22}$$

The subsidiary equation for (22) with zero initial temperature is

$$\frac{d^2\bar{v}}{dx^2} - \frac{p}{\kappa}\bar{v} = -A\bar{C}, \tag{23}$$

so that if $\bar{C}$ is known it may be inserted in (23), and C need not be evaluated.

As an example, consider the slab $0 < x < l$, with zero initial temperature and concentration, and with boundary conditions

$$\frac{\partial C}{\partial x} = 0, \qquad \frac{\partial v}{\partial x} = 0, \qquad x = 0 \tag{24}$$

$$C = C_0, \qquad v = 0, \qquad x = l. \tag{25}$$

Then, from (21) with these boundary conditions:

$$\bar{C} = \frac{C_0 \cosh x[(p+k)/D]^{\frac{1}{2}}}{p \cosh l[(p+k)/D]^{\frac{1}{2}}}.$$

Using this value of $\bar{C}$ in (23), gives

$$\bar{v} = \frac{AC_0\kappa D}{(\kappa - D)p[p + k\kappa/(\kappa - D)]}\left\{\frac{\cosh x(p/\kappa)^{\frac{1}{2}}}{\cosh l(p/\kappa)^{\frac{1}{2}}} - \frac{\cosh x[(p+k)/D]^{\frac{1}{2}}}{\cosh l[(p+k)/D]^{\frac{1}{2}}}\right\},$$

$$v = \frac{AC_0 D}{k}\left\{1 - \frac{\cosh x(k/D)^{\frac{1}{2}}}{\cosh l(k/D)^{\frac{1}{2}}}\right\} +$$

$$+\frac{16 AC_0 Dl^2}{\pi}\sum_{n=0}^{\infty}\frac{(-1)^n\cos(2n+1)\pi x/2l}{(2n+1)[(2n+1)^2\pi^2(\kappa - D) - 4kl^2]}e^{-\kappa(2n+1)^2\pi^2 t/4l^2} -$$

$$-16 AC_0\pi\kappa l^2 D\sum_{n=0}^{\infty}\frac{(-1)^n(2n+1)\cos(2n+1)\pi x/2l}{[D(2n+1)^2\pi^2 + 4kl^2][(2n+1)^2\pi^2(\kappa - D) - 4kl^2]}e^{-kt - (2n+1)^2\pi^2 Dt/4l^2}.$$

Many problems of this type may conveniently be discussed by the present method.†

15.8. Systems for automatic temperature control

In practice it is often desired to maintain constant temperature‡ in some region, such as an oven which loses heat both by steady flow through its walls and by artificial disturbances such as the opening of its door. In order to do this the rate of supply of heat to the oven is regulated by a thermometer in the oven, the two commonest methods of control being: (i) *'on-off'* control, in which the heating current (or part of it) is switched on when the thermometer falls to a certain temperature, and off when it has risen to another fixed temperature; (ii) *proportional control*, in which the rate of supply of heat is proportional to the departure of the thermometer reading from the desired temperature.

In both cases an essential difficulty is that there is always a time-lag between the temperature of the region and the appropriate changes in the heating current, caused, among other things, by the time taken for changes of temperature to penetrate through the shield of the thermometer and in the material of the thermometer itself. Thus the simplest idealization of an actual furnace would be the following: mass M of well-stirred fluid (the heating element, contents of the furnace, etc.) which has heat supplied to it at a prescribed rate $Q(t)$ and loses heat at a rate proportional to its temperature (by conduction through the furnace walls) is in contact at $x = 0$ with the slab $0 < x < l$ (the shield of the thermometer, etc.) with no loss of heat from $x = l$. It is the temperature v_l at $x = l$ which determines the rate of supply of heat $Q(t)$. This ideal system has been studied in § 3.13 (vii) and (viii) and

† Cf. Danckwerts, *Appl. Sci. Res.* A, **3** (1953) 385.

‡ For the systems used in practice see, e.g., Griffiths, *Thermostats* (Griffin, edn. 2, 1943); Rhodes, *Industrial Instruments for Measurement and Control* (McGraw-Hill, 1941). The theory is part of the general theory of servomechanisms, cf. Thaler and Brown, *Servomechanism Analysis* (McGraw-Hill, 1953).

other cases may be treated similarly. Thus for on-off systems, in which Q has always one of two constant values, the behaviour of v_l is easily studied for any special system (general results are too complicated to give) and in particular the behaviour of v_l for a Q which varies periodically in square wave form may be found by the methods of §§ 3.6, 15.5.

In the case of proportional control in the above system we have $Q(t) = C[V-v_l]$, where C and V are constants, and the theory presents interesting new features. To illustrate these with the simplest possible algebra, we consider the following system in detail.

Infinite solid $-\infty < x < \infty$ is to be heated by supply of heat in the plane $x = 0$. The initial temperature of the solid is zero, and the rate of supply of heat, Q per unit area per unit time, is given by

$$Q = 2C(V-v_l), \tag{1}$$

where C and V are constants, and v_l is the temperature at $x = l$.

Thus the boundary condition for the semi-infinite solid $x > 0$ (into which half the total heat supply flows) is

$$-K\frac{\partial v}{\partial x} = C(V-v_l), \quad x = 0. \tag{2}$$

The boundary condition for the subsidiary equation is therefore

$$-K\frac{d\bar{v}}{dx} = \frac{CV}{p} - C\bar{v}_l, \quad x = 0. \tag{3}$$

The solution of the subsidiary equation which remains finite as $x \to \infty$ is $\bar{v} = Ae^{-qx}$, so that $\bar{v}_l = Ae^{-ql}$. Substituting in (3) gives A, and we get finally

$$\bar{v} = \frac{Vke^{-qx}}{p(q+ke^{-ql})}, \tag{4}$$

where $k = C/K$ is a positive constant.

To determine v we may expand (4) as in § 12.5, and obtain

$$\bar{v} = kV \sum_{n=0}^{\infty} \frac{(-k)^n}{pq^{n+1}} e^{-q(x+nl)}. \tag{5}$$

Therefore

$$v = V \sum_{n=0}^{\infty} (-1)^n [2k\sqrt{(\kappa t)}]^{n+1} i^{n+1} \mathrm{erfc}\, \frac{x+nl}{2\sqrt{(\kappa t)}}. \tag{6}$$

This series is useful only for small values of k and t, and gives no information about the nature of v: the Inversion Theorem is more useful. This gives

$$v = \frac{kV}{2\pi i} \int_{\gamma-i\infty}^{\gamma+i\infty} \frac{e^{\lambda t - \mu x}\, d\lambda}{\lambda[\mu+ke^{-\mu l}]}. \tag{7}$$

The integrand of (7) has a branch point at $\lambda = 0$. Its poles are at the roots of

$$z + kle^{-z} = 0$$

for which $\mathbf{R}(z) \geqslant 0$, where we have put $z = \mu l$. Writing $z = \xi \pm i\eta$, $\eta > 0$, we have at these roots

$$\xi + kle^{-\xi} \cos \eta = 0, \tag{8}$$

$$\eta - kle^{-\xi} \sin \eta = 0. \tag{9}$$

The following properties of the roots appear immediately:

(i) There are no real roots, $\eta = 0$.

(ii) Since in (8) $\cos \eta$ must be negative, and in (9) $\sin \eta$ must be positive, η can only lie in the regions $(2n + \frac{1}{2})\pi \leqslant \eta \leqslant (2n+1)\pi$, $n = 0, 1,\dots$.

(iii) If $kl = (4n+1)\pi/2$, $n = 0, 1, 2,\dots$, there is a pure imaginary root, $\eta = (4n+1)\pi/2$.

(iv) If $kl = (8n+3)\pi 2^{-\frac{3}{2}} e^{(8n+3)\pi/4}$, $n = 0, 1, 2,\dots$, there is a root

$$\xi = \eta = (8n+3)\pi/4. \tag{10}$$

(v) Since $\xi = -\eta \cot \eta$, and $kl = \eta e^{\xi} \operatorname{cosec} \eta$, it follows that in each of the regions (ii), ξ and kl both increase steadily with η.

The way in which the roots vary with kl is shown in Fig. 47, in which the positions of the roots for various values of kl (marked on the curves) are shown. There is one root in the first region for all values of kl greater than $\pi/2$; one in the first region and one in the second for all values of kl between $5\pi/2$ and $9\pi/2$; and so on, there being n roots for kl in the region $2(n-1)\pi + \frac{1}{2}\pi < kl < 2n\pi + \frac{1}{2}\pi$.

Of these roots, those with real part greater than imaginary part (i.e. to the right of OP in Fig. 47) will correspond to an unstable oscillation, while those with real parts less than imaginary parts (to the left of OP) correspond to a damped oscillation. The roots (10) which lie on OP correspond to a steady maintained oscillation.

The line $(\gamma - i\infty, \gamma + i\infty)$ is to lie to the right of all these poles. Since the integrand of (7) has a branch point at $\lambda = 0$ we use the contour of Fig. 40, and, in the usual way, the line integral in (7) is found to be equal to the contributions from CD, EF, and the small circle about the origin, together with $2\pi i$ times the sum of the residues at the poles of the integrand. We write down the solution for the case

$$2(n-1)\pi + \tfrac{1}{2}\pi < kl < 2n\pi + \tfrac{1}{2}\pi$$

in which there are n pairs of poles at the points $\mu l = \alpha_r \pm i\beta_r$, $r = 1,\dots, n$; if $kl < \frac{1}{2}\pi$ there is no term of this type.

The poles $\lambda = \kappa[\alpha_r^2 - \beta_r^2 \pm 2i\alpha_r\beta_r]/l^2$ give a contribution

$$\frac{2lkV}{(\alpha_r+i\beta_r)(1+\alpha_r+i\beta_r)}\exp\left\{-\frac{x(\alpha_r+i\beta_r)}{l}+\frac{\kappa(\alpha_r^2-\beta_r^2)t}{l^2}+\frac{2i\kappa t\alpha_r\beta_r}{l^2}\right\}+\text{conjugate}$$

$$=\frac{4lkVe^{-\alpha_r x/l+\kappa(\alpha_r^2-\beta_r^2)t/l^2}}{\sqrt{\{(\alpha_r^2+\beta_r^2)[(1+\alpha_r)^2+\beta_r^2]\}}}\cos\left\{\frac{2\kappa\alpha_r\beta_r t}{l^2}-\frac{\beta_r x}{l}-\phi_r-\psi_r\right\},$$

where $\phi_r = \tan^{-1}(\beta_r/\alpha_r)$, $\psi_r = \tan^{-1}\{\beta_r/(1+\alpha_r)\}$.

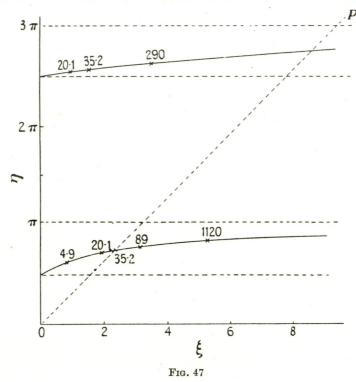

Fɪɢ. 47

The small circle about the origin gives V. And CD and EF give in the limit

$$-\frac{2kV}{\pi}\int_0^\infty \frac{e^{-\kappa u^2 t}[u\cos ux - k\sin u(l-x)]\,du}{u\{u^2+k^2-2ku\sin ul\}}.$$

Thus in this case

$$v = V + \sum_{r=1}^{n}\frac{4lkVe^{-\alpha_r x/l+\kappa(\alpha_r^2-\beta_r^2)t/l^2}}{\sqrt{\{(\alpha_r^2+\beta_r^2)[(1+\alpha_r)^2+\beta_r^2]\}}}\cos\left\{\frac{2\kappa\alpha_r\beta_r t}{l^2}-\frac{\beta_r x}{l}-\phi_r-\psi_r\right\}-$$

$$-\frac{2kV}{\pi}\int_0^\infty \frac{e^{-\kappa u^2 t}[u\cos ux - k\sin u(l-x)]\,du}{u\{u^2+k^2-2ku\sin ul\}}. \tag{11}$$

In Fig. 48 some values of v_l calculated from this formula are shown. In curve I, for $kl = 4\cdot9$, the temperature tends slowly to its final value V; for larger values of k, the final value is approached more rapidly, and a damped oscillation executed about it as in curve II, for $kl = 20\cdot1$; for $kl = 35\cdot2$, the critical value (10) with $n = 1$, there is a steady maintained oscillation about the final value, curve III; for still larger values of kl there would be an oscillation of increasing amplitude.

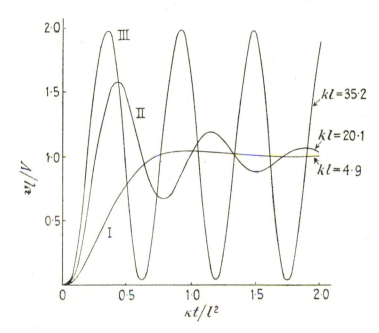

Fig. 48. Behaviour of a system with automatic temperature control for various values of the control parameters.

This type of behaviour, and the possibility of a steady 'hunting' of temperature, are characteristic of such systems. The simple example above, which is of no physical interest, was chosen because the zeros of the denominator of (7) could be discussed briefly; in practical systems the denominators are more complicated and need a tedious numerical and graphical study.†

The simplest extension of the case considered above is that of the semi-infinite solid $x > 0$ which loses heat from the plane $x = 0$ at a

† Callender, Hartree, and Porter, *Trans. Roy. Soc.* A, **235** (1936) 415; Hartree, Porter, Callender, and Stevenson, *Proc. Roy. Soc.* A, **161** (1937) 460. Hull and Wolfe, *Canad. J. Phys.* **32** (1954) 72–80, have developed a method of successive approximations for inverting transforms which appear in problems of the present type.

rate H times its temperature. We suppose heat to be supplied in the plane $x = 0$ at the rate

$$Q_0 + C(V_0 - v_l),\tag{12}$$

where v_l is the temperature at $x = l$, V_0 is the required final temperature, C is a constant, and $Q_0 = HV_0$ is the steady rate of supply which would be needed to maintain the surface $x = 0$ at V_0. Then, if initially the solid is at zero, the transform of the solution is found as above to be

$$\bar{v} = \frac{(h+k)V_0\, e^{-qx}}{p(q+h+ke^{-ql})},\tag{13}$$

where $h = H/K$ and $k = C/K$. The corresponding problem for a slab $0 < x < l$ with no loss of heat at $x = l$ can be solved in the same way: steady periodic solutions in this case have been studied by Turner.[†]

15.9. Non-homogeneous solids

If the thermal properties of a solid vary with position, exact solutions can be found in a limited number of special cases. Taking the problem of linear flow, the differential equation is[‡]

$$\frac{\partial}{\partial x}\left(K\frac{\partial v}{\partial x}\right) - \rho c\frac{\partial v}{\partial t} = 0,\tag{1}$$

and the corresponding subsidiary equation for zero initial temperature becomes

$$\frac{d}{dx}\left(K\frac{d\bar{v}}{dx}\right) - \rho c p\bar{v} = 0,$$

or

$$\frac{d^2\bar{v}}{dx^2} + \frac{1}{K}\frac{dK}{dx}\frac{d\bar{v}}{dx} - \frac{\rho c}{K}p\bar{v} = 0.\tag{2}$$

The simplest cases are those in which (2) can be transformed into Bessel's equation, and in this connexion the result ($W.B.F.$, § 4.31 (3)) that the general solution of

$$\frac{d^2u}{dz^2} + \frac{2\alpha - 2\beta\nu + 1}{z}\frac{du}{dz} + \frac{\beta^2\gamma^2 z^{2\beta} + \alpha(\alpha - 2\beta\nu)}{z^2}u = 0\tag{3}$$

† Turner, *Proc. Camb. Phil. Soc.* **32** (1936) 663. Cf. also Hopper, *Proc. Phys. Soc.* **54** (1942) 55.

‡ Problems on equations of this type go back to Kepinski, *Math. Ann.* **61** (1905) 397, and Gevrey, *J. Math. pures appl.* (6) **10** (1914) 105. They have recently become of interest in connexion with transport of heat by turbulent fluid, cf. Sutton, *Proc. Roy. Soc.* A, **146** (1934) 701; Brunt, *Physical and Dynamical Meteorology* (Cambridge, 1939); Sutton, *Atmospheric Turbulence* (Methuen, 1949). Explicit solutions related to those of Case I, below, have been given by Huber, *Z. angew. Math. Mech.* **7** (1927) 469; Sutton, *Proc. Roy. Soc.* A, **182** (1943) 48; Jaeger, *Quart. Appl. Math.* **3** (1945) 210; Calder, *Quart. J. Mech. Appl. Math.* **2** (1949) 153–76; Knighting, ibid. **5** (1952) 423–31; Davies, ibid. **7** (1954) 462–71; Davies and Bourne, *Quart. J. Mech. Appl. Math.* **9** (1956) 457–68. The equation $a\partial^2(vx)/\partial x^2 - \partial[(bx+c)v]/\partial x - \partial v/\partial t = 0$, which occurs in some biological problems, is discussed by Feller, *Ann. Math. Princeton*, **54** (1951) 173–82.

is
$$z^{\beta\nu-\alpha}C_\nu(\gamma z^\beta), \tag{4}$$

where $C_\nu(x)$ is the cylinder function $AJ_\nu(x)+BY_\nu(x)$, is frequently useful.
A number of typical soluble cases will now be considered.

I. $K = K_0 x^n$, $\rho c =$ constant $\qquad\qquad\qquad\qquad\qquad$ (5)

In this case the solutions of (2) are

$$x^{\frac{1}{2}(1-n)}I_\nu\left(\frac{2qx^{1-\frac{1}{2}n}}{2-n}\right) \quad\text{and}\quad x^{\frac{1}{2}(1-n)}K_\nu\left(\frac{2qx^{1-\frac{1}{2}n}}{2-n}\right), \tag{6}$$

where $\qquad \kappa = K_0/\rho c, \quad q^2 = p/\kappa, \quad \nu = (1-n)/(2-n),$ $\qquad$ (7)

and $n \neq 2$. If $n = 2$, the solutions are

$$x^{[-1\pm\sqrt{(1+4q^2)}]/2}. \tag{8}$$

As a definite example consider *the region $x > 0$ with zero initial temperature, and with the plane $x = 0$ raised to unit temperature at $t = 0$.*

For definiteness we suppose also $0 \leqslant n \leqslant 1$ which is the case of the greatest practical interest. The solution of (2) which is bounded at infinity becomes in this case

$$\bar{v} = Ax^{\frac{1}{2}(1-n)}K_\nu\left(\frac{2qx^{1-\frac{1}{2}n}}{2-n}\right), \tag{9}$$

where ν is given by (7), and A is to be chosen so that $\bar{v} \to 1/p$ as $x \to 0$.
Now as $z \to 0$, $\qquad\qquad z^\nu K_\nu(z) \to 2^{\nu-1}\Gamma(\nu).$ $\qquad\qquad$ (10)

Using this result we find A, and hence finally

$$\bar{v} = \frac{2x^{\frac{1}{2}(1-n)}}{\Gamma(\nu)(2-n)^\nu\kappa^{\frac{1}{2}\nu}}\, p^{\frac{1}{2}\nu-1}K_\nu\left(\frac{2x^{1-\frac{1}{2}n}}{(2-n)\kappa^{\frac{1}{2}}}\, p^{\frac{1}{2}}\right). \tag{11}$$

Thus, using Appendix V (26),†

$$v = \frac{1}{\Gamma(\nu)}\int_X^\infty e^{-u}u^{\nu-1}\,du, \tag{12}$$

where $\qquad\qquad\qquad X = \dfrac{x^{2-n}}{(2-n)^2\kappa t}.$ $\qquad\qquad$ (13)

Problems in which there is an arbitrary initial temperature, and problems on the finite region $0 < x < l$ may be treated in the usual way.

---

† Goldstein, *Proc. Lond. Math. Soc.* (2) **34** (1932) 104, equations (15) and (24). He also shows that the function whose transform is $p^n K_\nu(zp^{\frac{1}{2}})$ is a Whittaker function. Alternatively the Inversion Theorem may be used with the contour of Fig. 40.

II. *If $K = K_0(1+\alpha x)$ and $q = (p\rho c/K_0)^{\frac{1}{2}}$,* $\qquad$ (14)

the solutions of (2) are

$$I_0[2q(1+\alpha x)^{\frac{1}{2}}/\alpha] \quad \text{and} \quad K_0[2q(1+\alpha x)^{\frac{1}{2}}/\alpha], \qquad (15)$$

so that the solutions of problems in which the thermal conductivity is of this form may be derived from those of problems on regions bounded by circular cylinders. Thus, for example, for the semi-infinite region $x > 0$ with conductivity given by (14), zero initial temperature, and the surface $x = 0$ maintained at constant temperature V for $t > 0$, the temperature at x and t is the same as that at radius $r = a(1+\alpha x)^{\frac{1}{2}}$ and time $\alpha^2 a^2 t/4$ in the problem of § 13.5 I and is given by 13.5 (6) and can be read off from Fig. 41.

If α in (14) is negative, $\alpha = -\beta$, say, corresponding to conductivity decreasing linearly with distance to zero when $x = 1/\beta$, the temperature is the same as that in the problem of 7.6 (8) with $r = a(1-\beta x)^{\frac{1}{2}}$ at time $\beta^2 a^2 t/4$, and may be obtained from Fig. 24.

III. *If $K = K_0 x^n$, $c = c_0 x^n$,* $\qquad$ (16)

(2) becomes $\qquad \dfrac{d^2\bar{v}}{dx^2} + \dfrac{n}{x}\dfrac{d\bar{v}}{dx} - q^2\bar{v} = 0,$ $\qquad$ (17)

where $q^2 = p/\kappa_0$, and $\kappa_0 = K_0/\rho c_0$. The solutions of (17) are

$$x^{\frac{1}{2}(1-n)}K_{\frac{1}{2}(1-n)}(qx) \quad \text{and} \quad x^{\frac{1}{2}(1-n)}I_{\frac{1}{2}(1-n)}(qx). \qquad (18)$$

Solutions may be obtained as in I.

IV. *If $K = K_0(1+\alpha x)$, $c = c_0(1+\alpha x)$,* $\qquad$ (19)

(2) becomes $\qquad \dfrac{d^2\bar{v}}{dx^2} + \dfrac{\alpha}{1+\alpha x}\dfrac{d\bar{v}}{dx} - q^2\bar{v} = 0,$ $\qquad$ (20)

where $q^2 = p/\kappa_0$ and $\kappa_0 = K_0/\rho c_0$. The solutions of (20) are

$$I_0\left[\frac{q}{\alpha}(1+\alpha x)\right] \quad \text{and} \quad K_0\left[\frac{q}{\alpha}(1+\alpha x)\right]. \qquad (21)$$

For the region $x > 0$, initially at zero and with $x = 0$ maintained at V for $t > 0$, the temperature at x and t is the same as that in the cylindrical region $r > a$ at the point $a(1+\alpha x)$ at time $\alpha^2 a^2 t$.

V. *If $K = K_0 x^n$, $c = c_0 x^m$,* $\qquad$ (22)

(2) becomes $\qquad \dfrac{d^2\bar{v}}{dx^2} + \dfrac{n}{x}\dfrac{d\bar{v}}{dx} - \dfrac{q^2}{x^{n-m}}\bar{v} = 0,$ $\qquad$ (23)

where $q^2 = K_0/\rho c_0$. By (3), the solutions of (23) are

$$x^{-\frac{1}{2}(n-1)}K_\nu\left\{\frac{2q}{m-n+2}x^{\frac{1}{2}(m-n+2)}\right\} \quad \text{and} \quad x^{-\frac{1}{2}(n-1)}I_\nu\left\{\frac{2q}{m-n+2}x^{\frac{1}{2}(m-n+2)}\right\} \qquad (24)$$

where $\qquad\qquad\qquad \nu = (1-n)/(m-n+2),$

and solutions may be obtained as before.

VI. *Cylindrical and spherical cases*

The subsidiary equations in these cases take the forms

$$\frac{d}{dr}\left(rK\frac{d\bar{v}}{dr}\right) - \rho crp\bar{v} = 0, \qquad (25)$$

and $\qquad\qquad \dfrac{d}{dr}\left(r^2K\dfrac{d\bar{v}}{dr}\right) - \rho cr^2 p\bar{v} = 0.$ $\qquad$ (26)

and thus the theory of power law variations of K and c is contained in that given in V above.

15.10. The heating of a chain of radiating slabs. Laminated materials

This problem, which dates back to Fourier, is of interest in connexion with conduction of heat in laminated materials,[†] and has other practical applications.

Suppose we have a number of infinite parallel plane sheets so thin that the temperature may be assumed constant across their cross-section. Let M_r be the mass per unit area of the rth strip, c_r its specific heat, v_r its temperature, and

$$H_r(v_r - v_{r+1}) \tag{1}$$

the rate of heat transfer per unit time per unit area to the $(r+1)$th strip. Then the temperature v_r satisfies

$$M_r c_r \frac{dv_r}{dt} = H_{r-1}(v_{r-1} - v_r) - H_r(v_r - v_{r+1}). \tag{2}$$

If the initial temperature is zero, the subsidiary equation corresponding to (2) is

$$(M_r c_r p + H_{r-1} + H_r)\bar{v}_r - H_{r-1}\bar{v}_{r-1} - H_r \bar{v}_{r+1} = 0. \tag{3}$$

The system of equations (3) constitutes a set of difference equations connecting the transforms of the temperatures of the successive slabs. We consider only the case in which all the H_r, M_r, and c_r are equal, so that (3) becomes

$$\bar{v}_{r+1} - (2 + 2kp)\bar{v}_r + \bar{v}_{r-1} = 0, \tag{4}$$

where

$$k = Mc/2H. \tag{5}$$

The solution of (4) is

$$\bar{v}_r = Ae^{r\theta} + Be^{-r\theta}, \tag{6}$$

where A and B are constants to be found from the conditions at the beginning and end of the chain, and θ is given by

$$\cosh\theta = 1 + kp. \tag{7}$$

If there are n slabs, initially at zero, of which the first receives radiation from medium at V, and the last radiates into medium at zero, we have

$$\bar{v}_0 = \frac{V}{p}, \qquad \bar{v}_{n+1} = 0,$$

and (6) becomes

$$\bar{v}_r = \frac{V \sinh(n-r+1)\theta}{p \sinh(n+1)\theta}. \tag{8}$$

† Cf. Störmer, *Wiss. Veröff. Siemens-Werken*, **17** (1938), 90.

Using the Inversion Theorem we get finally for $r = 1,..., n$

$$v_r = \frac{V(n-r+1)}{n+1} - \sum_{s=1}^{n} \frac{V \sin[rs\pi/(n+1)]\sin[s\pi/(n+1)]}{(n+1)[1-\cos\{s\pi/(n+1)\}]} e^{-(t/k)[1-\cos s\pi/(n+1)]}.$$

(9)

If there is an infinite number of slabs, initially at zero, $r = 1, 2,...$, with the first receiving radiation from medium at V, we find in the same way

$$\bar{v}_r = \frac{V}{p} e^{-r\theta} = \frac{Vk^r}{p}\left\{\left(p+\frac{1}{k}\right) - \left[\left(p+\frac{1}{k}\right)^2 - \frac{1}{k^2}\right]^{\frac{1}{2}}\right\}^r.$$

(10)

Therefore, using Appendix V (27)

$$v_r = Vr \int_0^t e^{-\tau/k} I_r(\tau/k)\frac{d\tau}{\tau}.$$

(11)

15.11. Direct application of the Laplace transformation method to two- and three-dimensional problems

It was remarked in § 6.1 that, of the various methods available, the Laplace transformation provided the most direct solution of problems with zero initial temperature† and prescribed surface temperature. In this section some important problems of this type are solved in this way.

I. *The rectangular parallelepiped $0 < x < a$, $0 < y < b$, $0 < z < c$.* *Zero initial temperature. Surface temperature v_1, constant, on the face $x = 0$ and zero on the other faces*

The subsidiary equation is

$$\frac{\partial^2\bar{v}}{\partial x^2} + \frac{\partial^2\bar{v}}{\partial y^2} + \frac{\partial^2\bar{v}}{\partial z^2} - \frac{p}{\kappa}\bar{v} = 0.$$

(1)

This has to be solved with

$$\bar{v} = \frac{v_1}{p}, \quad x = 0, \quad 0 < y < b, \quad 0 < z < c,$$

(2)

and with $\bar{v}$ zero on the other faces.

To solve these we proceed precisely as in § 6.2, the only change being that a term (p/κ) is to be added to the right-hand side of 6.2 (3). Thus with this modification (and $v_2 = 0$) we can quote the solution 6.2 (9), and therefore after some changes of notation,

$$\bar{v} = \frac{16v_1}{\pi^2}\sum_{m=0}^{\infty}\sum_{n=0}^{\infty}\frac{\sin[(2m+1)\pi y/b]\sin[(2n+1)\pi z/c]\sinh q'(a-x)}{(2m+1)(2n+1)p\sinh q'a},$$

(3)

† If the initial temperature is not zero the method can, of course, still be applied but it has no advantage over the use of the Green's function.

where
$$q'^2 = \frac{(2m+1)^2\pi^2}{b^2} + \frac{(2n+1)^2\pi^2}{c^2} + \frac{p}{\kappa}. \tag{4}$$

Then, using the Inversion Theorem, we find

$$v = \frac{16v_1}{\pi^2} \sum_{m=0}^{\infty} \sum_{n=0}^{\infty} \frac{\sin[(2m+1)\pi y/b]\sin[(2n+1)\pi z/c]}{(2m+1)(2n+1)} \times$$

$$\times \frac{1}{2\pi i} \int_{\gamma-i\infty}^{\gamma+i\infty} \frac{e^{\lambda t}\sinh\mu'(a-x)\,d\lambda}{\lambda\sinh\mu'a}, \tag{5}$$

where
$$\mu'^2 = \frac{(2m+1)^2\pi^2}{b^2} + \frac{(2n+1)^2\pi^2}{c^2} + \frac{\lambda}{\kappa}.$$

The integrand of (5) has a simple pole at $\lambda = 0$, and simple poles at the values of λ corresponding to

$$\mu' = \frac{il\pi}{a}, \quad l = 1, 2,....$$

Evaluating the residues we get finally

$$v = \frac{16v_1}{\pi^2} \sum_{m=0}^{\infty} \sum_{n=0}^{\infty} \frac{\sin[(2m+1)\pi y/b]\sin[(2n+1)\pi z/c]\sinh(a-x)\sqrt{\alpha_{m,n,0}}}{(2m+1)(2n+1)\sinh a\sqrt{\alpha_{m,n,0}}} -$$

$$- \frac{32v_1}{\pi a^2} \sum_{m=0}^{\infty} \sum_{n=0}^{\infty} \sum_{l=1}^{\infty} \frac{l\sin[(2m+1)\pi y/b]\sin[(2n+1)\pi z/c]\sin[l\pi x/a]}{(2m+1)(2n+1)\alpha_{m,n,l}} e^{-\kappa t\alpha_{m,n,l}}, \tag{6}$$

where
$$\alpha_{m,n,l} = \frac{l^2\pi^2}{a^2} + \frac{(2m+1)^2\pi^2}{b^2} + \frac{(2n+1)^2\pi^2}{c^2}. \tag{7}$$

The first term in (6), which is derived from the pole $\lambda = 0$ in (5), is the old steady state solution, 6.2 (9).

II. *The problem of I except that $x = 0$ is kept at temperature $v_1\sin(\omega t+\epsilon)$*

The only change is that the integrand in (5) is to be multiplied by

$$\frac{\lambda(\omega\cos\epsilon+\lambda\sin\epsilon)}{\lambda^2+\omega^2}.$$

It then has poles at $\lambda = \pm i\omega$ which give the steady periodic solution,

$$\frac{16v_1}{\pi^2} \sum_{m=0}^{\infty} \sum_{n=0}^{\infty} \frac{\sin[(2m+1)\pi y/b]\sin[(2n+1)\pi z/c]}{(2m+1)(2n+1)} M_{m,n}\sin(\omega t+\epsilon+\phi_{m,n}), \tag{8}$$

where

$$M_{m,n}e^{i\phi_{m,n}} = \frac{\sinh(a-x)[(2m+1)^2\pi^2/b^2+(2n+1)^2\pi^2/c^2+i\omega/\kappa]^{\frac{1}{2}}}{\sinh a[(2m+1)^2\pi^2/b^2+(2n+1)^2\pi^2/c^2+i\omega/\kappa]^{\frac{1}{2}}}. \tag{9}$$

There are also poles at $-\kappa\alpha_{m,n,l}$ which give the transient part of the solution,

$$\frac{32v_1\kappa}{\pi a^2} \sum_{l=1}^{\infty} \sum_{m=0}^{\infty} \sum_{n=0}^{\infty} \frac{\begin{array}{c}l\sin[(2m+1)\pi y/b]\sin[(2n+1)\pi z/c]\sin[l\pi x/a] \times \\ \times (\omega\cos\epsilon - \kappa\alpha_{m,n,l}\sin\epsilon)\end{array}}{(\omega^2+\kappa^2\alpha_{m,n,l}^2)(2m+1)(2n+1)} e^{-\kappa t\alpha_{m,n,l}}.$$

If all surfaces are maintained at $\sin(\omega t + \epsilon)$, the solution is obtained by superposing six expressions of the above type. Clearly the results are so complicated as to be practically useless for calculation.

III. *The rectangular parallelepiped* $0 < x < a$, $-b < y < b$, $-c < z < c$. *Zero initial temperature. Surface temperature* v_1, *constant, on the face* $x = 0$, v_2, *constant, on the face* $x = a$. *Radiation at the other surfaces into medium at zero*

Here, as in I, we find, using 6.2 (23),

$$\bar{v} = \sum_{r=1}^{\infty} \sum_{s=1}^{\infty} \frac{4h^2[v_1 \sinh q'(a-x) + v_2 \sinh q'x]\cos \alpha_r y \cos \beta_s z}{p[(\alpha_r^2+h^2)b+h][(\beta_s^2+h^2)c+h]\cos \alpha_r b \cos \beta_s c \sinh q'a}, \tag{10}$$

where the α_r and β_s are the positive roots of 6.2 (16), (17), and

$$q'^2 = \alpha_r^2 + \beta_s^2 + (p/\kappa).$$

The solution v is found to be the sum of the steady state part, 6.2 (23), and the transient part

$$\frac{8\pi h^2}{a^2} \sum_{r=1}^{\infty} \sum_{s=1}^{\infty} \sum_{n=1}^{\infty} \frac{n[(-1)^n v_2 - v_1]\cos \alpha_r y \cos \beta_s z \sin[n\pi x/a]}{\gamma_{r,s,n}[(\alpha_r^2+h^2)b+h][(\beta_s^2+h^2)c+h]\cos \alpha_r b \cos \beta_s c} e^{-\kappa t \gamma_{r,s,n}}, \tag{11}$$

where

$$\gamma_{r,s,n} = \alpha_r^2 + \beta_s^2 + (n^2\pi^2/a^2). \tag{12}$$

IV. *The finite cylinder* $0 \leqslant r < a$, $0 < z < l$. *Zero initial temperature.* $z = 0$ *kept at temperature* V, *constant, and* $z = l$ *at temperature zero. Radiation at* $r = a$ *into medium at zero*

Here, using § 8.3 I as above,

$$v = V \sum_{n=1}^{\infty} \frac{2hJ_0(r\alpha_n)\sinh(l-z)\alpha_n}{a(h^2+\alpha_n^2)J_0(a\alpha_n)\sinh l\alpha_n} - \frac{4\pi h V}{a} \sum_{n=1}^{\infty} \sum_{m=1}^{\infty} \frac{mJ_0(r\alpha_n)\sin(m\pi z/l)}{(\alpha_n^2 l^2 + m^2\pi^2)(h^2+\alpha_n^2)J_0(a\alpha_n)} \times$$
$$\times e^{-\kappa t(\alpha_n^2 + m^2\pi^2/l^2)}, \tag{13}$$

where the α_n are the positive roots of

$$\alpha J_1(a\alpha) - hJ_0(a\alpha) = 0.$$

V. *The finite cylinder* $0 \leqslant r < a$, $0 < z < l$. *Zero initial temperature.* $z = 0$ *maintained at constant temperature* V *for* $t > 0$. *Radiation into medium at zero at the other faces*

$$v = \frac{2hV}{a} \sum_{n=1}^{\infty} \frac{J_0(r\alpha_n)[\alpha_n \cosh \alpha_n(l-z) + h \sinh \alpha_n(l-z)]}{(h^2+\alpha_n^2)J_0(a\alpha_n)[\alpha_n \cosh l\alpha_n + h \sinh l\alpha_n]} -$$
$$- \frac{4hV}{a} \sum_{n=1}^{\infty} \sum_{m=1}^{\infty} \frac{\beta_m(h^2+\beta_m^2)J_0(r\alpha_n)\sin z\beta_m}{(h^2+\alpha_n^2)[l(h^2+\beta_m^2)+h](\alpha_n^2+\beta_m^2)J_0(a\alpha_n)} e^{-\kappa(\alpha_n^2+\beta_m^2)t}, \tag{14}$$

where the α_n are the positive roots of

$$\alpha J_1(a\alpha) = hJ_0(a\alpha), \tag{15}$$

and the β_m are the positive roots of

$$\beta \cot \beta l + h = 0. \tag{16}$$

VI. *The semi-infinite cylinder $z > 0$, $0 \leqslant r < a$. Zero initial temperature. The surface $z = 0$ maintained at constant temperature V for $t > 0$. Radiation at $r = a$ into medium at zero*

$$v = \frac{2hV}{a} \sum_{n=1}^{\infty} \frac{J_0(r\alpha_n)}{(h^2+\alpha_n^2)J_0(a\alpha_n)} e^{-\alpha_n z} +$$

$$+ \frac{hV}{a} \sum_{n=1}^{\infty} \frac{J_0(r\alpha_n)}{(h^2+\alpha_n^2)J_0(a\alpha_n)} \left\{ e^{z\alpha_n} \operatorname{erfc}\left[\alpha_n\sqrt{(\kappa t)}+\frac{z}{2\sqrt{(\kappa t)}}\right] - \right.$$

$$\left. - e^{-z\alpha_n} \operatorname{erfc}\left[\alpha_n\sqrt{(\kappa t)}-\frac{z}{2\sqrt{(\kappa t)}}\right] \right\}. \quad (17)$$

Here we have used 8.3 (33) in the usual way, and Appendix V (19) and 12.2 (6) to find v from $\bar{v}$. The α_n are the positive roots of (15).

VII. *The semi-infinite cylinder $z > 0$, $0 \leqslant r < a$. Zero initial temperature. The surface $r = a$ maintained at constant temperature V for $t > 0$, and the surface $z = 0$ at zero*

We could use § 8.3 XIII, as above, but the following procedure is simpler. It is easy to verify that

$$\bar{v} = V\frac{I_0(qr)}{pI_0(qa)} - \frac{2V}{a} \sum_{m=1}^{\infty} \frac{\alpha_m J_0(r\alpha_m)}{p(q^2+\alpha_m^2)J_1(a\alpha_m)} e^{-z(\alpha_m^2+q^2)^{\frac{1}{2}}}, \quad (18)$$

where $q = \sqrt{(p/\kappa)}$, and the α_m are the roots of $J_0(a\alpha) = 0$, satisfies the subsidiary equation and boundary conditions.

It follows, using Appendix V (19), that

$$v = V - \frac{V}{a} \sum_{m=1}^{\infty} \frac{J_0(r\alpha_m)}{\alpha_m J_1(a\alpha_m)} \left\{ 2e^{-\kappa\alpha_m^2 t} \operatorname{erf}\frac{z}{2\sqrt{(\kappa t)}} + e^{z\alpha_m} \operatorname{erfc}\left[\frac{z}{2\sqrt{(\kappa t)}}+\alpha_m\sqrt{(\kappa t)}\right] + \right.$$

$$\left. + e^{-z\alpha_m} \operatorname{erfc}\left[\frac{z}{2\sqrt{(\kappa t)}}-\alpha_m\sqrt{(\kappa t)}\right] \right\}. \quad (19)$$

The first term of (18) is the value of $\bar{v}$ for the infinite cylinder with surface temperature V, so that the series in (18) may be regarded as an end correction. This device† is useful when surfaces extending to infinity are maintained at constant temperature so that Fourier's integral theorem is not applicable, though, in fact, results obtained by its use are usually correct.

VIII. *The wedge‡ $r > 0$, $0 < \theta < \theta_0$. Zero initial temperature. The surface $\theta = 0$ kept at V, and $\theta = \theta_0$ at zero, for $t > 0$*

In this case the subsidiary equation is

$$\frac{\partial^2 \bar{v}}{\partial r^2} + \frac{1}{r}\frac{\partial \bar{v}}{\partial r} + \frac{1}{r^2}\frac{\partial^2 \bar{v}}{\partial \theta^2} - q^2\bar{v} = 0, \quad 0 < \theta < \theta_0, \quad r > 0. \quad (20)$$

† Results for the region $z > 0, r > a$, and $r > a, 0 < z < l$, as well as the semi-infinite cylinder with constant surface temperature, have been given by Jaeger, *Bull. Amer. Math. Soc.* **47** (1941) 734.

‡ Bock, *J. Phys. Rad.* **10** (1939) 241. Jaeger, *Phil. Mag.* (7) **33** (1942) 527, gives some numerical results for this problem, and for the cylinder whose cross-section is a sector of a circle. The corresponding problem for the semi-infinite cylinder is discussed by Craggs, *Phil. Mag.* (7) **36** (1945) 220.

We seek a solution of (20) of the form

$$\bar{v} = \frac{V\theta}{p\theta_0} + \sum_{n=1}^{\infty} \sin s\theta \int_0^{\infty} f_n(u) J_s(ur) \, du, \tag{21}$$

where $s = n\pi/\theta_0$. This consists of a term corresponding to the steady state solution, and a series of terms which vanish on both boundaries $\theta = 0$ and $\theta = \theta_0$. The unknown function $f_n(u)$ is to be chosen to make (21) satisfy (20). Substituting (21) in (20), and using the result ($W.B.F.$, § 13.24 (1))

$$\int_0^{\infty} \frac{J_s(ur) \, du}{u} = \frac{1}{s},$$

we find

$$f_n(u) = \frac{2(-1)^n V}{\kappa u \theta_0 (q^2 + u^2)}.$$

Thus, finally,

$$\bar{v} = \frac{V\theta}{p\theta_0} + V \sum_{n=1}^{\infty} \frac{2(-1)^n}{\kappa \theta_0} \sin s\theta \int_0^{\infty} \frac{J_s(ur) \, du}{u(q^2 + u^2)}, \tag{22}$$

and

$$\frac{v}{V} = \frac{\theta}{\theta_0} + \frac{2}{\theta_0} \sum_{n=1}^{\infty} (-1)^n \sin s\theta \int_0^{\infty} e^{-\kappa u^2 t} \frac{J_s(ur) \, du}{u}. \tag{23}$$

The integrals in (23) may be expressed as confluent hypergeometric functions by $W.B.F.$, § 13.3 (2).

If both faces of the wedge are kept at unity for $t > 0$ and the initial temperature is zero, we get

$$v = 1 - \frac{4}{\theta_0} \sum_{n=0}^{\infty} \sin \frac{(2n+1)\pi\theta}{\theta_0} \int_0^{\infty} e^{-\beta u^2} \frac{J_s(u) \, du}{u}, \tag{24}$$

where $s = (2n+1)\pi/\theta_0$, and $\beta = \kappa t/r^2$.

IX. *The cone $0 < \theta < \theta_0$. Zero initial temperature. The surface kept at unity for $t > 0$*

Using the method of VIII we find

$$v = 1 + \sqrt{2} \sum_n \frac{(2n+1)\Gamma[\frac{1}{2}(n+1)]P_n(\mu)}{n\Gamma(\frac{1}{2}n)[dP_n(\mu)/dn]_{\mu=\mu_0}} \int_0^{\infty} e^{-\kappa u^2 t/r^2} \frac{J_{n+\frac{1}{2}}(u) \, du}{u^{\frac{1}{2}}}, \tag{25}$$

where the n are the positive roots of $P_n(\mu_0) = 0$.

XVI

STEADY TEMPERATURE

16.1. Introductory

PROBLEMS on steady temperature are of great practical importance: not only do most methods of measuring thermal conductivity employ the steady state, but in industrial applications it is the steady flow which is of the greatest interest. Unfortunately the regions in which the flow of heat is important in practice are rarely of the simple shapes discussed earlier; in this chapter we discuss other methods which give some information about more complicated systems.

As remarked in § 1.6 the steady temperature v satisfies Laplace's equation
$$\nabla^2 v = 0 \tag{1}$$
in a region in which no heat is generated. If heat is generated at a rate $A(x, y, z)$ per unit time per unit volume of the region the steady temperature v must satisfy Poisson's equation
$$\nabla^2 v = -\frac{1}{K} A(x, y, z), \tag{2}$$
provided, of course, that the medium is homogeneous and isotropic.

Thus the solutions of problems on steady flow of heat may be inferred from known results in Potential Theory, Electrostatics, Hydrodynamics, Flow of Electric Current, and other subjects in which these equations occur.

16.2. Sources and sinks in steady temperature

We suppose flow of heat in an infinite solid to be caused by steady supply of heat at certain points and its withdrawal at others. These points may be called Sources and Sinks of Heat.

In this case, if we describe a small sphere of radius r about a point at which heat is being supplied, when $r \to 0$ the rate at which heat flows out over the sphere must be equal to the rate at which heat is supplied at the source. Hence the solution of the equation
$$\nabla^2 v = 0$$
must take the form
$$\frac{Q}{4\pi r K} + \phi, \tag{1}$$

where ϕ is a solution of 16.1 (1) which remains finite at the source, and Q is the quantity of heat introduced there per unit time.

Similarly for a line source in which heat is introduced at the constant rate Q per unit length of the line per unit time, the part of v which tends to infinity as the distance r' from the line tends to zero is

$$-\frac{Q}{2\pi K}\ln r'. \tag{2}$$

If heat is being emitted steadily throughout a finite portion of an infinite solid, the temperature at any point is obtained by integration of (1) or (2).

If heat is produced in a bounded region whose surface is kept at zero (or is thermally insulated) the method of images is available, or more generally the Green's function of potential theory may be used.

The Green's function of potential theory for a point source at (x', y', z') in a given region with given boundary conditions is defined as the solution $u(x, y, z; x', y', z')$ of Laplace's equation which satisfies the boundary conditions and is finite within the region except at the point (x', y', z') where it is infinite in such a way that

$$\lim_{r\to 0}\left\{u-\frac{1}{r}\right\}$$

is finite, where $\qquad r^2 = (x-x')^2+(y-y')^2+(z-z')^2.$

Thus if u is the Green's function as defined above, the temperature due to a point source at (x', y', z') which emits Q units of heat per unit time is

$$\frac{Q}{4\pi K}u. \tag{3}$$

These Green's functions are well known† and may be found in works on potential theory. Here we remark only that for the regions considered in Chapter XIV they can be written down from the results of that chapter. Referring, for example, to § 14.10, it appears that, if we put $q = 0$, the $\bar{v}$ of that section satisfies Laplace's equation and the boundary conditions, and near the point (x', y', z') behaves like

$$\frac{1}{4\pi\kappa\sqrt{\{(x-x')^2+(y-y')^2+(z-z')^2\}}}.$$

Thus the Green's function u defined above is just

$$4\pi\kappa[\bar{v}]_{q=0}, \tag{4}$$

where $[\bar{v}]_{q=0}$ is the value when $q = 0$ of the Laplace transform of the

† For surfaces of the cylindrical coordinate system a full account, including alternative forms of the solutions, is given in *G. and M.*, Chap. IX.

solution of Chapter XIV for a unit instantaneous point source in the same region and with the same boundary conditions.

Thus, for example, if the planes $z = 0$ and $z = l$ are kept at zero temperature, and Q units of heat are emitted per second at the point whose cylindrical coordinates are (r', θ', z'), the temperature at (r, θ, z) is, by 14.10 (11),

$$\frac{Q}{\pi l K} \sum_{m=1}^{\infty} \sin\frac{m\pi z}{l} \sin\frac{m\pi z'}{l} K_0\left(\frac{m\pi R}{l}\right), \tag{5}$$

where

$$R^2 = r^2 + r'^2 - 2rr'\cos(\theta - \theta'). \tag{6}$$

Similarly, if the cylinder $r = a$ is kept at zero temperature and Q units of heat are emitted per second at the point $(r', \theta', 0)$ within it, the temperature at (r, θ, z) is, using 14.13 (5),

$$\frac{Q}{2\pi a^2 K} \sum_{n=-\infty}^{\infty} \cos n(\theta - \theta') \sum_{\alpha} \frac{e^{-\alpha|z|} J_n(\alpha r) J_n(\alpha r')}{\alpha [J'_n(\alpha a)]^2}, \tag{7}$$

where the summation in α is over the positive roots of $J_n(\alpha a) = 0$.

The corresponding result for a source at (r', θ', z') in the finite cylinder $0 < z < l$, $0 < r < a$, whose surface is kept at zero, is, using 14.15 (1),

$$\frac{Q}{\pi l K} \sum_{m=1}^{\infty} \sin\frac{m\pi z}{l} \sin\frac{m\pi z'}{l} \sum_{n=0}^{\infty} \epsilon_n \frac{I_n(m\pi r/l)}{I_n(m\pi a/l)} F_n\left(\frac{m\pi a}{l}; \frac{m\pi r'}{l}\right) \cos n(\theta - \theta'),$$
$$0 < r < r', \tag{8}$$

where

$$F_n(x, y) = I_n(x)K_n(y) - K_n(x)I_n(y), \tag{9}$$

and $\quad \epsilon_n = 1$, if $n = 0$, and $\quad \epsilon_n = 2$, if $n = 1, 2, 3, \dots$.

If $r' < r < a$ we interchange r and r' in (8).

As an example of the use of these solutions we determine the temperature in the cylinder $0 \leqslant r < a$, $0 < z < l$, whose surface is kept at zero, and in which there is steady emission of heat at the rate Q per unit length per unit time along a line parallel to the axis of the cylinder running between the points (r', θ', b) and $(r', \theta', l-b)$. This temperature is just

$$\int_b^{l-b} u\, dz',$$

where u is given by (8). Therefore, for $0 < r < r'$, it is

$$\frac{2Q}{K\pi^2} \sum_{m=1}^{\infty} \frac{1}{(2m-1)} \cos\frac{(2m-1)\pi b}{l} \sin\frac{(2m-1)\pi z}{l} \times$$

$$\times \sum_{n=0}^{\infty} \frac{I_n[(2m-1)\pi r/l]}{I_n[(2m-1)\pi a/l]} F_n\left\{\frac{(2m-1)\pi a}{l}; \frac{(2m-1)\pi r'}{l}\right\} \epsilon_n \cos n(\theta - \theta').$$

This is the temperature in a cylinder heated by a wire parallel to its axis.†

16.3. Steady flow to a nearly plane surface. Topographic corrections for the geothermal flux

As remarked in § 2.13 there is an approximately linear increase in temperature with depth below the Earth's surface which is modified by local irregularities of the surface. It is of considerable geophysical interest‡ to determine the effect of such irregularities and to correct observed temperature gradients for them.

Two methods of approach have been used: in the first of these,§ the surface of the Earth at sea-level is considered and its temperature at each point assumed to be $h(g-g')$, where h is the height above sea-level at that point, g is the geothermal gradient, and g' is the adiabatic lapse rate in the atmosphere, so that the mean surface temperature at height h above sea-level is less than that at sea-level by approximately hg'.

The problem then becomes that of *steady temperature in a semi-infinite solid* $z > 0$ *whose surface temperature is* $F(x,y)$, determined as above. Letting $t \to \infty$ in 14.9 (3) gives|| for the temperature v at $(0, 0, z)$

$$v = \frac{1}{2\pi} \int\limits_{-\infty}^{\infty} \int\limits_{-\infty}^{\infty} \frac{zF(x',y')\,dx'dy'}{R^3}, \tag{1}$$

where

$$R^2 = z^2+r'^2, \quad \text{and} \quad r'^2 = x'^2+y'^2. \tag{2}$$

In the present connexion we require the value of $\partial v/\partial z$ which is

$$\frac{\partial v}{\partial z} = \frac{1}{2\pi} \int\limits_{-\infty}^{\infty} \int\limits_{-\infty}^{\infty} \frac{(R^2-3z^2)}{R^5} F(x',y')\,dx'dy'. \tag{3}$$

Putting $F(x',y')$ into polar coordinates (r', θ') and writing

$$V = \frac{1}{2\pi} \int\limits_{0}^{\infty} F(r'\cos\theta', r'\sin\theta')\,d\theta' \tag{4}$$

for the mean value of $F(x',y')$ round a circle of radius r' about the origin, (3) becomes

$$\frac{\partial v}{\partial z} = \int\limits_{0}^{\infty} R^{-5}(R^2-3z^2)Vr'\,dr' = \int\limits_{z}^{\infty} R^{-4}(R^2-3z^2)V\,dR$$

$$= \left[V\left(\frac{z^2}{R^3}-\frac{1}{R}\right)\right]_z^{\infty} + \int\limits_{z}^{\infty} \left(\frac{1}{R}-\frac{z^2}{R^3}\right)\frac{dV}{dR}\,dR$$

$$= \int\limits_{z}^{\infty} \frac{r'^2}{R^3}\frac{dV}{dR}\,dR = \int\limits_{0}^{\infty} \frac{r'^2}{R^3}\frac{dV}{dr'}\,dr'. \tag{5}$$

---

† Lees, *Phil. Trans. Roy. Soc.* A, **204** (1905) 433.

‡ A full account, including an historical survey, is given by Birch, *Bull. Geol. Soc. Amer.* **61** (1950) 567–630.

§ Jeffreys, *Mon. Not. R. Astr. Soc. Geophys. Suppl.* **4** (1940) 309–12; Bullard, ibid. **4** (1940) 360–2.

|| Birch, loc. cit., generalizes this to include linear variation of surface temperature caused by erosion or uplift.

When $z = 0$ so that $R = r'$, this becomes

$$\left[\frac{\partial v}{\partial z}\right]_{z=0} = \int_0^\infty \frac{dV}{dr'} \frac{dr'}{r'}, \tag{6}$$

which is easily evaluated for any terrain. This quantity must be subtracted from the observed value of g to get the true value of the geothermal gradient.

Earlier writers have made exact calculations of the isothermals and temperature gradient for various nearly-plane surfaces whose form approximates to that of common surface features.[†] Considering the two-dimensional case only and taking the z-axis vertically downwards, it is necessary to find a solution of Laplace's equation

$$\frac{\partial^2 v}{\partial x^2} + \frac{\partial^2 v}{\partial z^2} = 0, \tag{7}$$

such that $v \to gz$ as $z \to \infty$, and that

$$v = V_0 + g'z \text{ on the surface } z = f(x), \tag{8}$$

where, as before, g is the geothermal gradient and g' the adiabatic lapse rate. The most interesting of these calculations is that of Lees,[‡] who noted that

$$v = V_0 + gz + A\frac{d}{dz}\ln[x^2 + (z+a)^2]^{\frac{1}{2}} = V_0 + gz + \frac{A(z+a)}{x^2 + (z+a)^2} \tag{9}$$

satisfies (7), behaves as required as $z \to \infty$, and satisfies (8) if the surface of the solid is

$$(g'-g)z = \frac{A(z+a)}{x^2 + (z+a)^2}. \tag{10}$$

The surface (10) has a minimum at $x = 0$, and tends monotonically to zero as $x \to \infty$; it provides a reasonable representation of a single mountain chain. If H is the height of the mountain and $2b$ its width at half its height, so that (10) passes through $(0, -H)$ and $(b, -\frac{1}{2}H)$, the parameters A and a are given by

$$a = H + (\tfrac{1}{4}H^2 + b^2)^{\frac{1}{2}}, \qquad A = (g - g')H(\tfrac{1}{4}H^2 + b^2)^{\frac{1}{2}}. \tag{11}$$

The isothermals are then given by (9) with various values of v, and the temperature gradient is easily calculated.

16.4. Steady flow in composite media

The problem of the disturbance of steady linear flow of heat in a uniform medium by an object of different conductivity buried in it is of considerable technical importance. Mathematically, it is precisely the same as that of induced magnetization in a body of the same shape placed in a uniform external field, and solutions will be found in text-books on electricity and magnetism, but, because of their importance, the principal ones are given briefly here. The solutions for spheres and ellipsoids may be used to estimate the modifications to the geothermal gradient caused by a buried mass of different conductivity and are thus

[†] Volterra, *Nuovo Cim.* (6) **4** (1912) 111–26, Thoma, *Diss. Karlsruhe* (1906), and Andreae, *Ann. Ponts et Chaussées* **128** (1958) 37, discuss a periodic hill and valley system.

[‡] Lees, *Proc. Roy. Soc.* A, **83** (1910) 339–46. The last term in (9) is the temperature due to a line doublet. A series of type $\sum A_n(d/dz)^n \ln[x^2 + (z+a_n)^2]$ may also be used, and an additional series with x replaced by $x+c$ may be added to give a representation of parallel mountain ranges.

fundamental for thermal methods of prospecting. Also, the exact result for a single sphere or ellipsoid is used statistically in calculations of the thermal conductivity of granular materials, these being regarded as a number of particles of one material embedded in a matrix of another; a simple example of the method is given in IV below.

The study of the behaviour of the flux and temperature near a plane contact between two materials is of importance in connexion with the geothermal flux. Idealized problems with rectangular boundaries can be treated by the methods of Chapter V, see V below. Simple polynomial solutions of Laplace's equation have also been used, cf. VI.

I. *The region within the sphere $0 \leqslant r < a$ is of conductivity K' and the region outside of conductivity K. The temperature v tends to Vz at great distances*

We assume for the temperatures v' and v, within and outside the sphere $r = a$, the forms

$$v = Vr\cos\theta + \frac{B}{r^2}\cos\theta, \tag{1}$$

$$v' = Ar\cos\theta, \tag{2}$$

where A and B are unknown constants and r, θ are spherical polar coordinates. These satisfy Laplace's equation, v' is finite as $r \to 0$, and $v \to Vz$ as $r \to \infty$, as required. The boundary conditions at $r = a$, namely

$$v = v', \qquad K'\frac{\partial v'}{\partial r} = K\frac{\partial v}{\partial r}, \qquad r = a, \qquad 0 \leqslant \theta \leqslant \pi,$$

give
$$Va^3 + B = Aa^3,$$
$$K(Va^3 - 2B) = K'Aa^3.$$

Solving for A and B, we get finally

$$v = Vr\cos\theta + \frac{Va^3(K-K')\cos\theta}{r^2(2K+K')}, \tag{3}$$

$$v' = \frac{3KVr\cos\theta}{2K+K'} = \frac{3KVz}{2K+K'}. \tag{4}$$

The temperature gradient in the sphere is $3KV/(2K+K')$.

II. *The cylinder $0 \leqslant r < a$ is of conductivity K' and has its axis perpendicular to the z-axis. The region outside the cylinder is of conductivity K and the temperature in it is Vz at great distances*

The temperatures v' inside, and v outside, the cylinder are

$$v' = \frac{2VKz}{K+K'}, \tag{5}$$

$$v = Vz - \frac{(K'-K)Va^2z}{(K'+K)r^2}. \tag{6}$$

III. *An ellipsoid of conductivity K' in a medium of conductivity K*

Suppose the ellipsoid is
$$\frac{x^2}{a^2}+\frac{y^2}{b^2}+\frac{z^2}{c^2} = 1, \tag{7}$$

and that the temperature at large distances from it tends to the value
$$V_1 x + V_2 y + V_3 z. \tag{8}$$

For any point (x, y, z), let λ be the positive root of
$$\frac{x^2}{a^2+\lambda}+\frac{y^2}{b^2+\lambda}+\frac{z^2}{c^2+\lambda} = 1. \tag{9}$$

Then it is known† that xA_λ, yB_λ, zC_λ, where
$$A_\lambda = \tfrac{1}{2}abc \int_\lambda^\infty \frac{du}{(a^2+u)\Delta(u)}, \qquad B_\lambda = \tfrac{1}{2}abc \int_\lambda^\infty \frac{du}{(b^2+u)\Delta(u)},$$

$$C_\lambda = \tfrac{1}{2}abc \int_\lambda^\infty \frac{du}{(c^2+u)\Delta(u)}, \tag{10}$$

$$\Delta(u) = [(a^2+u)(b^2+u)(c^2+u)]^{\frac{1}{2}}, \tag{11}$$

all satisfy Laplace's equation. Then, by arguments similar to those in I, it is found that the temperatures v_i inside and v_0 outside the ellipsoid are
$$v_i = \frac{V_1 x}{1+A_0(\epsilon-1)}+\frac{V_2 y}{1+B_0(\epsilon-1)}+\frac{V_3 z}{1+C_0(\epsilon-1)}, \tag{12}$$

$$v_0 = V_1 x + V_2 y + V_3 z - \frac{(\epsilon-1)V_1 A_\lambda x}{1+A_0(\epsilon-1)}-\frac{(\epsilon-1)V_2 B_\lambda y}{1+B_0(\epsilon-1)}-\frac{(\epsilon-1)V_3 C_\lambda z}{1+C_0(\epsilon-1)}, \tag{13}$$

where
$$\epsilon = K'/K, \tag{14}$$

and A_0, B_0, C_0 are the integrals A_λ, B_λ, C_λ with $\lambda = 0$. It may be noted that
$$A_0+B_0+C_0 = 1. \tag{15}$$

For the various spheroids in which two axes are equal, the integrals (10) can be expressed in terms of elementary functions. The results are as follows:‡

Prolate spheroid, $b = c < a$; $e' = [(a^2-b^2)/(a^2+\lambda)]^{\frac{1}{2}}$ is the eccentricity of the confocal ellipse through the external point considered, and e' for $\lambda = 0$ is e, the eccentricity of the generating ellipse.
$$A_\lambda = \frac{(1-e^2)}{e^3}\left\{\tfrac{1}{2}\ln\frac{1+e'}{1-e'}-e'\right\}, \tag{16}$$

$$B_\lambda = C_\lambda = \frac{(1-e^2)}{2e^3}\left\{\frac{e'}{1-e'^2}-\tfrac{1}{2}\ln\frac{1+e'}{1-e'}\right\}. \tag{17}$$

Oblate spheroid, $a = b > c$.
$$A_\lambda = B_\lambda = \frac{(1-e^2)^{\frac{1}{2}}}{2e^3}\left\{\cot^{-1}\nu-\frac{\nu}{\nu^2+1}\right\}, \tag{18}$$

$$C_\lambda = \frac{(1-e^2)^{\frac{1}{2}}}{e^3}\left\{\frac{1}{\nu}-\cot^{-1}\nu\right\}. \tag{19}$$

---

† The integrals are discussed in most works on hydrodynamics and electricity. Cf. Livens, *Theory of Electricity* (edn. 1, 1918) § 117. The analysis goes back at least to Lamé, *Liouville's J.* **2** (1837); cf. also Maxwell, *Electricity and Magnetism* (edn. 2) Vol. 2, § 437.

‡ For the evaluation, see Besant and Ramsey, *Hydrodynamics* (edn. 2, 1920) p. 169.

where $\qquad \nu = [(c^2+\lambda)/(a^2-c^2)]^{\frac{1}{2}} = (1-e'^2)^{\frac{1}{2}}/e'.$

As a simple example, consider a long prolate spheroid $a \gg b$ (which might be a thermal conductivity probe or a piece of well-casing), of conductivity K', with its axis in the direction of the temperature gradient in material of conductivity K. The temperature gradient in the spheroid is, by (12), $V_1/[1+A_0(K'-K)/K]$, also, from (16), since b/a is small we have approximately

$$A_0 = \frac{b^2}{a^2}\Big\{\ln\frac{2a}{b}-1\Big\}.$$

Then the temperature gradient within the spheroid is approximately

$$V_1\Big\{1-\frac{b^2(K'-K)}{Ka^2}\Big(\ln\frac{2a}{b}-1\Big)\Big\}. \tag{20}$$

IV. *The thermal conductivity of a simple granular medium*

Suppose that the material consists of a fraction α by volume of spheres of conductivity K' embedded in a matrix of conductivity K. It is assumed that the spheres are so far apart as to have no influence on one another. Suppose that the spheres are of radius a, and consider a larger sphere of radius b containing n of them so that $na^3 = \alpha b^3$. By (3), the temperature at great distances due to the n spheres in a linear temperature gradient is

$$Vz+\frac{na^3(K-K')}{r^3(2K+K')}Vz,$$

while, if K_{av} is the average conductivity of the material in the sphere of radius b, this must also be

$$Vz+\frac{b^3(K-K_{\mathrm{av}})}{r^3(2K+K_{\mathrm{av}})}Vz.$$

Equating these two expressions gives†

$$K_{\mathrm{av}} = \frac{3KK'\alpha+(2K+K')K(1-\alpha)}{3K\alpha+(2K+K')(1-\alpha)}. \tag{21}$$

V. *Steady flow in the composite infinite strip* $0 < y < l$, *of which* $x > 0$ *is of conductivity* K_1 *and* $x < 0$ *is of conductivity* K_2. *Boundary conditions*

$$v = 0, \qquad y = 0, \qquad -\infty < x < \infty,$$
$$v = V_1, \qquad y = l, \qquad x > 0,$$
$$v = V_2, \qquad y = l, \qquad x < 0.$$

If v_1 and v_2 are the temperatures for $x > 0$ and $x < 0$, respectively, it follows as in § 5.2 that

$$lv_1 = yV_1-\frac{2l(V_2-V_1)K_2}{\pi(K_1+K_2)}\sum_{n=1}^{\infty}\frac{(-1)^n}{n}\sin\frac{n\pi y}{l}e^{-n\pi x/l}, \tag{22}$$

$$lv_2 = yV_2+\frac{2l(V_2-V_1)K_1}{\pi(K_1+K_2)}\sum_{n=1}^{\infty}\frac{(-1)^n}{n}\sin\frac{n\pi y}{l}e^{n\pi x/l}. \tag{23}$$

† This simple discussion derives from Maxwell, *Electricity and Magnetism*, Vol. 1, § 314. More elaborate theories have been given by many writers, in particular, Lord Rayleigh, *Phil. Mag.* (5) 34 (1892) 481–502, who considers a cubic lattice of spheres. For a review and references see de Vries, 'The thermal conductivity of granular materials' (Inst. International du Froid, Paris).

If V_1 and V_2 are such that $K_1 V_1 = K_2 V_2 = -Fl$, so that the flux in both regions tends to F as $|x| \to \infty$, the flux across the plane $y = 0$ is

$$F \mp \frac{2(K_1 - K_2)F}{(K_1 + K_2)} \sum_{n=1}^{\infty} (-1)^n e^{-n\pi|x|/l}, \tag{24}$$

the positive sign being taken if $x < 0$, and the negative sign if $x > 0$. This shows the disturbance in the flux near the boundary of the two media.

VI. *Polynomial solutions*

It is known that polynomials $ax+by$, $ax+by+cxy+d(y^2-x^2)$,... of the first, second,... degrees satisfy Laplace's equation. These may be useful in simple problems on composite media. For example

$$v_1 = y(x+a), \qquad v_2 = y\{(K_1 x/K_2) + a\} \tag{25}$$

satisfy Laplace's equation and the following conditions:

$$v_1 = v_2 = 0, \quad \text{when } y = 0;$$

$$v_1 = v_2, \qquad K_1 \frac{\partial v_1}{\partial x} = K_2 \frac{\partial v_2}{\partial x}, \quad \text{when } x = 0, y > 0.$$

They thus provide an elementary solution for the composite region in which the material to the right of $x = 0$, $y > 0$ has conductivity K_1, and that to the left has conductivity K_2. The isothermals are arcs of rectangular hyperbolas.†

16.5. Practical problems

As remarked in § 16.1 the regions in which steady flow takes place in engineering or laboratory problems are more complicated than those hitherto considered.

Simple examples of regions of practical importance which have not yet been studied in this book are the following:

(i) Flow between a cylinder and a plane, for example between a buried pipe or cable and the surface of the ground.

(ii) Flow between two eccentric cylinders.

(iii) Flow from a grid of pipes.

(iv) Flow between two strips in an infinite medium.

(v) Flow through the walls of a furnace.

Of these the first three have geometrically simple boundaries so that a complete theoretical solution might be hoped for. The last two are more complicated.

A great deal of attention has recently been paid to the development of numerical methods for the solution of such problems; for them, see Chapter XVIII.

We remark first that, if the surfaces are isothermal, the rate of flow

† This method has been developed by Sbrana and Bossolasco, *Geofis. Pura é Appl.* **23** (1952) 3–8, for studying isothermals near the Earth's surface.

of heat between them follows immediately from a theoretical or experimental determination of the electrostatic capacity of the surfaces.[†] Suppose heat flows from a surface S_1 at v_1 to a surface S_2 at v_2 through medium of conductivity K. Then, defining the thermal resistance R of the system as the temperature difference between the surfaces divided by the heat transferred between them per unit time, we find

$$R = \frac{v_1 - v_2}{K \iint\limits_{S_2} (\partial v / \partial n) \, dS_2}, \tag{1}$$

where $\partial/\partial n$ denotes differentiation along the normal to S_2 drawn into the medium.

The temperature v in this case, being a solution of Laplace's equation which takes the values v_1 and v_2 on the boundaries, gives the electrostatic potential if the surfaces S_1 and S_2 are separated by material of unit dielectric constant and charged to potentials v_1 and v_2. The capacity of the condenser so formed is

$$C = \frac{1}{4\pi(v_1 - v_2)} \iint\limits_{S_2} \frac{\partial v}{\partial n} \, dS_2. \tag{2}$$

Comparing (1) and (2) it follows that

$$R = \frac{1}{4\pi K C}. \tag{3}$$

In the same way, if the space between the two surfaces is filled with electrolyte and the electric current between them caused by a known potential difference is measured, the thermal resistance of the system can be inferred. Langmuir, Adams, and Meikle[‡] used this method to determine empirical formulae for the thermal resistance of a furnace wall of thickness l surrounding a rectangular parallelepiped of sides $2a$, $2b$, $2c$, for the important case in which l is of the same order as a, b, and c.

Three-dimensional problems such as this are quite intractable. When they are further simplified to two dimensions, the method of conformal representation which has proved so important in other fields is useful. In the present connexion this has applications of three different types. Firstly, it provides exact solutions of simple problems such as (i) and (ii) above; secondly, it gives approximate solutions of some problems of type (iii), in the sense that it gives exact solutions for gratings of

† For some simple systems see Rudenberg, *Elektrotech. Z.* **46** (1925) 1342.
‡ *Trans. Amer. Electrochem. Soc.* **24** (1913) 53. A brief account is given in McAdams, *Heat Transmission* (edn. 2, 1942) p. 14.

oval curves which are nearly, but not accurately, circular; finally, it may be used to study certain simple types of region which occur very frequently in practice, for example, a right-angled bend in a wall, a change in thickness of a wall, a guard ring, etc.

This method is discussed in detail in text-books on electricity and hydrodynamics. Here we give a brief introduction *ab initio*, applying the method first to problems of steady flow of heat when the surface temperatures are arbitrary functions of position, and subsequently to the simpler case of flow between isothermal surfaces.

16.6. The use of conjugate functions in problems of steady temperature

Let ξ, η be real functions of x and y such that

$$\xi + i\eta = f(x+iy) = f(z).$$

Then ξ, η are called conjugate functions of x and y. Also we have

$$\frac{\partial \xi}{\partial x} + i\frac{\partial \eta}{\partial x} = f'(z),$$

$$\frac{\partial \xi}{\partial y} + i\frac{\partial \eta}{\partial y} = if'(z).$$

Therefore

$$\frac{\partial \xi}{\partial x} = \frac{\partial \eta}{\partial y}, \tag{1}$$

$$\frac{\partial \eta}{\partial x} = -\frac{\partial \xi}{\partial y}. \tag{2}$$

It follows that the curves $\xi = $ constant and $\eta = $ constant are orthogonal.

Again, since

$$\frac{\partial^2 \xi}{\partial x^2} = \frac{\partial^2 \eta}{\partial x \partial y},$$

and

$$\frac{\partial^2 \xi}{\partial y^2} = -\frac{\partial^2 \eta}{\partial x \partial y},$$

it follows that

$$\frac{\partial^2 \xi}{\partial x^2} + \frac{\partial^2 \xi}{\partial y^2} = 0, \tag{3}$$

and similarly

$$\frac{\partial^2 \eta}{\partial x^2} + \frac{\partial^2 \eta}{\partial y^2} = 0. \tag{4}$$

Further, if v is a function of ξ and η such that

$$\frac{\partial^2 v}{\partial \xi^2} + \frac{\partial^2 v}{\partial \eta^2} = 0, \tag{5}$$

we can show that
$$\frac{\partial^2 v}{\partial x^2} + \frac{\partial^2 v}{\partial y^2} = 0.$$

For
$$\frac{\partial v}{\partial x} = \frac{\partial v}{\partial \xi}\frac{\partial \xi}{\partial x} + \frac{\partial v}{\partial \eta}\frac{\partial \eta}{\partial x}$$

and
$$\frac{\partial^2 v}{\partial x^2} = \frac{\partial^2 v}{\partial \xi^2}\left(\frac{\partial \xi}{\partial x}\right)^2 + 2\frac{\partial^2 v}{\partial \xi \partial \eta}\frac{\partial \xi}{\partial x}\frac{\partial \eta}{\partial x} + \frac{\partial^2 v}{\partial \eta^2}\left(\frac{\partial \eta}{\partial x}\right)^2 + \frac{\partial v}{\partial \xi}\frac{\partial^2 \xi}{\partial x^2} + \frac{\partial v}{\partial \eta}\frac{\partial^2 \eta}{\partial x^2}.$$

Similarly
$$\frac{\partial^2 v}{\partial y^2} = \frac{\partial^2 v}{\partial \xi^2}\left(\frac{\partial \xi}{\partial y}\right)^2 + 2\frac{\partial^2 v}{\partial \xi \partial \eta}\frac{\partial \xi}{\partial y}\frac{\partial \eta}{\partial y} + \frac{\partial^2 v}{\partial \eta^2}\left(\frac{\partial \eta}{\partial y}\right)^2 + \frac{\partial v}{\partial \xi}\frac{\partial^2 \xi}{\partial y^2} + \frac{\partial v}{\partial \eta}\frac{\partial^2 \eta}{\partial y^2}.$$

Adding these two results, and using (1), (2), (3), (4), and (5), we see that
$$\frac{\partial^2 v}{\partial x^2} + \frac{\partial^2 v}{\partial y^2} = 0.$$

Thus, if we can obtain a solution of the equation
$$\frac{\partial^2 v}{\partial \xi^2} + \frac{\partial^2 v}{\partial \eta^2} = 0$$

satisfying certain boundary conditions at the curves
$$\xi = \xi_1, \qquad \xi = \xi_2,$$
$$\eta = \eta_1, \qquad \eta = \eta_2,$$

this solution in the $\xi\eta$-plane may be transferred to the xy-plane, the boundaries being the curves in the xy-plane which correspond by the transformation
$$\xi + i\eta = f(x + iy)$$

to the curves $\xi = \xi_1$, etc., while the temperatures at these boundaries correspond to the temperatures at the boundaries in the $\xi\eta$-plane.

Suppose that we have the case of the rectangle in the $\xi\eta$-plane given by
$$\xi = \xi_1, \qquad \xi = \xi_2,$$
$$\eta = \eta_1, \qquad \eta = \eta_2,$$

and that
$$v = f_1(\eta) \text{ at } \xi = \xi_1 \quad (\eta_1 < \eta < \eta_2),$$
$$v = f_2(\eta) \text{ at } \xi = \xi_2 \quad (\eta_1 < \eta < \eta_2),$$
$$v = F_1(\xi) \text{ at } \eta = \eta_1 \quad (\xi_1 < \xi < \xi_2),$$
$$v = F_2(\xi) \text{ at } \eta = \eta_2 \quad (\xi_1 < \xi < \xi_2).$$

The solution of this problem is obtained by breaking it up into four cases, in each of which three of the boundaries are kept at zero temperature. In this way we find, as in § 5.3 I,

$$v = \sum_{1}^{\infty} \frac{a_n \sinh \dfrac{n\pi(\xi_2-\xi)}{(\eta_2-\eta_1)} + a'_n \sinh \dfrac{n\pi(\xi-\xi_1)}{(\eta_2-\eta_1)}}{\sinh \dfrac{n\pi(\xi_2-\xi_1)}{(\eta_2-\eta_1)}} \sin \frac{n\pi(\eta-\eta_1)}{(\eta_2-\eta_1)} +$$

$$+ \sum_{1}^{\infty} \frac{b_n \sinh \dfrac{n\pi(\eta_2-\eta)}{(\xi_2-\xi_1)} + b'_n \sinh \dfrac{n\pi(\eta-\eta_1)}{(\xi_2-\xi_1)}}{\sinh \dfrac{n\pi(\eta_2-\eta_1)}{(\xi_2-\xi_1)}} \sin \frac{n\pi(\xi-\xi_1)}{(\xi_2-\xi_1)},$$

where a_n, a'_n, b_n, and b'_n are the coefficients in the sine series,

$$f_1(\eta) = \sum_{1}^{\infty} a_n \sin \frac{n\pi(\eta-\eta_1)}{(\eta_2-\eta_1)}, \qquad f_2(\eta) = \sum_{1}^{\infty} a'_n \sin \frac{n\pi(\eta-\eta_1)}{(\eta_2-\eta_1)},$$

$$F_1(\xi) = \sum_{1}^{\infty} b_n \sin \frac{n\pi(\xi-\xi_1)}{(\xi_2-\xi_1)}, \qquad F_2(\xi) = \sum_{1}^{\infty} b'_n \sin \frac{n\pi(\xi-\xi_1)}{(\xi_2-\xi_1)}.$$

Substituting for ξ, η from the relation

$$\xi + i\eta = f(x+iy),$$

we have the temperature in the region bounded by the curves which correspond to $\xi = \xi_1$, etc., these curves being kept at the temperatures corresponding to $f_1(\eta)$, etc.

16.7. Applications of this method†

I. *The sector of a circle*

Consider the transformation

$$\xi + i\eta = -\frac{i\pi}{\alpha} \ln \frac{x+iy}{a}. \tag{1}$$

In this case, writing $z = re^{i\theta}$,

$$\xi = \frac{\pi}{\alpha}\theta, \qquad \eta = \frac{\pi}{\alpha}\ln\left(\frac{a}{r}\right), \tag{2}$$

and the sector of radius a and angle α corresponds to the region

$$0 < \eta, \qquad 0 < \xi < \pi,$$

in the $\xi\eta$-plane.

Thus the equations

$$\frac{\partial^2 v}{\partial \xi^2} + \frac{\partial^2 v}{\partial \eta^2} = 0, \qquad 0 < \xi < \pi, \, 0 < \eta,$$

† Cf. Mathieu, *Cours de physique mathématique*, Chap. III; Kober, *Dictionary of Conformal Representations* (Dover, 1952).

$$v = 0, \quad \text{when } \xi = 0 \text{ and } \xi = \pi,$$

and
$$v = 1, \quad \text{when } \eta = 0,$$

lead to
$$\frac{\partial^2 v}{\partial x^2} + \frac{\partial^2 v}{\partial y^2} = 0 \quad \text{over the sector,}$$

$$v = 0, \quad \text{when } \theta = 0 \text{ and } \theta = \alpha,$$

and
$$v = 1, \quad \text{when } r = a.$$

These equations in (ξ, η) we have already discussed when dealing with the infinite rectangular solid, and their solution is, by 5.2 (11),

$$v = \frac{2}{\pi} \tan^{-1}\left(\frac{\sin \xi}{\sinh \eta}\right).$$

Therefore the temperature in the sector is given by

$$v = \frac{2}{\pi} \tan^{-1}\left\{\frac{\sin(\pi\theta/\alpha)}{\sinh[(\pi/\alpha)\ln(a/r)]}\right\}.$$

If the boundary $r = a$ had been kept at $v = f(\theta)$, the problem would have reduced to the solution of the equations

$$\frac{\partial^2 v}{\partial \xi^2} + \frac{\partial^2 v}{\partial \eta^2} = 0,$$

$$v = 0, \quad \text{when } \xi = 0 \text{ and } \xi = \pi,$$

and
$$v = f\left(\frac{\alpha\xi}{\pi}\right), \quad \text{when } \eta = 0,$$

and this has been discussed in § 5.2.

Similarly, if the surface $\theta = 0$ is maintained at prescribed temperature $f(r)$, the other faces being at zero, the solution follows from 5.2 (18).

In the same way (1) transforms the wedge $0 < \theta < \alpha$, $r > 0$, into the doubly infinite strip $-\infty < \eta < \infty$, $0 < \xi < \pi$, so that the solution for steady temperature in the wedge with arbitrary surface temperature follows immediately from 5.2 (19).

II. *The circle*

Consider the transformation

$$\xi + i\eta = -i \ln \frac{x + iy}{a}. \tag{3}$$

Then
$$\xi = \theta, \qquad \eta = \ln \frac{a}{r},$$

and the interior of the circle $r = a$ corresponds to the region

$$0 < \eta, \quad 0 < \xi < 2\pi$$

of the $\xi\eta$-plane.

Thus a solution $v(\xi, \eta)$ of the equations

$$\left.\begin{array}{c} \dfrac{\partial^2 v}{\partial \xi^2}+\dfrac{\partial^2 v}{\partial \eta^2} = 0 \quad \text{over this region,} \\[2mm] v = f(\xi), \quad \text{when } \eta = 0, \\[2mm] v(0, \eta) = v(2\pi, \eta), \quad \eta > 0, \end{array}\right\} \tag{4}$$

leads to

$$\left.\begin{array}{c} \dfrac{\partial^2 v}{\partial x^2}+\dfrac{\partial^2 v}{\partial y^2} = 0 \quad \text{in the circle,} \\[2mm] v = f(\theta), \quad \text{when } r = a. \end{array}\right\} \tag{5}$$

The solution of (4) is given by

$$v = \sum_0^\infty e^{-n\eta}(a_n \cos n\xi+b_n \sin n\xi),$$

where

$$a_0 = \frac{1}{2\pi} \int_0^{2\pi} f(\xi')\, d\xi',$$

$$a_n = \frac{1}{\pi} \int_0^{2\pi} f(\xi')\cos n\xi'\, d\xi',$$

$$b_n = \frac{1}{\pi} \int_0^{2\pi} f(\xi')\sin n\xi'\, d\xi'.$$

Thus we have

$$v = \frac{1}{2\pi} \int_0^{2\pi} \left\{1+2 \sum_1^\infty e^{-n\eta} \cos n(\xi-\xi')\right\} f(\xi')\, d\xi'$$

$$= \frac{1}{2\pi} \int_0^{2\pi} \frac{1-e^{-2\eta}}{1-2e^{-\eta} \cos(\xi-\xi')+e^{-2\eta}} f(\xi')\, d\xi'.$$

Therefore the temperature in the circle is given by

$$v = \frac{1}{2\pi} \int_0^{2\pi} f(\theta') \frac{a^2-r^2}{a^2-2ar \cos(\theta-\theta')+r^2}\, d\theta'. \tag{6}$$

The integral (6) is Poisson's integral, cf. *F.S.*, § 100.

III. *Two concentric circles*

This may be obtained from the same transformation. The solution can readily be found in the form

$$v = \sum_0^\infty \frac{\sinh n(\eta_2 - \eta)}{\sinh n(\eta_2 - \eta_1)} (a_n \cos n\xi + b_n \sin n\xi) +$$

$$+ \sum_0^\infty \frac{\sinh n(\eta - \eta_1)}{\sinh n(\eta_2 - \eta_1)} (a_n' \cos n\xi + b_n' \sin n\xi),$$

where

$$f_1(\xi) = \sum_0^\infty (a_n \cos n\xi + b_n \sin n\xi),$$

$$f_2(\xi) = \sum_0^\infty (a_n' \cos n\xi + b_n' \sin n\xi)$$

are the Fourier series for $f_1(\xi)$ and $f_2(\xi)$ in the interval 0 to 2π.

IV. *Two intersecting or non-intersecting circles*

Consider the transformation

$$\xi + i\eta = \ln \frac{x+1-iy}{x-1-iy}. \tag{7}$$

Then

$$\xi = \ln \frac{r_2}{r_1}, \qquad \eta = \theta_1 - \theta_2,$$

where r_1 and r_2 are the distances from the points $A(1, 0)$, $B(-1, 0)$, to the point $P(x, y)$, and θ_1, θ_2 are the angles AP and BP make with the positive direction of the axis of x, cf. Fig. 49.

Thus $\xi =$ constant represents the system of coaxal circles with A, B as limiting points, and $\eta =$ constant represents the system of circles passing through A, B, these two sets of curves, as in all cases of conjugate functions, being orthogonal. With this notation the xy-plane is given by $-\pi < \eta < \pi$ and $-\infty < \xi < \infty$; the lower side of the portion BA of the real axis is $\eta = -\pi$; the lines Ax and Bx' are $\eta = 0$; and the upper side of BA is $\eta = \pi$; the region $y > 0$ corresponds to positive values of η and the region $y < 0$ to negative values. Also A is the point $\xi = +\infty$, B is the point $\xi = -\infty$, and the line $y'Oy$ is $\xi = 0$.

We proceed to apply this transformation to several cases in which the region in the xy-plane is bounded by arcs of these circles.

(i) Consider the region bounded by

$$\xi = \xi_1 \quad \text{and} \quad \xi = \xi_2 \quad (0 < \eta < \pi),$$

$$\eta = 0 \quad \text{and} \quad \eta = \pi \quad (\xi_1 < \xi < \xi_2),$$

shown by the heavy lines in Fig. 49.

Let $\qquad v = 0 \qquad$ over $\quad \xi = \xi_2, \, \eta = 0, \text{ and } \eta = \pi,$

and $\qquad v = f(\eta) \quad$ over $\quad \xi = \xi_1.$

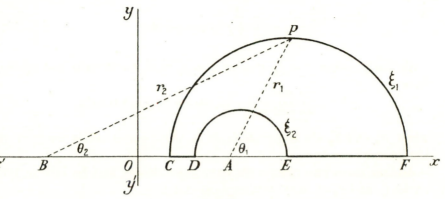

Fig. 49

Then we have

$$v = \sum_{1}^{\infty} a_n \frac{\sinh n(\xi_2 - \xi)}{\sinh n(\xi_2 - \xi_1)} \sin n\eta,$$

where $\qquad\qquad a_n = \frac{2}{\pi} \int_0^{\pi} f(\eta') \sin n\eta' \, d\eta'.$

It is easy to extend this solution to the case where

$$v = f_1(\eta) \quad \text{over} \quad \xi = \xi_1 \quad (0 < \eta < \pi),$$
$$v = f_2(\eta) \quad \text{over} \quad \xi = \xi_2 \quad (0 < \eta < \pi),$$
$$v = F_1(\xi) \quad \text{over} \quad \eta = 0 \quad (\xi_1 < \xi < \xi_2),$$
$$v = F_2(\xi) \quad \text{over} \quad \eta = \pi \quad (\xi_1 < \xi < \xi_2).$$

(ii) Consider the region bounded by the two complete circles $\xi = \xi_1$ and $\xi = \xi_2$ surrounding the limiting point A.

Let $v = f_1(\eta)$ over $\xi = \xi_1$, and $v = 0$ over $\xi = \xi_2$. Then the solution is obviously

$$v = \sum_{0}^{\infty} \frac{\sinh n(\xi_2 - \xi)}{\sinh n(\xi_2 - \xi_1)} (a_n \cos n\eta + b_n \sin n\eta), \qquad (8)$$

where

$$a_0 = \frac{1}{2\pi} \int_{-\pi}^{\pi} f_1(\eta') \, d\eta',$$

$$a_n = \frac{1}{\pi} \int_{-\pi}^{\pi} f_1(\eta') \cos n\eta' \, d\eta',$$

$$b_n = \frac{1}{\pi} \int_{-\pi}^{\pi} f_1(\eta') \sin n\eta' \, d\eta'.$$

Similarly, when $v = f_2(\eta)$ over $\xi = \xi_2$ and $v = 0$ over $\xi = \xi_1$, we have

$$v = \sum_{0}^{\infty} \frac{\sinh n(\xi - \xi_1)}{\sinh n(\xi_2 - \xi_1)} (a_n' \cos n\eta + b_n' \sin n\eta),$$

where a_n' and b_n' are the coefficients in the Fourier series for $f_2(\eta)$ in the interval $-\pi$ to π.

Adding these two results we have the solution for the case of the circles ξ_1 and ξ_2 at temperatures $f_1(\eta)$ and $f_2(\eta)$.

It is clear that if $f_1(\eta)$ and $f_2(\eta)$ are constant and equal to v_1 and v_2, respectively, we have only to solve the equations

$$\frac{\partial^2 v}{\partial \xi^2} = 0,$$

$$v = v_1, \quad \text{when } \xi = \xi_1,$$

$$v = v_2, \quad \text{when } \xi = \xi_2.$$

The solution is
$$v = v_1 \left(\frac{\xi_2 - \xi}{\xi_2 - \xi_1} \right) + v_2 \left(\frac{\xi - \xi_1}{\xi_2 - \xi_1} \right). \tag{9}$$

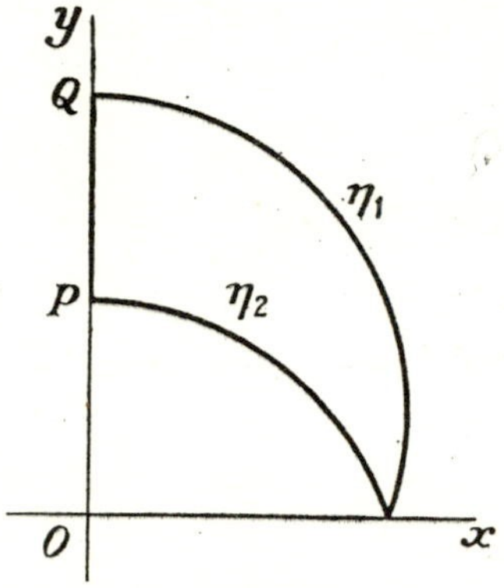

FIG. 50

Steady flow of heat between two given circles will be discussed again in § 16.9.

(iii) Consider the region bounded by

$$\eta = \eta_1 \quad (0 < \xi < \infty),$$
$$\eta = \eta_2 \quad (0 < \xi < \infty),$$
$$\xi = 0 \quad (\eta_1 < \eta < \eta_2),$$

as in Fig. 50.

Suppose first that

$$v = f_1(\eta) \quad \text{over} \quad \xi = 0 \quad (\eta_1 < \eta < \eta_2),$$
$$v = 0 \quad \text{over} \quad \eta = \eta_1 \quad (0 < \xi < \infty),$$
$$v = 0 \quad \text{over} \quad \eta = \eta_2 \quad (0 < \xi < \infty).$$

Then, as in § 5.2, the solution is

$$v = \sum_{n=1}^{\infty} a_n e^{-n\pi\xi/(\eta_2 - \eta_1)} \sin \frac{(\eta - \eta_1)n\pi}{(\eta_2 - \eta_1)}, \tag{10}$$

where
$$a_n = \frac{2}{\eta_2 - \eta_1} \int_{\eta_1}^{\eta_2} f_1(\eta') \sin \frac{(\eta' - \eta_1)n\pi}{(\eta_2 - \eta_1)} \, d\eta'.$$

Next suppose that

$$v = f_2(\xi) \quad \text{over} \quad \eta = \eta_1 \quad (0 < \xi < \infty),$$
$$v = 0 \quad \text{over} \quad \eta = \eta_2 \quad (0 < \xi < \infty),$$
$$v = 0 \quad \text{over} \quad \xi = 0 \quad (\eta_1 < \eta < \eta_2).$$

It follows from 5.2 (18) that

$$v = \frac{1}{2(\eta_2 - \eta_1)} \sin \frac{\pi(\eta - \eta_1)}{(\eta_2 - \eta_1)} \times$$

$$\times \int_0^\infty f_2(\xi') \, d\xi' \left\{ \frac{1}{\cos[\pi(\eta_2 - \eta)/(\eta_2 - \eta_1)] + \cosh[\pi(\xi - \xi')/(\eta_2 - \eta_1)]} - \right.$$
$$\left. - \frac{1}{\cos[\pi(\eta_2 - \eta)/(\eta_2 - \eta_1)] + \cosh[\pi(\xi + \xi')/(\eta_2 - \eta_1)]} \right\}. \qquad (11)$$

By adding (10) and two solutions of type (11) we find the solution when all three surfaces of Fig. 50 are kept at prescribed temperatures.

(iv) Consider the region bounded by

$$\eta = \eta_1 \quad (-\infty < \xi < \infty),$$
$$\eta = \eta_2 \quad (-\infty < \xi < \infty).$$

Let
$$v = f(\xi) \quad \text{over} \quad \eta = \eta_1,$$
and
$$v = 0 \quad \text{over} \quad \eta = \eta_2.$$

Then we find, as above, using 5.2 (19),

$$v = \frac{1}{2(\eta_2 - \eta_1)} \sin \frac{\pi(\eta - \eta_1)}{(\eta_2 - \eta_1)} \times$$

$$\times \int_{-\infty}^\infty \frac{f(\xi') \, d\xi'}{\cos[\pi(\eta_2 - \eta)/(\eta_2 - \eta_1)] + \cosh[\pi(\xi - \xi')/(\eta_2 - \eta_1)]}. \qquad (12)$$

V. *Confocal ellipses or hyperbolas*

Consider the transformation

$$\xi + i\eta = \cosh^{-1} \frac{x + iy}{c},$$

or
$$x + iy = c \cosh(\xi + i\eta).$$

Then
$$x = c \cosh \xi \cos \eta, \qquad y = c \sinh \xi \sin \eta,$$

and
$$\frac{x^2}{\cosh^2 \xi} + \frac{y^2}{\sinh^2 \xi} = c^2,$$

$$\frac{x^2}{\cos^2 \eta} - \frac{y^2}{\sin^2 \eta} = c^2.$$

Thus the curves ξ = constant, and η = constant, are a set of confocal ellipses and hyperbolas, and the xy-plane is given by $-\pi < \eta < \pi$ and $0 < \xi < \infty$, the lower part of the xy-plane having negative values of η and the upper part positive values.

(i) *Two confocal ellipses.* Consider the region bounded by $\xi = \xi_1$ and $\xi = \xi_2$.

Let
$$v = f_1(\eta) \quad \text{over} \quad \xi = \xi_1,$$
$$v = f_2(\eta) \quad \text{over} \quad \xi = \xi_2.$$

Then, as above,

$$v = \sum_{0}^{\infty} \frac{\sinh n(\xi_2 - \xi)}{\sinh n(\xi_2 - \xi_1)} (a_n \cos n\eta + b_n \sin n\eta) +$$

$$+ \sum_{0}^{\infty} \frac{\sinh n(\xi - \xi_1)}{\sinh n(\xi_2 - \xi_1)} (a'_n \cos n\eta + b'_n \sin n\eta),$$

where a_n, b_n, a'_n, and b'_n are the coefficients in the Fourier series for $f_1(\eta)$ and $f_2(\eta)$ in the interval $-\pi$ to π.

(.) *Two semi-ellipses and the part of the major axis between them.* In this case the region is bounded by

$$\xi = \xi_1 \quad \text{and} \quad \xi = \xi_2 \quad (0 < \eta < \pi),$$
$$\eta = 0 \quad \text{and} \quad \eta = \pi \quad (\xi_1 < \xi < \xi_2).$$

Let
$$v = f_1(\eta) \quad \text{over} \quad \xi = \xi_1,$$
$$v = f_2(\eta) \quad \text{over} \quad \xi = \xi_2,$$

and
$$v = 0 \quad \text{over} \quad \eta = 0 \text{ and } \eta = \pi.$$

It is clear that the solution is

$$v = \sum_{1}^{\infty} a_n \sin n\eta \, \frac{\sinh n(\xi_2 - \xi)}{\sinh n(\xi_2 - \xi_1)} + \sum_{1}^{\infty} a'_n \sin n\eta \, \frac{\sinh n(\xi - \xi_1)}{\sinh n(\xi_2 - \xi_1)},$$

where a_n and a'_n are the coefficients in the sine series for $f_1(\eta)$ and $f_2(\eta)$.

(iii) *A semi-ellipse.* In this case the region is bounded by

$$\xi = 0 \quad \text{and} \quad \xi = \xi_1 \quad (0 < \eta < \pi),$$
$$\eta = 0 \quad \text{and} \quad \eta = \pi \quad (0 < \xi < \xi_1).$$

Let $v = f(\eta)$ over $\xi = \xi_1$, and let the major axis be at zero temperature. Then

$$v = \sum_{1}^{\infty} a_n \sin n\eta \, \frac{\sinh n\xi}{\sinh n\xi_1},$$

where a_n is the coefficient in the sine series for $f(\eta)$.

(iv) *A complete ellipse.* In this case we have to satisfy

$$\frac{\partial^2 v}{\partial \xi^2} + \frac{\partial^2 v}{\partial \eta^2} = 0, \quad 0 < \xi < \xi_1, \ -\pi < \eta < \pi,$$

$$v = f(\eta), \quad \text{when } \xi = \xi_1 \quad (-\pi < \eta < \pi).$$

Also there must be no discontinuity in the temperature or the flow of heat as we cross the major axis or pass along it.

All these conditions are satisfied by the expression

$$v = \sum_{n=0}^{\infty} \left(a_n \frac{\cosh n\xi}{\cosh n\xi_1} \cos n\eta + b_n \frac{\sinh n\xi}{\sinh n\xi_1} \sin n\eta \right),$$

where a_n and b_n are the coefficients in the Fourier series for $f(\eta)$ in the interval $-\pi$ to π.

(v) *A quadrilateral bounded by the arcs of two confocal ellipses and hyperbolas.* This reduces to the rectangle in the $\xi\eta$-plane, and the solution follows.

16.8. Steady flow of heat in a polygon

The results of § 16.7 have been found by using well-known transformations suitable for the regions under consideration: there has been no general method of finding the transformation. For two-dimensional steady flow in the region bounded by a polygon it is theoretically possible to find the proper transformation by the use of the theorem given below.

The Schwarz–Christoffel theorem† states that any polygon bounded by straight lines in the z-plane ($z = x+iy$) can be transformed into the axis of ξ in the t-plane ($t = \xi+i\eta$), and that points inside the polygon in the z-plane transform into points on one side of the axis of ξ. The transformation which does this is obtained from the relation

$$\frac{dz}{dt} = C(t-\xi_1)^{\alpha_1/\pi-1}(t-\xi_2)^{\alpha_2/\pi-1}...(t-\xi_n)^{\alpha_n/\pi-1}, \tag{1}$$

where $\alpha_1, \alpha_2,..., \alpha_n$ are the interior angles of the polygon, $\xi_1,..., \xi_n$ are the points on the real axis into which the angular points transform, C is a constant, and, if one of the angular points corresponds to an infinite value of ξ, the corresponding factor in (1) is omitted.‡

Whatever the values of $\xi_1,..., \xi_n$ may be, the transformation (1) transforms the axis of ξ into a polygon with internal angles $\alpha_1,..., \alpha_n$; if this is to be similar to a given polygon, only three of $\xi_1,..., \xi_n$ can be chosen arbitrarily and the remainder must be determined from the dimensions of the given polygon.

When the transformation is known, the temperature at any point can be inferred from the known solution for the temperature in the half-plane $\eta > 0$ due to arbitrary temperature $f(\xi)$ in the surface $\eta = 0$. This is, by 5.2 (20),

$$\frac{\eta}{\pi} \int_{-\infty}^{\infty} \frac{f(\xi')\, d\xi'}{(\xi-\xi')^2+\eta^2}.$$

† Christoffel, 'Sul problema delle temperature stazionarie', *Ann. Mat. pura appl.* **1** (1867) 89; Schwarz, *Crelle*, **70** (1869) 105.

‡ Proofs are given in most treatises on electricity or hydrodynamics. Also in Bateman, *Partial Differential Equations of Mathematical Physics*, § 4.62; Carter, *J. Instn. Elect. Engrs.* **64** (1926) 1115; Thomson, *Recent Researches in Electricity and Magnetism*, Chap. III. The latter contain many examples of interest in the present connexion.

The proof depends on the fact that from (1) the argument of dz/dt is constant when t lies between any two of the points $\xi_1, \xi_2,...$, so that these regions of the t-plane correspond to straight lines in the z-plane. Also, as t passes from $\xi_r-\epsilon$ to $\xi_r+\epsilon$ round a semicircle of infinitesimal radius ϵ, the argument of dz/dt increases by $\pi-\alpha_r$, corresponding to an internal angle α_r of the polygon.

If $f(\xi) = V$, constant, for $\xi > 0$, and $f(\xi) = 0$, $\xi < 0$, this reduces to

$$V\left\{\frac{1}{2} + \frac{1}{\pi}\tan^{-1}\frac{\xi}{\eta}\right\}. \tag{2}$$

The transformation may be used to study flow of heat in a rectangle, also in the region between concentric squares.† In general the results will involve elliptic or more complicated types of function. The simplest

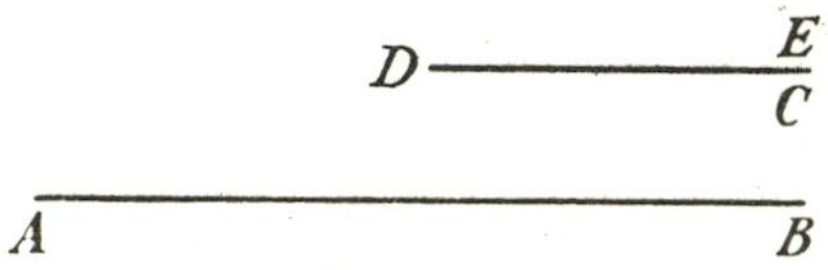

FIG. 51

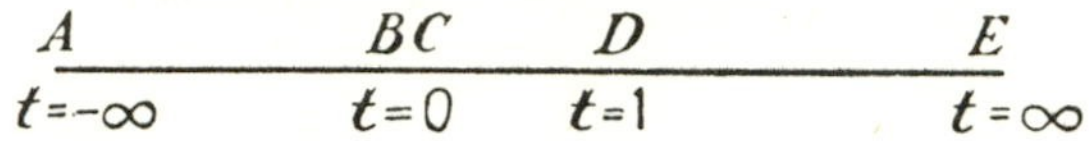

FIG. 52

and most important applications are to degenerate polygons with some corners at infinity; in a number of such cases the result involves only elementary functions and makes possible a study of two-dimensional problems which are idealizations of such commonly occurring systems as a bend in a wall, a change in thickness of a wall, a projection into a wall, a guard ring, etc. We consider briefly some examples of these types.‡

I. *Flow between an infinite plate AB and a semi-infinite plate CD* (Fig. 51)

The polygon in the z-plane is $ABCDEA$ which has a zero angle at BC at infinity, an internal angle of 2π at D, and, if the points corresponding to A and E are taken to be at infinity in the t-plane, the angle here need not be discussed. Since we can choose three points arbitrarily in the t-plane, we let A and E be at infinity; B, C at $t = 0$; and D at $t = 1$. The figure in the t-plane is shown in Fig. 52.

Here (1) gives

$$\frac{dz}{dt} = C\frac{t-1}{t}. \tag{3}$$

† Moulton, *Proc. Lond. Math. Soc.* (2) **3** (1905) 104; Bowman, ibid. (2) **39** (1935) 211, (2) **41** (1936) 271.

‡ For other examples of the method see the works of Thomson and Carter referred to above, also Love, *Proc. Lond. Math. Soc.* (2) **22** (1924) 337; Davy, *Phil. Mag.* (7) **35** (1944) 819; ibid. **36** (1945) 153; Bowman, *Introduction to Elliptic Functions* (London, 1953). Problems of the types given below, but with some of the corners rounded, are discussed by Cockcroft, *J. Instn. Elect. Engrs.* **66** (1928) 385; other problems with rounded corners are given by Page, *Proc. Lond. Math. Soc.* (2) **11** (1912) 314. The Schwarz method is extended to polygons with curved sides by Richmond, ibid. **22** (1924) 483.

The short method using internal angles of 0 and 2π used above to write down (3) is the most convenient. If preferred, the polygon of Fig. 51 may be regarded as the limit of the polygon $ABCDD'EFGA$ of Fig. 53 as $D' \rightarrow D$ and the corners A, B, C, E, F, G tend to infinity. By writing down the value of (1) for the polygon of Fig. 53 and taking the limit we again find (3).

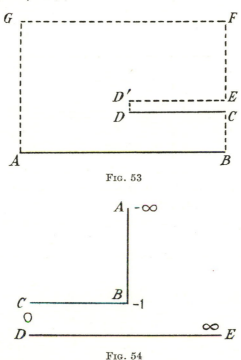

FIG. 53

FIG. 54

Integrating (3) gives

$$z = C(t - \ln t) + k. \qquad (4)$$

The arbitrary constants C and k are found by specifying precisely the polygon in the z-plane. We may always assume C to be real; giving it a complex value simply rotates the polygon. If AB is to be the axis of x, z must be real for t real and negative. This requires that $k = i\pi C$, and thus (4) becomes

$$z = C\{t - \ln t + i\pi\}. \qquad (5)$$

Now at the point D, $t = 1$, and thus $z = C(1 + i\pi)$. It follows that the distance between the plates is πC.

II. *Flow between an infinite plate and a right-angled corner*

Here the polygon is $ABCDE$, Fig. 54, and we take for A, B, CD, and E the points $-\infty$, -1, 0, and $+\infty$ on the axis of ξ. The internal angles at B and CD are $3\pi/2$ and 0, respectively, and (1) becomes

$$\frac{dz}{dt} = C\frac{(t+1)^{\frac{1}{2}}}{t},$$

giving
$$z = C\left\{2\surd(t+1)+\ln\frac{(t+1)^{\frac{1}{2}}-1}{(t+1)^{\frac{1}{2}}+1}\right\}+k. \qquad (6)$$

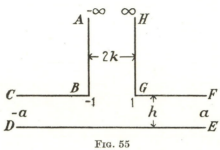

FIG. 55

To determine the arbitrary constants C and k we notice that as t increases through 0, z decreases by $i\pi C$. Also for $t > 0$, z is real: that is DE is the real axis in the z-plane. Thus, if the distance between BC and DE is h, we must have $\pi C = h$. Also when $t = -1$, $z = iC\pi + k$, so taking $k = 0$ fixes the origin at the intersection of DE and AB produced.

III. A guard ring

The polygon of Fig. 55 may be used to study the effect of a guard ring on the isothermals between parallel planes. We take $\xi = -\infty$ for A; $\xi = -1$ for B; $\xi = -a$ for CD, where a is to be determined later; $\xi = a$ for EF; $\xi = 1$ for G, and $\xi = +\infty$ for H. The internal angles are zero at CD and EF, and $3\pi/2$ at B and G. Thus

$$\frac{dz}{dt} = C\frac{\surd(t^2-1)}{t^2-a^2}.$$

Integrating we find

$$z = -iC\left\{\frac{\surd(1-a^2)}{2a}\ln\frac{(t+a)\{1-at+\surd(1-a^2)(1-t^2)\}}{(t-a)\{1+at+\surd(1-a^2)(1-t^2)\}}+\sin^{-1}t\right\}+$$
$$+\pi C\frac{\surd(1-a^2)}{2a}, \quad (7)$$

where the constant of integration has been chosen to make $z = 0$ when $t = 0$. DE is the axis of y, and the axis of x is midway between AB and HG. To find a and C, suppose $2k$ is the distance between the planes

AB and HG, and that h is the distance between the planes GF and DE, so that the point G is $z = h - ik$. Putting $t = 1$, $z = h - ik$, in (7) we find

$$h - ik = -\frac{iC\pi}{2} + \frac{\pi C\sqrt{(1-a^2)}}{2a}.$$

Therefore

$$C = 2k/\pi, \qquad a^2 = k^2/(k^2 + h^2).$$

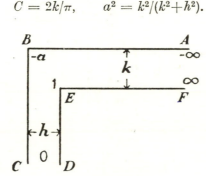

FIG. 56

IV. *A right-angled bend in a wall*

The bend is shown in Fig. 56. The points A, B, CD, E, F are to transform to $-\infty$, $-a$, 0, 1, $+\infty$ in the t-plane. The transformation is given by

$$\frac{dz}{dt} = \frac{C(t-1)^{\frac{1}{2}}}{t(t+a)^{\frac{1}{2}}}.$$

To integrate, put

$$\beta = \left(\frac{t-1}{t+a}\right)^{\frac{1}{2}}, \tag{8}$$

and we find

$$z = -\frac{2C}{a^{\frac{1}{2}}}\tan^{-1}\beta a^{\frac{1}{2}} + C\ln\frac{1+\beta}{1-\beta}, \tag{9}$$

where the constant of integration has been made zero by choosing $z = 0$ when $t = 1$, that is, E to be the origin in the z-plane.

·Alternatively, (9) may be written

$$z = -\frac{2C}{a^{\frac{1}{2}}}\sin^{-1}\left\{\frac{a(t-1)}{t(a+1)}\right\}^{\frac{1}{2}} + C\ln\frac{\sqrt{(t+a)}+\sqrt{(t-1)}}{\sqrt{(t+a)}-\sqrt{(t-1)}}. \tag{10}$$

Putting $t = -a$ in (10) we find the value of z for the point B to be $-\pi Ca^{-\frac{1}{2}} + i\pi C$. Now, if the distance between the vertical faces of the wall (Fig. 56) is h and that between the horizontal faces is k, we have for the point B, $z = -h + ik$, $t = -a$. Therefore from (10)

$$k = \pi C, \qquad h = \pi Ca^{-\frac{1}{2}}, \qquad a = k^2/h^2. \tag{11}$$

V. *A sudden change in the thickness of a wall*†

The surfaces are shown in Fig. 57, the values of ξ in the t-plane being marked at the angular points. The transformation is given by

$$\frac{dz}{dt} = \frac{C(t+1)^{\frac{1}{2}}}{t(t+a)^{\frac{1}{2}}}.$$

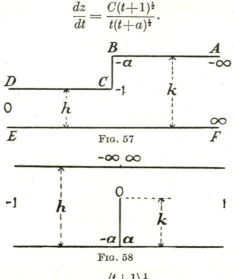

Fig. 57

Fig. 58

Putting
$$\beta = \left(\frac{t+1}{t+a}\right)^{\frac{1}{2}},$$

we find
$$z = C\ln\frac{1+\beta}{1-\beta} + Ca^{-\frac{1}{2}}\ln\frac{\beta a^{\frac{1}{2}}-1}{\beta a^{\frac{1}{2}}+1}, \tag{12}$$

the arbitrary constant of integration being chosen so that when $t = -a$, $z = i\pi C$. Also, putting $t = -1$ in (12), we find $z = i\pi Ca^{-\frac{1}{2}}$.

Finally for $t > 0$, z is real. Thus EF is the real axis in the z-plane and the origin is at the intersection of BC and EF. If h and k are the two thicknesses of the wall we have

$$h = \pi Ca^{-\frac{1}{2}}, \qquad k = \pi C, \qquad a = k^2/h^2. \tag{13}$$

VI. *A projection into a wall*‡

The values of ξ corresponding to the angular points are marked in Fig. 58. The transformation is

$$\frac{dz}{dt} = \frac{Ct}{(t^2-1)(t^2-a^2)^{\frac{1}{2}}}.$$

† Lees, *Phil. Mag.* (6) **16** (1908), 734; *Proc. Roy. Soc.* A, **91** (1915) 440. Castoldi, *Geofis. Pura é Appl.* **23** (1952) 27.

‡ Lees, *Proc. Phys. Soc.* **23** (1911) 361. A large number of solutions of problems of this type is given by Schofield, *Phil. Mag.* (7) **6** (1928) 567. Radiation at the plane boundary is discussed by Awbery., ibid. **7** (1929) 1143. A projection from a plane is discussed by Lees, *Proc. Roy. Soc.* A, **91** (1915) 440, and by Langton and Davy, *Brit. J. Appl. Phys.* **5** (1954) 405.

$$z = -\frac{C}{2\sqrt{(1-a^2)}} \ln \frac{\sqrt{(1-a^2)}+\sqrt{(t^2-a^2)}}{\sqrt{(1-a^2)}-\sqrt{(t^2-a^2)}}, \qquad (14)$$

where the additive constant has been chosen to give $z = 0$ when $t = \pm a$; that is, the origin in the z-plane is at the base of the projection. Now from (14):

$$\text{when } t \to \pm\infty, \quad z \to -\frac{i\pi C}{2\sqrt{(1-a^2)}};$$

$$\text{when } t = 0, \qquad z = -\frac{iC}{\sqrt{(1-a^2)}} \tan^{-1}\frac{a}{\sqrt{(1-a^2)}}.$$

Thus, if the distance between the walls is h and the length of the projection is k, we must have

$$\frac{\pi C}{2\sqrt{(1-a^2)}} = h, \qquad a = \sin\frac{\pi k}{2h}. \qquad (15)$$

16.9. Flow between isothermal surfaces

The method of §§ 16.6–16.8 is a very general one suitable for the study of problems in which the surface temperature is an arbitrary function of position: the only use which is made of the theory of conformal representation is in the transformation of the given region in the z-plane into a simpler region in the t-plane for which the solution can be written down.

The most important practical problems are rather easier ones in which the boundaries of the region in the z-plane are kept at constant, instead of arbitrary, temperature, and the solution in the t-plane takes a very simple form such as 16.7 (9) or 16.8 (2). Instead of writing down solutions in this way, we develop here a slightly different approach to such problems, which is precisely that used in electricity† or hydrodynamics. The temperature v itself is taken as one of a pair of conjugate functions, then, by studying also the function u conjugate to v we can find the value of the rate of flow of heat across any portion of one of the bounding surfaces, and thus an expression for the thermal resistance (16.5 (1)) to steady flow between the boundaries at constant temperature.

As in § 16.6 let
$$w = u+iv = f(x+iy) = f(z),$$

then all the results of § 16.6 hold with ξ, η replaced by u, v. In particular

$$\frac{\partial^2 v}{\partial x^2}+\frac{\partial^2 v}{\partial y^2} = 0. \qquad (1)$$

† Cf. Jeans, *Electricity and Magnetism* (Cambridge, edn. 5, 1925) Chaps. VIII, X.

Thus the imaginary part (or in the same way the real part) of any function $f(z)$ is a solution of the equation of steady temperature, and, in particular, represents the temperature in the region between the curves $v = v_1$ and $v = v_2$, constants, in the z-plane.

For example, if

$$u + iv = \frac{1}{\pi} \ln z = \frac{1}{\pi} \ln r + \frac{i\theta}{\pi}, \tag{2}$$

we have $v = \theta/\pi$, and this gives the steady temperature in the region between the plane $\theta = 0$, maintained at temperature zero, and the plane $\theta = \pi$, maintained at unit temperature (cf. 16.8 (2)).

Next we remark that the curves $u = $ const. and $v = $ const. in the z-plane cut orthogonally (by 16.6 (1), (2)). Thus, since $v = $ const. are the isothermals, the curves $u = $ const. are the lines of flow of heat. Also, again from 16.6 (1), (2),

$$|f'(z)| = \left\{\left(\frac{\partial u}{\partial y}\right)^2 + \left(\frac{\partial v}{\partial y}\right)^2\right\}^{\frac{1}{2}} = \left\{\left(\frac{\partial v}{\partial x}\right)^2 + \left(\frac{\partial v}{\partial y}\right)^2\right\}^{\frac{1}{2}}. \tag{3}$$

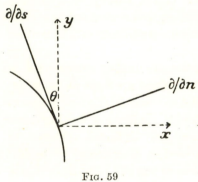

FIG. 59

Therefore the magnitude of the flux vector† at any point is

$$K\left\{\left(\frac{\partial v}{\partial x}\right)^2 + \left(\frac{\partial v}{\partial y}\right)^2\right\}^{\frac{1}{2}} = K\left|\frac{dw}{dz}\right|. \tag{4}$$

The rate at which heat flows across any portion of an isothermal surface can also be written down very simply. We write $\partial/\partial n$ for differentiation in the direction of the normal to the surface and $\partial/\partial s$ for differentiation along the surface (Fig. 59).

The flux of heat at any point across the surface is

$$-K\frac{\partial v}{\partial n} = -K\left\{\frac{\partial v}{\partial x}\cos\theta + \frac{\partial v}{\partial y}\sin\theta\right\}$$

$$= -K\left\{-\frac{\partial u}{\partial y}\cos\theta + \frac{\partial u}{\partial x}\sin\theta\right\}$$

$$= K\frac{\partial u}{\partial s}.$$

† Cf. § 1.3.

Thus the rate of flow of heat across the portion of the surface from s_1 to s_2 is

$$K \int_{s_1}^{s_2} \frac{\partial u}{\partial s} ds = K(u_2 - u_1), \tag{5}$$

where u_1 and u_2 are the values of u at s_1 and s_2.

It appears that the introduction of the function u conjugate to v greatly simplifies the calculation of the rate of flow of heat across an isothermal surface, and therefore of the thermal resistance between isothermal surfaces.

From the present point of view the $t = \xi + i\eta$ plane of § 16.7 may be regarded as the w-plane of this section. In the more complicated problems of § 16.8 we make the transformation from the z-plane to the t-plane as before: in the simple case of two isothermal boundaries we write down the solution for the t-plane by the relation (2)

$$w = \frac{1}{\pi} \ln t, \tag{6}$$

or an appropriate modification of it, and the solution in the z-plane follows. When the magnitude of the flux vector at any point is needed we have the relation

$$\left|\frac{dw}{dz}\right| = \left|\frac{dw}{dt}\right| \times \left|\frac{dt}{dz}\right| = \left|\frac{dw}{dt}\right| \Big/ \left|\frac{dz}{dt}\right|. \tag{7}$$

We now discuss the thermal resistance between some of the boundaries considered earlier.

I. *Steady flow of heat between circular cylinders*

The appropriate transformation has been given in § 16.7, we have to put the results in the simplest form for practical calculations. Following 16.7 (7) but with some changes of notation, we take

$$u + iv = i \ln \frac{(x+c) + iy}{(x-c) + iy}, \tag{8}$$

so that

$$v = \ln \frac{r}{r'} \quad \text{and} \quad u = \theta' - \theta, \tag{9}$$

where r and r' are the distances of the point P, (x, y) from the points $(\mp c, 0)$, and θ and θ' are the angles PCX and PDX (Fig. 60).

Suppose, now, we wish to find the thermal resistance between two cylinders of radii a_1 and a_2 whose centres are at distance d apart. We discuss in detail the case in which the cylinder of radius a_1 encloses that of radius a_2. Let OX in Fig. 60 be the line of centres: we have to find the value of c, and the ratio (r/r') for each of the circles, in terms of the given quantities a_1, a_2, d.

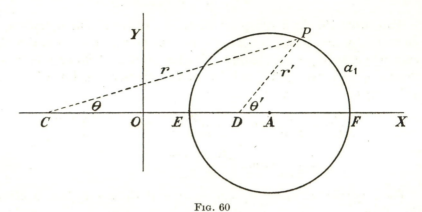

Fig. 60

Suppose the circle of radius a_1 has centre A distant d_1 from O. Let k_1 be the constant value of (r/r') for this circle, then, writing down the value of this ratio for the points E and F, respectively, we get

$$k_1 = \frac{c+(d_1-a_1)}{c-(d_1-a_1)} = \frac{(d_1+a_1)+c}{(d_1+a_1)-c}. \tag{10}$$

It follows that
$$d_1^2-a_1^2 = c^2. \tag{11}$$

For the circle of radius a_2 with centre distant d_2 from O we have in the same way
$$d_2^2-a_2^2 = c^2. \tag{12}$$

Subtracting (11) from (12) we get
$$d_1^2-d_2^2 = a_1^2-a_2^2. \tag{13}$$

Also, since the distance between the centres of the circles is d,
$$d_1-d_2 = d. \tag{14}$$

It follows that
$$d_1 = \frac{a_1^2-a_2^2+d^2}{2d}, \qquad d_2 = \frac{a_1^2-a_2^2-d^2}{2d}. \tag{15}$$

Also, using (11) in (10) gives
$$k_1 = \frac{\sqrt{(d_1+a_1)}+\sqrt{(d_1-a_1)}}{\sqrt{(d_1+a_1)}-\sqrt{(d_1-a_1)}} = \frac{d_1+\sqrt{(d_1^2-a_1^2)}}{a_1}. \tag{16}$$

Thus, from (9), for this cylinder
$$v_1 = \ln k_1 = \cosh^{-1}\frac{d_1}{a_1} = \cosh^{-1}\left\{\frac{a_1^2-a_2^2+d^2}{2a_1 d}\right\}, \tag{17}$$

and for the cylinder of radius a_2

$$v_2 = \ln k_2 = \cosh^{-1}\left\{\frac{a_1^2 - a_2^2 - d^2}{2a_2 d}\right\}. \tag{18}$$

Also, from (9), in a complete circuit of either of the cylinders, u increases by 2π. Thus, by (5), the rate of flow of heat between the cylinders is $2\pi K$ per unit length. The temperature difference between them is $v_2 - v_1$ given by (17) and (18), and the thermal resistance between unit lengths of the cylinders is thus

$$\frac{(v_2 - v_1)}{2\pi K} = \frac{1}{2\pi K}\left\{\cosh^{-1}\frac{a_1^2 - a_2^2 - d^2}{2a_2 d} - \cosh^{-1}\frac{a_1^2 - a_2^2 + d^2}{2a_1 d}\right\}$$

$$= \frac{1}{2\pi K}\cosh^{-1}\frac{a_1^2 + a_2^2 - d^2}{2a_1 a_2}. \tag{19}$$

For a cylinder of radius a_1 with its axis at distance d_1 from a plane, the cylinder and plane being isothermals,† (17) holds as before, and we get for the thermal resistance between unit length of the cylinder and the plane

$$\frac{1}{2\pi K}\ln\frac{d_1 + \sqrt{(d_1^2 - a_1^2)}}{a_1} = \frac{1}{2\pi K}\cosh^{-1}\frac{d_1}{a_1}. \tag{20}$$

For *the case in which the cylinders are external to one another* we suppose the cylinder of radius a_1 to surround the limiting point D and that of radius a_2 to enclose C. We proceed as before except that (14) is replaced by $d_1 + d_2 = d$. It follows that v_1 is given by (17), and v_2 by

$$v_2 = -\ln\frac{d_2 + \sqrt{(d_2^2 - a_2^2)}}{a_2} = -\cosh^{-1}\frac{d_2}{a_2} = -\cosh^{-1}\left\{\frac{d^2 - a_1^2 + a_2^2}{2a_2 d}\right\}. \tag{21}$$

Thus the thermal resistance between unit length of the two cylinders is

$$\frac{1}{2\pi K}\left\{\cosh^{-1}\frac{a_1^2 - a_2^2 + d^2}{2a_1 d} + \cosh^{-1}\frac{d^2 - a_1^2 + a_2^2}{2a_2 d}\right\} = \frac{1}{2\pi K}\cosh^{-1}\frac{d^2 - a_1^2 - a_2^2}{2a_1 a_2}. \tag{22}$$

II. *Other problems involving circular boundaries*

Systems of obvious practical importance which are next in order of difficulty to those of I are the cylinder between parallel planes, and the grating of cylinders regularly spaced. Exact solutions of these problems have not been obtained by conformal transformation, but solutions are

---

† This solution is of use in finding the steady loss of heat from a buried cable, cf. Melsom and Booth, *J. Instn. Elect. Engrs.* **52** (1915) 779. The case of radiation at the plane surface has been discussed by Awbery, *Phil. Mag.* (7) **7** (1929) 1143, and Charny, *C.R. (Doklady) Acad. Sci. U.R.S.S.* **48** (1945) 27. Other problems with the radiation boundary condition are studied by Schofield, *Phil. Mag.* (7) **31** (1941) 471.

known for oval curves which by choice of parameters may be made to approximate very closely to circles.†

III. *Flow below a non-conducting strip*

Suppose the region $-\infty < x < -c$ is maintained at constant temperature π, and $x > c$ at zero temperature, there being no flow across the strip $-c < x < c$. We take

$$u+iv = \cosh^{-1}(x+iy)/c,$$

so that
$$x = c \cosh u \cos v,$$

$$y = c \sinh u \sin v.$$

The equipotentials are the hyperbolas

$$\frac{x^2}{c^2\cos^2 v} - \frac{y^2}{c^2\sin^2 v} = 1,$$

and the lines of flow are ellipses. The equipotential $v = 0$ is the line $x > c$; $-\infty < u < 0$ corresponds to the underside and $0 < u < \infty$ to the upper side. Similarly, the equipotential $v = \pi$ is the line $x < -c$, with $-\infty < u < 0$ corresponding to the underside and $0 < u < \infty$ to the upper side. The rate of flow of heat Q from the part $-X < x < -c$ of the underside of the x-axis is by (5)

$$K[u]_{x=-X}^{x=-c} = K \cosh^{-1}\frac{X}{c}.$$

IV. *Flow between the plates AB and CD of Fig. 51, AB being at temperature unity, and CD at zero temperature*

Here, as in (2) or (6) we take

$$w = \frac{1}{\pi}\ln t,$$

also, by 16.8 (5),

$$z = \frac{d}{\pi}\{t-\ln t+i\pi\} = \frac{d}{\pi}\{e^{\pi w}-\pi w+i\pi\}, \tag{23}$$

where d is the distance between the plates.

The magnitude of the flux vector at any point is

$$K\left|\frac{dw}{dz}\right| = K\left|\frac{dw}{dt}\right| \bigg/ \left|\frac{dz}{dt}\right| = \frac{K}{d|t-1|}.$$

† The cylinder between parallel planes is discussed by Schofield, *Phil. Mag.* (7) **12** (1931) 329; the grating by Richmond, *Proc. Lond. Math. Soc.* (2) **22** (1924) 389. Using another method, the cylinder between parallel planes is discussed by Knight, ibid. **39** (1935) 272; the row of cylinders between parallel planes by Knight and McMullen, *Phil. Mag.* (7) **24** (1937) 35; and other problems on cylinders between parallel planes by Howland and McMullen, *Proc. Camb. Phil. Soc.* **32** (1936) 402.

As $t \to 0$, that is near the points BC, this tends to K/d, which is the steady flux between two infinite parallel planes. At the point D, $t = 1$, the flux is infinite.

On CD we have, since here $v = 0$,

$$z = \frac{d}{\pi}\{e^{\pi u} - \pi u + i\pi\}.$$

Also, since D is the point $d(1+i\pi)/\pi$, the value of u at D is 0. Its values at the points on CD and DE distant l from D are the roots of

$$\frac{\pi l}{d} + 1 = e^{\pi u} - \pi u, \tag{24}$$

the negative root corresponding to a point on the lower side CD (since here $0 < t < 1$, cf. Fig. 52) and the positive root to the upper side DE.

If l is large the negative root of (24) is approximately

$$-\left(\frac{l}{d} + \frac{1}{\pi}\right),$$

and thus the rate at which heat flows across length l of the underside DC of the strip is

$$K\left(\frac{l}{d} + \frac{1}{\pi}\right) \tag{25}$$

per unit time per unit length normal to the plane of the figure. This is an increase of K/π over the corresponding value for flow between infinite planes.

For large values of l the positive root of (24) is approximately

$$\frac{1}{\pi}\ln\left(\frac{\pi l}{d} + 1\right).$$

Thus the rate of flow of heat across this length l of the upper side of DE is

$$\frac{K}{\pi}\ln\left(\frac{\pi l}{d} + 1\right). \tag{26}$$

V. *Flow of heat in the right-angled bend of Fig.* 56

We suppose the surface DEF to be at zero temperature and ABC at unit temperature. Then, as before,

$$w = \frac{1}{\pi}\ln t. \tag{27}$$

At the point E, $t = 1$, we have $u = 0$. We proceed to find the rate of flow of heat, per unit length along the corner, over length x of EF

measured from the corner E, assuming that x is large. For this we need the value of t which corresponds to x by 16.8 (8), (9), (11), namely

$$x = -\frac{2h}{\pi}\tan^{-1}\frac{k\beta}{h}+\frac{2k}{\pi}\ln(1+\beta)-\frac{k}{\pi}\ln(1-\beta^2), \qquad (28)$$

where $\qquad \beta = \left(\frac{t-1}{t+a}\right)^{\frac{1}{2}}, \qquad 1-\beta^2 = \frac{1+a}{t+a}, \qquad a = \frac{k^2}{h^2}. \qquad (29)$

Now if x is large, t is large, and β is nearly unity. Putting $\beta = 1$ in the first two terms of the right-hand side of (28) which are small compared with x and $\ln(1-\beta^2)$, and using (29) in the latter, we find as a first approximation

$$\ln\frac{1+a}{t+a} = -\frac{\pi x}{k}-\frac{2h}{k}\tan^{-1}\frac{k}{h}+2\ln 2. \qquad (30)$$

Since t is large we may replace $\ln(t+a)$ by $\ln t$ with an error of the order of a/t, and (30) gives approximately

$$\ln t = \frac{\pi x}{k}+\frac{2h}{k}\tan^{-1}\frac{k}{h}+\ln\frac{k^2+h^2}{4h^2}. \qquad (31)$$

Since $v = 0$ on EF, we have $\pi u = \ln t$ by (27), and thus the rate of flow of heat across length x of EF measured from E is

$$K\left\{\frac{x}{k}+\frac{2h}{\pi k}\tan^{-1}\frac{k}{h}+\frac{1}{\pi}\ln\frac{k^2+h^2}{4h^2}\right\}. \qquad (32)$$

The first term is that corresponding to steady flow between planes distant k apart. Adding the corresponding term for DE we find that *the effect of the right-angled bend is to increase the rate of flow of heat, per unit time per unit length along the corner,* by

$$K\left\{\frac{2h}{\pi k}\tan^{-1}\frac{k}{h}+\frac{2k}{\pi h}\tan^{-1}\frac{h}{k}+\frac{2}{\pi}\ln\frac{k^2+h^2}{4hk}\right\}. \qquad (33)$$

If the walls are of equal thickness, $k = h$, this reduces to

$$K\left\{1-\frac{2}{\pi}\ln 2\right\} = 0{\cdot}559K. \qquad (34)$$

XVII

INTEGRAL TRANSFORMS

17.1. Introductory

THE classical method of solution of problems in conduction of heat consists in assuming a solution in the form of a series of elementary solutions of the differential equation and some of the boundary conditions, the coefficients in this series being determined from the theory of Fourier or similar series. This method is completely adequate for problems on finite regions, but when infinite regions are considered, the corresponding procedure, using Fourier integrals, must usually be regarded as purely formal because of convergence difficulties (essential functions such as unity do not possess Fourier transforms). Nevertheless, this formal theory does give correct results which may be verified *a posteriori*, and it has been made rigorous by its extension† to the theory of Fourier transforms in the complex plane. Also, there has been an increasing tendency to use the equivalent idea of a Fourier transform rather than a Fourier integral in the actual analysis (cf. § 2.3).

The past three decades have seen the growth of the Laplace transformation method which, when applied to problems involving only one space variable, has conspicuous advantages over the older Fourier methods, namely, (i) it provides a routine procedure which is applied in the same way to all problems, (ii) it applies indifferently to all boundary conditions, so that the necessity for developing a new theory for each type of boundary condition is avoided, (iii) it has a large body of simple theorems such as those of § 12.2 which may be used to produce new results and transforms, (iv) convergence difficulties are, in the main, avoided and the analysis for simple, special problems such as constant initial or surface temperature can usually be regarded as completely rigorous. For problems involving more than one space variable, the situation is not so satisfactory, and the method used throughout this book has been to use the classical Fourier methods in the space-variables after removing the time-variable by the Laplace transformation.

Recently, the practice has grown up of rewriting all the series and integral theorems used in this type of work in a transform notation:‡

---

† Titchmarsh, *Theory of Fourier Integrals* (Oxford, 1937).

‡ A good short account is given by Tranter, *Integral Transforms in Mathematical Physics* (Methuen, 1951).

this has the great advantage of treating all variables in the same manner. It has, in fact, to a greater degree, the advantage (i) of the Laplace transform referred to above, namely, that by setting up a standard routine procedure or 'drill' applicable to all cases, it avoids the necessity for seeking special solutions and slightly lowers the level of manipulative skill needed. On the other hand, these other transforms do not possess the advantages (ii) –(iv) of the Laplace transform: they do not treat all boundary conditions equally, but a new transform and theory must be developed for each region and boundary condition; they are not in general so well tabulated as the Laplace transform, and their theory tends to be more sophisticated; for the integral theorems there are usually serious convergence difficulties and most work with them is purely formal. Finally, it must be emphasized that, at this stage, the integral transform theory contains nothing more than the theory of the original integral or series and thus is no more powerful.

Thus, while it is possible to solve problems on conduction of heat in one space variable by transformation with regard to this variable, the Laplace transformation with regard to the time must be regarded as the more powerful and satisfactory method. On the other hand, for problems in steady flow, the neatness of the integral transform methods is attractive. However, they are likely to assume greatest importance in problems involving several space variables when successive transformations (possibly including Laplace transformations) can be made very elegantly.

In this chapter a brief account of some of the simpler integral transforms will be given to indicate the type of analysis used and to enable a comparison to be made with the classical Fourier methods used earlier.

17.2. Integral transforms and their inversion formulae

In this and the following sections, v will be a function of x or r, its transform will be indicated by a capital V, and the nature of the transformation either by a suffix or by a characteristic new variable ω, s, or σ. In all cases, it will be assumed without comment that the integrals in question exist, and that, if necessary, the functions and their derivatives tend to zero as the variable tends to infinity.

I. *The complex Fourier transform*†

Writing $F[v(x)]$ to represent the operation of taking the complex

† The systematic use of the Fourier transform may be said to date from Campbell and Foster's large table of *Fourier Integrals for Practical Applications* (Bell Telephone System Monograph B-584, 1931). The accurate mathematical theory, as well as many

Fourier transform, this is defined by

$$F[v(x)] \equiv V(\omega) = \frac{1}{(2\pi)^{\frac{1}{2}}} \int_{-\infty}^{\infty} e^{i\omega x} v(x) \, dx. \tag{1}$$

By 2.3 (7), the inversion formula is

$$v(x) = \frac{1}{(2\pi)^{\frac{1}{2}}} \int_{-\infty}^{\infty} e^{-i\omega x} V(\omega) \, d\omega. \tag{2}$$

Integrating by parts gives

$$F\left[\frac{\partial v}{\partial x}\right] = -i\omega V(\omega), \tag{3}$$

$$F\left[\frac{\partial^2 v}{\partial x^2}\right] = -\omega^2 V(\omega), \tag{4}$$

provided that $v(x)$ and $\partial v/\partial x \to 0$ as $x \to \pm\infty$.

II. *The Fourier sine transform*

Writing $F_s[v(x)]$ to represent the operation of taking the sine transform, and $V_s(\omega)$ for its value, this is defined by

$$F_s[v(x)] \equiv V_s(\omega) = \left(\frac{2}{\pi}\right)^{\frac{1}{2}} \int_0^{\infty} v(x)\sin \omega x \, dx, \tag{5}$$

and the inversion formula is, by 2.3 (10),

$$v(x) = \left(\frac{2}{\pi}\right)^{\frac{1}{2}} \int_0^{\infty} V_s(\omega)\sin \omega x \, d\omega. \tag{6}$$

Also, if both $v(x)$ and $\partial v/\partial x \to 0$ as $x \to \infty$, integration by parts gives

$$F_s\left[\frac{\partial v}{\partial x}\right] = -\omega V_c(\omega), \tag{7}$$

$$F_s\left[\frac{\partial^2 v}{\partial x^2}\right] = -\omega^2 V_s(\omega) + \omega(2/\pi)^{\frac{1}{2}}v(0), \tag{8}$$

where $V_c(\omega)$ is defined in (9) below.

applications, are given in Titchmarsh, loc. cit. A great many practical applications are given by Sneddon, *Fourier Transforms* (McGraw-Hill, 1951); Chap. V of this work, pp. 159–205, is devoted to problems in conduction of heat. Both Titchmarsh and Sneddon discuss the other integral transforms of this section: it may be said here that there are no tables of these other transforms comparable in usefulness to that of Campbell and Foster or to existing tables of Laplace transforms. The most complete recent table of integral transforms is that of Erdélyi, Magnus, Oberhettinger, and Tricomi, *Tables of Integral Transforms* (McGraw-Hill, 1954).

III. *The Fourier cosine transform*

Writing $F_c[v(x)]$ to represent the operation of taking the cosine transform, and $V_c(\omega)$ for its value, we have

$$F_c[v(x)] \equiv V_c(\omega) = \left(\frac{2}{\pi}\right)^{\frac{1}{2}} \int_0^\infty v(x)\cos \omega x \, dx, \qquad (9)$$

and by 2.3 (13) the inversion formula is

$$v(x) = \left(\frac{2}{\pi}\right)^{\frac{1}{2}} \int_0^\infty V_c(\omega)\cos \omega x \, d\omega. \qquad (10)$$

Also, if both $v(x)$ and $\partial v/\partial x \to 0$ as $x \to \infty$,

$$F_c\left[\frac{\partial v}{\partial x}\right] = \omega V_s(\omega) - (2/\pi)^{\frac{1}{2}} v(0), \qquad (11)$$

$$F_c\left[\frac{\partial^2 v}{\partial x^2}\right] = -\omega^2 V_c(\omega) - (2/\pi)^{\frac{1}{2}} \left[\frac{\partial v}{\partial x}\right]_{x=0} \qquad (12)$$

IV. *The Hankel transform*

The Hankel transform of order $\nu > -\frac{1}{2}$, $H_\nu[v(r)]$ or $V_\nu(\sigma)$, of a function $v(r)$ is defined as

$$H_\nu[v(r)] \equiv V_\nu(\sigma) = \int_0^\infty rJ_\nu(\sigma r)v(r) \, dr, \qquad (13)$$

and its inversion formula† is

$$v(r) = \int_0^\infty \sigma J_\nu(\sigma r)V_\nu(\sigma) \, d\sigma. \qquad (14)$$

Also, integrating by parts twice and using Appendix III (2) gives

$$H_\nu\left[\frac{1}{r}\frac{\partial}{\partial r}\left(r\frac{\partial v}{\partial r}\right) - \frac{\nu^2 v}{r^2}\right] = -\sigma^2 V_\nu(\sigma), \qquad (15)$$

provided that $rvJ_\nu'(\sigma r)$ and $rJ_\nu(\sigma r)(\partial v/\partial r)$ tend to zero as $r \to 0$ and as $r \to \infty$.

V. *The Mellin transform*

The Mellin transform, $M[v(r)]$ or $V_m(s)$, of $v(r)$ is defined as

$$M[v(r)] \equiv V_m(s) = \int_0^\infty r^{s-1}v(r) \, dr, \qquad (16)$$

---

† This is Hankel's integral theorem, *W.B.F.*, § 14.4; *G. and M.*, Chap. VIII.

and its inversion formula† is

$$v(r) = \frac{1}{2\pi i} \int_{\gamma-i\infty}^{\gamma+i\infty} V_m(s)r^{-s}\, ds. \tag{17}$$

Also, integrating by parts,

$$M\left[r\frac{\partial}{\partial r}\left(r\frac{\partial v}{\partial r}\right)\right] = s^2 V_m(s), \tag{18}$$

provided that $r^s v$ and $r^{s+1}(\partial v/\partial r)$ vanish as $r \to 0$ and as $r \to \infty$.

17.3. Application to variable flow of heat

This may be illustrated by a number of typical problems.

I. *The region* $-\infty < x < \infty$ *with initial temperature* $f(x)$

The equation to be solved is

$$\frac{\partial^2 v}{\partial x^2} - \frac{1}{\kappa}\frac{\partial v}{\partial t} = 0, \tag{1}$$

in the region $-\infty < x < \infty$, $t > 0$. The complex Fourier transform of (1) is by 17.2 (4)

$$\frac{dV(\omega)}{dt} + \kappa\omega^2 V(\omega) = 0, \qquad t > 0,$$

with $V(\omega) = F(\omega)$, when $t = 0$. It follows that

$$V(\omega) = F(\omega)e^{-\kappa\omega^2 t},$$

and, by the Inversion Theorem, 17.2 (2),

$$v = \frac{1}{(2\pi)^{\frac{1}{2}}} \int_{-\infty}^{\infty} e^{-i\omega x - \kappa\omega^2 t} F(\omega)\, d\omega,$$

which is just 2.3 (17).

II. *The region* $x > 0$ *with zero initial temperature and* $v = v_0$, *when* $x = 0$, $t > 0$

Here the differential equation to be solved is (1) in $x > 0$, and we notice that, of the sine and cosine transforms which are both available for the region $x > 0$, the former introduces the temperature v_0 in 17.2 (8) and the latter the unknown value of $[\partial v/\partial x]_{x=0}$ in 17.2 (12). We therefore use a sine transform which, when applied to (1), gives

$$\frac{dV_s}{dt} + \kappa\omega^2 V_s = \kappa\omega v_0 (2/\pi)^{\frac{1}{2}}$$

† Titchmarsh, loc. cit., § 1.29. The result is related to the inversion formulae for the Fourier and Laplace transforms, indeed the latter is often called the 'Fourier–Mellin theorem'.

to be solved with $V_s = 0$ when $t = 0$. It follows that

$$V_s = \frac{v_0 2^{\frac{1}{2}}}{\omega \pi^{\frac{1}{2}}} (1 - e^{-\kappa \omega^2 t}).$$

Then, by the Inversion Formula, 17.2 (6),

$$v = \frac{2v_0}{\pi} \int_0^\infty (1 - e^{-\kappa \omega^2 t}) \sin \omega x \frac{d\omega}{\omega}, \tag{2}$$

which reduces to the usual result $v_0 \operatorname{erfc}[x/2(\kappa t)^{\frac{1}{2}}]$ after some manipulation.

This problem illustrates very clearly the comparison between the Fourier and Laplace transformation methods in one-variable problems of this type. Firstly, if adequate tables of transforms are available, the amount of work involved in the two methods is the same. Secondly, if tables of transforms are not available, a certain amount of manipulation of the integrals obtained from the inversion theorem is necessary in either case. The essential superiority of the Laplace transformation for this type of problem appears in connexion with the boundary conditions since it treats all boundary conditions in the same way. On the other hand, it was necessary to use a sine transform above since v was specified at $x = 0$; if the flux had been specified a cosine transformation would have had to be used; for a radiation boundary condition neither of these is suitable and a new type of transformation has to be developed; for the boundary condition 1.9 F, yet another, and so on.

III. *The infinite region $r > 0$, $-\infty < z < \infty$, $0 < \theta < 2\pi$. Initial temperature a function $f(r)$ of r only*

Here we have to solve

$$\frac{\partial^2 v}{\partial r^2} + \frac{1}{r} \frac{\partial v}{\partial r} - \frac{1}{\kappa} \frac{\partial v}{\partial t} = 0 \tag{3}$$

with $v = f(r)$ when $t = 0$. The Hankel transform of order zero of (3) is, by 17.2 (15),

$$\frac{dV_0(\sigma)}{dt} + \kappa \sigma^2 V_0(\sigma) = 0 \tag{4}$$

to be solved with $V_0(\sigma) = F_0(\sigma)$ when $t = 0$. This gives

$$V_0(\sigma) = e^{-\kappa \sigma^2 t} F_0(\sigma),$$

and, by the inversion formula, 17.2 (14),

$$v(r) = \int_0^\infty e^{-\kappa \sigma^2 t} F_0(\sigma) \sigma J_0(\sigma r) \, d\sigma,$$

which reduces to 10.3 (11) on evaluating an integral.

17.4. Application to steady flow of heat

Probably the most important property of the transformations listed in § 17.2 is that they reduce Laplace's equation in two variables to an ordinary differential equation in many important special cases.

Thus, if
$$\frac{\partial^2 v}{\partial x^2} + \frac{\partial^2 v}{\partial y^2} = 0, \tag{1}$$

applying the complex Fourier transform in x gives

$$\frac{d^2 V}{dy^2} - \omega^2 V = 0, \tag{2}$$

with similar results for the sine and cosine transforms.

Applying the Hankel transform of order zero to

$$\frac{1}{r}\frac{\partial}{\partial r}\left(r\frac{\partial v}{\partial r}\right) + \frac{\partial^2 v}{\partial z^2} = 0, \tag{3}$$

gives
$$\frac{d^2 V_0}{dz^2} - \sigma^2 V_0 = 0. \tag{4}$$

Applying the Mellin transform to

$$\frac{1}{r}\frac{\partial}{\partial r}\left(r\frac{\partial v}{\partial r}\right) + \frac{1}{r^2}\frac{\partial^2 v}{\partial \theta^2} = 0, \tag{5}$$

gives
$$\frac{d^2 V_m}{d\theta^2} + s^2 V_m = 0. \tag{6}$$

As examples, we consider the following:

I. *The region $z > 0$, $0 \leqslant r < \infty$, with constant flux Q over $0 \leqslant r < a$ and zero flux over $r > a$*

We have to solve (3) with the boundary conditions

$$-K\left[\frac{\partial v}{\partial z}\right]_{z=0} = Q, \quad 0 \leqslant r < a \atop = 0, \quad r > a \bigg\}. \tag{7}$$

Taking the Hankel transform of order zero, V_0 has to satisfy (4) and to vanish as $z \to \infty$; its boundary condition at $z = 0$ is the Hankel transform of (7), that is

$$-K\left[\frac{dV_0}{dz}\right]_{z=0} = Q\int_0^a rJ_0(\sigma r)\,dr = \frac{aQ}{\sigma}J_1(\sigma a).$$

It follows that
$$V_0 = (aQ/K\sigma^2)J_1(\sigma a)e^{-\sigma z},$$

and, by the inversion formula 17.2 (14),

$$v = \frac{aQ}{K} \int_0^\infty e^{-\sigma z} J_0(\sigma r) J_1(\sigma a) \frac{d\sigma}{\sigma}, \tag{8}$$

in agreement with 8.2 (7). The results of § 8.2 were derived by the use of the integrals 8.2 (3), (4) to draw attention to an important and simple method: if Hankel's integral theorem had been used in the classical manner the analysis would have been very similar to that above.

II. *The wedge* $-\alpha < \theta < \alpha$, $r > 0$, *with* $v = 1$, $0 < r < a$, *and* $v = 0$, $r > a$, *on* $\theta = \pm\alpha$.

We have to solve (5) with the above boundary conditions. Taking the Mellin transform, V_m has to satisfy (6) with

$$V_m = \int_0^a r^{s-1}\, dr = a^s/s, \quad \text{when } \theta = \pm\alpha.$$

It follows that

$$V_m = \frac{a^s \cos s\theta}{s \cos s\alpha}.$$

Using the inversion formula 17.2 (17) then gives

$$v = \frac{1}{2\pi i} \int_{\gamma-i\infty}^{\gamma+i\infty} \frac{a^s r^{-s} \cos s\theta\, ds}{s \cos s\alpha}$$

$$= \frac{1}{2} + \frac{1}{\pi} \int_0^\infty \frac{\sin[\zeta \ln(a/r)]\cosh \zeta\theta\, d\zeta}{\zeta \cosh \zeta\alpha}.$$

17.5. Finite transforms

The same procedure may be used to rewrite the theory of Fourier and allied series in the transform notation.† Here, to conform with previous results, we shall take the interval to be $0 < x < l$, though $0 < x < \pi$ is more usual.

Writing $f_s[v(x)]$ for the operation of taking the finite sine transform and $V_s(n)$, $n = 1, 2, 3,...$ for its value, we define

$$f_s[v(x)] \equiv V_s(n) = \int_0^l v(x)\sin\frac{n\pi x}{l}\, dx. \tag{1}$$

The inversion formula, which is Fourier's sine series, is

$$v(x) = \frac{2}{l} \sum_{n=1}^\infty V_s(n)\sin\frac{n\pi x}{l}. \tag{2}$$

† For the finite Fourier transform see: Kniess, *Math. Z.* **44** (1939) 266–91; Brown, *Bull. Amer. Math. Soc.* **48** (1942) 522; ibid. **50** (1944) 376; *J. Appl. Phys.* **14** (1943) 609; Koschmieder, *Dtsch. Math.* **5** (1945) 521–45; Roettinger, *Quart. Appl. Math.* **5** (1947) 298–319; Jacobson, *Bull. Amer. Math. Soc.* **55** (1949) 804–9, *Quart. Appl. Math.* **7** (1949) 293–302. For the finite Hankel transform see Sneddon, *Phil. Mag.* (7) **37** (1946) 17–25.

Integrating by parts gives

$$f_s\left[\frac{\partial^2 v}{\partial x^2}\right] = \frac{n\pi}{l}\{v(0)-(-1)^n v(l)\}-\frac{n^2\pi^2}{l^2}V_s(n). \tag{3}$$

Similarly, writing $f_c[v(x)]$ for the operation of taking the finite cosine transform, and $V_c(n)$, $n = 0, 1, 2,\ldots$, for its value, we have

$$f_c[v(x)] \equiv V_c(n) = \int_0^l v(x)\cos\frac{n\pi x}{l}\,dx, \tag{4}$$

and the inversion formula is

$$v(x) = \frac{1}{l}V_c(0)+\frac{2}{l}\sum_{n=1}^{\infty}V_c(n)\cos\frac{n\pi x}{l}. \tag{5}$$

Also

$$f_c\left[\frac{\partial^2 v}{\partial x^2}\right] = (-1)^n\left[\frac{\partial v}{\partial x}\right]_{x=l}-\left[\frac{\partial v}{\partial x}\right]_{x=0}-\frac{n^2\pi^2}{l^2}V_c(n). \tag{6}$$

It appears from (3) that the sine transform will be useful for problems involving prescribed temperature at the boundaries, and from (6) that the cosine transform is available for problems involving flux. For the radiation boundary condition, a new type of transform based on the expansion of § 3.9 has to be developed. In the same way, for radial flow, finite Hankel transforms can be defined for the regions $0 < r < a$, and $a < r < b$, yet another set being needed for the radiation boundary condition.

I. *The slab $0 < x < l$ with zero initial temperature and the surfaces $x = 0$ and $x = l$ kept at unity for $t > 0$*

Since the surface temperatures are involved, we take the sine transform of

$$\frac{\partial^2 v}{\partial x^2}-\frac{1}{\kappa}\frac{\partial v}{\partial t} = 0, \quad 0 < x < l,$$

which is, by (3),

$$\frac{n\pi}{l}[1-(-1)^n]-\frac{n^2\pi^2}{l^2}V_s(n)-\frac{1}{\kappa}\frac{dV_s(n)}{dt} = 0,$$

to be solved with $V_s = 0$ when $t = 0$. The solution is

$$V_s(n) = \frac{l[1-(-1)^n]}{n\pi}\{1-e^{-\kappa n^2\pi^2 t/l^2}\}.$$

Therefore, by (2),

$$v = \frac{4}{\pi}\sum_{m=0}^{\infty}\frac{[1-e^{-\kappa(2m+1)^2\pi^2 t/l^2}]}{(2m+1)}\sin\frac{(2m+1)\pi x}{l},$$

which reduces to 3.4 (1) on using the sine series for unity.

II. *Steady flow in the rectangle $0 < x < a$, $0 < y < b$. $y = 0$ kept at $u(x)$, the other surfaces at zero*

Taking the sine transform with respect to x we get

$$\frac{d^2 V_s}{dy^2}-\frac{n^2\pi^2}{a^2}V_s = 0,$$

to be solved with $V_s = U_s$ when $y = 0$, and $V_s = 0$ when $y = b$. The solution is

$$V_s = U_s\frac{\sinh n\pi(b-y)/a}{\sinh n\pi b/a}.$$

Then by (2)

$$v = \frac{2}{a} \sum_{n=1}^{\infty} U_s(n) \frac{\sinh n\pi(b-y)/a}{\sinh n\pi b/a} \sin \frac{n\pi x}{a}$$

$$= \frac{2}{a} \sum_{n=1}^{\infty} \frac{\sinh n\pi(b-y)/a}{\sinh n\pi b/a} \sin \frac{n\pi x}{a} \int_0^a u(x')\sin \frac{n\pi x'}{a} \, dx'$$

in agreement with 5.3 (9).

17.6. Successive transformations

In problems involving several variables, a great economy in notation can be achieved by making integral transformations successively with regard to several variables. Either a Laplace transformation may be used first to remove the time-variable, followed by other integral transforms on the space-variables, or successive integral transforms may be used on the space-variables.[†]

To illustrate the method we consider a problem already solved in § 15.11, VIII which is used as an example by J. C. Cooke,[‡] namely, *conduction of heat in the wedge* $r > 0, 0 < \theta < \theta_0$, *with zero initial temperature and unit surface temperature.*

The equation to be solved is

$$\frac{\partial^2 v}{\partial r^2} + \frac{1}{r}\frac{\partial v}{\partial r} + \frac{1}{r^2}\frac{\partial^2 v}{\partial \theta^2} - \frac{1}{\kappa}\frac{\partial v}{\partial t} = 0, \quad r > 0, \quad 0 < \theta < \theta_0, \tag{1}$$

with

$$v = 0, \quad t = 0, \quad r > 0, \quad 0 < \theta < \theta_0, \tag{2}$$

$$v = 1, \quad \theta = 0, \quad \text{and} \quad \theta = \theta_0, \quad t > 0. \tag{3}$$

Writing V_s for the finite sine transform of v with respect to θ, it follows from 17.5 (3) that V_s satisfies

$$\frac{\partial^2 V_s}{\partial r^2} + \frac{1}{r}\frac{\partial V_s}{\partial r} - \frac{\nu^2}{r^2}V_s - \frac{1}{\kappa}\frac{\partial V_s}{\partial t} = -\frac{\nu[1-(-1)^n]}{r^2}, \tag{4}$$

where

$$\nu = n\pi/\theta_0. \tag{5}$$

Writing $V_{s,\nu}$ for the Hankel transform of order ν of V_s, it follows from 17.2 (15) that $V_{s,\nu}$ satisfies

$$\frac{1}{\kappa}\frac{dV_{s,\nu}}{dt} + \sigma^2 V_{s,\nu} = \nu[1-(-1)^n]\int_0^{\infty} \frac{J_\nu(\sigma r) \, dr}{r} = [1-(-1)^n], \tag{6}$$

since

$$\int_0^{\infty} \frac{J_\nu(z)}{z} \, dz = \frac{1}{\nu}, \tag{7}$$

by *W.B.F.*, 13.24 (1). The solution of this with $V_{s,\nu} = 0$ when $t = 0$ is

$$V_{s,\nu} = \frac{[1-(-1)^n]}{\sigma^2}\{1-e^{-\kappa\sigma^2 t}\}. \tag{8}$$

---

† Some examples of the use of repeated transformations are given by Delavault, *C.R. Acad. Sci. Paris*, **236** (1953) 2484–6; ibid. **237** (1953) 1067–8; Sneddon, *Proc. Glasgow Math. Ass.* **1** (1952) 21–27.

‡ *Amer. Math. Mon.* **62** (1955) 331–4.

Inverting the Hankel transform by 17.2 (14), and using (7) and (5), gives

$$V_s = \frac{\theta_0[1-(-1)^n]}{n\pi} - [1-(-1)^n] \int_0^\infty \frac{1}{\sigma} e^{-\kappa\sigma^2 t} J_\nu(\sigma r)\, d\sigma, \tag{9}$$

and, finally, inverting the finite sine transform by 17.5 (2), we get

$$v = \frac{2}{\pi} \sum_{n=1}^\infty \frac{[1-(-1)^n]}{n} \sin\frac{n\pi\theta}{\theta_0} - \frac{2}{\theta_0} \sum_{n=1}^\infty [1-(-1)^n] \sin\frac{n\pi\theta}{\theta_0} \int_0^\infty \frac{1}{\sigma} e^{-\kappa\sigma^2 t} J_\nu(\sigma r)\, d\sigma$$

$$= 1 - \frac{4}{\theta_0} \sum_{n=0}^\infty \sin\frac{(2n+1)\pi\theta}{\theta_0} \int_0^\infty \frac{1}{\sigma} e^{-\kappa\sigma^2 t} J_{(2n+1)\pi/\theta_0}(\sigma r)\, d\sigma, \tag{10}$$

on using the sine series for unity. This is the result 15.11 (24). The comparison between the two methods is interesting; they are of about the same length, they both require the result (7), but, while the method of § 15.11 assumed a form 15.11 (21) for $\bar{v}$, the present method proceeds directly without any assumptions.

XVIII

NUMERICAL METHODS

18.1. Introductory

In the past few years a great deal of attention has been paid to the development of numerical methods for the solution of problems in conduction of heat. This has been due partly to the increasing interest in numerical analysis, and partly to the possibility of solving important practical problems by the use of electronic and other calculating machines.

It is clear from the preceding chapters that the exact solutions available are practically confined to linear problems on regions of simple shapes. If bodies of complicated shapes, or non-linear boundary conditions, have to be considered, recourse must be had to numerical methods.† It is obviously impossible to give anything like a complete treatment here, but it seems desirable to give a survey of the present position and to indicate methods which can easily be employed. The worker with exact solutions often comes to a stage where he wishes to check the adequacy of approximations (such as linearization) or to solve simple problems for which no exact solution is available. It is, in fact, extremely easy to use simple numerical methods such as those of § 18.3 since these do not need an apprenticeship to numerical analysis. For this reason, most space has been given to the simpler step-by-step methods: in fact, because of their application to machine calculation, it is also these which have been most studied theoretically. Finally, the point must be made that, while the methods described in this chapter are the most obvious ones from the point of view of finite differences, they are far from being the only ones.‡

18.2. Finite differences§

Suppose that we know the values v_m of a function $v(x)$ at regular intervals $x = m\epsilon, m = ..., -2, -1, 0, 1, 2,...,$ of its argument. Then the

† From the practical point of view, a full account is given in Dusinberre, *Numerical Analysis of Heat Flow* (McGraw-Hill, 1949). Shorter accounts are given in most works on heat transfer, and in Ingersoll, Zobel, and Ingersoll, *Heat Conduction* (University of Wisconsin Press, 1954). Methods appropriate to problems on variable diffusivity are described in *M.D.*, Chaps. IX, X.

‡ For example, Monte Carlo methods are used by King, *Ind. Eng. Chem.* **43** (1951) 2475; an interesting method using steps in v has been developed by Philip, *Trans. Faraday Soc.* **51** (1955) 885, and other methods for variable diffusivity are described in *M.D.*

§ For the theory of finite differences see Hartree, *Numerical Analysis* (Oxford, 1952);

first, second, third,... forward differences, Δv_m, $\Delta^2 v_m$, $\Delta^3 v_m$,... of v_m are defined to be

$$\Delta v_m = v_{m+1} - v_m, \tag{1}$$

$$\Delta^2 v_m = \Delta v_{m+1} - \Delta v_m = v_{m+2} - 2v_{m+1} + v_m, \tag{2}$$

$$\Delta^3 v_m = \Delta^2 v_{m+1} - \Delta^2 v_m = v_{m+3} - 3v_{m+2} + 3v_{m+1} - v_m, \tag{3}$$

and so on. These are called 'forward' differences because they involve successive values v_m, v_{m+1},... working forward, or in the direction of m increasing, from v_m. In the same way, backward differences (working backwards from v_m) may be defined, but these will not be needed here. Clearly, however, it may be convenient to have a notation which involves points symmetrical about m, and for this reason the central difference notation is introduced. In this, the first, second,... *central* differences are defined as

$$\delta^1 v_{m+\frac{1}{2}} = \Delta v_m = v_{m+1} - v_m, \tag{4}$$

$$\delta^2 v_m = \delta^1 v_{m+\frac{1}{2}} - \delta^1 v_{m-\frac{1}{2}} = \Delta v_m - \Delta v_{m-1} = v_{m+1} - 2v_m + v_{m-1}, \tag{5}$$

and so on. It will be seen that, as in (4), differences of odd order involve points distributed symmetrically about $m+\frac{1}{2}$ and not about m; to get formulae with symmetry about m, the so-called *mean differences* are introduced, defined by

$$\mu\delta^1 v_m = \tfrac{1}{2}\{\delta^1 v_{m+\frac{1}{2}} + \delta^1 v_{m-\frac{1}{2}}\} = \tfrac{1}{2}(v_{m+1} - v_{m-1}), \tag{6}$$

$$\mu\delta^3 v_m = \tfrac{1}{2}\{\delta^3 v_{m+\frac{1}{2}} + \delta^3 v_{m-\frac{1}{2}}\}, \tag{7}$$

where μ denotes the operation of taking the mean of the values for $m\pm\frac{1}{2}$.

The whole of the theory is now based on the assumption that the values ..., v_{-2}, v_{-1}, v_0, v_1, v_2,... are known and that a 'difference table' of them and their successive differences (1), (2),... is formed. In principle, a polynomial can then be found which passes through any specified number of these points (its coefficients can be expressed in terms of the v_m or their differences) and any operations such as interpolation, differentiation or integration are performed on this polynomial. Thus the first derivative $[dv/dx]_m$ implies the first derivative of the interpolating polynomial at the point $m\epsilon$. In this sense, if the values of a function are known at the points $m\epsilon$, a great many formulae are available which express its derivatives at any point in terms of either its tabulated values

Milne, *Numerical Calculus* (Princeton, 1949). The classical papers on the application of differences to the solution of partial differential equations are Richardson, *Phil. Trans. Roy. Soc.* A, **210** (1910) 307; ibid. A, **226** (1927) 299. It may be remarked that while a background knowledge of finite differences is very desirable, it is not essential for the present purposes: the essential results, such as 18.3 (4), can be obtained immediately from Taylor's theorem.

or their differences. For example

$$\epsilon\left[\frac{dv}{dx}\right]_m = \Delta v_m - \tfrac{1}{2}\Delta^2 v_m + \tfrac{1}{3}\Delta^3 v_m + \cdots, \tag{8}$$

$$= \tfrac{1}{2}(-3v_m + 4v_{m+1} - v_{m+2}) + \tfrac{1}{3}\Delta^3 v_m + \cdots, \tag{9}$$

$$= \mu\delta^1 v_m - \tfrac{1}{6}\mu\delta^3 v_m + \tfrac{1}{30}\mu\delta^5 v_m - \cdots, \tag{10}$$

$$= \tfrac{1}{2}(v_{m+1} - v_{m-1}) - \tfrac{1}{6}\mu\delta^3 v_m + \cdots, \tag{11}$$

$$\epsilon^2\left[\frac{d^2v}{dx^2}\right]_m = \Delta^2 v_m - \Delta^3 v_m + \cdots, \tag{12}$$

$$= \delta^2 v_m - \tfrac{1}{12}\delta^4 v_m + \tfrac{1}{90}\delta^6 v_m - \cdots, \tag{13}$$

$$= (v_{m-1} - 2v_m + v_{m+1}) - \tfrac{1}{12}\delta^4 v_m + \cdots. \tag{14}$$

Proofs of these and other results may be found in the texts referred to. In the present context, in which ϵ is regarded as small, the differences of the nth order are of order ϵ^n. Thus neglecting an nth difference implies an error which is $O(\epsilon^n)$. In the simple treatments given below, higher differences will be neglected, only the first terms on the right-hand sides of (8) to (14) being used. The error introduced in this way will depend on the order of the first term neglected, and thus a formula such as (11) in which only third differences are neglected and the error in dv/dx is $O(\epsilon^2)$ is to be regarded as preferable to (8) in which a second difference is neglected and so the error in dv/dx is $O(\epsilon)$. The higher differences are always retained and made use of in serious numerical analysis: in this way the time taken to solve a problem may be reduced and the accuracy increased, but at the expense of an increase of mathematical sophistication.

Two other results, which are needed for work in cylindrical or spherical coordinates, may be given here. It follows from (11) and (14), neglecting third and higher differences, that

$$\left[\frac{d^2v}{dx^2} + \frac{1}{x}\frac{dv}{dx}\right]_m = \frac{1}{\epsilon^2}\{v_{m-1} - 2v_m + v_{m+1}\} + \frac{1}{2m\epsilon^2}(v_{m+1} - v_{m-1})$$

$$= \frac{1}{2m\epsilon^2}\{(2m-1)v_{m-1} - 4mv_m + (2m+1)v_{m+1}\}, \tag{15}$$

provided $m \neq 0$. If $(dv/dx) = 0$ when $x = 0$, the result is

$$\left[\frac{d^2v}{dx^2} + \frac{1}{x}\frac{dv}{dx}\right]_0 = \frac{4}{\epsilon^2}(v_1 - v_0). \tag{16}$$

Similarly

$$\left[\frac{d^2v}{dx^2} + \frac{2}{x}\frac{dv}{dx}\right]_m = \frac{1}{m\epsilon^2}\{(m-1)v_{m-1} - 2v_m + (m+1)v_{m+1}\}, \tag{17}$$

if $m \neq 0$, and if $(dv/dx) = 0$ when $x = 0$

$$\left[\frac{d^2v}{dx^2} + \frac{2}{x}\frac{dv}{dx}\right]_0 = \frac{6}{\epsilon^2}(v_1 - v_0). \tag{18}$$

When functions of two or more variables are involved, partial differences may be defined in the same way. For example, if $v(x, y)$ is a function of x and y and we choose the same interval ϵ in x and y, so that $v_{m,n} = v(m\epsilon, n\epsilon)$, it follows from (14) that

$$\left[\frac{\partial^2 v}{\partial x^2}\right]_{m,n} = \epsilon^{-2}(v_{m-1,n} - 2v_{m,n} + v_{m+1,n}) + O(\epsilon^2), \tag{19}$$

$$\left[\frac{\partial^2 v}{\partial y^2}\right]_{m,n} = \epsilon^{-2}(v_{m,n-1} - 2v_{m,n} + v_{m,n+1}) + O(\epsilon^2). \tag{20}$$

Therefore

$$\left[\frac{\partial^2 v}{\partial x^2} + \frac{\partial^2 v}{\partial y^2}\right]_{m,n} = \epsilon^{-2}(v_{m-1,n} + v_{m+1,n} + v_{m,n-1} + v_{m,n+1} - 4v_{m,n}) + O(\epsilon^2), \tag{21}$$

a result which is needed for the study of Laplace's equation in two dimensions.

18.3. Linear flow of heat in an infinite region

The equation to be solved is

$$\frac{\partial^2 v}{\partial x^2} - \frac{1}{\kappa}\frac{\partial v}{\partial t} = 0, \qquad -\infty < x < \infty, \tag{1}$$

with v a prescribed function of x when $t = 0$. Here, κ may be a function of v; it was shown in § 1.6 that the general case in which K is a function of v can be reduced to the form (1) by a change of variable.

Suppose we now choose an interval ϵ in x, and write $v_m(t)$ for the value of $v(x, t)$ at the point $x = m\epsilon$, $m = ..., -2, -1, 0, 1, 2,...$. Replacing† $\partial^2 v/\partial x^2$ in (1) by a difference, using 18.2 (14) and neglecting higher differences, gives

$$v_{m+1}(t) - 2v_m(t) + v_{m-1}(t) - \frac{\epsilon^2}{\kappa}\frac{dv_m(t)}{dt} = 0, \tag{2}$$

to be solved with known values of $v_m(0)$. This is precisely the system 15.10 (2) with $M_r c_r = \rho c \epsilon$, $H_r = K/\epsilon$, and (for constant κ) may be solved by the methods of that section. The physical significance of the approximation being made can now be seen—the set (2) corresponds to dividing

---

† Alternatively the time-derivative in (1) can be replaced by a difference. This yields a set of differential equations in x which can be solved successively, cf. Hartree and Womersley, *Proc. Roy. Soc.* A, **161** (1937) 353.

the solid into slabs of thickness ϵ, and replacing these by slabs of perfect conductor of the same heat capacity separated by thermal resistance†
ϵ/K which is that of a slab of the original material of thickness ϵ.

If κ is constant, a comparison of solutions of (2) obtained by the methods of § 15.10 with the accurate results allows the effects of the approximation to be studied. For variable κ, and also non-linear boundary conditions, (2) has been solved by the use of the differential analyser.‡ One result of these investigations is that it has been found that in many cases good results can be obtained with surprisingly large values of ϵ: for a slab of thickness a, quite good results are obtained with $\epsilon = a/6$.

To get a completely numerical method, it is necessary also to replace the time-derivative in (2) by a difference. To do this, we choose an interval of time τ, and write $v_{m,n}$ for $v(m\epsilon, n\tau)$. Both ϵ and τ are as yet unspecified, and their choice will be discussed later.

There are various formulae for $\partial v/\partial t$ which may be used and, of these, the simplest is 18.2 (8), which, neglecting the second difference, is

$$\left[\frac{\partial v_m}{\partial t}\right]_{t=n\tau} = \frac{v_{m,n+1}-v_{m,n}}{\tau}. \tag{3}$$

Using this in (2) gives

$$v_{m,n+1} = M(v_{m+1,n}+v_{m-1,n})-(2M-1)v_{m,n}, \tag{4}$$

where
$$M = \kappa\tau/\epsilon^2 \tag{5}$$

is sometimes called the modulus.§ It will be observed that (4) gives the values of v at time $(n+1)\tau$ immediately in terms of those at time $n\tau$, and thus may be used to work forward from the known values at $t = 0$. Since a second difference has been neglected in (3), it might be expected that the method would not be a very accurate one: it does, however, prove to be surprisingly accurate and to be the only really simple method available; it is almost the only one which will be discussed here.

It is now necessary to discuss the size of ϵ and τ, and their relation to the reliability of the method. Clearly, ϵ is to some extent fixed by the problem in hand, for a slab of thickness a, it was remarked that $\epsilon = \tfrac{1}{6}a$

† The same approximation is made in many analogue machines for studying heat flow. The most important of these is the resistance-capacity network system known as the 'heat and mass flow analyser', Paschkis and Baker, *Trans. Amer. Soc. Mech. Engrs.* **64** (1942) 105–12, on which a great deal of work is now done. See also Paschkis and Heisler, *Elect. Engng.* **63** (1944) 165; *J. Appl. Phys.* **17** (1946) 246–54. Moore, *Ind. Eng. Chem.* **28** (1936) 704, has constructed a hydraulic analogue.

‡ Eyres, Hartree, *et al.*, 'The calculation of variable heat flow in solids', *Phil. Trans. Roy. Soc.* A, **240** (1946) 1–57, give a survey of the important problems and methods.

§ Frequently, the case $\kappa = 1$ is considered and τ/ϵ^2 is called the mesh-ratio.

gave quite good results for the system (2): in any case, ϵ will not be too small a fraction of a. ϵ and τ enter together in M in (4) and (5), so, to get a solution in a reasonable number of steps, it is desirable to choose τ as large as is safe, but that some restriction on its value is necessary may be seen by the following argument.† Suppose that when $t = n\tau$ the maximum error in any of the $v_{m,n}$ is η, then by (4) if $dv_{m,n+1}$ is the small change in $v_{m,n+1}$ caused by small changes $dv_{m,n}$ in the $v_{m,n}$, and if M is constant,

$$|dv_{m,n+1}| \leqslant M|dv_{m+1,n}| + M|dv_{m-1,n}| + |2M-1||dv_{m,n}|$$
$$\leqslant \{2M + |2M-1|\}\eta. \tag{6}$$

Now if the method is to be useful, errors must not grow,‡ that is, we must have $|dv_{m,n+1}| \leqslant \eta$, and by (6) this requires

$$M = \kappa\tau/\epsilon^2 \leqslant \tfrac{1}{2}. \tag{7}$$

This is the required restriction on τ, usually called the stability condition. It is obviously a sufficient condition, but by no means a necessary one. Many elaborate discussions of stability based on solutions of the difference equations have been given.§

Another simple method of investigating the stability and accuracy of numerical solutions is by studying the case $v_{0,0} = 1$, $v_{m,0} = 0$, $m \neq 0$. This may be regarded either as showing how the effect of a unit error is propagated through the system, or as the Green's function of the system corresponding to the liberation of a quantity of heat $\epsilon\rho c$ at the origin

† Due to Price and Slack, *Brit. J. Appl. Phys.* **3** (1952) 379–84, where the effect of boundary conditions on stability is also discussed.

‡ Apart from accidental errors, there are 'round-off' errors in the last place of decimals used.

§ A great deal of accurate analytical work has been done on this problem mostly with reference to (4) with linear boundary conditions. It must be emphasized that there are several different but interrelated questions, in particular, firstly, convergence, that is, the convergence of the solutions of the difference equation to the solution of the partial differential equation as the steps in space and time are made smaller, and secondly, stability, that is, the question of whether numerical errors and round-off errors die out or increase as the time increases. Fowler, *Quart. Appl. Math.* **3** (1946) 361–76, discusses convergence by studying exact solutions of the difference equation (4). O'Brien, Hyman, and Kaplan, *J. Math. Phys.* **29** (1950) 223–51, study the stability of (4) by a method due to von Neumann and point out the inherent superiority of 'implicit' relationships such as (11). Leutert, *Proc. Amer. Math. Soc.* **2** (1951) 433–9, *J. Math. Phys.* **30** (1951) 245–51, points out that convergence is possible in some cases in which the stability condition is not satisfied. The relationship between convergence and stability is further discussed by Hildebrand, *J. Math. Phys.* **31** (1952) 35–41, and Evans, Brousseau, and Keirstead, *J. Math. Phys.* **34** (1955) 267–85; the latter paper gives a very full discussion with interesting numerical examples. Most of these papers emphasize the fact that convergence and stability depend on the form of the initial and boundary conditions. They illustrate the difficulties involved in dealing with linear problems: such questions for non-linear problems have, as yet, hardly been touched.

at $t = 0$. The exact solution for this case is, by 10.3 (4),

$$v_{m,n} = \frac{1}{2(\pi Mn)^{\frac{1}{2}}} e^{-m^2/4Mn}. \tag{8}$$

To illustrate this method, a portion of the tabulation of $v_{m,n}$ for this case is set out below for the values 0·25, 0·5, and 0·6 of M. In all cases only the portion $0 \leqslant m \leqslant 5$ of the table is shown, results are, of course, symmetrical about $m = 0$.

$M = 0{\cdot}25$	1	0	0	0	0	0
	0·5	0·25	0	0	0	0
	0·375	0·25	0·062	0	0	0
	0·312	0·234	0·094	0·016	0	0
	0·273	0·219	0·109	0·031	0·004	0
	0·246	0·205	0·117	0·044	0·010	0·001
	0·226	0·193	0·121	0·054	0·016	0·003
	0·209	0·183	0·122	0·061	0·022	0·006
	0·196	0·175	0·122	0·067	0·028	0·009
	0·185	0·167	0·121	0·071	0·033	0·012
	0·176	0·160	0·120	0·074	0·037	0·015

$M = 0{\cdot}5$	1	0	0	0	0	0
	0	0·5	0	0	0	0
	0·5	0	0·25	0	0	0
	0	0·375	0	0·125	0	0
	0·375	0	0·25	0	0·062	0
	0	0·312	0	0·156	0	0·031

$M = 0{\cdot}6$	1	0	0	0	0	0
	−0·2	0·6	0	0	0	0
	0·76	−0·24	0·36	0	0	0
	−0·44	0·72	−0·216	0·216	0	0
	0·952	−0·538	0·605	−0·173	0·130	0
	−0·836	1·042	−0·548	0·476	−0·130	0·078
	1·418	−1·039	1·020	−0·502	0·358	−0·093

Considering first the case $M = 0·6$, it appears that an error of unity in one figure will ultimately lead to large errors which oscillate in sign. Such an effect rapidly becomes obvious in practical computing. This case, according to the criterion (7), is unstable.

In the cases $M = 0·5$ and $M = 0·25$ which, by (7), should be stable, it appears from (8) that the values in the sixth row of the former, and the eleventh row of the latter, should both be $(10\pi)^{-\frac{1}{2}}\exp(-m^2/10)$, that is

| 0·178 | 0·161 | 0·120 | 0·073 | 0·036 | 0·015 |

It appears that the values for $M = 0·25$ are in excellent agreement with these, there being only small discrepancies in the last figures. The values for $M = 0·5$ are alternately zero and approximately twice the accurate values (this may be regarded as being due to the same total quantity of heat being concentrated in alternate slabs) so that, in fact, a curve drawn through half the values gives an increasingly good approximation to the correct result as n increases. If the $v_{m,0}$ are points on a continuous curve, this oscillation is smoothed out.

In this case, $M = 0·5$, (4) takes the particularly simple form

$$v_{m,n+1} = \tfrac{1}{2}(v_{m-1,n}+v_{m+1,n}) \tag{9}$$

so that the value of v at $m\epsilon$, $(n+1)\tau$ is just the arithmetic mean of the values at $(m\pm1)\epsilon$, $n\tau$. This process of taking the arithmetic mean can be carried out either numerically or graphically, and in the latter form has been familiar for many years in works on heat transfer under the name of *Schmidt's method*.†

Both (4) and (9) are adequate for most practical purposes where an accuracy of a few per cent is sufficient. They have the advantage of involving repetition of extremely simple operations, and are thus well adapted to the use of either unskilled computers or electronic calculating machines.

Considering now the possibility of finding difference equations which represent the differential equation more accurately, it is natural to replace the crude 18.2 (8) by 18.2 (11) in which only third differences have to be neglected. Using 18.2 (11) in (2) gives

$$v_{m,n+1} = v_{m,n-1}+2M(v_{m-1,n}-2v_{m,n}+v_{m+1,n}), \tag{10}$$

where, as before, $M = \kappa\tau/\epsilon^2$. This, like (4), determines $v_{m,n+1}$ immediately in terms of known quantities, though these are now in the two preceding rows. Unfortunately, this system proves to be unstable. Thus, if we calculate the progress of a unit error as before we get for $M = \tfrac{1}{2}$

1	0	0	0	0
−2	1	0	0	0
7	−4	1	0	0

† Schmidt, *Föppl's Festschrift* (Springer, 1924). Nussbaum, *Z. angew. Math. Mech.* **8** (1928) 133–42. Nessi and Nisolle, *Chal. et Industr.* **9** (1928) 193, and Patton, *Ind. Eng. Chem.* **36** (1944) 990–6, extend the method to cylindrical and spherical bodies.

18.2 (11) is used in a particularly satisfactory method due to Crank and Nicholson.† They replace the differential equation by a difference equation at $(n+\tfrac{1}{2})\tau$, using 18.2 (11) for the value of $(\partial v/\partial t)$ at $(n+\tfrac{1}{2})\tau$, and the mean of the values 18.2 (14) at $n\tau$ and $(n+1)\tau$ for the value of $(\partial^2 v/\partial x^2)$ at $t = (n+\tfrac{1}{2})\tau$. This gives

$$v_{m,n+1} - v_{m,n}$$

$$= \frac{\kappa\tau}{2\epsilon^2}\{(v_{m-1,n}+v_{m-1,n+1}) - 2(v_{m,n}+v_{m,n+1}) + (v_{m+1,n}+v_{m+1,n+1})\}. \quad (11)$$

This method has the advantage that, in cases of variable diffusivity, heat production, or boundary conditions, this variation can be taken into account in a physically satisfactory way, namely, by giving (say) κ a mean value for the region instead of the value corresponding to $v_{m,n}$. On the other hand, (11) is not so easy to use as (4) since it does not immediately give the $v_{m,n+1}$ in terms of the $v_{m,n}$, but only a set of algebraic equations which have to be solved. This solution can be done by a routine based on relaxation or similar methods; it is fully described in *M.D.* A further refinement of the method, using higher order differences, has been made by Douglas.‡

Finally, it should be remarked that cases of radial flow in spheres or cylinders may be treated in much the same way, using 18.2 (17) or 18.2 (15).

18.4. Boundary conditions

We shall only consider a left-hand boundary, so that the region in question is $x > 0$ and we are concerned with $v_{m,n}$, $m = 0, 1, 2,...$. A right-hand boundary may be treated similarly.

In the case of *prescribed temperature*, $v_{0,n}$ is given and introduced into the set of difference equations. As remarked in § 18.3, the nature of this temperature function may affect the stability of the equations.

All other boundary conditions involve the flux

$$-K\frac{\partial v}{\partial x} = F(v,t). \quad (1)$$

If the difference equation 18.3 (4) is used in the body of the solid and we know $v_{m,n}$, $m = 0, 1, 2,...$ for any n, the values of $v_{m,n+1}$, $m = 1, 2,...$

† Crank and Nicholson, *Proc. Camb. Phil. Soc.* **43** (1947) 50–67; *M.D.*, Chap. X.
‡ Douglas, *J. Math. Phys.* **35** (1956) 145–51; Crandall, *Quart. Appl. Math.* **13** (1955) 318–20.

can be found from 18.3 (4), and it remains to find $v_{0,n+1}$ from the surface condition (1). This may be done in a number of ways:

I. Using 18.2 (8), neglecting the second difference, (1) becomes

$$v_{0,n+1}-v_{1,n+1} = \frac{\epsilon}{K}f\{v_{0,n+1},\ (n+1)\tau\}. \tag{2}$$

This is an equation which can be solved for $v_{0,n+1}$.

II. A more accurate result of the same type may be obtained by using 18.2 (9) in place of 18.2 (8). This gives

$$3v_{0,n+1}-4v_{1,n+1}+v_{2,n+1} = \frac{2\epsilon}{K}f\{v_{0,n+1},\ (n+1)\tau\}. \tag{3}$$

Neither of these methods can be regarded as particularly satisfactory, since they involve extrapolation of a function in a region in which it may be varying rapidly.

III. This method may most simply be stated by using the device due to Schmidt of introducing the set of temperatures $v_{-1,n}$ at the fictitious point $x = -\epsilon$. If $v_{m,n}$ is known for $m \geqslant -1$, $v_{0,n+1}$ can then be found from 18.3 (4). To find $v_{-1,n}$, 18.2 (11) is used in (1), neglecting third differences, and gives

$$v_{-1,n} = v_{1,n}+\frac{2\epsilon}{K}f(v_{0,n},\ n\tau). \tag{4}$$

In this case, $v_{-1,n}$ and hence $v_{0,n+1}$, is determined immediately without the solution of an equation such as (2) or (3).

For the case of no flow of heat at the boundary, (4) becomes

$$v_{-1,n} = v_{1,n}. \tag{5}$$

IV. It may be noticed that I and II involve the value of the flux at $(n+1)\tau$ while III contains its value at $n\tau$. If the flux varies rapidly with time or temperature, these might be expected to give different results. This variation can be taken into account by using the value of the flux at $(n+\frac{1}{2})\tau$ in (4), which gives, approximately,

$$v_{-1,n} = v_{1,n}+\frac{2\epsilon}{K}\left\{f(v_{0,n},\ n\tau)+\tfrac{1}{2}(v_{0,n+1}-v_{0,n})\left[\frac{\partial f}{\partial v}\right]_{0,n}\right\}. \tag{6}$$

This, combined with 18.3 (4), still gives a linear equation for $v_{0,n+1}$.

These formulae and others are discussed by Price and Slack (loc. cit.) who investigate their stability and accuracy for the case of linear heat transfer.

One further practical point of some importance must now be mentioned, namely the use of a *starting solution*. When, for example, boundary conditions are suddenly imposed on a solid at constant temperature, there will be high values of the flux near the surface, and the values of $\partial v/\partial x$ computed during the first few steps in time will be very inaccurate. It is thus desirable, instead of starting from constant values at $t = 0$, to start from a computed solution at (say) $t = \tau$ or 2τ, even if this solution has to be computed from an approximate (linearized) boundary condition. Such solutions can be obtained from the exact solutions given earlier. New ones can probably be found for important boundary conditions, for example, a solution useful for small values of the time is known for the case in which the surface flux is a power of the surface temperature.†

In the case of linear heat transfer a number of comparisons of the accurate solutions with various methods of treating the boundary condition have been made by Price and Slack (loc. cit.).

18.5. Heat production, variable diffusivity, and latent heat

In the case of heat production, a term $A(v, x, t)/K$ has to be added to the left-hand side of 18.3 (1). In the finite difference approximations, a term $\kappa\tau A(v_{m,n}, m\epsilon, n\tau)/K$, or, to a better approximation

$$\frac{\kappa\tau}{K}\left\{A(v_{m,n}, m\epsilon, n\tau) + \tfrac{1}{2}(v_{m,n+1} - v_{m,n})\left[\frac{\partial A}{\partial v}\right]_{m,n}\right\}, \tag{1}$$

has to be added to the right-hand side of 18.3 (4). In the Crank-Nicholson formula 18.3 (11), a term

$$\frac{\kappa\tau}{2K}\{A(v_{m,n}, m\epsilon, n\tau) + A(v_{m,n+1}, m\epsilon, (n+1)\tau)\} \tag{2}$$

is to be added to the right-hand side.

Little work on problems of this type has yet been done. Crank and Nicholson (loc. cit.) discuss a case of heat production by a chemical reaction. Blanch‡ discusses stability and gives a number of numerical examples.

The case of variable diffusivity has been briefly referred to in § 18.3. To take it into account, κ in 18.3 (5) has to be given the value corresponding to $v_{m,n}$, while in 18.3 (11) it has to be given the value corresponding to $v_{m,n+\frac{1}{2}}$. This case is of great importance in diffusion and is fully treated in M.D.

In conduction of heat, the most important problems of this type are (i) the case in which the diffusivity is a step-function of the temperature (which also corresponds to the liberation of latent heat over a melting range), and (ii) the related problem of liberation of latent heat at a melting-point. These are of great technical importance; also, while exact solutions are known for such problems in the semi-infinite solid, there are none for the slab or cylinder. These latter must be treated by numerical methods, but the exact solutions of Chapter XI are extremely useful§

† Jaeger, *Proc. Camb. Phil. Soc.* **46** (1950) 634–41. In particular, black-body radiation and natural convection are discussed.

‡ Blanch, *J. Res. Nat. Bur. Stand.* **50** (1953) 343–56.

§ Jaeger, *Amer. J. Sci.* **255** (1957) 306.

as 'starting solutions'. The effect of latent heat is studied by Price and Slack.[†] They point out that, in problems of this type, it is more satisfactory to work with Q, the heat content per unit mass of the solid which satisfies the differential equation

$$\rho \, \frac{\partial Q}{\partial t} = K \, \frac{\partial^2 v}{\partial x^2}, \tag{3}$$

of which the finite difference form corresponding to 18.3 (4) is

$$Q_{m,n+1} = Q_{m,n} + \frac{K\tau}{\rho \epsilon^2} (v_{m+1,n} + v_{m-1,n} - 2v_{m,n}). \tag{4}$$

The calculation then proceeds as before, there being a discontinuity in Q at a melting-point, or in c at the end of a melting range.

18.6. Relaxation methods

The preceding sections have dealt with step-by-step methods for solving equations in the open region $t > 0$. Relaxation methods[‡] are applicable to problems stated in a closed region with conditions at all points of the boundary. They have been in use for many years for the solution of Laplace's and Poisson's equations, and are particularly useful for problems in conduction of heat, such as steady flow through a bend in a wall or a corner of a hollow cube, which are difficult or impossible by analytical methods. The method is too well known to need description here: many full accounts are given in the literature.[§]

In steady flow of heat in two dimensions, it is necessary to solve

$$\frac{\partial^2 v}{\partial x^2} + \frac{\partial^2 v}{\partial y^2} = 0, \tag{1}$$

in a given region with prescribed conditions over its boundary. Using 18.2 (21), neglecting higher differences,[||] this is replaced by the difference equation

$$v_{m,n+1} + v_{m,n-1} + v_{m-1,n} + v_{m+1,n} - 4v_{m,n} = 0, \tag{2}$$

which has to be solved with prescribed conditions at the mesh points on or near the boundary. It was for such problems that the relaxation methods were originally devised, and it is only recently that attempts have been made to apply them to initial value problems such as those of variable flow of heat.

To apply relaxation methods to the differential equation

$$\frac{\partial^2 v}{\partial x^2} - \frac{1}{\kappa} \frac{\partial v}{\partial t} = 0, \tag{3}$$

it is necessary to consider a closed region, say, $0 < x < l$, $0 < t < T$, where T

[†] Price and Slack, *Brit. J. Appl. Phys.* **5** (1954) 285–7. See also, Crank, *Quart. J. Mech. Appl. Math.* **10** (1957) 220.

[‡] Southwell, *Relaxation Methods in Engineering Science* (Oxford, 1940); *Relaxation Methods in Theoretical Physics* (Oxford, 1946).

[§] e.g. Hartree, loc. cit. From the point of view of conduction of heat, short accounts are given in most works on heat transfer. The method is discussed at length in Dusinberre, loc. cit. A useful short introduction is given by Emmons, *Trans. Amer. Soc. Mech. Engrs.* **65** (1943) 607–15. The solution of Poisson's equation in three dimensions is discussed by Allen and Dennis, *Quart. J. Mech. Appl. Math.* **4** (1951) 199–208. The combined use of relaxation methods and integral transforms is discussed by Tranter, *Quart. J. Mech. Appl. Math.* **1** (1948) 125–30. The application of relaxation methods to problems involving latent heat is discussed by Allen and Severn, *Quart. J. Mech. Appl. Math.* **5** (1952) 447–54.

[||] For the use of higher differences see Fox, *Proc. Roy. Soc.* A, **190** (1947) 31–59.

is the time during which the solution is required,† and to have conditions on all four lines, $x = 0$, $x = l$, $t = 0$, $t = T$. One method of ensuring this is by introducing‡ a new function w defined by

$$v = \frac{\partial w}{\partial t} + \kappa \frac{\partial^2 w}{\partial x^2},$$ (4)

with $w = 0$ when $x = 0$; $w = 0$ when $x = l$; and $\partial w/\partial t = 0$ when $t = T$. These conditions are not inconsistent, and imply no restrictions on v. Using (4) in (3) gives

$$\frac{\partial^2 w}{\partial t^2} - \kappa^2 \frac{\partial^4 w}{\partial x^4} = 0,$$ (5)

with a set of boundary conditions to which relaxation methods are applicable.

A number of methods is also in use which are described as relaxation methods, but which are in fact step-by-step methods in which relaxation methods are used to solve the set of equations for the values of v at any particular time. Thus, relaxation methods have been used§ for the solution of the set of equations 18.3 (11) which gives the $v_{m,n+1}$ in terms of the $v_{m,n}$ in the Crank-Nicholson method. An important method of this type has been introduced by Liebmann,‖ who in place of 18.3 (3) uses the equally valid

$$\left[\frac{\partial v_m}{\partial t}\right]_n = \frac{1}{\tau}(v_{m,n} - v_{m,n-1})$$ (6)

in 18.3 (2) to get the difference equation

$$M v_{m-1,n} - (2M+1)v_{m,n} + M v_{m+1,n} + v_{m,n-1} = 0,$$ (7)

where $M = \kappa\tau/\epsilon^2$. This set of equations for $v_{m,n}$ in terms of $v_{m,n-1}$ is then solved by relaxation methods. It has the advantage of being stable for all values of M.

---

† Gilmour, *Brit. J. Appl. Phys.* **2** (1951) 199–204, remarks that the final steady state, if one exists, may be used for this purpose.

‡ Allen and Severn, *Quart. J. Mech. Appl. Math.* **4** (1951) 209–22.

§ Mitchell, *Appl. Sci. Res.* A, **4** (1954) 109–19. A term $A(x, t)$ corresponding to heat production is included.

‖ Liebmann, *Brit. J. Appl. Phys.* **6** (1955) 129–35. The method is readily applicable to problems in two space variables.

APPENDIX I

Contour Integrals and the Verification of Solutions obtained by the Laplace Transformation

IT was remarked in § 12.3 that from the point of view of strict pure mathematics the solutions which have been obtained by the Laplace transformation process† were essentially 'formal'; that is to say, various operations, such as the inversion of order of limiting processes, were made without justification in the course of the analysis. It was added that it could be verified that all the one-variable solutions given here did satisfy the differential equations and initial and boundary conditions of their problems.

The procedure is as follows: in all cases‡ we have found $\bar{v}$, and derived v from it by the use of the Inversion Theorem in the form 12.3 (8),

$$v = \frac{1}{2\pi i} \int_{\gamma - i\infty}^{\gamma + i\infty} e^{\lambda t} \bar{v}(\lambda) \, d\lambda, \tag{1}$$

where γ is so large that all singularities of $\bar{v}(\lambda)$ lie to the left of the line $(\gamma - i\infty, \gamma + i\infty)$. In all the cases we have met, these singularities consist of a finite number of poles and branch points, together with possibly an infinite number of poles at isolated points on the negative real axis. Thus we can always choose a new path L', Fig. 61, which begins at infinity in the direction $\arg \lambda = -\beta$, where $\pi > \beta > \frac{1}{2}\pi$, passes to the right of the origin, keeping all singularities of the integrand to the left, and ends in the direction $\arg \lambda = \beta$.

For the values of $\bar{v}(\lambda)$ met earlier it can always be shown that

$$\int_{L'} e^{\lambda t} \bar{v}(\lambda) \, d\lambda = \int_{\gamma - i\infty}^{\gamma + i\infty} e^{\lambda t} \bar{v}(\lambda) \, d\lambda. \tag{2}$$

This follows by considering the closed contour $ABB''A''A$ of Fig. 62, consisting of portion AB of the line $\mathbf{R}(\lambda) = \gamma$, portion $A''B''$ of the contour L', and arcs $BB'B''$ and $A''A'A$ of a large circle of centre the origin and radius R. There are no singularities within the contour. The result (2) will follow by Cauchy's theorem, if in the limit as $R \to \infty$ the integrals over the circular arcs $AA'A''$ and $BB'B''$ tend to zero. A little discussion of the integrand shows that this is true in any of the special cases we consider.§

---

† The same remark applies to almost all the general solutions obtained by the earlier methods : indeed they are much more open to criticism on the ground that they usually assume that an arbitrary function can be expanded in a series of functions without verifying that this series is complete : it is thus easy to omit portion of the solution (cf. § 7.8, footnote). Actually very few rigorous investigations even of quite simple problems involving arbitrary functions have been made, cf. Moore's remarks, *Bull. Amer. Math. Soc.* **51** (1945) 650, on the problem of radial flow of heat in a circular cylinder.

‡ From the present point of view the use of the Table of Transforms, when possible, may be regarded as a matter of convenience : all the results in it may be obtained by applying the Inversion Theorem.

§ Here and elsewhere it is necessary to study the order of magnitude of the integrand on a circle of large radius R. Theoretically this is to be done for each special problem ; actually large classes of problems can be treated together. Details for the problem of

Thus in these cases we arrive at the result

$$v = \frac{1}{2\pi i} \int_{L'} e^{\lambda t} \bar{v}(\lambda) \, d\lambda. \tag{3}$$

The integral (3) can then be shown to be uniformly convergent in the space-variable in the given region for fixed t, and uniformly convergent in t for $t \geqslant 0$ when the space-variable is fixed. Differentiation under the integral sign is

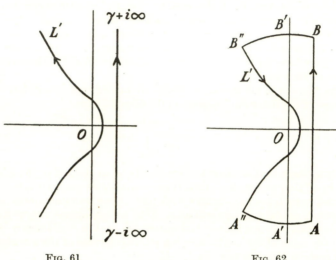

FIG. 61   FIG. 62

justifiable, and thus it can be shown that the given differential equation is satisfied. And in the same way the initial and boundary conditions will be found to be satisfied. The advantage of the path L' over $(\gamma - i\infty, \gamma + i\infty)$ is that on the former we have a factor of type

$$\exp\{-k|\lambda|^{\frac{1}{2}} \cos \tfrac{1}{2}\beta + t|\lambda| \cos \beta\},$$

where $\pi > \beta > \frac{1}{2}\pi$, and k is a positive quantity involving the space variable. This factor provides the required uniform convergence.

If in (3) we change the variable to $\alpha = \lambda^{\frac{1}{2}} e^{i\pi/2}$ we get the path P of Fig. 63 which begins at $\arg \alpha = \theta$, where $0 < \theta < \frac{1}{4}\pi$, and ends in the direction $\arg \alpha = \pi - \theta$. Clearly integrals along this path may be discussed in the same way as those along the path L', and we shall have

$$v = -\frac{1}{\pi i} \int_{P} e^{-\alpha^2 t} \bar{v}(\alpha^2 e^{-i\pi}) \alpha \, d\alpha. \tag{4}$$

Contour integrals of this type have been made fundamental in *C.H.* and many

§ 12.6 I are given by the authors in *Operational Methods in Applied Mathematics*, §§ 41, 58; for the problem of § 12.7 in ibid. § 47; for a problem on a composite spherical solid, *Proc. Camb. Phil. Soc.* **35** (1939) 394; for some problems on the circular cylinder, *Proc. Lond. Math. Soc.* (2) **46** (1940) 361; and for a complete set of problems on the cylindrical regions $0 \leqslant r < a$, $a < r < b$, and $r > a$, with boundary conditions 1.9 (14), *Proc. Roy. Soc. N.S.W.* **75** (1942) 130. There is some advantage in using a parabolic contour instead of the circle, cf. Churchill, *Math. Z.* **43** (1938) 743.

problems are solved there by their use. The method is seen to be the same as that used here except for the above change of variable.

In verifying the Green's functions of Chapter XIV we proceed in the same way except that we have to verify that w satisfies the differential equation and vanishes at $t = 0$, and that v satisfies the boundary conditions.

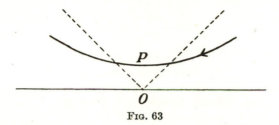

FIG. 63

The final forms of the solutions we have usually given have been deduced as series or integrals by the use of the contours of Figs. 39 and 40 respectively. By the verification process above we know that (1) satisfies all the conditions of the problem, and to make the final solutions rigorous the reduction of (1) to its final form must be carefully justified. This requires a proof that in the limit, as the radii of the large circles of the contours tend to infinity (if necessary by a discrete set of values, so that the circles pass through no pole of the integrand), the integrals over these circles tend to zero. This can be proved for all the solutions given here by the type of analysis used above: sufficient details are given in the references.

APPENDIX II

The Error Function and Related Functions

WE write†

$$\operatorname{erf} x = \frac{2}{\sqrt{\pi}} \int_0^x e^{-\xi^2}\, d\xi, \tag{1}$$

so that

$$\operatorname{erf} \infty = 1, \tag{2}$$

and

$$\operatorname{erf}(-x) = -\operatorname{erf} x.$$

Also we use

$$\operatorname{erfc} x = 1 - \operatorname{erf} x = \frac{2}{\sqrt{\pi}} \int_x^\infty e^{-\xi^2}\, d\xi. \tag{3}$$

We proceed to find approximations for $\operatorname{erf} x$ for large and for small values of x. For small values of x we use the series for $e^{-\xi^2}$ in (1), and thus obtain

$$\operatorname{erf} x = \frac{2}{\sqrt{\pi}} \int_0^x e^{-\xi^2}\, d\xi = \frac{2}{\sqrt{\pi}} \int_0^x d\xi \sum_{n=0}^\infty \frac{(-1)^n \xi^{2n}}{n!}.$$

Since the series is uniformly convergent it may be integrated term by term, and therefore

$$\operatorname{erf} x = \frac{2}{\sqrt{\pi}} \sum_{n=0}^\infty \frac{(-1)^n x^{2n+1}}{(2n+1)n!}. \tag{4}$$

For large values of x we proceed as follows: a single integration by parts gives

$$\int_x^\infty e^{-\xi^2}\, d\xi = \frac{e^{-x^2}}{2x} - \frac{1}{2} \int_x^\infty \frac{e^{-\xi^2}\, d\xi}{\xi^2},$$

and repeating the process n times we find

$$\frac{\sqrt{\pi}}{2} \operatorname{erfc} x = \int_x^\infty e^{-\xi^2} d\xi = \tfrac{1}{2} e^{-x^2}\left(\frac{1}{x} - \frac{1}{2x^3} + \frac{1.3}{2^2 x^5} - \ldots + (-1)^{n-1} \frac{1.3\ldots(2n-3)}{2^{n-1} x^{2n-1}}\right) +$$

$$+ (-1)^n \frac{1.3\ldots(2n-1)}{2^n} \int_x^\infty \frac{e^{-\xi^2}\, d\xi}{\xi^{2n}}.$$

† The first table of these integrals was published by Encke in a paper on the 'Method of Least Squares' in the Berlin *Astronomisches Jahrbuch* for 1834, giving the value of $\operatorname{erf} x$ for $x = 0$ to $x = 2$ at intervals of 0.01 computed to seven decimal places. De Morgan extended this to $x = 3$ in his 'Essay on Probabilities' (1838). A new table, to fifteen places, from $x = 0$ to $x = 3$ at intervals of 0.0001 was given by Burgess in his paper 'On the Definite Integral $\frac{2}{\sqrt{\pi}} \int_0^t e^{-t^2}\, dt$ with extended Tables of Values', *Trans. Roy. Soc. Edin.* **39** (1899) 257. The most recent tables are those of Sheppard, 'The Probability Integral', *Brit. Ass. Adv. Science: Math. Tables*, Vol. VII (1939), and the Works Project Administration 'Tables of the Probability Function' (New York, 1941). Shorter tables are to be found in, e.g., Milne-Thomson and Comrie, *Standard Four-Figure Mathematical Tables* (Macmillan, 1931), and in Jahnke–Emde, *Tables of Functions* (Teubner, edn. 2, 1933); the latter give values of the derivatives also.

Many different notations have been used for the function (1); for a discussion of these see Jeffreys, *Operational Methods in Mathematical Physics* (Cambridge, edn. 2, 1931), p. 110.

This series does not converge, since the ratio of the nth term to the $(n-1)$th does not remain less than unity, as n increases. However, if we take n terms of the series, the remainder, namely

$$\frac{1.3...(2n-1)}{2^n} \int_x^\infty \frac{e^{-\xi^2} d\xi}{\xi^{2n}},$$

is less than the nth term, since

$$\int_x^\infty \frac{e^{-\xi^2} d\xi}{\xi^{2n}} < e^{-x^2} \int_x^\infty \frac{d\xi}{\xi^{2n}}.$$

We can thus stop at any term, and take the sum of the terms up to this term as an approximation for the function, the error being less in absolute value than the last term we have retained. In this sense erfc x may, for large x, be calculated numerically from the formula

$$\pi^{-\frac{1}{2}} e^{-x^2} \left(\frac{1}{x} - \frac{1}{2x^3} + \frac{1.3}{2^2 x^5} - \frac{1.3.5}{2^3 x^7} + ... \right). \tag{5}$$

The function e^{x^2} erfc x which appears in this discussion often arises in problems on conduction of heat.

Other important integrals leading to error functions are†

$$\int_0^\infty e^{-\xi^2} \frac{\sin 2\xi y}{\xi} d\xi = \tfrac{1}{2}\pi \operatorname{erf} y, \tag{6}$$

$$\int_0^\infty e^{-\xi^2} \sin 2\xi y \, d\xi = \tfrac{1}{2}\sqrt{\pi} e^{-y^2} \operatorname{erf} y. \tag{7}$$

Derivatives and integrals of error functions

Successive derivatives of the error function,

$$\Phi_n(x) = \frac{d^n}{dx^n} \operatorname{erf} x, \tag{8}$$

so that

$$\Phi_1(x) = \frac{2}{\sqrt{\pi}} e^{-x^2},$$

$$\Phi_2(x) = -\frac{4}{\sqrt{\pi}} x e^{-x^2},$$

etc., are tabulated in Jahnke–Emde (loc. cit.). In problems on conduction of heat the repeated integrals of the error function are more important.‡ Write

$$i^n \operatorname{erfc} x = \int_x^\infty i^{n-1} \operatorname{erfc} \xi \, d\xi, \quad n = 1, 2,..., \tag{9}$$

with

$$i^0 \operatorname{erfc} x = \operatorname{erfc} x; \tag{10}$$

† *F.S.*, p. 213, Ex. 13. For others see Horenstein, *Quart. Appl. Math.* **3** (1945) 183.
‡ These have been studied and tabulated by Hartree, *Mem. Manchr. Lit. Phil. Soc.* **80** (1935) 85. The values given in Table I below are taken from this paper. Kaye, *J. Math. Phys.* **34** (1955) 119–25, gives values for $n = 1$ to 11 and $x = 0$, (0·01), 0·2, (0·05), 1, (0·1), 3.

we also often write ierfc x in place of i^1erfc x for shortness. Then, integrating by parts, we have

$$\text{ierfc } x = \frac{1}{\sqrt{\pi}} e^{-x^2} - x \text{ erfc } x, \tag{11}$$

and also

$$i^2\text{erfc } x = \frac{1}{4}\left[(1+2x^2)\text{erfc } x - \frac{2}{\sqrt{\pi}} xe^{-x^2}\right] \tag{12}$$

$$= \tfrac{1}{4}[\text{erfc } x - 2x \text{ ierfc } x]. \tag{13}$$

The general recurrence formula

$$2n \, i^n\text{erfc } x = i^{n-2}\text{erfc } x - 2x \, i^{n-1}\text{erfc } x, \tag{14}$$

of which (13) is the case $n = 2$, is easily established by induction.

It follows from (14) that

$$i^n\text{erfc } 0 = \frac{1}{2^n \Gamma(\tfrac{1}{2}n+1)}. \tag{15}$$

Again, it follows from (14) that $y = i^n\text{erfc } x$ satisfies the differential equation

$$\frac{d^2y}{dx^2} + 2x \, \frac{dy}{dx} - 2ny = 0. \tag{16}$$

The error function of complex argument

This function, which is of great importance in conduction of heat, has only recently been tabulated by Faddeeva and Terentév.† They define

$$w(z) = u(x,y) + iv(x,y) = e^{-z^2}\left\{1 + \frac{2i}{\pi^{\frac{1}{2}}} \int_0^z e^{t^2} dt\right\}, \tag{17}$$

where $z = x + iy$,

so that $w(iz) = e^{z^2} \text{erfc } z.$ $\qquad\qquad\qquad$ (18)

Some values of $u(x,y)$ and $v(x,y)$ are given in Tables II and III.

The results (4) and (5), proved above for real z, are true also if z is complex.

---

† Faddeeva and Terentév, *Tables of Values of the Function w(z) for a Complex Argument* (Gosudarstv. Izdat. Tehn. Teor. Lit., Moscow, 1954), give values to 6D at intervals of 0·02 in the arguments. Karpov, *Tables of the Function w(z) in a Complex Region* (Izdat. Akad. Nauk SSSR, Moscow, 1954), gives a similar table with arguments in polar co-ordinates. For real z, (17) has been tabulated by Dawson, *Proc. Lond. Math. Soc.* (1) **29** (1897–8) 519, and by Miller and Gordon, *J. Phys. Chem.* **35** (1931) 2785. If $x = y$, the integral can be expressed in terms of Fresnel's integrals and has been tabulated by Clemmow and Munford, *Phil. Trans. Roy. Soc.* A, **245** (1952) 189–211.

TABLE I. THE ERROR FUNCTION AND ITS DERIVATIVES AND INTEGRALS

x	e^{x^2} erfc x	$4\pi^{-\frac{1}{2}} x e^{-x^2}$	$2\pi^{-\frac{1}{2}} e^{-x^2}$	erf x	erfc x	2 ierfc x	4 i²erfc x	6 i³erfc x	8 i⁴erfc x	10 i⁵erfc x	12 i⁶erfc x
0	1·0	0	1·1284	0	1·0	1·1284	1·0	0·5642	0·25	0·0940	0·0313
0·05	0·9460	0·1126	1·1256	0·056372	0·943628	1·0312	0·8921	0·4933	0·2148	0·0795	0·0261
0·1	0·8965	0·2234	1·1172	0·112463	0·887537	0·9396	0·7936	0·4301	0·1841	0·0671	0·0217
0·15	0·8509	0·3310	1·1033	0·167996	0·832004	0·8537	0·7040	0·3740	0·1573	0·0564	0·0180
0·2	0·8090	0·4336	1·0841	0·222703	0·777297	0·7732	0·6227	0·3243	0·1341	0·0474	0·0149
0·25	0·7703	0·5300	1·0600	0·276326	0·723674	0·6982	0·5491	0·2805	0·1139	0·0396	0·0123
0·3	0·7346	0·6188	1·0313	0·328627	0·671373	0·6284	0·4828	0·2418	0·0965	0·0331	0·0101
0·35	0·7015	0·6988	0·9983	0·379382	0·620618	0·5639	0·4233	0·2079	0·0816	0·0275	0·0083
0·4	0·6708	0·7692	0·9615	0·428392	0·571608	0·5043	0·3699	0·1782	0·0687	0·0228	0·0068
0·45	0·6423	0·8294	0·9215	0·475482	0·524518	0·4495	0·3223	0·1522	0·0577	0·0189	0·0055
0·5	0·6157	0·8788	0·8788	0·520500	0·479500	0·3993	0·2799	0·1297	0·0484	0·0156	0·0045
0·55	0·5909	0·9172	0·8338	0·563323	0·436677	0·3535	0·2423	0·1101	0·0404	0·0128	0·0036
0·6	0·5678	0·9447	0·7872	0·603856	0·396144	0·3119	0·2090	0·0932	0·0336	0·0105	0·0029
0·65	0·5462	0·9614	0·7395	0·642029	0·357971	0·2742	0·1798	0·0787	0·0279	0·0086	0·0024
0·7	0·5259	0·9678	0·6913	0·677801	0·322199	0·2402	0·1541	0·0662	0·0231	0·0070	0·0019
0·75	0·5069	0·9644	0·6429	0·711156	0·288844	0·2097	0·1316	0·0555	0·0190	0·0057	0·0015
0·8	0·4891	0·9520	0·5950	0·742101	0·257899	0·1823	0·1120	0·0464	0·0156	0·0046	0·0012
0·85	0·4723	0·9314	0·5479	0·770668	0·229332	0·1580	0·0950	0·0386	0·0128	0·0037	0·0010
0·9	0·4565	0·9035	0·5020	0·796908	0·203092	0·1364	0·0803	0·0321	0·0104	0·0030	0·0008
0·95	0·4416	0·8695	0·4576	0·820891	0·179109	0·1173	0·0677	0·0265	0·0085	0·0024	0·0006
1·0	0·4276	0·8302	0·4151	0·842701	0·157299	0·1005	0·0568	0·0218	0·0069	0·0019	0·0005
1·1	0·4017	0·7403	0·3365	0·880205	0·119795	0·0729	0·0396	0·0147	0·0045	0·0012	0·0003
1·2	0·3785	0·6416	0·2673	0·910314	0·089686	0·0521	0·0272	0·0097	0·0029	0·0007	0·0002
1·3	0·3576	0·5413	0·2082	0·934008	0·065992	0·0366	0·0184	0·0063	0·0019	0·0004	0·0002
1·4	0·3387	0·4450	0·1589	0·952285	0·047715	0·0253	0·0122	0·0041	0·0011	0·0003	0·0001
1·5	0·3216	0·3568	0·1189	0·966105	0·033895	0·0172	0·0080	0·0026	0·0007	0·0002	0·0001
1·6	0·3060	0·2791	0·0872	0·976348	0·023652	0·0115	0·0052	0·0016	0·0004	0·0001	
1·7	0·2917	0·2132	0·0627	0·983790	0·016210	0·0076	0·0033	0·0010	0·0003		
1·8	0·2786	0·1591	0·0442	0·989091	0·010909	0·0049	0·0021	0·0006	0·0002		
1·9	0·2665	0·1160	0·0305	0·992790	0·007210	0·0031	0·0013	0·0003	0·0001		
2·0	0·2554	0·0827	0·0207	0·995322	0·004678	0·0020	0·0008	0·0002	0·0001		
2·1	0·2451	0·0576	0·0137	0·997021	0·002979	0·0012	0·0005	0·0001			
2·2	0·2356	0·0393	0·0089	0·998137	0·001863	0·0007	0·0003				
2·3	0·2267	0·0262	0·0057	0·998857	0·001143	0·0004	0·0002				
2·4	0·2185	0·0171	0·0036	0·999311	0·000689	0·0002	0·0001				
2·5	0·2108	0·0109	0·0022	0·999593	0·000407	0·0001					
2·6	0·2036	0·0068	0·0013	0·999764	0·000236	0·0001					
2·7	0·1969	0·0042	0·0008	0·999866	0·000134						
2·8	0·1905	0·0025	0·0004	0·999925	0·000075						
2·9	0·1846	0·0015	0·0003	0·999959	0·000041						
3·0	0·1790	0·0008	0·0001	0·999978	0·000022						

TABLE II. $u(x,y)$

y \ x	0	0.1	0.2	0.4	0.6	0.8	1.0	1.2	1.4	1.6	1.8	2.0	2.5	3	4	5
0	1.0000	0.9900	0.9608	0.8521	0.6977	0.5273	0.3679	0.2369	0.1409	0.0773	0.0392	0.0183	0.0019	0.0001	0.0000	0.0000
0.1	0.8965	0.8885	0.8650	0.7773	0.6511	0.5093	0.3732	0.2574	0.1684	0.1058	0.0651	0.0402	0.0147	0.0079	0.0039	0.0024
0.2	0.8090	0.8026	0.7835	0.7121	0.6083	0.4897	0.3732	0.2709	0.1892	0.1289	0.0871	0.0595	0.0268	0.0156	0.0078	0.0048
0.3	0.7346	0.7293	0.7138	0.6552	0.5692	0.4695	0.3694	0.2792	0.2047	0.1473	0.1055	0.0764	0.0382	0.0231	0.0117	0.0072
0.4	0.6708	0.6665	0.6537	0.6053	0.5336	0.4492	0.3630	0.2834	0.2157	0.1617	0.1208	0.0909	0.0488	0.0303	0.0155	0.0096
0.5	0.6157	0.6121	0.6015	0.5613	0.5011	0.4294	0.3549	0.2846	0.2233	0.1728	0.1333	0.1034	0.0584	0.0371	0.0192	0.0119
0.6	0.5678	0.5648	0.5560	0.5222	0.4715	0.4103	0.3456	0.2835	0.2280	0.1812	0.1434	0.1138	0.0672	0.0436	0.0229	0.0142
0.7	0.5259	0.5234	0.5160	0.4876	0.4444	0.3919	0.3357	0.2807	0.2306	0.1872	0.1514	0.1226	0.0751	0.0497	0.0264	0.0165
0.8	0.4891	0.4870	0.4807	0.4566	0.4198	0.3745	0.3254	0.2767	0.2314	0.1914	0.1576	0.1298	0.0821	0.0553	0.0298	0.0187
0.9	0.4565	0.4547	0.4494	0.4288	0.3972	0.3580	0.3151	0.2718	0.2308	0.1940	0.1623	0.1356	0.0883	0.0605	0.0331	0.0209
1.0	0.4276	0.4260	0.4215	0.4038	0.3766	0.3425	0.3047	0.2662	0.2292	0.1954	0.1657	0.1402	0.0938	0.0653	0.0363	0.0230
1.1	0.4017	0.4004	0.3965	0.3812	0.3576	0.3279	0.2946	0.2602	0.2268	0.1957	0.1680	0.1438	0.0985	0.0697	0.0393	0.0251
1.2	0.3785	0.3774	0.3740	0.3608	0.3402	0.3142	0.2847	0.2540	0.2237	0.1952	0.1694	0.1465	0.1025	0.0736	0.0422	0.0271
1.3	0.3576	0.3566	0.3537	0.3422	0.3242	0.3013	0.2752	0.2476	0.2202	0.1941	0.1700	0.1485	0.1060	0.0772	0.0449	0.0290
1.4	0.3387	0.3379	0.3353	0.3252	0.3095	0.2892	0.2660	0.2412	0.2163	0.1923	0.1700	0.1497	0.1088	0.0804	0.0475	0.0309
1.5	0.3216	0.3208	0.3186	0.3097	0.2958	0.2779	0.2571	0.2349	0.2123	0.1902	0.1695	0.1504	0.1112	0.0832	0.0499	0.0327
1.6	0.3060	0.3053	0.3033	0.2955	0.2832	0.2672	0.2487	0.2286	0.2080	0.1878	0.1685	0.1506	0.1132	0.0857	0.0521	0.0344
1.7	0.2917	0.2911	0.2893	0.2824	0.2715	0.2572	0.2406	0.2224	0.2037	0.1851	0.1672	0.1504	0.1147	0.0879	0.0542	0.0361
1.8	0.2786	0.2780	0.2765	0.2703	0.2606	0.2479	0.2329	0.2164	0.1993	0.1822	0.1656	0.1499	0.1159	0.0897	0.0562	0.0377
1.9	0.2665	0.2660	0.2646	0.2592	0.2505	0.2390	0.2255	0.2106	0.1949	0.1792	0.1637	0.1490	0.1167	0.0914	0.0580	0.0392
2.0	0.2554	0.2550	0.2537	0.2488	0.2410	0.2307	0.2185	0.2049	0.1906	0.1761	0.1617	0.1480	0.1172	0.0927	0.0597	0.0406
2.2	0.2356	0.2353	0.2343	0.2303	0.2240	0.2155	0.2055	0.1942	0.1821	0.1697	0.1573	0.1452	0.1176	0.0947	0.0626	0.0433
2.4	0.2185	0.2182	0.2174	0.2142	0.2090	0.2020	0.1936	0.1842	0.1740	0.1633	0.1526	0.1420	0.1172	0.0960	0.0651	0.0457
2.6	0.2036	0.2034	0.2027	0.2000	0.1957	0.1899	0.1829	0.1749	0.1662	0.1571	0.1477	0.1384	0.1162	0.0966	0.0670	0.0478
2.8	0.1905	0.1904	0.1898	0.1876	0.1840	0.1791	0.1731	0.1663	0.1589	0.1510	0.1428	0.1346	0.1147	0.0967	0.0686	0.0496
3.0	0.1790	0.1788	0.1784	0.1765	0.1734	0.1693	0.1643	0.1584	0.1520	0.1451	0.1380	0.1308	0.1129	0.0964	0.0698	0.0512
3.2	0.1687	0.1686	0.1682	0.1666	0.1640	0.1605	0.1562	0.1511	0.1456	0.1396	0.1333	0.1269	0.1108	0.0957	0.0707	0.0526
3.4	0.1595	0.1594	0.1591	0.1577	0.1555	0.1525	0.1487	0.1444	0.1395	0.1343	0.1288	0.1231	0.1086	0.0948	0.0712	0.0537
3.6	0.1513	0.1512	0.1509	0.1497	0.1478	0.1452	0.1419	0.1381	0.1339	0.1293	0.1244	0.1193	0.1063	0.0937	0.0715	0.0547
3.8	0.1438	0.1437	0.1434	0.1424	0.1408	0.1385	0.1357	0.1323	0.1286	0.1245	0.1202	0.1157	0.1039	0.0924	0.0717	0.0554
4.0	0.1370	0.1369	0.1367	0.1358	3.1344	0.1324	0.1299	0.1270	0.1237	0.1201	0.1162	0.1121	0.1016	0.0909	0.0716	0.0560
4.2	0.1308	0.1307	0.1305	0.1298	0.1285	0.1267	0.1246	0.1220	0.1191	0.1158	0.1124	0.1087	0.0992	0.0894	0.0713	0.0564
4.4	0.1251	0.1251	0.1249	0.1242	0.1231	0.1216	0.1196	0.1173	0.1147	0.1119	0.1088	0.1055	0.0968	0.0879	0.0710	0.0567
4.6	0.1199	0.1199	0.1197	0.1191	0.1181	0.1168	0.1150	0.1130	0.1107	0.1081	0.1053	0.1024	0.0945	0.0862	0.0705	0.0569
4.8	0.1151	0.1151	0.1150	0.1144	0.1135	0.1123	0.1108	0.1089	0.1069	0.1045	0.1020	0.0994	0.0922	0.0846	0.0699	0.0570
5.0	0.1107	0.1107	0.1105	0.1101	0.1093	0.1082	0.1068	0.1052	0.1033	0.1012	0.0989	0.0965	0.0899	0.0830	0.0692	0.0570

TABLE III. $v(x, y)$

$y \backslash x$	0	0·1	0·2	0·4	0·6	0·8	1·0	1·2	1·4	1·6	1·8	2	2·5	3	4	5
0	0	0·1121	0·2198	0·4062	0·5357	0·6004	0·6072	0·5724	0·5151	0·4513	0·3913	0·3400	0·2517	0·2012	0·1460	0·1152
0·1	0	0·0943	0·1853	0·3447	0·4597	0·5229	0·5386	0·5183	0·4765	0·4262	0·3762	0·3316	0·2500	0·2007	0·1458	0·1152
0·2	0	0·0800	0·1574	0·2947	0·3939	0·4576	0·4790	0·4695	0·4400	0·4008	0·3597	0·3213	0·2471	0·1997	0·1455	0·1150
0·3	0	0·0684	0·1347	0·2536	0·3446	0·4022	0·4272	0·4257	0·4058	0·3759	0·3425	0·3098	0·2430	0·1980	0·1450	0·1148
0·4	0	0·0589	0·1161	0·2197	0·3010	0·3551	0·3822	0·3864	0·3741	0·3518	0·3250	0·2975	0·2381	0·1957	0·1442	0·1144
0·5	0	0·0510	0·1008	0·1915	0·2643	0·3148	0·3429	0·3513	0·3449	0·3288	0·3076	0·2848	0·2324	0·1930	0·1433	0·1140
0·6	0	0·0445	0·0880	0·1679	0·2332	0·2803	0·3085	0·3199	0·3180	0·3070	0·2906	0·2719	0·2262	0·1898	0·1421	0·1134
0·7	0	0·0391	0·0773	0·1480	0·2068	0·2505	0·2784	0·2919	0·2935	0·2865	0·2742	0·2590	0·2195	0·1862	0·1408	0·1128
0·8	0	0·0345	0·0682	0·1311	0·1842	0·2248	0·2520	0·2668	0·2710	0·2674	0·2584	0·2464	0·2126	0·1824	0·1393	0·1120
0·9	0	0·0306	0·0606	0·1167	0·1648	0·2024	0·2288	0·2443	0·2505	0·2496	0·2434	0·2341	0·2055	0·1782	0·1376	0·1112
1·0	0	0·0272	0·0540	0·1044	0·1480	0·1829	0·2082	0·2242	0·2319	0·2330	0·2292	0·2222	0·1983	0·1739	0·1358	0·1103
1·1	0	0·0244	0·0484	0·0938	0·1335	0·1659	0·1900	0·2061	0·2149	0·2177	0·2159	0·2108	0·1911	0·1694	0·1339	0·1094
1·2	0	0·0219	0·0435	0·0845	0·1208	0·1509	0·1739	0·1899	0·1994	0·2035	0·2033	0·1999	0·1839	0·1649	0·1319	0·1083
1·3	0	0·0198	0·0393	0·0765	0·1098	0·1377	0·1595	0·1753	0·1853	0·1904	0·1915	0·1895	0·1769	0·1602	0·1298	0·1072
1·4	0	0·0180	0·0357	0·0695	0·1000	0·1260	0·1467	0·1621	0·1724	0·1783	0·1804	0·1797	0·1700	0·1556	0·1276	0·1060
1·5	0	0·0163	0·0325	0·0634	0·0914	0·1156	0·1352	0·1502	0·1607	0·1671	0·1701	0·1704	0·1632	0·1509	0·1253	0·1048
1·6	0	0·0149	0·0296	0·0580	0·0838	0·1064	0·1250	0·1394	0·1499	0·1568	0·1605	0·1616	0·1567	0·1462	0·1229	0·1035
1·7	0	0·0136	0·0272	0·0532	0·0771	0·0981	0·1157	0·1297	0·1401	0·1472	0·1515	0·1533	0·1504	0·1416	0·1205	0·1022
1·8	0	0·0125	0·0249	0·0489	0·0711	0·0907	0·1074	0·1208	0·1311	0·1384	0·1431	0·1455	0·1442	0·1371	0·1181	0·1008
1·9	0	0·0115	0·0230	0·0451	0·0657	0·0841	0·0998	0·1128	0·1229	0·1303	0·1352	0·1381	0·1384	0·1327	0·1157	0·0994
2·0	0	0·0107	0·0212	0·0417	0·0609	0·0781	0·0930	0·1054	0·1153	0·1227	0·1279	0·1312	0·1327	0·1283	0·1132	0·0980
2·2	0	0·0092	0·0183	0·0360	0·0526	0·0678	0·0812	0·0926	0·1019	0·1093	0·1148	0·1186	0·1221	0·1199	0·1083	0·0950
2·4	0	0·0079	0·0158	0·0313	0·0459	0·0593	0·0713	0·0818	0·0906	0·0977	0·1033	0·1074	0·1124	0·1120	0·1034	0·0920
2·6	0	0·0070	0·0139	0·0274	0·0403	0·0522	0·0631	0·0727	0·0809	0·0877	0·0933	0·0976	0·1036	0·1046	0·0986	0·0890
2·8	0	0·0061	0·0122	0·0242	0·0356	0·0463	0·0561	0·0649	0·0726	0·0791	0·0845	0·0888	0·0956	0·0977	0·0939	0·0859
3·0	0	0·0054	0·0108	0·0215	0·0317	0·0413	0·0502	0·0582	0·0654	0·0716	0·0768	0·0811	0·0883	0·0912	0·0893	0·0828
3·2	0	0·0048	0·0097	0·0192	0·0283	0·0370	0·0451	0·0525	0·0592	0·0650	0·0700	0·0743	0·0817	0·0853	0·0850	0·0798
3·4	0	0·0044	0·0087	0·0172	0·0255	0·0334	0·0408	0·0476	0·0537	0·0592	0·0640	0·0682	0·0757	0·0797	0·0808	0·0768
3·6	0	0·0039	0·0078	0·0156	0·0230	0·0302	0·0370	0·0432	0·0490	0·0541	0·0587	0·0627	0·0702	0·0746	0·0768	0·0739
3·8	0	0·0036	0·0071	0·0141	0·0209	0·0275	0·0337	0·0395	0·0448	0·0497	0·0540	0·0578	0·0653	0·0699	0·0730	0·0711
4·0	0	0·0032	0·0065	0·0128	0·0191	0·0251	0·0308	0·0361	0·0411	0·0457	0·0498	0·0535	0·0608	0·0656	0·0694	0·0683
4·2	0	0·0030	0·0059	0·0117	0·0175	0·0230	0·0282	0·0332	0·0379	0·0421	0·0460	0·0496	0·0567	0·0616	0·0660	0·0656
4·4	0	0·0027	0·0054	0·0108	0·0160	0·0211	0·0260	0·0306	0·0349	0·0390	0·0427	0·0460	0·0530	0·0579	0·0627	0·0630
4·6	0	0·0025	0·0050	0·0099	0·0148	0·0195	0·0240	0·0283	0·0323	0·0361	0·0396	0·0429	0·0496	0·0545	0·0597	0·0605
4·8	0	0·0023	0·0046	0·0092	0·0136	0·0180	0·0222	0·0262	0·0300	0·0336	0·0369	0·0400	0·0465	0·0513	0·0568	0·0581
5·0	0	0·0021	0·0043	0·0085	0·0126	0·0167	0·0206	0·0244	0·0279	0·0313	0·0344	0·0374	0·0436	0·0484	0·0541	0·0558

APPENDIX III

Note on Bessel Functions

WE collect here for reference some properties of Bessel functions which have been required in the text.†

The Bessel function $J_\nu(z)$ is defined by the equation

$$J_\nu(z) = \sum_{r=0}^{\infty} \frac{(-1)^r(\tfrac{1}{2}z)^{\nu+2r}}{r!\,\Gamma(\nu+r+1)}, \tag{1}$$

where ν is real and z may be complex, its argument being given its principal value. The function $J_\nu(z)$ satisfies Bessel's equation of order ν:

$$\frac{d^2y}{dz^2}+\frac{1}{z}\frac{dy}{dz}+\left(1-\frac{\nu^2}{z^2}\right)y = 0. \tag{2}$$

If ν is not an integer, $J_\nu(z)$ and $J_{-\nu}(z)$ are independent solutions of (2), but if ν is an integer, n,

$$J_n(z) = (-1)^n J_{-n}(z).$$

In order to have a second solution of (2) which is available for all values of ν, the function

$$Y_\nu(z) = \frac{J_\nu(z)\cos\nu\pi - J_{-\nu}(z)}{\sin\nu\pi} \tag{3}$$

is defined, the function of integral order $Y_n(z)$ being defined as $\lim_{\nu\to n} Y_\nu(z)$. With this definition

$$\tfrac{1}{2}\pi Y_0(z) = \{\ln(\tfrac{1}{2}z)+\gamma\}J_0(z)+(\tfrac{1}{2}z)^2-(1+\tfrac{1}{2})\frac{(\tfrac{1}{2}z)^4}{(2!)^2}+(1+\tfrac{1}{2}+\tfrac{1}{3})\frac{(\tfrac{1}{2}z)^6}{(3!)^2}-\dots, \tag{4}$$

where $\gamma = 0\cdot5772\dots$ is Euler's constant.

Also, when n is any positive integer,

$$\pi Y_n(z) = 2\{\ln(\tfrac{1}{2}z)+\gamma\}J_n(z)- \sum_{r=0}^{\infty} (-1)^r \frac{(\tfrac{1}{2}z)^{n+2r}}{r!\,(n+r)!}\left[\sum_{m=1}^{n+r} m^{-1}+ \sum_{m=1}^{r} m^{-1}\right]-$$
$$- \sum_{r=0}^{n-1} (\tfrac{1}{2}z)^{-n+2r}\frac{(n-r-1)!}{r!}, \tag{5}$$

where for $r = 0$ we replace $\left(\sum_{m=1}^{n+r} m^{-1}+ \sum_{m=1}^{r} m^{-1}\right)$ by $\sum_{m=1}^{n} m^{-1}$.

The modified Bessel equation

$$\frac{d^2y}{dz^2}+\frac{1}{z}\frac{dy}{dz}-\left(1+\frac{\nu^2}{z^2}\right)y = 0 \tag{6}$$

is satisfied by

$$I_\nu(z) = \sum_{r=0}^{\infty} \frac{(\tfrac{1}{2}z)^{\nu+2r}}{r!\,\Gamma(\nu+r+1)}. \tag{7}$$

† For full information see *W.B.F.*, *G. and M.*, or McLachlan, *Bessel Functions for Engineers* (Oxford, 1934).

If ν is not an integer, $I_{-\nu}(z)$ is an independent solution of (6), but to obtain a second solution available for all values of ν we define

$$K_\nu(z) = \tfrac{1}{2}\pi \frac{I_{-\nu}(z)-I_\nu(z)}{\sin\nu\pi}, \tag{8}$$

the function $K_n(z)$ of integral order n being defined as $\lim_{\nu\to n} K_\nu(z)$. With this definition

$$K_0(z) = -\{\ln(\tfrac{1}{2}z)+\gamma\}I_0(z)+(\tfrac{1}{2}z)^2+(1+\tfrac{1}{2})\frac{(\tfrac{1}{2}z)^4}{(2!)^2}+(1+\tfrac{1}{2}+\tfrac{1}{3})\frac{(\tfrac{1}{2}z)^6}{(3!)^2}+\cdots \tag{9}$$

and, when n is any positive integer,

$$K_n(z) = (-1)^{n+1}\{\ln(\tfrac{1}{2}z)+\gamma\}I_n(z)+\tfrac{1}{2}(-1)^n \sum_{r=0}^{\infty} \frac{(\tfrac{1}{2}z)^{n+2r}}{r!\,(n+r)!}\left[\sum_{m=1}^{n+r} m^{-1}+ \sum_{m=1}^{r} m^{-1}\right]+$$

$$+\tfrac{1}{2}\sum_{r=0}^{n-1}(-1)^r(\tfrac{1}{2}z)^{-n+2r}\frac{(n-r-1)!}{r!}, \tag{10}$$

where, for $r = 0$, $\left(\sum_{m=1}^{n+r} m^{-1}+ \sum_{m=1}^{r} m^{-1}\right)$ is replaced by $\sum_{m=1}^{n} m^{-1}$.

For large values of z

$$K_\nu(z) = \left(\frac{\pi}{2z}\right)^{\tfrac{1}{2}}e^{-z}\left\{1+\frac{4\nu^2-1^2}{1!\,8z}+\frac{(4\nu^2-1^2)(4\nu^2-3^2)}{2!\,(8z)^2}+O\left(\frac{1}{z^3}\right)\right\}, \tag{11}$$

$$I_\nu(z) = \frac{e^z}{\sqrt{(2\pi z)}}\left\{1-\frac{4\nu^2-1^2}{1!\,8z}+\frac{(4\nu^2-1^2)(4\nu^2-3^2)}{2!\,(8z)^2}+O\left(\frac{1}{z^3}\right)\right\}+\frac{e^{-z\pm(\nu+\tfrac{1}{2})\pi i}}{\sqrt{(2\pi z)}}\left\{1+O\left(\frac{1}{z}\right)\right\}, \tag{12}$$

the positive sign being taken if $-\tfrac{1}{2}\pi < \arg z < \tfrac{3}{2}\pi$, and the negative sign if

$$-\tfrac{3}{2}\pi < \arg z < \tfrac{1}{2}\pi.$$

The properties of the Bessel functions most frequently needed are the following:

$$zI_\nu'(z)+\nu I_\nu(z) = zI_{\nu-1}(z), \tag{13}$$

$$zI_\nu'(z)-\nu I_\nu(z) = zI_{\nu+1}(z), \tag{14}$$

$$zK_\nu'(z)+\nu K_\nu(z) = -zK_{\nu-1}(z), \tag{15}$$

$$zK_\nu'(z)-\nu K_\nu(z) = -zK_{\nu+1}(z), \tag{16}$$

$$zJ_\nu'(z)+\nu J_\nu(z) = zJ_{\nu-1}(z), \tag{17}$$

$$zJ_\nu'(z)-\nu J_\nu(z) = -zJ_{\nu+1}(z). \tag{18}$$

$Y_\nu(z)$ satisfies the same relations (17) and (18) as $J_\nu(z)$. In the case $\nu = 0$ these become

$$I_0'(z) = I_1(z);\quad K_0'(z) = -K_1(z);\quad J_0'(z) = -J_1(z);\quad Y_0'(z) = -Y_1(z). \tag{19}$$

$$J_\nu(z)Y_\nu'(z)-Y_\nu(z)J_\nu'(z) = \frac{2}{\pi z}, \tag{20}$$

$$I_\nu(z)K_\nu'(z)-K_\nu(z)I_\nu'(z) = -\frac{1}{z}, \tag{21}$$

$$I_\nu(z)K_{\nu+1}(z)+K_\nu(z)I_{\nu+1}(z) = \frac{1}{z}, \tag{22}$$

$$J_\nu(ze^{m\pi i}) = e^{m\nu\pi i}J_\nu(z), \tag{23}$$

$$Y_\nu(ze^{m\pi i}) = e^{-m\nu\pi i}Y_\nu(z) + 2i\sin m\nu\pi \cot\nu\pi \, J_\nu(z), \tag{24}$$

$$K_\nu(ze^{\pm\frac{1}{2}\pi i}) = \pm\tfrac{1}{2}\pi i e^{\mp\frac{1}{2}\nu\pi i}[-J_\nu(z)\pm iY_\nu(z)], \tag{25}$$

$$I_\nu(ze^{\pm\frac{1}{2}\pi i}) = e^{\pm\frac{1}{2}\nu\pi i}J_\nu(z), \tag{26}$$

$$K_{-\nu}(z) = K_\nu(z), \tag{27}$$

$$K_{\frac{1}{2}}(z) = \left(\frac{\pi}{2z}\right)^{\frac{1}{2}}e^{-z}, \tag{28}$$

$$\int_0^\infty \xi J_0(r\xi)e^{-b\xi^2}\,d\xi = \frac{1}{2b}\,e^{-r^2/4b}, \tag{29}$$

$$\tfrac{1}{2}\int_0^\infty \exp\left\{-\xi-\frac{z^2}{4\xi}\right\}\frac{d\xi}{\xi} = K_0(z), \quad \mathbf{R}(z^2) > 0. \tag{30}$$

APPENDIX IV

The Roots of certain Transcendental Equations

TABLE I

The first six roots,† α_n, of

$$\alpha \tan \alpha = C.$$

C	α_1	α_2	α_3	α_4	α_5	α_6
0	0	3·1416	6·2832	9·4248	12·5664	15·7080
0·001	0·0316	3·1419	6·2833	9·4249	12·5665	15·7080
0·002	0·0447	3·1422	6·2835	9·4250	12·5665	15·7081
0·004	0·0632	3·1429	6·2838	9·4252	12·5667	15·7082
0·006	0·0774	3·1435	6·2841	9·4254	12·5668	15·7083
0·008	0·0893	3·1441	6·2845	9·4256	12·5670	15·7085
0·01	0·0998	3·1448	6·2848	9·4258	12·5672	15·7086
0·02	0·1410	3·1479	6·2864	9·4269	12·5680	15·7092
0·04	0·1987	3·1543	6·2895	9·4290	12·5696	15·7105
0·06	0·2425	3·1606	6·2927	9·4311	12·5711	15·7118
0·08	0·2791	3·1668	6·2959	9·4333	12·5727	15·7131
0·1	0·3111	3·1731	6·2991	9·4354	12·5743	15·7143
0·2	0·4328	3·2039	6·3148	9·4459	12·5823	15·7207
0·3	0·5218	3·2341	6·3305	9·4565	12·5902	15·7270
0·4	0·5932	3·2636	6·3461	9·4670	12·5981	15·7334
0·5	0·6533	3·2923	6·3616	9·4775	12·6060	15·7397
0·6	0·7051	3·3204	6·3770	9·4879	12·6139	15·7460
0·7	0·7506	3·3477	6·3923	9·4983	12·6218	15·7524
0·8	0·7910	3·3744	6·4074	9·5087	12·6296	15·7587
0·9	0·8274	3·4003	6·4224	9·5190	12·6375	15·7650
1·0	0·8603	3·4256	6·4373	9·5293	12·6453	15·7713
1·5	0·9882	3·5422	6·5097	9·5801	12·6841	15·8026
2·0	1·0769	3·6436	6·5783	9·6296	12·7223	15·8336
3·0	1·1925	3·8088	6·7040	9·7240	12·7966	15·8945
4·0	1·2646	3·9352	6·8140	9·8119	12·8678	15·9536
5·0	1·3138	4·0336	6·9096	9·8928	12·9352	16·0107
6·0	1·3496	4·1116	6·9924	9·9667	12·9988	16·0654
7·0	1·3766	4·1746	7·0640	10·0339	13·0584	16·1177
8·0	1·3978	4·2264	7·1263	10·0949	13·1141	16·1675
9·0	1·4149	4·2694	7·1806	10·1502	13·1660	16·2147
10·0	1·4289	4·3058	7·2281	10·2003	13·2142	16·2594
15·0	1·4729	4·4255	7·3959	10·3898	13·4078	16·4474
20·0	1·4961	4·4915	7·4954	10·5117	13·5420	16·5864
30·0	1·5202	4·5615	7·6057	10·6543	13·7085	16·7691
40·0	1·5325	4·5979	7·6647	10·7334	13·8048	16·8794
50·0	1·5400	4·6202	7·7012	10·7832	13·8666	16·9519
60·0	1·5451	4·6353	7·7259	10·8172	13·9094	17·0026
80·0	1·5514	4·6543	7·7573	10·8606	13·9644	17·0686
100·0	1·5552	4·6658	7·7764	10·8871	13·9981	17·1093
∞	1·5708	4·7124	7·8540	10·9956	14·1372	17·2788

† The roots of this equation are all real if $C > 0$.

TABLE II

The first six roots,† α_n, of

$$\alpha \cot \alpha + C = 0.$$

C	α_1	α_2	α_3	α_4	α_5	α_6
−1·0	0	4·4934	7·7253	10·9041	14·0662	17·2208
−0·995	0·1224	4·4945	7·7259	10·9046	14·0666	17·2210
−0·99	0·1730	4·4956	7·7265	10·9050	14·0669	17·2213
−0·98	0·2445	4·4979	7·7278	10·9060	14·0676	17·2219
−0·97	0·2991	4·5001	7·7291	10·9069	14·0683	17·2225
−0·96	0·3450	4·5023	7·7304	10·9078	14·0690	17·2231
−0·95	0·3854	4·5045	7·7317	10·9087	14·0697	17·2237
−0·94	0·4217	4·5068	7·7330	10·9096	14·0705	17·2242
−0 93	0·4551	4·5090	7·7343	10·9105	14·0712	17·2248
−0·92	0·4860	4·5112	7·7356	10·9115	14·0719	17·2254
−0·91	0·5150	4·5134	7·7369	10·9124	14·0726	17·2260
−0·90	0·5423	4·5157	7·7382	10·9133	14·0733	17·2266
−0·85	0·6609	4·5268	7·7447	10·9179	14·0769	17·2295
−0·8	0·7593	4·5379	7·7511	10·9225	14·0804	17·2324
−0·7	0·9208	4·5601	7·7641	10·9316	14·0875	17·2382
−0·6	1·0528	4·5822	7·7770	10·9403	14·0946	17·2440
−0·5	1·1656	4·6042	7·7899	10·9499	14·1017	17·2498
−0·4	1·2644	4·6261	7·8028	10·9591	14·1088	17·2556
−0·3	1·3525	4·6479	7·8156	10·9682	14·1159	17·2614
−0·2	1·4320	4·6696	7·8284	10·9774	14·1230	17·2672
−0·1	1·5044	4·6911	7·8412	10·9865	14·1301	17·2730
0	1·5708	4·7124	7·8540	10·9956	14·1372	17·2788
0·1	1·6320	4·7335	7·8667	11·0047	14·1443	17·2845
0·2	1·6887	4·7544	7·8794	11·0137	14·1513	17·2903
0·3	1·7414	4·7751	7·8920	11·0228	14·1584	17·2961
0·4	1·7906	4·7956	7·9046	11·0318	14·1654	17·3019
0·5	1·8366	4·8158	7·9171	11·0409	14·1724	17·3076
0·6	1·8798	4·8358	7·9295	11·0498	14·1795	17·3134
0·7	1·9203	4·8556	7·9419	11·0588	14·1865	17·3192
0·8	1·9586	4·8751	7·9542	11·0677	14·1935	17·3249
0·9	1·9947	4·8943	7·9665	11·0767	14·2005	17·3306
1·0	2·0288	4·9132	7·9787	11·0856	14·2075	17·3364
1·5	2·1746	5·0037	8·0385	11·1296	14·2421	17·3649
2·0	2·2889	5·0870	8·0962	11·1727	14·2764	17·3932
3·0	2·4557	5·2329	8·2045	11·2560	14·3434	17·4490
4·0	2·5704	5·3540	8·3029	11·3349	14·4080	17·5034
5·0	2·6537	5·4544	8·3914	11·4086	14·4699	17·5562
6·0	2·7165	5·5378	8·4703	11·4773	14·5288	17·6072
7·0	2·7654	5·6078	8·5406	11·5408	14·5847	17·6562
8·0	2·8044	5·6669	8·6031	11·5994	14·6374	17·7032
9·0	2·8363	5·7172	8·6587	11·6532	14·6870	17·7481
10·0	2·8628	5·7606	8·7083	11·7027	14·7335	17·7908
15·0	2·9476	5·9080	8·8898	11·8959	14·9251	17·9742
20·0	2·9930	5·9921	9·0019	12·0250	15·0625	18·1136
30·0	3·0406	6·0831	9·1294	12·1807	15·2380	18·3018
40·0	3·0651	6·1311	9·1987	12·2688	15·3417	18·4180
50·0	3·0801	6·1606	9·2420	12·3247	15·4090	18·4953
60·0	3·0901	6·1805	9·2715	12·3632	15·4559	18·5497
80·0	3·1028	6·2058	9·3089	12·4124	15·5164	18·6209
100·0	3·1105	6·2211	9·3317	12·4426	15·5537	18·6650
∞	3·1416	6·2832	9·4248	12·5664	15·7080	18·8496

† The roots of this equation are all real if $C > -1$. These negative values of C arise in connexion with the sphere, § 9.4.

TABLE III

The first six roots, α_n, of

$$\alpha J_1(\alpha) - C J_0(\alpha) = 0.$$

C	α_1	α_2	α_3	α_4	α_5	α_6
0	0	3·8317	7·0156	10·1735	13·3237	16·4706
0·01	0·1412	3·8343	7·0170	10·1745	13·3244	16·4712
0·02	0·1995	3·8369	7·0184	10·1754	13·3252	16·4718
0·04	0·2814	3·8421	7·0213	10·1774	13·3267	16·4731
0·06	0·3438	3·8473	7·0241	10·1794	13·3282	16·4743
0·08	0·3960	3·8525	7·0270	10·1813	13·3297	16·4755
0·1	0·4417	3·8577	7·0298	10·1833	13·3312	16·4767
0·15	0·5376	3·8706	7·0369	10·1882	13·3349	16·4797
0·2	0·6170	3·8835	7·0440	10·1931	13·3387	16·4828
0·3	0 7465	3·9091	7·0582	10·2029	13·3462	16·4888
0·4	0·8516	3·9344	7·0723	10·2127	13·3537	16·4949
0·5	0·9408	3·9594	7·0864	10·2225	13·3611	16·5010
0·6	1·0184	3·9841	7·1004	10·2322	13·3686	16·5070
0·7	1·0873	4·0085	7·1143	10·2419	13·3761	16·5131
0·8	1·1490	4·0325	7·1282	10·2516	13·3835	16·5191
0·9	1·2048	4·0562	7·1421	10·2613	13·3910	16·5251
1·0	1·2558	4·0795	7·1558	10·2710	13·3984	16·5312
1·5	1·4569	4·1902	7·2233	10·3188	13·4353	16·5612
2·0	1·5994	4·2910	7·2884	10·3658	13·4719	16·5910
3·0	1·7887	4·4634	7·4103	10·4566	13·5434	16·6499
4·0	1·9081	4·6018	7·5201	10·5423	13·6125	16·7073
5·0	1·9898	4·7131	7·6177	10·6223	13·6786	16·7630
6·0	2·0490	4·8033	7·7039	10·6964	13·7414	16·8168
7·0	2·0937	4·8772	7·7797	10·7646	13·8008	16·8684
8·0	2·1286	4·9384	7·8464	10·8271	13·8566	16·9179
9·0	2·1566	4·9897	7·9051	10·8842	13·9090	16·9650
10·0	2·1795	5·0332	7·9569	10·9363	13·9580	17·0099
15·0	2·2509	5·1773	8·1422	11·1367	14·1576	17·2008
20·0	2·2880	5·2568	8·2534	11·2677	14·2983	17·3442
30·0	2·3261	5·3410	8·3771	11·4221	14·4748	17·5348
40·0	2·3455	5·3846	8·4432	11·5081	14·5774	17·6508
50·0	2·3572	5·4112	8·4840	11·5621	14·6433	17·7272
60·0	2·3651	5·4291	8·5116	11·5090	14·6889	17·7807
80·0	2·3750	5·4516	8·5466	11·6461	14·7475	17·8502
100·0	2·3809	5·4652	8·5678	11·6747	14·7834	17·8931
∞	2·4048	5·5201	8·6537	11·7915	14·9309	18·0711

TABLE IV

The first five roots, α_n, of

$$J_0(\alpha) Y_0(k\alpha) - Y_0(\alpha) J_0(k\alpha) = 0.$$

k	α_1	α_2	α_3	α_4	α_5
1·2	15·7014	31·4126	47·1217	62·8302	78·5385
1·5	6·2702	12·5598	18·8451	25·1294	31·4133
2·0	3·1230	6·2734	9·4182	12·5614	15·7040
2·5	2·0732	4·1773	6·2754	8·3717	10·4672
3·0	1·5485	3·1291	4·7038	6·2767	7·8487
3·5	1·2339	2·5002	3·7608	5·0196	6·2776
4·0	1·0244	2·0809	3·1322	4·1816	5·2301

APPENDIX V

Table of Laplace Transforms

$$\bar{v}(p) = \int_0^\infty e^{-pt} v(t)\, dt.$$

WE write $q = \sqrt{(p/\kappa)}$. κ and x are always real and positive. α and h are unrestricted.

	$\bar{v}(p)$	$v(t)$
1.	$\dfrac{1}{p}$	1
2.	$\dfrac{1}{p^{\nu+1}}, \quad \nu > -1$	$\dfrac{t^\nu}{\Gamma(\nu+1)}$
3.	$\dfrac{1}{p+\alpha}$	$e^{-\alpha t}$
4.	$\dfrac{\omega}{p^2+\omega^2}$	$\sin \omega t$
5.	$\dfrac{p}{p^2+\omega^2}$	$\cos \omega t$
6.	e^{-qx}	$\dfrac{x}{2\sqrt{(\pi\kappa t^3)}} e^{-x^2/4\kappa t}$
7.	$\dfrac{e^{-qx}}{q}$	$\left(\dfrac{\kappa}{\pi t}\right)^{\frac{1}{2}} e^{-x^2/4\kappa t}$
8.	$\dfrac{e^{-qx}}{p}$	$\operatorname{erfc} \dfrac{x}{2\sqrt{(\kappa t)}}$
9.	$\dfrac{e^{-qx}}{pq}$	$2\left(\dfrac{\kappa t}{\pi}\right)^{\frac{1}{2}} e^{-x^2/4\kappa t} - x \operatorname{erfc} \dfrac{x}{2\sqrt{(\kappa t)}}$
10.	$\dfrac{e^{-qx}}{p^2}$	$\left(t+\dfrac{x^2}{2\kappa}\right) \operatorname{erfc} \dfrac{x}{2\sqrt{(\kappa t)}} - x\left(\dfrac{t}{\pi\kappa}\right)^{\frac{1}{2}} e^{-x^2/4\kappa t}$
11.	$\dfrac{e^{-qx}}{p^{1+\frac{1}{2}n}}, \quad n = 0, 1, 2, \ldots$	$(4t)^{\frac{1}{2}n}\, i^n \operatorname{erfc} \dfrac{x}{2\sqrt{(\kappa t)}}$
12.	$\dfrac{e^{-qx}}{q+h}$	$\left(\dfrac{\kappa}{\pi t}\right)^{\frac{1}{2}} e^{-x^2/4\kappa t} - h\kappa e^{hx+\kappa th^2} \times$ $\times \operatorname{erfc}\left\{\dfrac{x}{2\sqrt{(\kappa t)}}+h\sqrt{(\kappa t)}\right\}$
13.	$\dfrac{e^{-qx}}{q(q+h)}$	$\kappa e^{hx+\kappa th^2} \operatorname{erfc}\left\{\dfrac{x}{2\sqrt{(\kappa t)}}+h\sqrt{(\kappa t)}\right\}$
14.	$\dfrac{e^{-qx}}{p(q+h)}$	$\dfrac{1}{h} \operatorname{erfc} \dfrac{x}{2\sqrt{(\kappa t)}} - \dfrac{1}{h} e^{hx+\kappa th^2} \times$ $\times \operatorname{erfc}\left\{\dfrac{x}{2\sqrt{(\kappa t)}}+h\sqrt{(\kappa t)}\right\}$

	$\bar{v}(p)$	$v(t)$
15.	$\dfrac{e^{-qx}}{pq(q+h)}$	$\dfrac{2}{h}\left(\dfrac{\kappa t}{\pi}\right)^{\frac{1}{2}}e^{-x^2/4\kappa t}-\dfrac{(1+hx)}{h^2}\operatorname{erfc}\dfrac{x}{2\sqrt{(\kappa t)}}+$ $+\dfrac{1}{h^2}e^{hx+\kappa th^2}\operatorname{erfc}\left\{\dfrac{x}{2\sqrt{(\kappa t)}}+h\sqrt{(\kappa t)}\right\}$
16.	$\dfrac{e^{-qx}}{q^{n+1}(q+h)}$	$\dfrac{\kappa}{(-h)^n}e^{hx+\kappa th^2}\operatorname{erfc}\left\{\dfrac{x}{2\sqrt{(\kappa t)}}+h\sqrt{(\kappa t)}\right\}-$ $-\dfrac{\kappa}{(-h)^n}\sum_{r=0}^{n-1}[-2h\sqrt{(\kappa t)}]^r\,i^r\mathrm{erfc}\dfrac{x}{2\sqrt{(\kappa t)}}$
17.	$\dfrac{e^{-qx}}{(q+h)^2}$	$-2h\left(\dfrac{\kappa^3 t}{\pi}\right)^{\frac{1}{2}}e^{-x^2/4\kappa t}+\kappa(1+hx+2h^2\kappa t)e^{hx+\kappa th^2}$ $\times\operatorname{erfc}\left\{\dfrac{x}{2\sqrt{(\kappa t)}}+h\sqrt{(\kappa t)}\right\}$
18.	$\dfrac{e^{-qx}}{p(q+h)^2}$	$\dfrac{1}{h^2}\operatorname{erfc}\dfrac{x}{2\sqrt{(\kappa t)}}-\dfrac{2}{h}\left(\dfrac{\kappa t}{\pi}\right)^{\frac{1}{2}}e^{-x^2/4\kappa t}-$ $-\dfrac{1}{h^2}\{1-hx-2h^2\kappa t\}e^{hx+\kappa th^2}\times$ $\times\operatorname{erfc}\left\{\dfrac{x}{2\sqrt{(\kappa t)}}+h\sqrt{(\kappa t)}\right\}$
19.	$\dfrac{e^{-qx}}{p-\alpha}$	$\tfrac{1}{2}e^{\alpha t}\left\{e^{-x\sqrt{(\alpha/\kappa)}}\operatorname{erfc}\left[\dfrac{x}{2\sqrt{(\kappa t)}}-\sqrt{(\alpha t)}\right]+\right.$ $\left.+e^{x\sqrt{(\alpha/\kappa)}}\operatorname{erfc}\left[\dfrac{x}{2\sqrt{(\kappa t)}}+\sqrt{(\alpha t)}\right]\right\}$
20.	$\dfrac{1}{p^{\frac{3}{4}}}e^{-qx}$	$\dfrac{1}{\pi}\left(\dfrac{x}{2t\kappa^{\frac{1}{2}}}\right)^{\frac{1}{2}}e^{-x^2/8\kappa t}K_{\frac{1}{4}}\left(\dfrac{x^2}{8\kappa t}\right)$
21.	$\dfrac{1}{p^{\frac{1}{2}}}K_{2\nu}(qx)$	$\dfrac{1}{2\sqrt{(\pi t)}}e^{-x^2/8\kappa t}K_\nu\left(\dfrac{x^2}{8\kappa t}\right)$
22.	$\left.\begin{array}{l}I_\nu(qx')K_\nu(qx),\quad x>x'\\ I_\nu(qx)K_\nu(qx'),\quad x<x'\end{array}\right\}$	$\dfrac{1}{2t}e^{-(x^2+x'^2)/4\kappa t}I_\nu\left(\dfrac{xx'}{2\kappa t}\right),\quad \nu\geqslant 0$
23.	$K_0(qx)$	$\dfrac{1}{2t}e^{-x^2/4\kappa t}$
24.	$\dfrac{1}{p}e^{x/p}$	$I_0[2\sqrt{(xt)}]$
25.	$\dfrac{\exp\{xp-x[(p+a)(p+b)]^{\frac{1}{2}}\}}{[(p+a)(p+b)]^{\frac{1}{2}}}$	$e^{-\frac{1}{2}(a+b)(t+x)}I_0\{\tfrac{1}{2}(a-b)[t(t+2x)]^{\frac{1}{2}}\}$
26.	$p^{\frac{1}{2}\nu-1}K_\nu(x\sqrt{p})$	$x^{-\nu}2^{\nu-1}\displaystyle\int_{x^2/4t}^{\infty}e^{-u}u^{\nu-1}\,du$
27.	$[p-\sqrt{(p^2-x^2)}]^\nu,\quad \nu>0.$	$\nu x^\nu I_\nu(xt)/t$
28.	$\dfrac{\exp\{x[(p+a)^{\frac{1}{2}}-(p+b)^{\frac{1}{2}}]^2\}}{(p+a)^{\frac{1}{2}}(p+b)^{\frac{1}{2}}[(p+a)^{\frac{1}{2}}+(p+b)^{\frac{1}{2}}]^{2\nu}},$ $\nu\geqslant 0$	$\dfrac{t^{\frac{1}{2}\nu}e^{-\frac{1}{2}(a+b)t}I_\nu[\tfrac{1}{2}(a-b)t^{\frac{1}{2}}(t+4x)^{\frac{1}{2}}]}{(a-b)^\nu(t+4x)^{\frac{1}{2}\nu}}$

	$\bar{v}(p)$	$v(t)$
29.	$\dfrac{e^{-qx}}{(p-\alpha)^2}$	$\tfrac{1}{2}e^{\alpha t}\Big\{\Big(t-\dfrac{x}{2\sqrt{(\kappa\alpha)}}\Big)\times$ $\times e^{-x\sqrt{(\alpha/\kappa)}}\mathrm{erfc}\Big[\dfrac{x}{2\sqrt{(\kappa t)}}-\sqrt{(\alpha t)}\Big]+$ $+\Big(t+\dfrac{x}{2\sqrt{(\kappa\alpha)}}\Big)e^{x\sqrt{(\alpha/\kappa)}}\mathrm{erfc}\Big[\dfrac{x}{2\sqrt{(\kappa t)}}+\sqrt{(\alpha t)}\Big]\Big\}$
30.	$\dfrac{e^{-qx}}{q(p-\alpha)}$	$\tfrac{1}{2}e^{\alpha t}\Big(\dfrac{\kappa}{\alpha}\Big)^{\frac{1}{2}}\Big\{e^{-x\sqrt{(\alpha/\kappa)}}\mathrm{erfc}\Big[\dfrac{x}{2\sqrt{(\kappa t)}}-\sqrt{(\alpha t)}\Big]-$ $-e^{x\sqrt{(\alpha/\kappa)}}\mathrm{erfc}\Big[\dfrac{x}{2\sqrt{(\kappa t)}}+\sqrt{(\alpha t)}\Big]\Big\}$
31.	$\dfrac{e^{-qx}}{(p-\alpha)(q+h)}, \quad \alpha \neq \kappa h^2$	$\tfrac{1}{2}e^{\alpha t}\Big\{\dfrac{\kappa^{\frac{1}{2}}}{h\kappa^{\frac{1}{2}}+\alpha^{\frac{1}{2}}}e^{-x\sqrt{(\alpha/\kappa)}}\mathrm{erfc}\Big[\dfrac{x}{2\sqrt{(\kappa t)}}-\sqrt{(\alpha t)}\Big]+$ $+\dfrac{\kappa^{\frac{1}{2}}}{h\kappa^{\frac{1}{2}}-\alpha^{\frac{1}{2}}}e^{x\sqrt{(\alpha/\kappa)}}\mathrm{erfc}\Big[\dfrac{x}{2\sqrt{(\kappa t)}}+\sqrt{(\alpha t)}\Big]\Big\}-$ $-\dfrac{h\kappa}{h^2\kappa-\alpha}e^{hx+h^2\kappa t}\mathrm{erfc}\Big[\dfrac{x}{2\sqrt{(\kappa t)}}+h\sqrt{(\kappa t)}\Big]$
32.	$\dfrac{1}{p}\ln p$	$-\ln(Ct), \quad \ln C = \gamma = 0\cdot5772...$
33.	$p^{\frac{1}{2}\nu}K_\nu(x\sqrt{p})$	$\dfrac{x^\nu}{(2t)^{\nu+1}}e^{-x^2/4t}$

APPENDIX VI

Thermal Properties of some Common Substances

THIS table is intended only to indicate the orders of magnitude likely to occur in practice; for fuller information, and for the variation of the thermal properties with temperature, the International Critical Tables, or other standard works giving physical constants, may be consulted. The values for non-metals are to be regarded as rough average values only, as there may be large differences between the thermal conductivities of different samples of the same substance. The units are c.g.s., calorie, and °C.

Substance	Density ρ	Specific heat c	Conductivity K	Diffusivity κ
Silver	10·49	0·0556	1·00	1·71
Gold	19·30	0·0308	0·70	1·18
Copper	8·94	0·0914	0·93	1·14
Magnesium	1·74	0·240	0·38	0·91
Aluminium	2·70	0·206	0·48	0·86
Zinc	7·14	0·0917	0·27	0·41
Tin	7·30	0·0534	0·15	0·38
Brass (70:30) . . .	8·5	0·09	0·25	0·33
Platinum . . .	21·46	0·0315	0·17	0·25
Lead	11·34	0·0302	0·084	0·25
Mild steel (0·1% C) . .	7·85	0·118	0·11	0·12
Cast iron . . .	7·4	0·136	0·12	0·12
Bismuth	9·80	0·0292	0·020	0·070
Mercury	13·55	0·0335	0·020	0·044
Non-metals				
Air	0·00129	0·240	0·000058	0·187
Granite	2·6	0·21	0·006	0·011
Limestone	2·5	0·22	0·004	0·007
Sandstone	2·3	0·23	0·006	0·011
Average rock† . . .	..	..	0·0042	0·0118
Ice	0·92	0·502	0·0053	0·0115
Glass (crown) . . .	2·4	0·20	0·0028	0·0058
Concrete (1:2:4) . . .	2·3	0·23	0·0022	0·0042
Brick (building) . . .	2·6	0·20	0·0020	0·0038
Snow (fresh) . . .	0·1	0·5	0·00025	0·0050
Soil (average) . . .	2·5	0·2	0·0023	0·0046
Soil (sandy, dry) . . .	1·65	0·19	0·00063	0·0020
Soil (sandy, 8% moist) .	1·75	0·24	0·0014	0·0033
Wood (spruce, with grain) .	0·41	0·30	0·00055	0·0045
Wood (spruce, across grain) .	0·41	0·30	0·00030	0·0024
Water	1·0	1·0	0·00144	0·00144
Ground cork . . .	0·15	0·48	0·0001	0·0014

† Kelvin, cf. § 2.14.

ᴋ k

AUTHOR INDEX

SUBJECT INDEX